GMP Audits in Pharmaceutical and Biotechnology Industries

The fact that good manufacturing practice (GMP) audits in the pharmaceutical and biotechnology industries have to be evaluated, and with very limited resources, has created a gap in this field. The lack of trained and qualified GMP auditors is on the rise in all organizations that are required to implement FDA, EMA, MHRA, WHO, TGA, and PIC/S regulations. This volume is an essential reference source for those organizations operating in the field of health and presents the basic knowledge needed to perform audits. The author also provides useful tips and a selection of samples about GMP audits that are indispensable for professionals and health inspectors working in industry and health authorities.

Features

- An essential reference source for those organizations operating in the field of health and presents the basic knowledge needed to perform audits.
- Anyone working in the manufacturing sector needs to be aware of GMP, be able to identify operational flaws as well as legal violations, and have a clear understanding of how to meet GMP standards.
- Assists readers in understanding the importance of GMP and how they can apply each aspect in their working environment.
- Covers a global regulatory landscape.
- Suitable for relevant degree courses including industrial pharmaceutics and pharmaceutical biotechnology.

Drugs and the Pharmaceutical Sciences

A Series of Textbooks and Monographs
Series Editor

Anthony J. Hickey

RTI International, Research Triangle Park, USA

The Drugs and Pharmaceutical Sciences series is designed to enable the pharmaceutical scientist to stay abreast of the changing trends, advances and innovations associated with therapeutic drugs and that area of expertise and interest that has come to be known as the pharmaceutical sciences. The body of knowledge that those working in the pharmaceutical environment have to work with, and master, has been, and continues, to expand at a rapid pace as new scientific approaches, technologies, instrumentations, clinical advances, economic factors and social needs arise and influence the discovery, development, manufacture, commercialization and clinical use of new agents and devices.

Recent Titles in Series

Biotechnology: The Science, the Products, the Government, the Business Ronald P. Evens

Filtration and Purification in the Biopharmaceutical Industry, Third Edition, Maik W. Jornitz

Handbook of Drug Metabolism, Third Edition, Paul G. Pearson and Larry Wienkers

The Art and Science of Dermal Formulation Development, Marc Brown and Adrian C. Williams

Pharmaceutical Inhalation Aerosol Technology, Third Edition, Anthony J. Hickey and Sandro R. da Rocha

Good Manufacturing Practices for Pharmaceuticals, Seventh Edition, Graham P. Bunn

Pharmaceutical Extrusion Technology, Second Edition, Isaac Ghebre-Sellassie, Charles E. Martin, Feng Zhang, and James Dinunzio

Biosimilar Drug Product Development, Laszlo Endrenyi, Paul Declerck, and Shein-Chung Chow

High Throughput Screening in Drug Discovery, Amancio Carnero

Generic Drug Product Development: International Regulatory Requirements for Bioequivalence, Second Edition, Isadore Kanfer and Leon Shargel

Aqueous Polymeric Coatings for Pharmaceutical Dosage Forms, Fourth Edition, Linda A. Felton

Good Design Practices for GMP Pharmaceutical Facilities, Second Edition, Terry Jacobs and Andrew A. Signore

Handbook of Bioequivalence Testing, Second Edition, Sarfaraz K. Niazi

FDA Good Laboratory Practice Requirements, First Edition, Graham Bunn

Continuous Pharmaceutical Processing and Process Analytical Technology, Ajit Narang and Atul Dubey

Project Management for Drug Developers, Joseph P. Stalder

Emerging Drug Delivery and Biomedical Engineering Technologies: Transforming Therapy, Dimitrios Lamprou

RNA-seq in Drug Discovery and Development, Feng Cheng and Robert Morris

Patient Safety in Developing Countries: Education, Research, Case Studies, Yaser Al-Worafi

Industrial Hygiene in the Pharmaceutical and Consumer Healthcare Industries, Casey Cosner

Cancer Targeting Therapies: Conventional and Advanced Perspectives, Muhammad Yasir Ali, Shazia Bukhari

Molecular Recognition in Pharmacology, Mikhail Darkhovskiy

GMP Audits in Pharmaceutical and Biotechnology Industries, Mustafa EDİK

For more information about this series, please visit: www.crcpress.com/Drugs-and-the-Pharmaceutical-Sciences/book-series/IHCDRUPHASCI

GMP Audits in Pharmaceutical and Biotechnology Industries

Mustafa Edik

Quality Academia Training & Consulting Vice President, Turkey

CRC Press
Taylor & Francis Group
Boca Raton London New York

CRC Press is an imprint of the
Taylor & Francis Group, an **informa** business

First edition published 2024
by CRC Press
2385 Executive Center Drive, Suite 320, Boca Raton, FL 33431

and by CRC Press
4 Park Square, Milton Park, Abingdon, Oxon, OX14 4RN

CRC Press is an imprint of Taylor & Francis Group, LLC

Library of Congress Cataloging-in-Publication Data
Names: Edik, Mustafa, author.
Title: GMP audits in pharmaceutical and biotechnology industries/authored
by Mustafa Edik.
First edition. | Boca Raton: CRC Press, 2024. | Series: Drugs and the pharmaceutical
sciences | Includes bibliographical references. |
Identifiers: LCCN 2023029896 (print) | LCCN 2023029897 (ebook) | ISBN
9781032257303 (hardback) | ISBN 9781032257761 (paperback) | ISBN
9781003284970 (ebook)
Subjects: LCSH: Pharmaceutical industry--Auditing. | Pharmaceutical
industry--Quality control.
Classification: LCC HD9665.9.A1 E45 2024 (print) | LCC HD9665.9.A1
(ebook) | DDC 657/.450246151--dc23/eng/20231017
LC record available at https://lccn.loc.gov/2023029896
LC ebook record available at https://lccn.loc.gov/2023029897

ISBN: 978-1-032-25730-3 (hbk)
ISBN: 978-1-032-25776-1 (pbk)
ISBN: 978-1-003-28497-0 (ebk)

DOI: 10.1201/9781003284970

Typeset in Times
by Deanta Global Publishing Services, Chennai, India

To the meaning of my life, my dear wife Deniz, who is always by my side in good times and bad, listens to me patiently, inspires me, and believes in and trusts me.

To my beloved brave-hearted father, Selahattin Edik, and to my tenderhearted mother, İftihar Edik, who raised me and gave me opportunities to become the most preferred consultant in Turkey.

To my inspiring sister Zuhal Ertuğrul, who encouraged and shaped me to become a successful man.

To my smart uncle Yaşar Ertaş, the secret hero of my life, who expresses his belief in me at every opportunity.

Contents

Preface

The pharmaceutical and biopharmaceutical industries are going through a more active period than ever before. With the rapid development of technology and the ever-increasing expectations and sanctions of health authorities, the desire of patients to reach safe and effective drugs is increasing day by day. In this dazzling environment, pharmaceutical and biopharmaceutical companies are trying to keep up with the developing technology, while providing a facility that is fully compatible with good manufacturing practices (GMP) and looking for ways to reduce costs as much as possible. For this reason, the need for examining, researching, and inspecting raw and packaging materials, which are the starting materials of pharmaceutical products, has gained more importance in order to respond to the GMP and quality expectations of health authorities. The tragic GMP disasters of the past continue today and show their terrible face at the first opportunity. Cough syrup incidents following active pharmaceutical ingredient (API) impurities are an indication that GMP noncompliance will continue unabated. Before I started writing this book, I thought everything I did would benefit the pharmaceutical and biotechnology industries. After nearly 25 years in the field, I decided to do what I know best as an expert and GxP auditor and write this book. Of course, the audience I am targeting in the book is not patients with direct or indirect access to drugs. I aim to reach all quality and GMP professionals, including senior executives in the health authority, pharmaceutical, and biotechnology sectors, and to blend my knowledge by adding examples of the activities they are conducting, and adding other perspectives.

Chapter 1, the emergence of the concept of quality and the ISO 9000 series with the words of international quality gurus on this subject, is aimed at the reader to realize what stages they have passed on the way to the pharmaceutical quality management system. Following the ISO 9000 system standard, other quality management standards used in various sectors are briefly explained. At the end of this section, an introduction to the quality management system established in the pharmaceutical and biotechnology sector is introduced.

Chapter 2 covers why GMP inspections should be done; the differences between audits from health authority inspections; why GMP audits are needed; how the role and task distribution should be in the audit; the audit plan; the selection of the auditors; the establishment of the audit team; the performance of the audit; the working staff, what points should be paid attention to in one-to-one interviews; why listening and note-taking should be considered important in audits; how audit findings should be classified according to health authority guidelines; what kind of information should be included in the audit report; how and by whom the necessary actions should be taken after the audits; examination in follow-up audits explaining the required parts and how GMP audits are handled in US Food and Drug Administration (FDA), Pharmaceutical Inspection and Co-operation Scheme (PIC/S), European Medicines Agency (EMA), UK Medicines and Healthcare products Regulatory Agency (MHRA), Therapeutic Goods Administration (TGA), World Health Organization (WHO), and Turkish Medicines and Medical Devices Agency (TMMDA) guidelines; and finally, what should be considered in FDA Form 483 and warning letters.

Chapter 3 is the longest part of the book and all the topics are explained in detail. In line with what is explained in the first and second chapters, how medical drug product production facilities, third-party manufacturing facilities, and due diligence audits should be done, and then the parts that need attention in a mock inspection, which was carried out in order to fully prepare for health authority inspections, are explained. After explaining the audits of API, excipient, and packaging materials manufacturers and suppliers, critical issues that should be considered in a computerized system and data integrity audits are mentioned. After detailing the tricks in the audits of suppliers providing logistics services, industry experiences on how GMP internal audits should be carried out are shared. After emphasizing the importance of documentation in GMP audits, how laboratory audits should be carried out is explained. In the most prominent part of the book, the things to be considered in the audits of biopharmaceutical facilities are explained. Important topics in the audits of protein drug substances, fractionators, cell–gene therapy, and vaccine production facilities are explained through examples, then radiopharmaceutical, medical gas, cleaning companies for cleanroom clothes, pest control services, pallet suppliers, software suppliers, disposable systems, sterilization services, clean rooms, disinfection, sterile pharmaceutical products, media fill operations, and lyophilization inspections are detailed.

In Chapter 4, audits of validation, qualification, calibration, complaints, change control, training, quality risk management, recalls, and deviations, which are components of the pharmaceutical quality management system, have been explained.

At the end of the chapter, audits of HVAC, water, steam, and gas systems, which are auxiliary businesses, are included.

In Chapter 5, methods that can be used as alternatives to on-site audits are emphasized and the features of remote audits, which have started to be preferred with COVID-19, and the parts that need attention are explained.

Chapter 6 includes checklists that the reader can use in various audits, and also briefly explains the lessons that the reader should learn from FDA Form 483 and warning letters. The book is completed by presenting the names of international guidelines and directives.

At the end of each chapter, questions have been added so that the reader can examine the topics more clearly and obtain some useful results.

One of the aims of this book is not to prepare a list of questions for each topic but rather for the reader to create their own list of questions at the end of each chapter based on what they understand. The list of questions presented at the end of the book are examples and do not include any topic in its entirety. The main purpose of presenting sample question lists is for the user to have an idea about the question types.

GMP auditors do not have all the features such as leadership, technical knowledge, communication skills, and dominance of guidelines and directives. Generally, GMP auditors who will perform the audits are expected to be experienced in the sector and to be managers in at least one department. Knowing that most auditors try to specialize in only one or a few areas, especially when the subject is technical skills, knowledge, and field experience, which are emphasized in the book, brief information about each topic is presented first.

Some chapters do not contain information about the overall flow of a GMP audit from start to finish, because the reader is expected to make an inference using their own imagination and to construct the audits in their head.

The EU Annex 1, PIC/S GMP Guide was revised in August 2022 and became a booklet of approximately 60 pages, much more detailed than the previous version. In this book, an attempt has been made to cover the newly published Annex 1 in a way that covers almost all items, and questions have been formed from every subject that should be asked or verified in GMP audits.

I hope this unique book will open a new door of opportunity for all GMP professionals.

About the Author

Mustafa Edik, after graduating as a chemist from university, began his 25-year-plus career as a laboratory supervisor at Bayer, a German pharmaceutical company. After 15 years of working as a quality assurance assistant manager, laboratory supervisor, pharmaceutical quality management systems, and GMP lead auditor, he decided to continue his career as a consultant. He served the Turkish Atomic Energy Authority (TAEA) as principal GMP auditor and consultant for six years. TAEA was audited by the Republic of Turkey's Ministry of Health and granted a GMP certificate for five radiopharmaceuticals. This success has won great acclaim from all health authorities and industry.

He has prepared and presented various training courses and workshops to more than 8,000 individuals from 150 International and local pharmaceutical, medical device, and cosmetics companies on GMP, GDP, and pharmaceutical quality management systems. He has taken part in several international pharmaceutical facility establishment projects as a GMP consultant and has also set up various quality management systems for local pharmaceutical and medical device companies.

While he was the vice president of quality and technical operations at Quality Academia Training and Consultancy firm, he acquired and converted it into a 100% Turkish company. As the only IRCA Certificated Pharmaceutical Quality Management Systems and GMP Lead Auditor in Turkey, he currently conducts API, excipient, packaging materials suppliers and manufacturers, third-party logistics service providers, sterile and non-sterile manufacturing facilities audits according to FDA, EMA, PIC/S, TMMDA, MHRA, TGA, Health Canada, and WHO regulations and guidelines.

He completed his second university degree in biopharmaceutical sciences BSc (Hons) at Atlantic Technological University–Ireland. He is the author of a chapter in the book *Good Distribution Practice: A Handbook for Healthcare Manufacturers and Suppliers*, published by the Parenteral Drug Association (PDA).

Abbreviations

ABEF	Australian Business Excellence Framework
ACN	acetonitrile
AHU	air handling unit
ALARA	as low as reasonably achievable
ANSI	American National Standards Institute
API	active pharmaceutical ingredient
APIC	Active Pharmaceutical Ingredients Committee
APR	annual product review
APS	aseptic process simulation
AQL	acceptable quality level
ASQ	American Society for Quality
ASTM	American Society for Testing and Materials
ASU	air separation unit
ATMPs	advanced therapy medicinal products
BFS	blow–fill–seal
BI	biological indicator
BMS	building management system
BP	British Pharmacopoeia
BS	British Standards
BSL	biosafety level
CAPA	corrective and preventive action
CAR	corrective action report
CCS	contamination control strategy
CDS	chromatography data system
CEN	European Committee for Standardization
CEP	Certification of Suitability to the monographs of the European Pharmacopoeia
CEO	chief executive officer
CFR	Code of Federal Regulations
CFU	colony-forming unit
CI	constant improvement
CI	chemical indicator
CIP	clean-in-place
cGLP	current good laboratory practice
cGMP	current good manufacturing practice
CMO	contract manufacturing organization
CMT	compliance management team
CNS	clean unclassified
CoA	Certificate of Analysis
COTS	commercial off-the-shelf
CPGM	Compliance Program Guidance Manual
CQA	Certified Quality Auditor
CRO	contract research organization
CTD	common technical document
DEG	diethylene glycol
DI	data integrity
DMF	Drug Master File
DMAIC	Define, Measure, Analyze, Improve, Control

DQ	design qualification
DS	drug substance
ECD	electrochemical detection
EDI	electrodeionization
EDMS	electronic document management system
EDQM	European Directorate for the Quality of Medicines & HealthCare
EFQM	European Foundation for Quality Management
EIR	Establishment Inspection Report
EM	environmental monitoring
EMA	European Medicines Agency
EO	ethylene oxide
ESTRI	Electronic Standards for the Transfer of Regulatory Information
EU	European Union
FAT	factory acceptance test
FDA	US Food and Drug Administration
FDCA	US Food, Drug, and Cosmetic Act
FFS	form–fill–seal
FID	flame ionization detector
FDG	fluorodeoxyglucose
FIFO	first in, first out
FFP	finished pharmaceutical product
FMEA	failure mode and effects analysis
FOIA	Freedom of Information Act
FTA	fault tree analysis
FT-IR	Fourier transform infrared
GC	gas chromatography
GCP	good clinical practice
GDP	good distribution practice
GMDP	good manufacturing and distribution practice
GMP	good manufacturing practice
GPVP	good pharmacovigilance practice
HACCP	hazard analysis of critical control points
HDPE	high-density polyethylene
HEPA	high-efficiency particulate absorbing (filter)
HETP	height equivalent to a theoretical plate
HIV	human immunodeficiency virus
HLS	high-level structure
HPLC	high-performance liquid chromatography
HT	heat treated
HVAC	Heating, ventilation, and air-conditioning
IaaS	Infrastructure as a Service
IATF	International Automotive Task Force
ICCR	International Cooperation on Cosmetics Regulation
ICH	International Council for Harmonisation
ID	identity
IM	intermediate
IMPs	investigational medicinal products
IPC	in-process control
IPEC	International Pharmaceutical Excipients Council
IPPC	International Plant Protection Convention
IQ	installation qualification

IQPP	International Quality Plasma Program
IRB	institutional review board
ISPE	International Society for Pharmaceutical Engineering
ISPM	International Standards for Phytosanitary Measures
ISO	International Organization for Standardization
IT	information technology
kDA	kilodalton
KF	Karl Fischer
LAF	laminar airflow
LAL	limulus amebocyte lysate
LES	lab execution system
LIMS	laboratory information management system
LIN	liquid nitrogen
LPS	lipopolysaccharide
LSS	Lean Six Sigma
MA	marketing authorization
mAb	monoclonal antibody
MAL	material airlock
MB	methyl bromide
MCB	Master Cell Bank
MES	manufacturing execution system
MHRA	Medicines and Healthcare products Regulatory Agency (UK)
MIA	Manufacturing/Importation Authorization
MRA	mutual recognition agreement
NCAs	national competent authorities
NDA	nondisclosure agreement
NDMA	N-nitrosodimethylamine
NPPO	national plant protection organization
OECD	Organization for Economic Co-operation and Development
OOC	out of calibration
OOP	out of parameters
OOS	out of specification
OOT	out of trend
OPEX	operational excellence
OQ	operational qualification
OSCS	oversulfated chondroitin sulfate
OSD	oral solid dosage
PAI	preapproval inspection
PAL	personnel airlock
PaaS	Platform as a Service
PAT	process analytical technology
PC	personal computer
PDA	Parenteral Drug Association
PDCA	plan, do, check, act
PDMA	Prescription Drug Marketing Act
PET	positron emission tomography
PET	polyethylene terephthalate
Ph. Eur.	European Pharmacopoeia
PIC/S	Pharmaceutical Inspection Co-operation Scheme
PLC	programmable logic controller
PPE	personal protective equipment

PQ	performance qualification
PQG	Pharmaceutical Quality Group
PQR	product quality review
PQS	pharmaceutical quality system
PUPSIT	pre-use post sterilization integrity test
QA	quality assurance
QC	quality control
QMS	quality management system
QRM	quality risk management
QSEAL	Quality Standards of Excellence, Assurance and Leadership
RA	regulatory affairs
RABS	restricted access barrier system
R&D	research and development
SAL	sterility assurance level
SAP	Systemanalyse und Programmentwicklung
SaaS	Software as a Service
SCADA	supervisory control and data acquisition
SIP	sterilization in place
SM	starting material
SME	subject matter expert
SMF	Site Master File
SOP	standard operating procedure
SST	system suitability test
SUS	single-use system
SUT	single-use technology
SWOT	strengths, weaknesses, opportunities, threats
TBA	2,4,6-tribromoanisole
TC	technical committee
TGA	Therapeutic Goods Administration
TLC	thin-layer chromatography
TMMDA	Turkish Medicines and Medical Devices Agency
TOC	total organic carbon
TQM	total quality management
UF/DF	ultrafiltration/diafiltration
UK	United Kingdom
ULPA	ultra-low penetration air (filter)
UNC	unclassified
USA	United States of America
USB	Universal Serial Bus
USP	US Pharmacopoeia
URS	user requirement specification
UV	ultraviolet
VICH	International Cooperation on Harmonisation of Technical Requirements for Registration of Veterinary Medicinal Products
VLP	virus-like particle
VMF	Validation Master File
WCB	Working Cell Bank
WF	water for injection
WHO	World Health Organization

1 The Concept of Quality

DOI: 10.1201/9781003284970-1

1.1 FIVE WAYS OF LOOKING AT QUALITY DEFINITIONS

During the course of days when I decided to write this book, discussions were going on whether to be good manufacturing practices (GMP) compliant in the pharmaceutical and biotechnology industries or whether the concept of quality precedes GMP. As an expert, consultant, trainer, project leader, and GMP lead auditor, who has worked 25 years in the industry, I can easily express that I have encountered numerous events in which I have observed that neither GMP nor quality management is sufficient for full compliance. Since the 2000 revision of ISO 9001, those who have followed the standard have witnessed how the standard has evolved into the latest risk-oriented management. While I remember information management and information security in the 2008 version of the standard, the 2015 version of the standard emphasizes the need for the management to carry out every activity with a risk focus. Before ICH Q10, I had followed the US Food and Drug Administration's (FDA's) initiatives on quality and GMP compliance, and the quality metrics models of the International Society for Pharmaceutical Engineering (ISPE) and FDA with admiration. Although it is approaching the concept of quality management far behind other industries, it is obvious that pharmaceutical and biotechnology companies want to obtain quality products for the first time, accurately and effectively, as well as GMP compliance.

However, the desire to spread the quality concept from the shop floor level to the top management seems to have caused us to be bombarded with "quality culture" without a clear understanding of the concept. If you cannot internalize the concept of quality in the axis of the product you have produced, and if you cannot attach a meaning to quality that prioritizes patient health, quality will remain only a rough term for you. In this chapter, I hope that the health industry will gain its own conclusions and gain a completely different perspective by addressing the different perspectives of different quality gurus on quality through their own recipes.

The quality manual is one of the documents that I request to be prepared before visiting a facility, even during GMP audits. However, when the point of view is an ISO 9001-based approach, even the quality manuals can be dragged to another medium by shifting from the GMP axis. In my opinion, the most important move that companies should make is that they should declare what they get from quality without being too attached to quality as a word.

1. Customer based
    - "Quality consists of the capacity to satisfy wants." (C.D. Edwards, "The Meaning of Quality," *Quality Progress*, October 1968)
    - "Quality is fitness for use." (J.M. Juran, ed., *Quality Control Handbook*, 1988)

2. Manufacturing based
   - "Quality is the degree to which a specific product conforms to a design or specification." (H.L. Gilmore, "Product Conformance Cost," *Quality Progress*, June 1974)
   - "Quality [means] conformance to requirements." (P.B. Crosby, *Quality Is Free*)
3. Product based
   - "Quality refers to the amount of the unpriced attributes contained in each unit of the priced attribute." (K.B. Leifler, "Ambiguous Changes in Product Quality," *American Economic Review*, December 1982)
4. Value based
   - "Quality is the degree of excellence at an acceptable price and the control of variability at an acceptable cost." (R.A. Broh, *Managing Quality for Higher Profits*, 1982)
5. Transcendent
   - "Quality is neither mind nor matter, but a third entity independent of the two … even though quality cannot be defined, you know what it is." (R.M. Pirsig, *Zen and the Art of Motorcycle Maintenance*)

Table 1.1 shows examples of quality expectations in different industries. While fast delivery stands out in the food industry, defect-free is the biggest expectation in the automotive industry. When we look at the health sector, the most obvious expectation is the correct diagnosis of diseases. Therefore, expectations may vary according to the dynamics of each sector, but the common denominator in all of them is customer satisfaction.

John Ruskin, the Victorian art critic, suggested that "quality is never an accident, it is always the result of intelligent effort." In other words, "quality" doesn't just happen – we have to work at it!

Across the Atlantic, Ralph Waldo Emerson, the 19th-century American poet, essayist, and philosopher, proposed that a "quality product," even if only a mousetrap, was something worth seeking out! "If a man should paint a better picture, preach a better sermon, or even build a better mousetrap, then even though he builds his house in the middle of a forest, people will make a path to his door."

People will make a path to your door! And that is what an aspect of quality is about – getting people to make a path to your corporate door!

Ask any group of people what quality is and what it is about and you will get a myriad of answers. Quality is undoubtedly a movable feast – and what is quality for one person is not necessarily quality for another.

TABLE 1.1
Different Quality Expectations

Area	Example
Airlines	On-time, comfortable, low-cost service
Health care	Correct diagnosis, minimum wait time, lower cost, security
Food services	Good product, fast delivery, good environment
Postal services	Fast delivery, correct delivery, cost containment
Academia	Proper preparation for future, on-time knowledge delivery
Consumer products	Properly made, defect-free, cost effective
Insurance	Payoff on time, reasonable cost
Military	Rapid deployment, decreased wages, no graft
Automotive	Defect-free
Communications	Clearer, faster, cheaper service

Generally, though, such answers will center on well-rehearsed clichés such as "meeting specifications" or "the right product, at the right time and at the right price." "Getting it the right the first time, every time" and "making goods which do not come back for customers who do" would also surface in any discussion. We may even have definitions of quality thrown in, such as Joseph Juran's "fitness for purpose" and Philip Crosby's "conformance to requirements."

ISO, the International Organization for Standardization, in "ISO 9000:2005 – Fundamentals & Vocabulary," defines quality as "the degree to which a set of characteristics fulfills requirements."

The word "quality" has its origin in the Latin word *qualitas*. It means "general excellence" or "distinctive feature." The *Oxford Reference Dictionary* defines quality as "A standard of how good something is as measured against other similar things." If we try to analyze this definition we come across some keywords:

- Standard
- Goodness
- Measurement
- Comparison

Quality in the production process denotes exceptional levels of performance, or a point where there are no faults, deficiencies, or noticeable changes. This state is achieved by maintaining an unwavering commitment to specific standards that aim for product consistency, so as to meet unique needs of customers or users. The ISO 8402-1986 standard describes quality as "the sum of product or service features and properties that contribute to its capacity to meet expressed or implicit requirements."

1.2 QUALITY IN PHARMA AND BIOTECH INDUSTRIES

ICH Quality Guideline Q6A Specifications Quality: The suitability of either a drug substance or drug product for its intended use. This term includes such attributes as identity, strength, and purity.

ICH Quality Guideline Q9 Quality Risk Management Quality: The degree to which a set of inherent properties of a product, system, or process fulfills requirements.

1.3 WHAT ARE THE DIMENSIONS OF QUALITY?

- Performance
- Features
- Reliability
- Conformance
- Durability
- Serviceability
- Aesthetics
- Perceived Quality

These dimensions can serve as a framework for strategic analysis. Some of these are always mutually reinforcing; some are not. The challenge to managers is to compete on selected dimensions.

An application similar to the PDCA (plan, do, check, act) cycle is in Table 1.2. The CAMC cycle includes control, assurance, management, and culture. This cycle offers us an insight into how quality evolves.

TABLE 1.2
How Pharmaceutical Quality Evolved

Phase 1 Q-Control
- Testing
- Monitoring

Phase 2 Q-Assurance
- Methods
- Processes

Phase 3 Q-Management
- QA-Oversight
- PDCA
- Kaizen

Phase 4 Q-Culture
- Q-Control
- Q-Assurance
- Q-Management
- Human behavior

Quality culture combines the technical aspects of "what we have to do" with the psychological aspects of human behavior.

1.4 WHAT IS QUALITY BEHAVIOR?

Quality behaviors are observable or countable actions (in doing or saying) of an individual or a group that positively determine or influence the quality of a product or service.

Quality behavior is relevant in all tasks of our daily business:

- Perform a manufacturing step
- Perform an analytical test
- Write a standard operating procedure (SOP) or instruction
- Train a new colleague
- Investigate a deviation or complaint
- Recruit new employees
- Coach employees
- Chair a meeting or participate in a meeting
- Call someone on the phone
- Write an email

As John Ruskin said, "Quality is never an accident; it is always the result of intelligent effort."

1.5 WHAT IS QUALITY?

- A degree of excellence – The meaning used by the general public.
- Freedom from deficiencies or defects (Juran) – The meaning used by those making a product or delivering a service.
- Conformity to requirements (Crosby) – The meaning used by those designing a product or a service or assessing conformity.
- Fitness for use (Juran) – The meaning used by those accepting a product or service.

- Fitness for the purpose (Sales and Supply of Goods Act 1994) – The meaning used by those selling and purchasing goods.
- The degree to which a set of inherent characteristics fulfill requirements (ISO 9000:2005) – The meaning used by those managing or assessing the achievement of quality.
- Sustained satisfaction (Deming) – The meaning used by those in upper management using quality for competitive advantage.

The *Oxford English Dictionary* (*OED*) contains 17 meanings of "quality," most of which relate to personal characteristics, but the following three are relevant to the quality of the product:

- A particular class, kind, or grade of something, is determined by its character, especially its excellence. (This meaning is addressed further under the classification of products and services.)
- The standard or nature of something as measured against other things of a similar kind.
- The degree of excellence possessed by a thing.

1.5.1 Quality System

The quality of a system is the sum of all that is necessary to implement an organization's quality policy and meet quality objectives. It includes organizational structure, responsibilities, procedures, systems, processes, and resources. Typically these features will be addressed in different kinds of documents such as the quality manual and documented procedures, modus operandi, etc.

1.5.2 What Does Quality Mean to You?

If the quality risk management (QRM) applied in pharmaceutical companies is well adapted and clearly understood by the employees, it is successfully applied both in drug development and commercial production stages and can quickly resolve potential problems that may arise in the future, and the degree to which the supporting process and any supporting process meet predetermined criteria.

However, some quality experts argue that there cannot be a single definition of quality because strict formal thinking will always prevail. However, in my opinion, this approach may have its shortcomings, because in the research I have done during the preparation of this book, I have determined that quality is its main mission, such as meeting customer expectations, and therefore it should be clearly defined. I would like to express that defining quality with a single characteristic is a narrow viewpoint. According to Juran, quality means suitability for use and avoiding deficiencies. I was a little surprised to see two different definitions of quality on the International Council for Harmonisation (ICH) website. While quality in ICH Q6A is expressed as being fit for use, another definition caught my attention in the old version of ICH Q9, which has been used in almost all pharmaceutical and biotechnology industries, published in 2005. ICH Q9 presents quality as conformance to requirements. Measuring the quality of the pharmaceutical product, raw material, and packaging material, after its production, and valuing the related product has always seemed counterintuitive to me. Quality with a design seems more realistic to me and I am a firm believer that quality should be designed in the product when it is applied consciously by the right people.

So, how should it be evaluated from the patient's point of view when trying to define quality? As a result, the patient wants to receive the right dosage, right purity, right potency, and right production process drug product every time. For the patient, meeting these criteria may mean access to quality medicine. However, the patient can hardly distinguish the quality of the drug product according to the definition. The indicators of poor quality that may be encountered during use are color, appearance, taste, etc. It is a visual difference, however, it is not possible to notice the deterioration of the drug product, the use of the missing or excess active pharmaceutical ingredients (APIs), and the microbial risks it carries. In some cases, the end user may make the mistake of ordering the drug over the internet, which is never safe but inevitably unavoidable, for uninterrupted access to the drug, and the results can be tragic. In this case, the economic cost of quality is regulated under the supervision of governments, not taking into account the demands of the consumer. As a result, the primary goal of QRM applications should be to detect all the risks that the patient may face in advance and to seek ways to reduce them even if they cannot eliminate them. It is regrettable to say that low-quality medicines pose a very serious public health threat in both developed and developing countries. We can easily talk and express our opinion on the second or even the third quality of the products or services we buy in daily life, however, it is unthinkable to talk about the second quality when it comes to pharmaceutical products and the protection of patient health. But we know that there are enormous quality differences even between pharmaceutical products offered to the domestic market and exported. In the preparation phase of this book, in October 2022, a new diethylene glycol (DEG) contamination and a case in which nearly 80 children lost their lives hit our agenda as a bombshell. Previously, there was great concern and confusion in the global sense with the sartan case in 2018 and the incessant withdrawals. If so, why do these events occur in the industry where so many cases are experienced? Is the active substance used in the drug product incorrectly, the amount missing or excessive, have other components not passed the necessary tests, are the production areas not planned according to the GMP guidelines and directives, are the stability studies insufficient, are there packaging errors, are storage and shipment not carried out under the desired conditions?

We all know that the answer to most of these questions is yes. Even if you are in a developed country, you are either late or insufficient to monitor the problems at their source.

1.5.3 How to Build Customer Focus

One of the approaches that I think is one of the reasons for failing in the quality journey in pharmaceutical companies and is applied incompletely is the concept of customer focus. First, you need to build trust. If you are one of the companies that come to mind the most and people can remember at least a few of your products, it means you have built trust. Of course, the opposite may be the case. For some reason, when one thinks of data integrity, which is the trendy topic of today, Barr or Able Laboratories comes to mind, right? Or, if your name is one of the few companies that come to mind when it comes to vaccines since the COVID-19 pandemic, it means that you have built trust in one way or another. You not only produce quality pharmaceutical products but also create customer value.

1.5.4 How Valuable Is Your Brand?

When it comes to patient health, the desire to remedy their problems and to help them in every way is like proof that you are successfully implementing the pharmaceutical quality management system (QMS). A quality management system that you have established by using ICH Q10 not only affects the quality of your products but also is a factor in their rapid delivery to patients. Although a customer-oriented approach is displayed or it is desired to increase the brand value, it cannot be expected that every pharmaceutical company will do what a company like Apple does. Focusing on the customer in a multivariate and yet complex industry requires tough challenges.

1.5.5 How Loyal Are Your Customers?

Pressures faced by pharmaceutical companies include keeping up with changing market conditions, patent expiration, timely product launches, research and development (R&D), multivariate circumstances, and the changing role of the patient. With the introduction of pharmaceutical companies to the market with products that are intertwined with medical devices, and the introduction of different treatment options, interaction with patients has started to increase even more. At the same time, huge moves and advances in the life sciences have led to patients' tendencies toward more personalized care and service. If you are trying to reach the customer by claiming to be innovative, you have to get one step closer to the patient each time by focusing on R&D activities. If you can spare money, people, and time to adapt innovations such as big data, machine learning, digital twins, ChatGPT, artificial intelligence, blockchain, the Internet of Things, and the dissemination of digital technologies, it means that customer loyalty will begin to form by itself. For example, even developing a digital solution for rheumatoid arthritis patients can increase your sales by almost 10%.

1.5.6 Can You Involve Your Employees in Your Processes?

When quality is mentioned, if you think of a checklist where a right–wrong, pass–fail, or good–bad evaluation is made; or the indispensable goal of top management; or the only and main task of the quality unit, you are on the wrong track. Take the first exit and take the right road. Quality is the common responsibility of everyone, from the R&D director to the officer who cleans the production areas. Therefore, everyone has something to say when it comes to quality. When you include the staff who clean the production areas in quality meetings, you will be surprised to see that there is so much to learn. Edward Deming, one of the first names that comes to mind when it comes to quality, emphasizes that employees are always responsible for quality. If you have process development and improvement plans, you better listen to the employees. Using the ISO 9001 Quality Management System as a source, a study was carried out to identify the components that have an impact on quality.

At the end of the study, eight components were identified:

- Customer focus
- Leadership
- Employee engagement
- Process approach
- Systemic management
- Continuous improvement
- Evidence-based decision-making
- Mutually beneficial supplier relationships

Training, continuous open communication channels, transparent management, rewards, encouragement, elimination of fear of job loss, suggestion systems, valuation, and at least even addressing all employees by their names can create value on their own.

1.5.7 Is Your Approach Process Oriented?

If the way you do business in your company is based entirely on individual, department-specific responsibilities and duties that are not in communication and interaction with each other, you are again on the wrong path. However, when all processes interact with each other and the task sharing between departments is result and process oriented, you are more likely to produce a product without compromising quality. In fact, with the process-oriented management model, you will ensure continuous improvement, you will be able to predict the results, you will witness your performance increase, and you will spread a solid quality management system. The first and most important job you will do will be to write clear SOPs that employees can easily understand and perform their activities without errors. Years ago, pharmaceutical companies put more responsibility on stakeholders in the supply chain and scapegoated them in case of any problem. If the raw materials you purchased passed the quality control tests, you could easily offer your pharmaceutical product to the market. However, with the developing technology, and the latest updates in guidelines and directives, the responsibility is transferred to the marketing authorization holder throughout the drug life cycle. The process-oriented pharmaceutical quality management system model has to be applied carefully in pharmaceutical companies that are completely dependent on external resources. The quality system should provide transparency throughout the product life cycle and ensure that all stakeholders in the supply chain adopt a risk-based approach.

Pharmaceutical companies that are process oriented should constantly monitor the effective pharmaceutical quality management system.

In this context, the following are critically important:

- Effective risk assessment applications for suppliers
- Conducting on-site audits of suppliers
- Monitoring, documenting, and constantly reviewing a supplier's performance
- Establishing up-to-date procedures regarding the management of suppliers

1.5.8 Is Your Quality Management System Integrated?

Pharmaceutical quality management systems form a general framework for the management of interrelated processes. Whatever you do, transparency should be at the forefront of your processes. A pharmaceutical quality management system supported by GMP and ICH Q10 is well-founded and offers clues as to what an integrated system should look like. It has always been profitable to utilize systems such as ISO 9001 in the creation of ICH Q10.

With the establishment of an integrated QMS, it is possible to support the following:

- Quality policy
- Employees
- Development and implementation
- Generate Documentation
- Facilities and equipment
- Self-monitoring
- Management responsibilities
- Evaluation of suppliers and purchasing
- Complaints and withdrawals
- Corrective and preventive actions
- Measurement, analysis, development

1.5.9 Is the Strategic Approach Important?

For some reason, leadership is often evaluated only by its dictionary meaning and it is not understood how full its content is. Especially if you are developing and producing pharmaceuticals or biotechnology products that involve immediate access and use, unlike other industries, leadership may be an even more out-of-the-box approach. However, leadership is an integral element of the systemic and strategic approach in quality-oriented companies. The management team is responsible for developing the organization's vision, mission, strategy, tactics, and goals, involving everyone in this goal, and encouraging staff to pursue quality goals. The most important task of the leadership team in a pharmaceutical company is to develop a culture of quality and GMP and to be at the forefront of applications. The concept of leadership can also lead to misunderstandings as if it is only the responsibility of the top management. Through leadership, the pharmaceutical company will embrace being forward-thinking. It is important to apply the principles of Kaizen to take small but efficient steps toward long-term goals by acting more rationally and in a planned way, instead of only focusing on immediate goals and spending resources in that area. In this way, it will enable the possible risks that may come to the fore to be determined and eliminated in advance.

A successful pharmaceutical quality management system plays a key role in achieving a pharmaceutical company's short-, medium-, and long-term goals. Thanks to the leadership of the pharmaceutical quality management system, the coordination of the processes within the organization, real-time cooperation between the departments, and the training and education of the personnel can be kept up to date.

1.5.10 How Do You Ensure Continuous Improvement?

Continuous improvement means that a system is constantly readapted according to innovations and kept up to date. Continuous improvement and development activities are at the forefront of the success of a pharmaceutical company. The letter *c* in cGMP refers to the ability of a pharmaceutical company to constantly update itself and to keep up with the developing technology quickly. In this context, it is not just an effort to adapt the guidelines and directives published by health authorities. Pharmaceutical companies have to keep themselves up to date not only with advancing technology but also with changing guidelines and directives, and changing patient expectations and needs, while trying to meet ruthless competitive conditions, increasing costs, and strict regulatory expectations.

So, how and what can a pharmaceutical quality management system being implemented in a pharmaceutical company contribute to continuous improvement?

- Ensuring centralized management of improvement goals
- Raising employee training on SOPs and best practices
- Evaluating training performance
- Ensuring improvement collaborations among employees
- Continually monitoring and auditing progress

1.5.11 Are Your Decisions Based on Facts?

You will be successful and reach your goals to the extent that you can make your decisions based on data. Although the evidence-based decision-making approach is a concept that catches our attention in ISO 9001, it is an important ISO component that pharmaceutical companies need to adopt more widely in my opinion. Whether you take a step-by-step approach to quality goals or explore any deviations or noncompliance with the root cause, at the end of the day, you'll always have data to turn to. We now know clearly that pharmaceutical companies must continuously use data throughout the product life cycle. In a very simple statement, it is possible to decide where the changes will work, and when and how changes should be made to maintain product quality. Thanks to the data, it could be. Are your communication channels open?

How do you think the communication language you use throughout the company or to reach your patients communicates with your activities included in the quality management system? Effectively interacting with your employees, your processes, and your technology with your processes is possible thanks to an effective communication language. The communication language you use also supports your harmonious work with your suppliers and your valuable stakeholders and means more productive collaborations. Communication has become more important than ever, thanks to health authority expectations, sanctions, and changing patient needs. The pharmaceutical quality management system you are implementing must also ensure that the roles in communication are easy to understand.

1.6 QUALITY AND PRICE

What percentage of people who read this book do you think evaluate certain products and services based on price? Before we buy a product or service, we think that we can get the same features at a lower price. We can't help but make this comparison, hoping we'll get the same features.

"Quality is free." Could this statement be true? If we can eliminate all errors before, during, and after our production process and reach our zero-error target, then we can reduce costs and increase our customer satisfaction level. Doing the right things right the first time, not doing the right things right the first time.

1.7 QUALITY COSTS

Deming, Juran, Crosby, Ishikawa, and Taguchi, the important quintet of quality-guiding gurus, for 80 years, have promoted the use of quality tools and techniques to improve product quality and reduce operating costs. The contributions of the masters have made a significant impact on the world, improving not only production operations but also entire organizations, including the military and educational institutions. One of the tools and techniques, cost of quality, allows organizations to analyze and address the cost of poor quality.

As we manage day-to-day operations, do product testing, and engage with release or shipping, we try to meet the needs of various departments simultaneously – yet we know that each has seemingly different goals and expectations. On the one hand, increasing pressures on costs from the finance department and often the boardroom, and on the other hand, the increasing expectations for technical and validation requirements from health authorities. Finally, as long as the product reaches the market as quickly as possible, our sales and marketing colleagues may not care how much validation is done.

The "cost of quality" is a widely used but misunderstood phrase! The cost of quality is not the cost of producing a quality product or service. It is the cost of not creating a quality product or service. If we are doing what we are doing all over again, again, and again, the cost is doomed to be the cost of really low quality.

There is very little published data on quality costs, especially in the pharmaceutical, biotech, and medical device industries. The number of specific data published for the pharmaceutical industry can be counted on one hand, but the critical point to remember is that the pharmaceutical industry is scrutinizing quality costs very carefully and the available information is unofficial.

From the perspective of traditional quality costs, it is noteworthy that the quality costs for the pharmaceutical industry constitute approximately 10%–15% of the total production costs. Although the measurable costs are high, the hidden costs can be even greater. It is worth considering the cost of recalls and complaints. Recalls can push a product or company into a dark pit at worst and, at best, damage the product and the company's reputation, resulting in reduced sales and profits.

Permanent failure of the autoclave in a sterile production area in a facility can negatively affect morale among staff and create friction between departments,management, and employees. A validated and controlled process will result in fewer internal failures: less rejection, rework, retesting, reexamination, and waste.

As it is known and accepted by the authorities, validation makes it possible to do the job right the first time.

In addition, a scientifically studied and controlled process eliminates the possibility of sending defective products to the end-user patient; therefore, the rate of withdrawals or complaints will also decrease. Spending resources on prevention (validation) will also enable the drug manufacturer to reduce evaluation (testing) costs. Failure costs (also known as non-quality costs) are those associated with getting things done wrong. There may be intangible costs such as rejection, rework, loss of sales, damage to image, or sanctions from health authorities.

A striking example of the costs of failure encompassing many of these categories is an incident I experienced while consulting for an international pharmaceutical company. Contract manufacturers–suppliers carry out the production of secondary packaging materials. The remaining cardboard from the production of secondary packaging materials should be destroyed by the supplier by the mutually signed quality agreement, and the destruction form should be sent to the customer. However, if you receive a report that some of the secondary packaging materials are used as countertop protectors in hawker stalls in the public market of a province, the situation may turn in an unethical direction. The tracing work, which started from waste cardboard, took us to a local printing house. At an urgent meeting with company officials, my client was the quality director and general manager. His assistant expressed deep concern. Who do you think was billed?

The printing house could not verify the methods it claims to have applied for the disposal of waste packaging materials. The most striking sentence of the defense was as follows: "A pirate company took 20 sheets of waste cardboard from the disposal area with the help of a mole working in a printing house and used it to package and sell counterfeit pharmaceutical products."

1.7.1 Preventive Costs

Preventive costs are costs incurred to prevent failures or reduce evaluation costs. Preventive costs are costs associated with ensuring activities are done correctly.

a. Quality planning
b. Supplier approval system
c. Training and education

The goal of every manufacturer is to produce a quality product in a reasonable time with minimum cost. Compared to the loss of income that can result from inadequate education, the cost of education is minimal. Critical evaluation: "If you think education is expensive, try ignorance." Here are the list of documents such as SOPs, monographs:

a. Preventive maintenance and calibration
b. Hygiene and sanitation
c. Process validation
d. Internal audit
e. Annual review of data or trend analysis

1.7.2 Evaluation Costs

Evaluation costs are the costs of testing and quality assessment. Evaluation costs are related to checking that activities are done correctly. However, it should be emphasized that it is not an activity that adds value in itself. It can be taken as a "historical measurement of what has already happened."

Some examples of evaluation costs:

a. Testing of raw materials and packaging materials
b. Testing of in-process materials
c. Testing of finished products
d. Stability testing
e. Control of equipment and devices

1.7.3 Internal Failure Costs

Internal failure costs are costs associated with nonconforming material (material that does not meet quality standards) that is still in the company's possession, e.g., a solid dosage form that has not yet been released.

Some examples of internal failure costs are:

a. Rejects
b. Reworks

c. Rechecks
d. Retests
e. Waste and scrap
f. Troubleshooting
g. Substandard material research

1.7.4 External Failure Costs

External failure costs are costs associated with a nonconformity situation after the product has left the company's boundaries.

Some examples of external failure costs are:

a. Recalls
b. Complaints
c. Returns due to quality-related issues
d. Sale losses

Most experts agree that unless the defect is detected early, the cost of a defect will typically increase tenfold at each major machining point. Once a defect is put into production, the cost to the business increases tenfold through product rework or research time.

When the defective product is then shipped to the end-user patient, the costs increase exponentially to 100 and can result in an elaborate withdrawal and investigation process, in addition to the potentially devastating consequences of harming a patient.

1.8 ISO 9000 SERIES

The ISO 9000 series is a subset of the family of ISO/TC 176 standards.

Figure 1.1 shows the most used ISO standards and targets that we can use in GMP audits.

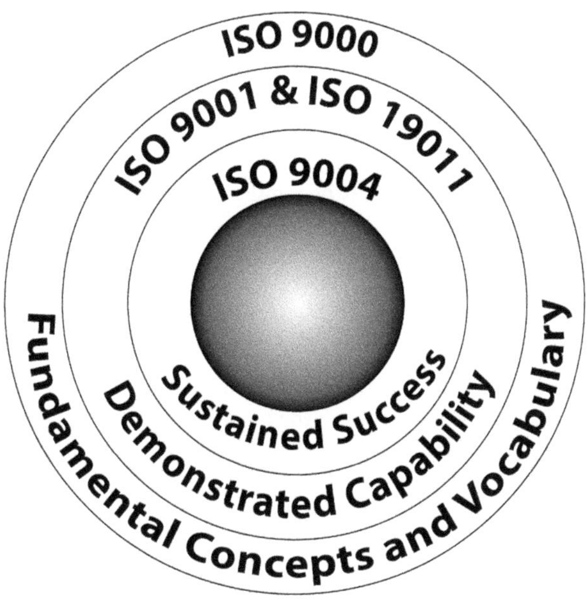

FIGURE 1.1 ISO family of standards.

1.8.1 Purpose of ISO 9001

ISO 9001 can be used to measure whether companies meet the criteria required by their customers or regulatory requirements.

1.8.2 Intent of ISO 9001

ISO 9001 is a standard with rules used to achieve quality assurance in products or services. For this reason, it is binding only in documentation or contracts.

1.8.3 Scope of ISO 9001

The main purpose of ISO 9001 is the combination of the requirements of a quality management system that allows companies offering products or services to manage customer satisfaction sustainably.

1.9 OTHER QUALITY MANAGEMENT STANDARDS

1.9.1 Aerospace Industry (AS 9100)

AS 9100 is the international quality management system standard for the aviation, space, and defense (AS&D) industry, created by the International Aerospace Quality Group (IAQG), and AS9100 Rev D (2016) is the most recent version. The standard provides suppliers with requirements for creating and maintaining a comprehensive quality system for providing safe and reliable products to the AS&D industry, as well as civil and military aviation requirements.

1.9.2 Automotive Industry (IATF16949:2016)

IATF 16949:2016, formerly known as ISO/TS 16949, is a global standard that concerns automotive manufacturers and suppliers. The aim of the standard is to gather standards in the USA and Europe under one roof and is an indispensable reference guide in the automotive supply chain.

With the development of this standard, common processes in the automotive industry were brought together. Almost all global automotive brands are among the members of IATF and, thanks to this standard, safe vehicles continue to be produced every day.

1.9.3 Computer Software (ISO/IEC/IEEE 90003:2018)

The ISO/IEC/IEEE 90003:2018 standard has become a standard used in the field of software engineering by integrating the main requirements of the ISO 9001:2015 standard for software developer companies.

1.9.4 Information Security (ISO/IEC 27001:2022)

The ISO 27001:2022 standard adopts a process approach for establishing, implementing, operating, monitoring, reviewing, maintaining, and improving an organization's information security management system. ISO 27001 was established by the ISO. It was first launched in 2005 as a replacement for BS 7799.

1.9.5 Cosmetics (GMP ISO 22716:2007)

ISO 22716:2007 is the international standard for GMP for cosmetics. The guidelines for the standard have been approved and accepted by many global regulatory bodies, including the International

Cooperation on Cosmetics Regulation (ICCR), US FDA, and the European Committee for Standardization (CEN).

1.9.6 Primary Packaging Materials for Medicinal Products (ISO 15378:2017)

ISO 15378:2017, "Primary packaging materials for medicinal products — Particular requirements for the application of ISO 9001:2015, with reference to good manufacturing practice (GMP)," specifies requirements for a QMS when an organization needs to demonstrate its ability to consistently provide products and services that meet customer and applicable statutory and regulatory requirements, to enhance customer satisfaction through effective systems, continuous improvement, and commitment to customer and applicable statutory and regulatory requirements. ISO 15378:2017's GMP principles in production and control of primary packaging materials are important for the safety of a patient receiving the medicinal product because of the direct contact between the packaging materials and the product.

1.9.7 Medical Devices – Quality Management Systems (ISO 13485:2016)

ISO 13485:2016 standard is a standard that is based on ISO 9001 and regulates many main rules, including the design, development, validation, and verification of medical devices. It provides guidance to companies providing medical devices and related services and assists companies providing these services at all stages, from the design and development stage to production, storage, and distribution. This standard is also adopted by organizations that act as suppliers to companies that provide medical device production or services.

1.9.8 Specification for Security Management Systems for the Supply Chain (ISO 28000:2022)

ISO 28000:2022 is an ISO 9001-based standard that tries to define all critical requirements regarding the security of the supply chain. Although the pharmaceutical and biotechnology industries, which constitute the majority of counterfeit and adulterated products today, are hesitant to implement this standard, the purposes of the standard encourage them to implement it.

1.10 OTHER QUALITY MODELS

1.10.1 Business Excellence

The EFQM model, preferred by many organizations, gives very successful results. Developed by the European Foundation for Quality Management, the model argues that the main reason for the existence of companies is to achieve remarkable success in the global market and that quality is the most determining factor in this goal.

No matter where your main purpose is, you may be surprised or upset to see how well your current situation fits with the EFQM model. Because the greater the distance, the longer and more troublesome your journey may be. While evaluating the results, your financial situation, your customer satisfaction score, your personnel satisfaction level, the impact you leave on society, and your success and agility in managing your processes are of great importance. First, after you plan how you will improve, you can move on to the step of what you will improve. So how does the top management view this situation?

If you successfully implement this model of excellence, you will witness your performance increase rapidly and you will focus more systematically on thought and action.

Many different models of business excellence are offered in the world

Although there are minor differences, almost all of the following models are focused on the same goal:

- Baldrige
- EFQM
- Japan Quality Award Model
- Canadian Business Excellence Model
- Australian Framework for Business Excellence (ABEF)

1.10.2 EFQM Model (New) 2020

The EFQM model, one of the most outstanding quality models, offers a detailed and understandable plan for the development and improvement of organizations in both Europe and other continents. Many companies, from the defense industry to the automotive industry to the nuclear industry to the pharmaceutical industry, are successfully applying the EFQM model. To successfully implement the EFQM model, you need the following seven criteria:

1. Goals, vision, and strategies
2. Company culture and leadership abilities
3. Full participation of all stakeholders
4. Sustainability of values
5. Performance and transformation movement
6. Stakeholder perceptions
7. High strategic and operational performance

1.10.3 Lean Six Sigma

The pharmaceutical industry is a strategic economic sector that gathers research, manufacturing, and marketing medicinal products for humans or veterinary medicines. It is one of the most profitable and economically important industries in the world.

Many pharma and biotech businesses implement Lean Six Sigma (LSS) practices to lower their operational costs and provide high-quality products to customers.

The principles of continuous improvement are certainly not new. From Stewart and Deming to Ohno and Shingo the benefits of Lean Manufacturing and Six Sigma are well recognized. However, despite the well-publicized benefits, the pharmaceutical and biotech industries have been slow to fully embrace these principles and benefits. The body of evidence has continued to grow as to the benefits of the methodology as companies including Southwest Airlines, Bank of America, and Pfizer have proclaimed the virtues of its structured methodology and tangible business improvement entitlements. However, the level of organizational inertia required to fully embrace Six Sigma to the same levels as General Electric (GE) requires significant corporate commitment. Lean Manufacturing, on the other hand, requires very little organizational inertia to realize efficiency improvements. However, the luster has diminished somewhat because of quality issues over the last decade. Given the other market challenges, our industry has lost its enthusiasm for LSS. However, as a Lean Six Sigma Master Black Belt, I confess that the benefits and principles of these methodologies have been firmly demonstrated. I believe what sets successful Lean corporate cultural transformations apart from the rest revolves around the integration of LSS philosophies into the drug development life cycle.

1.10.4 Operational Excellence

As per the Operational Excellence Society, operational excellence is defined as the constant and deliberate improvement of an organization's performance and the conditions of those who work there – to pursue operational excellence by design and not by accident. Operational excellence is required for all sectors. Operational excellence in the current business environment spreads beyond the organization's shop floor, incorporating the whole supply chain, and optimizing processes between the organization, suppliers, and customers. During the whole production, supply chain, and order-to-delivery cycle, operational excellence is about fulfilling customer requirements by knowing where everything is, how much is required, and what is supposed to be done to deliver a quality product. In addition to the supply chain, operational excellence supports organizations to constantly provide quality products with sustainable performance where potential risks are identified up front and proactively mitigated.

1.11 ISO 9001 STANDARD INTRODUCTION

ISO 9001 was initiated in 1987 with a primary focus on manufacturing. There were 20 requirements, or clauses, each stand-alone, and it covered not only manufacturing but also included requirements for purchasing, sales (contract review), and design. The 1994 revision of the standard was primarily just a continuation of the ISO 9001:1987 version of the standard, and the changes were incremental. In 2000, the standard made a radical change and became more of the standard it is today, that is, customer focused and process focused. Along with the customer and process orientation, the standard included requirements for customer satisfaction and continual improvement. Quality was thought of as "inspection" over 100 years ago and has slowly evolved from inspection to quality assurance to design to business strategy. For the first time, ISO 9001 is embracing risk, which is defined as the "effect of uncertainty" in the organization. If processes help meet objectives (definition of management), then risk analysis is conducted on the processes to ensure that objectives are met. The interaction of the "context" – a new term in ISO 9001 – i.e., business processes, objectives, goals, and leadership, will lead to some interesting auditing opportunities.

In 2015, the standard underwent another radical change. "Radical change" is a misnomer. Most importantly, it has taken on the concept of risk, which it defines as the "effect of uncertainty." The second major change in the 2015 revision was the reorganization of the standard into what ISO calls its high-level structure (HLS). The HLS change altered the overall structure of ISO 9001:2015 and of all ISO's management system standards (MSS). This is a conceptual change from the old process model, incorporating context, interested party expectations, and risk. Most importantly, it has leadership in the center ensuring that the PDCA cycle is working.

1.11.1 Pharmaceutical QMS and GMP

- PIC/S Guide to Good Manufacturing Practices – PE 009-16 (as of February 1, 2022)
- EU Guide to Good Manufacturing Practices (Eudralex Vol. 4)
- World Health Organization (WHO) cGMPs
- United States – FDA CFR Part 21
- 21 CFR Part 210/211 for Drugs – current GMPs
- 21 CFR Part 820 Quality Systems for Medical Devices – current GMPs
- 21 CFR Part 58 – cGLP
- ICH Q7 GMP for Active Pharmaceutical Ingredients
- Canadian cGMP (aligned with PIC/S)
- ISO 13485: 2016 – Medical Devices
- ICH Guidance Documents – Technical Standards
- EU/PICs/cGMPs – Chapter 1 – Quality Management
- ICH Q8 – Pharmaceutical Product Development
- ICH Q9 - Risk Management in Pharmaceuticals
- ICH Q10 – Pharmaceutical Quality System
- FDA Quality Systems Approach to Pharmaceutical cGMP Regulations (9/2006)

1.11.2 Pharmaceutical Quality Management System

There are several reasons for the development of the pharmaceutical quality management system concept in the field of pharmaceuticals and biotechnology. Until the 1980s, the only factor determining drug quality was testing the product at all costs. Even this concept alone claimed to provide evidence that a quality drug product was produced. However, as in other industries, adopting this approach in pharmaceutical products meant testing 100% of the product. In this case, there would be no drug product to be offered to the patient. In order to prevent this, although the approach of testing the samples representing the whole lot came to the fore, here again, it started to be considered that a small amount of quality does not mean that the whole lot is of good quality. In Figure 1.3, it is explained why the raw material from the supplier must pass quality control, why it must have a certificate of analysis, and that it must be usable for production after release.

Releasing the entire batch by checking only the samples can mean that serious defects are overlooked. Even if all microbiological tests of drug products on the market are successful, there is always a risk of contamination over time. In other words, it has been understood that testing alone cannot reliably determine whether a product meets the predefined specifications, and the concept of quality assurance has emerged in this way. In Figure 1.2, the stages of quality since the 1970s are explained. It can be seen that sterilization and validation were prominent concepts in those years. To

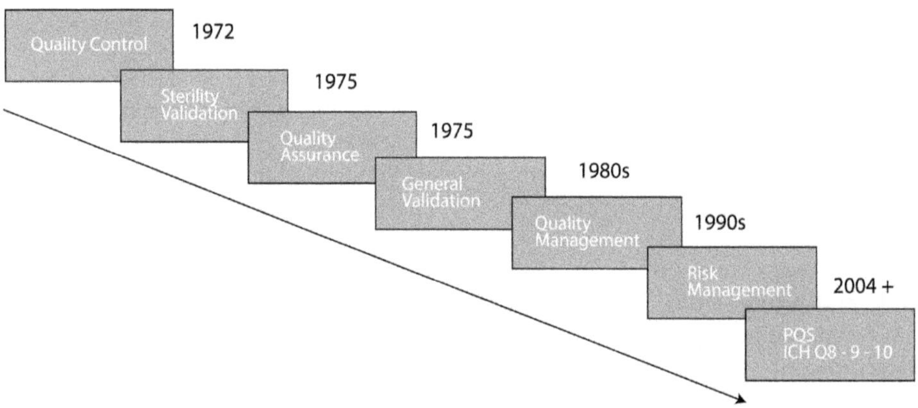

FIGURE 1.2 Evolution of the quality concept.

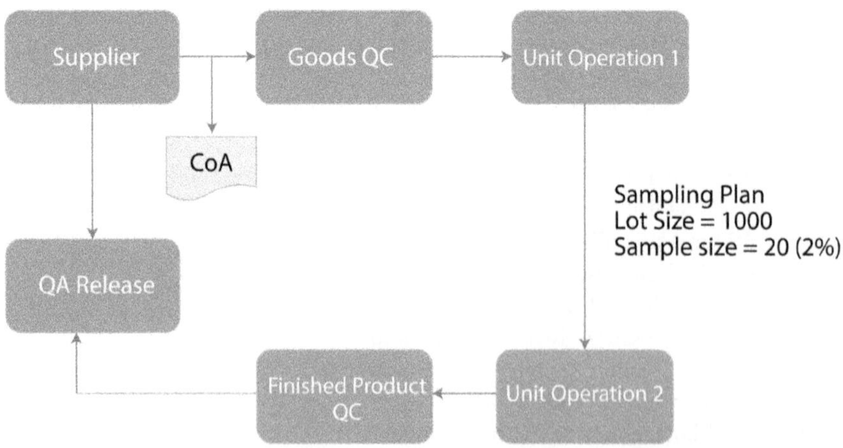

FIGURE 1.3 Quality control and sampling.

determine the quality of pharmaceutical or biotechnology products, additional data on how they are produced is needed. This insight has led to the development of measures such as the detailed review of batch records by quality units, the investigation of defects and the identification of root causes, and the approval of production documents.

In addition to all these developments, as you will remember, the FDA introduced the concept of efficiency for the pharmaceutical industry for the first time in 2001. Referring to the dizzying developments in the pharmaceutical and biotechnology industries, the FDA stated that manufacturers need to be more efficient to bring a product of sufficient quality to the market within the expected time frame. After three years, the FDA started to highlight the concept of a pharmaceutical quality management system in 2004. At this point, as an expert, consultant, and auditor who has been actively working in the pharmaceutical industry for 25 years, I am ashamed to state that the pharmaceutical industry is the industry that has adopted the quality management concept at the latest. However, the quality management system has been a partner in process robustness and accompanying cost-reduction efforts in other industries for many years. Thanks to the awareness-raising efforts and incentives of the FDA, a compact document such as ICH Q10 has finally been put into practice and made available to drug manufacturers. This move marked a formal paradigm shift in the pharmaceutical and biotech industries.

1.12 WHAT IS ICH Q10?

I believe that to be able to evaluate the ICH Q10, first, it is necessary to know ICH Q8 and Q9 very well and to absorb the close relationship between the three. Although ICH Q10 alone plays a key role in the development of any quality management system, it can become a more useful tool when used effectively with these two guides. As I have always defended, unfortunately, the pharmaceutical industry has witnessed the development of the ICH Q10 model, although it has lagged behind industries such as the aerospace, defense, and nuclear industries that make the concept of quality a daily function. Adopting the ICH Q10 model would be very cool to show your stakeholders, especially the health authority, that the quality of your products is at the highest level. While

TABLE 1.3
Quality and Compliance

Quality refers to:

- Product quality – meeting agreed specifications
- Quality systems – planned and deployed processes (systems) used to monitor, report, and take corrective actions

GMP compliance refers to:

- Identification, documentation and deployment of GMP obligations
- Ongoing verification that GMP obligations are being met, or not

TABLE 1.4
PQS/Compliance Gap Analysis

- Identify and assess compliance obligations and potential failures – not just GMP
- Include external and internal obligations in gap analysis
- Prioritize gaps using use risk principles
- Establish a compliance register or database
- Develop a PQS remediation/ improvement plan
- Establish a system to monitor external changes in PQS and compliance obligations, e.g., legislation, cGMP updates

Table 1.3 seeks answers to the question of quality or GMP compliance, one of the most debated topics today, Table 1.4 emphasizes the importance of gap analysis in the pharmaceutical quality system (PQS).

When we look at the internal structure of ICH Q10, we can say that it is a successful blend of ISO and GMP standards:

- Based on ISO 9000/ISO13485/CFR 820 systems model.
- Compliments ICH Q8 and ICH Q9.
- Applies across the product life cycle.
- Consistent with GMPs; not intended to add new expectations to regulations and compliance.
- Applies to APIs, drug products, and biotechnology.
- Strengthens the link between product development and manufacturing activities.
- The size and complexity of the company's activities should be taken into consideration when developing a new pharmaceutical quality system or modifying an existing one.
- While some aspects of the pharmaceutical quality system can be company-wide and others site-specific, the effectiveness of the implementation of the pharmaceutical quality system is normally demonstrated at the site level.

1.13 GMPS

- *EudraLex – Volume 1* – Pharmaceutical legislation for medicinal products for human use
- *EudraLex – Volume 4* – Good Manufacturing Practice (GMP) guidelines
- *EudraLex – Volume 5* – Pharmaceutical legislation for medicinal products for veterinary use

1.13.1 Eu Gmp

Volume 4 of "The rules governing medicinal products in the European Union" contains guidance for the interpretation of the principles and guidelines of good manufacturing practices for medicinal products for human and veterinary use laid down in Commission Directives 91/356/EEC, as amended by Directive 2003/94/EC, and 91/412/EEC respectively.

Commission Directive 2003/94/EC of 8 October 2003, laying down the principles and guidelines of good manufacturing practice in respect of medicinal products for human use and investigational medicinal products for human use. Replacement of Commission Directive 91/356/EC of 13 June 1991 to cover good manufacturing practice of investigational medicinal products.

Commission Directive 91/412/EEC of 23 July 1991 laid down the principles and guidelines of good manufacturing practice for veterinary medicinal products. Commission Delegated Regulation (EU) 2017/1569 of 23 May 2017 supplementing Regulation (EU) 536/2014 of the European Parliament and of the Council by specifying principles and guidelines for good manufacturing practice for investigational medicinal products for human use and arrangements for inspections (applicable as from the date of entry into application of Regulation (EU) No 536/2014 on Clinical Trials).

Commission Directive (EU) 2017/1572 of 15 September 2017 supplementing Directive 2001/83/EC of the European Parliament and of the Council as regards the principles and guidelines of good manufacturing practice for medicinal products for human use (applicable as from the date of entry into application of Regulation (EU) No 536/2014 on Clinical Trials).

Part I – Basic Requirements for Medicinal Products
Chapter 1 – Pharmaceutical Quality System (into operation since 31 January 2013)
Chapter 2 – Personnel (into operation since 16 February 2014)
Chapter 3 – Premise and Equipment (into operation since 1 March 2015)
Chapter 4 – Documentation (January 2011)
Chapter 5 – Production (into operation since 1 March 2015)
Chapter 6 – Quality Control (into operation since 1 October 2014)
Chapter 7 – Outsourced activities (into operation since 31 January 2013)
Chapter 8 – Complaints and Product Recall (into operation since 1 March 2015)
Chapter 9 – Self Inspection

Part II – Basic Requirements for Active Substances Used as Starting Materials (August 2014)

Part III – GMP-Related Documents

Site Master File
Q9 Quality Risk Management
Q10 Note for Guidance on Pharmaceutical Quality System
MRA Batch Certificate

Template for the "written confirmation" for active substances exported to the European Union for medicinal products for human use (Version 2, January 2013)
Guideline on setting health- based exposure limits for use in risk identification in the manufacture of different medicinal products in shared facilities
Guidelines of 19 March 2015 on the formalized risk assessment for ascertaining the appropriate manufacturing practice for excipients of medicinal products for human use
A risk assessment as set out in these guidelines should be carried out for excipients for authorized medicinal products for human use by 21 March 2016
Template for IMP batch release (applicable as from the date of entry into application of Regulation (EU) No 536/2014 on Clinical Trials)

Annexes

Annex 1 Manufacture of Sterile Medicinal Products [revised in August 2022]
Annex 2 Manufacture of Biological Active Substances and Medicinal Products for Human Use
Annex 3 Manufacture of Radiopharmaceuticals
Annex 4 Manufacture of Veterinary Medicinal Products other than Immunological Veterinary Medicinal Products [to be revised]
Annex 5 Manufacture of Immunological Veterinary Medical Products [to be revised]
Annex 6 Manufacture of Medicinal Gases
Annex 7 Manufacture of Herbal Medicinal Products
Annex 8 Sampling of Starting and Packaging Materials
Annex 9 Manufacture of Liquids, Creams and Ointments
Annex 10 Manufacture of Pressurised Metered Dose Aerosol Preparations for Inhalation
Annex 11 Computerised Systems [to be revised]
Annex 12 Use of Ionising Radiation in the Manufacture of Medicinal Products
Annex 13 Manufacture of Investigational Medicinal Products
Annex 14 Manufacture of Products Derived from Human Blood or Human Plasma
Annex 15 Qualification and Validation

Annex 16 Certification by a Qualified Person and Batch Release
Annex 17 Parametric Release
Annex 19 Reference and Retention Samples

Glossary

Part IV – GMP Requirements for Advanced Therapy Medicinal Products

Other Documents Related to GMP

- Compilation of Community Procedures on Inspection and Exchange of Information updated to include new EU formats and procedures.
- A revised version of the "Guidelines on Good Distribution Practice of Medicinal Products for Human Use" was published in the Official Journal and is applicable as of 24 November 2013 (OJ C 343/1, 23.11.2013).
- Guidelines of 19 March 2015 on principles of Good Distribution Practice of active substance for medicinal products for human use. These guidelines will come into operation on 21 September 2015.

1.13.2 FDA GMP

The FDA regulates several industries, including Pharmaceuticals, Medical Devices, Cosmetics, Food and Tobacco. When an auditor prepares for a GMP audit, he must follow the major codes of regulations:

- 21 CFR Part 11
- 21 CFR Part 210, 211
- 21 CFR Part 820
- 21 CFR Part 58

1.13.3 Medicines and Healthcare Products Regulatory Agency (MHRA)

The new 11th edition of the MHRA GMP Guide has been updated to include changes made after the UK left the European Union on 31 January 2020. The final version of the MHRA's GMP guide, also known as the "Orange Guide," was published after Brexit and the last GMP revisions.

1.13.4 The Role of the European Medicines Agency (EMA)

The European Medicines Agency (EMA) was established in 1995 to increase the effectiveness of market surveillance of medicinal products in Europe. It coordinates the scientific resources of the member states with regard to the evaluation and supervision of medicinal products for human and veterinary use. The EMA was originally based in London. However, as a result of the withdrawal of the United Kingdom from the EU (Brexit), it became necessary to relocate the agency's headquarters. Since 30 March 2019, the agency has had its new headquarters in Amsterdam.

1.13.4.1 Responsibilities

Marketing authorization. The EMA plays a central role in the authorization of medicinal products in the European Union (EU) and the European Economic Area (EEA) countries. Based on its scientific assessment, the European Commission issues an affirmative or negative decision on the marketing authorization applications submitted under the centralized procedure.

Pharmacovigilance. Another focus is on the continuous monitoring of the safety of medicinal products already authorized in the EU. The EMA defines safety standards and coordinates the pharmacovigilance system of the EU (EudraVigilance).

Coordination. In addition, the EMA coordinates the supervisory responsibilities of the member states. The main focus here is on the provision of information on medicinal products and the monitoring of compliance with good manufacturing practice (GMP), good laboratory practice (GLP), and good clinical practice (GCP).

1.13.4.2 Organization

The EMA is headed by an independent management board and an executive director.

Transparency, independence, and scientific expertise are declared goals of the EMA. To ensure that these are achieved, the agency has implemented various procedural rules, for which compliance is monitored on a regular basis.

The EMA also works closely with the competent authorities of the member states in a regulatory network. This means that the EMA has access to thousands of experts from the EU.

1.13.4.3 Committees and Working Groups

Within the agency, essential tasks relating to the evaluation and supervision of medicinal products are carried out by various committees. Their members are usually high-ranking staff of national medicinal product authorities. The committees are in turn supported by working groups, which provide a broad range of specialist expertise

In the area of the EMA's GMP/GDP supervision, the expertise of the member states is brought together in the GMP/GDP Inspectors Working Group. Among the most important responsibilities of this working group are:

- Redrafting or revision of GMP regulations
- Communication of inspection results within the EU
- Review of the exchange of information on the risks associated with medicinal products for efficiency and effectiveness
- Ongoing development of the "Compilation of Community Procedures on Inspections and Exchange of Information"

The GMP/GDP Inspectors Working Group cooperates with other GMP organizations and associations (such as PIC/S, EDQM, ISPE, and PDA).

The interface to the authorization system is essential for GMP supervision. GMP inspectors often require information about the development history of pharmaceutical processes and the background behind critical process attributes. Close cooperation with the assessors who have assessed the Common Technical Document is therefore helpful.

For the assessors, the counterpart to the GMP/GDP Inspectors Working Group within the EMA is the Quality Working Party. The GMP/GDP Inspectors Working Group and Quality Working Party come together at least once a year for a joint meeting. In addition, there is a regular exchange between GMP inspectors and GCP inspectors, particularly at the GMP/GCP interfaces, such as the production of investigational medicinal products.

1.14 ICH

According to the ICH website:

> The International Council for Harmonisation of Technical Requirements for Pharmaceuticals for Human Use (ICH) is unique in bringing together the regulatory authorities and pharmaceutical industry to discuss scientific and technical aspects of drug registration. Since its inception in 1990, ICH has gradually evolved, to respond to the increasingly global face of drug development.
>
> ICH's mission is to achieve greater harmonization worldwide to ensure that safe, effective, and high-quality medicines are developed and registered in the most resource-efficient manner. Harmonization is achieved through the development of ICH Guidelines via a process of scientific consensus with regulatory and industry experts working side by side. Key to the success of this process is the commitment of the ICH regulators to implement the final Guidelines.

Quality Guidelines

Harmonization achievements in the Quality area include pivotal milestones such as the conduct of stability studies, defining relevant thresholds for impurities testing and a more flexible approach to pharmaceutical quality based on Good Manufacturing Practice (GMP) risk management.

Safety Guidelines

ICH has produced a comprehensive set of safety Guidelines to uncover potential risks like carcinogenicity, genotoxicity, and reprotoxicity. A recent breakthrough has been a non-clinical testing strategy for assessing the QT interval prolongation liability: the single most important cause of drug withdrawals in recent years.

Efficacy Guidelines

The work carried out by ICH under the Efficacy heading is concerned with the design, conduct, safety and reporting of clinical trials. It also covers novel types of medicines derived from biotechnological processes and the use of pharmacogenetics/genomics techniques to produce better-targeted medicines.

Multidisciplinary Guidelines

Those are the cross-cutting topics that do not fit uniquely into one of the Quality, Safety and Efficacy categories. It includes the ICH medical terminology (MedDRA), the Common Technical Document (CTD) and the development of Electronic Standards for the Transfer of Regulatory Information (ESTRI).

Q1 A–Q1F Stability
Q2 Analytical Validation
Q3 A–Q3E Impurities
Q4 A–Q4B Pharmacopoeias
Q5 A–Q5E Quality of Biotechnological Products
Q6 A–Q6B Specifications
Q7 Good Manufacturing Practice
Q8 Pharmaceutical Development

Q9	Quality Risk Management
Q10	Pharmaceutical Quality System
Q11	Development and Manufacture of Drug Substances
Q12	Lifecycle Management
Q13	Continuous Manufacturing of Drug Substances and Drug Products
Q14	Analytical Procedure Development

1.15 QUESTIONS TO AUDITORS

- Do you think that the ISO 9001 system has an impact on the establishment of the quality management system in pharmaceutical and biotechnology companies?
- Assuming you're auditing a company with a pharmaceutical quality management system, which would be the top three SOPs you should review, and why?
- Although not among the mandatory documents, do you think the ICH guidelines work in GMP audits?

REFERENCES

Ann W. Phillips, *ISO 9001:2015 Internal Audits Made Easy*, Fourth Edition, 2015, ASQ Quality Press
Bikash Chatterjee, *Applying Lean Six Sigma in the Pharmaceutical Industry*, Gower, 2014
Carmen Medina, *Compliance Handbook for Pharmaceuticals, Medical Devices, and Biologics*, 2004, Taylor & Francis
Dennis R. Arter, *Quality Audits for Improved Performance*, Third Edition, 2013, ASQ Quality Press
Douglas C. Wood, *Principles of Quality Costs*, Fourth Edition, 2013, ASQ Quality Management Division
John Sharp, *Good Pharmaceutical Manufacturing Practice Rationale and Compliance*, 2005, Taylor & Francis
J.P. Russell, *The Internal Auditing Pocket Guide*, Second Edition, 2007, ASQ Quality Pres
Martin Hinsch, *ISO 9001:2015 for Everyday Operations*, Springer, 2019, Springer Wieveg

2 GMP Audits

DOI: 10.1201/9781003284970-2

2.1 WHAT IS A GMP AUDIT AND HOW DOES IT DIFFER FROM A REGULATORY INSPECTION?

2.1.1 GMP Audit

A good manufacturing practice (GMP) audit is defined as a documented, systematic tool used by management and done periodically by independent, qualified people to verify and evaluate an organization's use of the principles of GMP and compliance with applicable GMP regulations.

Are GMP Audits and Inspections different concepts? Can they be used interchangeably?

The word audit derives from the Latin word *audire* meaning "to hear." The definition can also be found in the US Food and Drug Administration (FDA) guidelines.

The purpose of the quality audit is to look carefully at the organization's quality system and to ensure that the outputs obtained from these activities are examined concerning the procedures established by the organization. In addition, the effective implementation of procedures also evaluates whether the organization has achieved the quality objectives that it has previously determined.

- Observation of a facility, a piece of equipment, a building, or a process.
- Verification that certain standards, regulations, or guidelines are being met.

2.1.2 Audit

- Evaluation and measurement of the performance of a facility, a piece of equipment, a building, or a process.
- Compared against the expectations outlined in a set of standards or guidance.

An audit is a systematic and unique approach that includes objective evaluation in order to determine how much of the predetermined criteria can be met by collecting as much objective evidence as possible.

The word "inspection" is also derived from Latin and means "to look at, to inspect."

In EMA E6(R2), inspection is defined as follows:

> It is the process of formally evaluating all documentation, facilities, records, and all resources accepted by the health authority that may be found at the site of the clinical trial that have been regonized by health authorities as being relevant to clinical trials.

As a result, both forms of expression appear to describe a systematic process for firms to evaluate and examine the systems in place and to judge the degree of compliance with accepted or legally required standards. Therefore, whereas the word "audit" is used to cover all kinds of evaluations, "inspection" is mostly used by the health authorities for detailed examinations in the pharmaceutical industry.

For audits, often the type of audit and the reason for the audit are further distinguished.

2.1.3 What Is An Audit?

Let's start with the definition of auditing ISO 9000, a member of the ISO family of standards:

> It is an independent, impartial, and documented activity of providing and evaluating impartial and objective evidence in order to determine how much of the predetermined criteria are met.

An audit examines in detail not only the products, but also the services purchased. In fact, it aims to determine whether the activities carried out by any organization are correct and at the same time are operating efficiently. Audits can be performed separately as product audits, process audits, and system audits according to their purpose and scope.

2.1.4 GMP Inspection (FDA Definition)

In order to audit whether a pharmaceutical or biotechnology company is producing in accordance with the applicable legal regulations, it is usually carried out by highly equipped inspectors/auditors who receive special training on GMP issues and some skills such as reading body language, detecting the person who is lying, acting like a detective. These are long and detailed investigations that are carried out and begin with a site visit.

While the inspectors/auditors learn about the general condition of the facility, they also learn about quality management systems, production procedures throughout the product life cycle, supplier qualifications and audits, validation of facilities, maintenance, cleaning qualification and validation processes, documentation, training, batch records, laboratories, and test methods. They evaluate numerous topics such as deviations, customer complaints, and withdrawals with the help of evidence documents.

Performing an inspection is like looking at what is on the surface of a lawn, while doing an audit is like looking down from a branch of a nearby tree to examine the lawn more broadly.

2.1.5 Documentation Should Come First

Before I elaborate on GMP audits, I should specifically point out an issue that I have not encountered very often in the numerous facility audits I have carried out so far.

I am in favor of companies having a policy for GMP audits in the pharmaceutical industry because, as it is known, policies are among the highest-level documents and each policy leads to a standard operation procedure (SOP) group at the bottom of the documentation pyramid. So if you have a GMP audit policy, it is definitely desirable to have a GMP audit SOP. The point that should not be overlooked here is that the firm does not have a separate audit SOP for an internal audit and a separate supplier audit.

Among the important components of the GMP audit SOP are the:

- SOP's name
- SOP's revision number
- SOP's page numbers
- SOP's validity date
- Aim
- Scope
- Internal audits, supplier and customer audits, health authority, and certification body audits
- Responsibilities
- Execution
- Abbreviations
- Definitions
- Referenced document names
- Document history
- Attachments

2.1.6 What Should Be Covered by the Audit SOP?

First, the SOP should strictly serve its purpose and convey all the activities to the user in simple language in detail. It should have some useful and guiding information such as audit requirements, audit plans, audit team selection, audit checklist, which areas should be audited and how, criteria that auditors should have, classification of audit findings, audit reports, and follow-up audits.

2.1.7 System-Based Audit

Before addressing system-based audits, if we touch on the short definition of a system; it is a collection of elements formed by interacting units in order to achieve a holistically determined purpose. If we are going to audit the systems, each component that makes up the system should be handled separately and their interactions with each other should be evaluated. In system audits,

TABLE 2.1
System-Based Audit Examples

System	What to Look For	Purpose
Water supply	Filters, UV units, ion exchangers, reverse osmosis units, pumps, holding tanks, pipes, valves, withdrawal stopcocks, disinfection equipment, etc.	Production of purified water
Ventilation system	Aspiration points, filters, flaps, motors, ducts, heating, humidifiers, grating, etc.	Compliance with temperature, humidity, and particle limits in the production rooms
Compressed air supply	Aspiration point, filters, compressors, water/oil separator, valves, withdrawal stopcocks, etc.	Production of compressed air intended to be in direct contact with the product and in accordance with a defined specification
Computer-assisted system	Servers, switches, hubs, coaxial cables, PCs, software, production environment, etc.	Electronic data processing, regulation and control of machines
Quality assurance system	Organization charts, job descriptions, QA manual, SOPs, etc.	Ensuring a defined quality

areas and procedures should also be examined in detail in order to see whether the functionality of the entire system is generally ensured. In Table 2.1, the purpose and points of system-based audits are explained with examples.

2.1.8 Product-Based Audit

Product-based audits evaluate stages from the production, storage, and use of components such as raw materials and packaging materials to the production of the finished product. One or more batches of the product are checked by adhering to the file in which the necessary permits are obtained for the marketing of the product and the procedures used can also be controlled simultaneously. This audit would usually be performed for contract manufacturers. Table 2.2 summarizes the points to be considered in product-based audits.

2.1.9 Procedure-Based Audit

Procedures are a documented version of activities performed in a particular way. Reminders of how the procedures should be followed are included in the instructions.

TABLE 2.2
Product-Based Audit Examples

- Specifications of the raw materials and packaging materials
- Standard operating procedures (SOPs)
- Manufacturing instructions/record
- Equipment logs
- Sampling procedures
- In-process controls of intermediate products and bulk
- Testing procedure/record
- Release procedure

TABLE 2.3
Procedure-Based Audit Examples

- Inward transfer and discharging staff from clean rooms
- Sampling of raw materials
- Manual cleaning of equipment
- Transfer of rejected goods
- Manual packaging of small batches
- Initial weight of raw materials
- Integrity tests of product-carrying sterile filters

One of the purposes of procedure-based audits is to determine whether the steps defined in the instructions are fulfilled. Although it is not always necessary to carry out this control, its necessity has always been proven in practice. The most critical issue to be considered here is how much control procedures and instructions are followed up to date. Even a small change in a procedure or instruction with change control means updating the relevant document. Table 2.3 summarizes the points to be considered in procedure-based audits.

In fact, another important purpose of procedure-based audits is to determine whether the steps specified in the procedures and instructions are applied correctly and on time by the relevant personnel each time. For example, while the maintenance, cleaning, calibration, or qualification status of equipment that needs to be used by the personnel is a requirement to reach the specified parameters, even the cleaning status alone will make it possible to decide whether the equipment will be used or not. Another issue is that the compulsory and periodic training that the personnel should receive is not given on time or that they are given incompletely, resulting in deviations from procedures. In addition, the practical experience of the personnel who perform the procedures with great care in the same order everyday without getting tired can enable the improvement, development, and even the terminology of the procedures or instructions, making them easy to understand. Therefore, it will always be an advantage to be in close contact with the personnel in procedure-based audits.

2.1.10 Area-Based Audit

Area-based audits, which aim to include areas such as the warehouse, shipping and packaging departments, raw material weighing stations, personnel airlocks, and toilet and shower areas, are essential because they provide the opportunity to audit the entire factory area in periodic audits. It should be noted that not only what is written in the procedures or instructions, but also the conditions of the areas should be taken into account. Table 2.4 summarizes the points to be considered in area-based audits.

2.2 INTERNAL AND EXTERNAL AUDITS

When evaluated from the GMP point of view and the relevant guides are examined, it should not be so easy to find very specific topics related to audits. For this reason, separating GMP audits as internal and external audits is more common in ISO-based quality management systems. It is even possible to classify internal and external audits as first- and second- or even third-party audits. In this case, whereas the firm's internal audit is a first-party audit, external audits consist of second- and third-party audits. While the second-party audit is mostly the audit in which the companies evaluate their suppliers, the third-party audit can be characterized as a certification body or health authority audit. Regardless of the party's audit, the health authority does not set very strict rules for the company to carry out the GMP audits by itself or to outsource it. In this case, there is no obstacle

TABLE 2.4
Area-Based Audit Examples

- Hygiene and cleanliness status of the rooms and facilities
- Status of floors, walls, and ceilings
- Labeling of material, equipment, and pipes
- Provision zones and material movements
- Machine logbooks (calibration, maintenance, malfunctions)
- Prevention of cross-contamination
- Required instructions and on-site records
- Responsibility regulations of the area
- Clothing and conduct of the employees at their place of work
- Staff training documents

to outsourcing external audits to companies with competent and certified auditors. If you ask me what is the trend in the industry or what is common practice, generally the organization is more willing to carry out GMP internal audits with in-house trained, experienced, and certified auditors. Figure 2.2 has been added to make the distinction between internal and external audits easier to understand and provides an explanation of how audits should be classified.

An audit can be a mix of different types of audits. External auditors (customers) may join third-party auditors (certification), second-party auditors (client auditors), or internal auditors.

2.2.1 First-Party Audit

As stated, first-party audits are audits performed by experienced and certified GMP auditors that the company has employed within its own body. Detailed information on internal audits is discussed in the following chapters. Thanks to this audit, the pharmaceutical company has the opportunity to see its strengths and weaknesses, as well as be prepared for health authority audits and customer

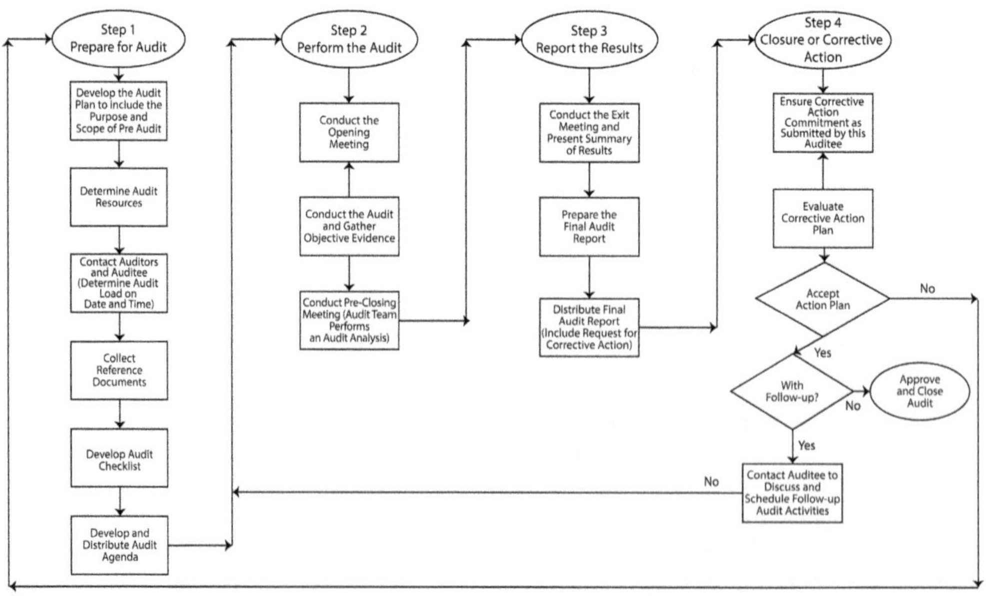

FIGURE 2.1 Audit process flow.

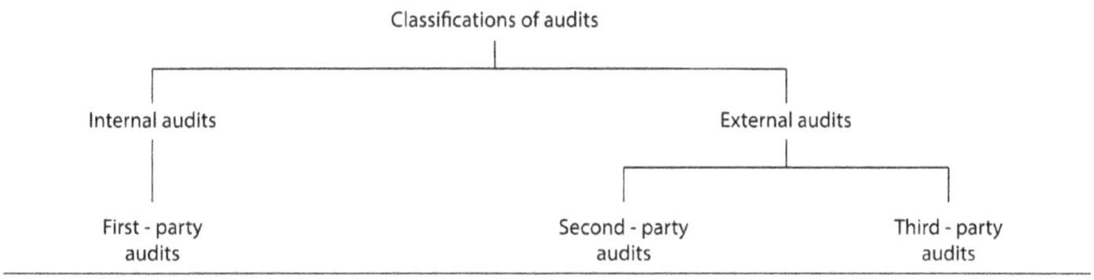

FIGURE 2.2 Classification of audits.

audits if it carries out contracted production. The part that should be noted here and that the health authority is happy to question is that in GMP internal audits, auditors do not audit their own departments. This issue is also particularly emphasized in the ISO 19011 standard. In fact, the idea is to establish an independent audit department within the company that will only deal with audits. This department can report directly to senior management, and situations such as conflict of interest are prevented in any departmental audit.

Another alternative, and in many cases a preferred approach due to time and personnel constraints, is for a GMP auditor to take on the role of GMP internal auditor as additional responsibilities and duties while working in any department. The application itself may have advantages and disadvantages. The GMP internal auditor will perform the audit and not disrupt daily work. In my opinion, all GMP internal auditors should have information about each department, and if the company expects an efficient and effective audit result, they should be assigned to departments within the company for a short time, and even on certain days of the week, the auditors should examine key activities as observers in these departments. A person working in the quality assurance department can witness all the activities in real-time, not only related to his/her job, for example, by spending one day a week in the quality control laboratory and production areas.

Another application is that the company wants to carry out its GMP internal audit activity entirely by outsourcing. Taking on this role by external GMP auditors also has its own advantages and disadvantages.

If a company with global operations audits its own facilities located in different countries, this audit can also be considered a GMP internal audit or a first-party audit. While working in Turkey at the factory of a German pharmaceutical company, one of the largest pharmaceutical companies in the world, I witnessed experienced and certified inspectors from Germany's head office come to the factory, stay for a minimum of five days, examine all the details, and leave the facility with a GMP compliance note. While I was waiting for these audits of the technical audit committee with great curiosity, and preparing my own department for the audit, I was also excited about what I could learn from these experienced, competent seniors.

2.2.2 Second-Party Audit

In this audit, the pharmaceutical company audits its own supplier according to GMP rules. If the pharmaceutical company also carries out its production or packaging operations in another GMP-certified facility, this audit can also be considered as a second-party. Contract law issues also come into play, as second-party auditing proceeds from the customer to the supplier. Second-party audits are performed in a more formal setting than first-party audits because the audit's outcome will directly impact the decision to purchase a product or service from the other party. In some cases, before the audit, the firm may want to send a list of preliminary assessment questions to its supplier. This assessment will provide information on critical aspects such as facilities, resources, systems, capacity, technical equipment, production capability, skilled workforce, and track record of the party that will take responsibility.

2.2.3 Third-Party Audit

The third-party audit is free of potential conflicts of interest between the customer and the supplier and is usually performed by an independent audit firm. The independence of the auditing organization is the most striking point and is the basis of such audits. The fact that a health authority such as the FDA performs an audit within 21 CFR Parts 210, 211, and 820 is sometimes defined as a third-party audit. Third-party audits may be performed on behalf of the auditee's potential clients who cannot afford to review or audit external entities themselves, or who see a third-party audit as a more cost-effective alternative.

2.2.4 Follow-Up Audit

Regardless of whether the GMP audit is internal or external, it will be an evaluation result and this evaluation will of course include some nonconformities. A corrective and preventive action (CAPA) will definitely be prepared for the findings and the implementation of the actions will be done within a time plan. Whether it's a product, a service, or a process audit, whether or not CAPAs are closed on time has always been an agenda item. Although the audited party claims to have taken all the actions after the audit, and even supports this claim with various visuals, the party performing the audit may request an on-site audit again in order to be completely sure. This audit is called a follow-up audit.

Considering the possibility that a follow-up audit alone would be too costly, pharmaceutical companies often want to kill two birds with one stone, preferring to evaluate this check at the beginning of the next inspection. But remember that a risk assessment definitely needs to be done in advance.

2.3 WHY CONDUCT GMP AUDITS?

2.3.1 Why Is a Pharmaceutical Company Audited for GMP Compliance?

To demonstrate that it is a GMP-compliant facility, the organization must first conduct an internal self-audit. This audit is carried out by a team under the leadership of the quality assurance department or by a third-party organization, as stated earlier.

Other GMP compliance inspections are carried out by a country's health authorities. The most important reason for the health authority to carry out these inspections is to protect and secure public health and safety. The health authority focuses on several very important issues such as the organization's risk management system, documentation practices, quality management system, validation, and data integrity. Of course, GMP inspection is not limited to checking these criteria. Figure 2.2 provides an example of what steps a GMP audit should follow.

The pharmaceutical establishment is supervised not only by the health authority to which it is affiliated but also by the health authorities of different countries. The best example of these audits can be given by the organization called the Pharmaceutical Inspection Co-operation Scheme (PIC/S). These regulatory agencies have highly trained auditors who conduct frequent GMP audits against the PIC/S scheme GMP requirements.

How long does a GMP audit take? A GMP audit can take from one day to several weeks, depending on the criticality of the facility, product variety, and the pharmaceutical company's audit history and size. Similar to the familiar quality management systems audits, the findings in the GMP audit provide the most important input to the audit report to be issued later. The health authority formally transmits the inspection report to the pharmaceutical company for the issuance of CAPAs after the inspection. Nonconformities detected in GMP audits are also called audit findings.

What is the primary and most important purpose of the GMP audit?

The purpose of an on-site or remote GMP audit by a health authority is to investigate compliance with GMP-related regulations and product-related directives and guidelines. When a health authority cannot go to the production facility to check the GMP compliance of a facility, for example, due to COVID-19, audits had to be done remotely, in this case, the remote audit option is considered.

2.4 AUDIT SCOPE

According to ISO 19011, the audit scope is the extent and boundaries of an audit. The audit scope normally includes a description of the physical locations, organizational units, activities and processes, and the time period covered. The audit scope indicates or fixes a limit or extent of the audit. The scope has been described as the breadth of the audit and may specify areas not to be included in the audit.

The scope or criteria of an audit can include:

- Physical locations
- Departments, areas, or units
- Products, processes, or systems
- Areas excluded from the audit
- Timeline for audit activities or events
- Relevant system and process policies, procedures, instructions, and plans
- Applicable standards, contracts, regulations, codes, and other legal documents

If the scope of the audit is to be changed in advance or during the audit, this change should be notified to the audit participants and the relevant revision should be made in the plan.

2.4.1 What Is Required for an Audit?

- Local and national statutes and regulations
- Industry codes and standards
- Guides, handbooks, and so on

Standards, codes, and regulations are issued by related industrial or professional associations, by national standards-writing organizations concerned with the intended marketplace, by local/state/national legislative bodies, and by international bodies.

2.4.2 Audit Bases

To perform an audit, an auditor must be aware of the audit basis, sometimes called reference standards, audit criteria, or performance standards. The compliance or adequacy of a system cannot be measured until those requirements are defined.

Regardless of the requirements, an audit must be performed against a basis for reference (for example, organization performance standards and/or national standards such as ISO 9001). These reference documents may include the following:

- Management system, product, or process standards
- Contracts
- Specifications
- Organization policies and objectives
- Laws or regulations

2.4.3 Standards

Certain international, national, and industry standards are mandated for many organizations. Audits verify compliance/conformance with the applicable management system standard, whether it be ISO 9001, 22716, or 14644.

2.4.4 Laws and Regulations

Auditors verify mandatory governmental standards such as the FDA's current good manufacturing practices (cGMPs). The audit criteria must be stipulated as part of the audit plan. There is no minimum or maximum limit to the amount or kinds of audit criteria. However, for an audit to be performed, there must be audit criteria. If there are no criteria to compare the organization with, the investigation may be called a survey or review.

2.5 AUDITING ROLES AND RESPONSIBILITIES

An audit involves three key participants who may interrelate in a number of ways. Described by function, these participants are the client, the auditor, and the auditee.

The client is the person or organization that has requested or commissioned the audit. The client is usually a member of senior management, and the audit is typically conducted of an organizational unit under the client's jurisdiction, of independent suppliers, or to support an application for third-party certification.

The auditor is the person who plans and carries out the audit. An auditing organization, that employs auditors to carry out audits, may be internal to a company or an independent organization, such as the auditing group of a quality or environmental program certification body or consulting organization. Table 2.5 provides a list of how roles and responsibilities should be allocated in audits.

The auditee is the organization to be audited. The auditee may be a division of the client's organization or an entirely separate entity, such as a supplier. In internal audits, the client is the top management and the auditee is the function or area to be audited. The audit process involves several participants. By its nature, an audit can cause stress among participants. Therefore, it is in everyone's best interest if the participants work together to ensure a successful and effective audit. The more contentious the relationship between participants (such as the auditor and the auditee), the more difficult it will be to achieve compliance, conformity, or improvement.

The following are audit process participants:

- Client – The person or organization that requested the audit
- Auditor – The person carrying out the audit
- Lead auditor or audit team leader – The auditor responsible for managing the audit
- Auditee – The person or organization to be audited
- Escort – The person assigned to escort the audit team members
- Coordinator – The person in contact with the lead auditor or the audit program manager in order to arrange for the audit
- Audit program manager – The person responsible for the audit program

2.5.1 Auditor Characteristics

Auditor characteristics are those qualities that make auditors effective. They become integrated into the auditor's behavior so that they appear to be a natural pattern of conduct. Characteristics such as interpersonal skills, problem-solving, and the ability to work independently as well as on a team are learned through study and practice. For that reason, most audit programs require knowledge and skill competencies gained through a combination of education, training, experience, and demonstration of personal traits.

Auditor competence is composed of several things, including:

- Education, training, and knowledge
- Work experience
- Audit skills
- Communication skills
- Personal traits

TABLE 2.5
List of Responsibilities and Duties

Function	Responsibilities and Duties
Auditor	• Understands the purpose and scope of the audit • Understands the audit criteria being audited against • Prepares for the audit • Performs the audit to collect evidence to verify conformance or nonconformance to the audit criteria • Records the results of the investigation (perhaps on a checklist) • Attends the opening and closing meetings • Reports findings to the lead auditor • Cooperates with the lead auditor • Verifies the correction of previous nonconformities if directed to do so • Provides input to the formal report if directed to do so by the lead auditor or client • Maintains confidentiality of the audit information • Reports conflicts of interest to the lead auditor • Is ethical and adheres to an organization code of conduct or the principles of auditing
Lead auditor	• Is responsible for communication with the client, auditor program management, and the auditee representative • Provides audit team selection input if requested to do so • Communicates audit plan and requirements to auditee • Ensures that necessary resources are available to audit team • Ensures the team has the appropriate working papers • Plans the audit and directs the audit team • Conducts audit process meetings • Prepares audit report • Manages the audit process and resolves conflicts of interest or other personnel issues • Ensures reports and records are properly filed and safeguarded
Auditee	• Coordinates audit with the lead auditor • Informs employees of the pending audit purpose and scope • Addresses logistical issues with the lead auditor • Provides adequate space and privacy for the opening and closing meetings • Attends the opening and closing meetings • Provides area for auditors to work and meet if requested • Cooperates with the auditors • Provides access to areas included in the audit scope • Acknowledges audit results • Takes corrective action on audit findings
Audit manager	• Assigns auditors to scheduled audits • Ensures availability of resources (budgeting) • Establishes a reporting relationship that ensures objective and impartial audits • Qualifies auditors (knowledge, experience, and skills) • Establishes controls (procedures, criteria, plans, and objectives) for an effective and efficient audit program • Creates, distributes, and maintains audit program schedules • Reports audit program progress to management • Monitors auditor performance • Determines audit program objectives and creates plans to accomplish the objectives • Keeps and safeguards audit program information • Promotes ethical behavior on the part of auditors and those involved in managing the audit program

Auditors must understand how to conduct an effective audit in the auditee organization environment and be able to apply the audit criteria. Auditors must possess the personal attributes necessary for conducting an effective and efficient audit.

ASK → LISTEN → OBSERVE → RECORD

2.5.1.1 Mental Stability

Because of the need to obtain the proper information and because of their educational role, the auditor should be able to prevent any conflict situation and create a positive, constructive environment. The audit should always be conducted in an amiable atmosphere, with tact, honesty, diplomacy, and persuasiveness. Therefore, the auditor should have a very stable character, and be able to restrain themself from emotional, aggressive, or discouraging declarations in the course of the audit.

During the audit, the auditor's role is to collect the information in an objective manner, keeping the auditee informed as the audit proceeds. In this way, serious misunderstandings in the closing meeting can be avoided. In the closing meeting or review of the audit report phase, the auditor should resist any pressure from the auditee to revise or reclassify the audit findings, unless of course a clearly explained error in the accuracy of the findings has occurred and is accepted by the lead auditor.

2.5.1.2 Conflict Situations and Compatibility

Whenever a conflict situation arises, the auditor should never resort to threats, intimidation, or strong-arm tactics. In most cases, a calm, patient, sympathetic, or persuasive attitude will overcome the person's reluctance or hostility. If these actions fail, further actions will depend on the nature of the audit.

2.5.1.3 Audit of an Aggressive Supplier

Appropriate actions may include notification of senior management of the supplier, directly or, as appropriate, through the auditor's purchasing department.

2.5.1.4 Audit of One of Your Facilities

Information provided with justification to senior plant management or corporate management of the situation, mentioning the possibility of discontinuing the audit if no satisfactory solution can be found.

2.5.1.5 Audit by a Customer

These audits are usually witnessed through the local sales agents whose role is also to smooth any conflict situation when these arise and who may notify the auditor's hierarchy or commercial departments whenever such situations arise.

2.5.1.6 Audit by an Authority

This situation should never arise as authorities are under strict behavioral instructions. In the event of professional misconduct, notification of their inspectorate may be appropriate by senior management of the auditee, usually the corporate quality director.

2.5.1.7 Social Competence

All auditors (lead auditors and co-auditors) should exhibit proper social behavior to ensure no conflict situations or improper attitudes arise in the course of the audit. Auditors should realize the auditees may be stressed and feel challenged, undermined, or aggressed by the attitude, remarks, and comments of the auditors. There are no educational requirements for social competence. However, management should assess the social behavior of candidates before nomination as auditor. Unsociable or unstable individuals do not qualify as auditors.

2.5.1.8 When a Conflict of Interest Exists

When there is an actual or potential conflict of interest with the organization or people being audited, the auditor must relay this information to audit program management or decline to conduct the audit, whichever is more appropriate. Actions that management and the audit team leader can take include:

- Ensuring that sufficient time has passed to eliminate the conflict
- Assigning a different auditor to cover the specific area of conflict
- Removing the auditor or the audit team leader from the team

By avoiding conflicts of interest, an auditor upholds the standards of independence, fairness, objectivity, and impartiality of the audit process.

2.5.2 Confidentiality

With processes, formulas, and equipment being developed by individual companies, the question of confidentiality of proprietary information has become a major concern during audits. Businesses could suffer great financial loss if customers or competitors were to gain access to proprietary processing knowledge, formulas, and trade secrets. Each auditor needs to be prepared to sign agreements or utilize techniques for working around a proprietary area.

2.5.3 Interpersonal Skills

Regardless of education or training, an auditor who lacks certain interpersonal skills will not be able to conduct an effective audit.

An auditor must be able to understand the technical materials presented during an audit.

The needs and objectives of the audit program should be carefully considered, and knowledge and skills identified in this section are common with auditors performing many audits. Critical points in communication are given in Table 2.6.

2.6 TECHNICAL QUALIFICATION REQUIREMENTS

Considering the legal regulations, there is no requirement for the qualifications required by the auditors whether in internal or third-party audits. Still the issues that must be present in the people performing the GMP audit are generally accepted characteristics in the pharmaceutical industry. In the pharmaceutical industry, it has been the preferred method to assign auditors to audits based on

TABLE 2.6
Communication Skills

• Listening	• Being clear and concise
• Questioning	• Good writing skills
• Probing	• Good oral skills
• Critiquing	• Appropriate body language
• Handling conflict	• Tact

their knowledge, skills, and experience. It is only possible for these people to carry out GMP audits effectively if they have these features.

2.7 KNOWLEDGE REQUIREMENTS FOR AUDITORS

- All national and international GMP guidelines, such as European Union (EU), PIC/S, Medicines and Healthcare products Regulatory Agency (MHRA), World Health Organization (WHO), FDA, and Therapeutic Goods Administration (TGA) GMP guidelines
- Principles regarding structuring and business processes in organizations
- Processes related to marketing and production authorization
- Microbiology, analytical chemistry, process systems, laboratory tools and equipment, solutions, reagents, standards, heating, ventilation, and air conditioning (HVAC), water, steam, air systems, qualification and validation, stability, computerized systems, data integrity
- Sampling, transport, storage testing, and release procedures
- Temperature mapping, storage conditions, safety, and security
- Pharmaceutical storage and shipping
- Audit techniques and current approaches
- Quality management standards
- International Society for Pharmaceutical Engineering (ISPE), Parental Drug Association (PDA), and ICH guidelines and regulations
- Written and verbal communication, using body language and foreign language knowledge

The GMP auditor needs to be aware of the technical characteristics of the relevant area to be audited and mindful of the legal regulations and ways of implementation in that area. In addition, sufficient expertise is sought for specific areas such as microbiology, sterile production, aseptic operations, computerized systems, stability, biotechnological products, and Active Pharmaceutical Ingredient (API) production techniques. GMP auditors should receive regular and up-to-date training as other pharmaceutical company personnel do. Successful GMP audits are inevitable when their interest in the areas to be audited is combined with the technical knowledge and skills they have acquired. One approach states that many years of professional experience in a pharmaceutical company is an opportunity for ideal auditors. These people have proven their professional competence and successfully managed soft skills such as time management skills, people management, and conflict management. If the pharmaceutical company determines that the GMP auditor still needs particular knowledge in any field, the auditor candidate must complete this training. Figure 2.3 summarizes the features that a GMP auditor should have. No one should be willing to consult in any GMP audit, whether a pharmaceutical company auditor, a third-party auditor, or a government authority auditor. The duties and responsibilities of GMP auditors and the limit they can go to are determined.

In the USA, FDA inspectors come from varied backgrounds and qualifications. Although few FDA inspectors have extensive industrial experience compared to EU inspectors, all FDA inspectors or investigators undergo extensive FDA training to perform their respective inspectional duties.

Particular attention is also paid to the inspector's personal characteristics in practice, in addition to the purely technical aspects of the inspector's qualifications. They should be able to approach the inspectors and communicate with them. The inspector should be able to recognize personal interactions and be able to react to them adequately.

During the GMP audit, the main task of the auditor is to collect as much information as possible about the object or product they are examining. The GMP auditor should always be open to new ideas and maintain impartiality towards the information. A GMP audit can also be troublesome and challenging for the auditor and the auditee. For this reason, the auditor who feels himself under pressure should be able to work under pressure without revealing this situation too much. The GMP auditor must remain unwavering and composed, even if they encounter conflict with any employee

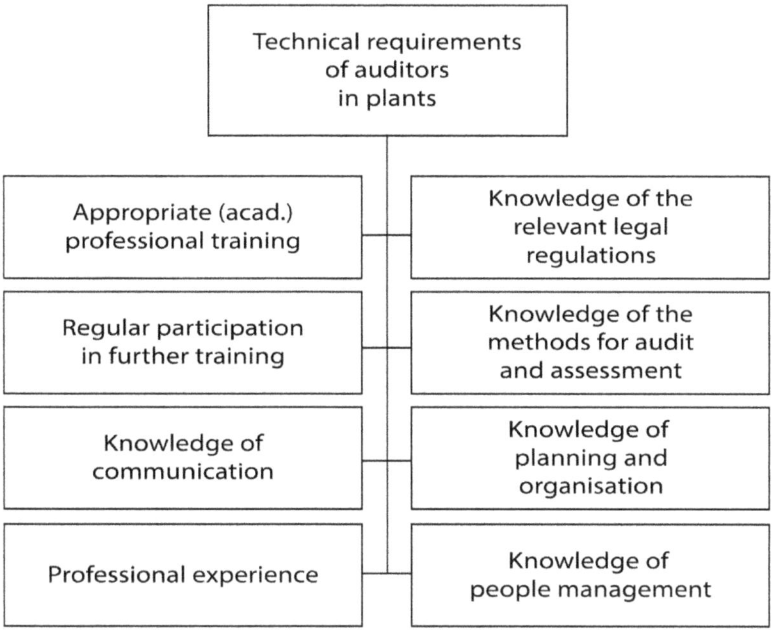

FIGURE 2.3 Technical requirements of auditors.

at the audited party's facility. They should be consistent in the decisions they have made and should be able to readily admit if they make the wrong decision. A question to consider is: How many auditors do you think can easily recognize their mistakes?

2.8 AUDITOR QUALIFICATION AND COMPETENCE

The processes for the selection and training of auditors have as their objective that the audit program can rely on persons having the required competence. It is generally believed that the qualification requirements for internal auditors are lower than those for external auditors, possibly resulting in internal auditors being less competent than external auditors. I strongly disagree with this view. Each type of auditor needs a different set of skills and knowledge.

Qualification is used for both the process of qualifying and the minimal result, whereas competence is the possession of attributes, skills, and knowledge to a variable degree.

2.8.1 Personal Attributes and Attitudes

Desirable attributes, which are the source of attitudes and behaviors, are independent of the internal audit's scope and objectives. Anyone wishing to be an auditor should possess:

- A high level of personal integrity or honesty, which will help in:
  - Making adequate judgments in various situations
  - Being sincere and honest in professional relations
  - Fully committing oneself to the purpose of the audit
- A sense of self-security that is independent of external factors; this self-esteem enables the auditor to:
  - Be able to really listen to others and to be open-minded about other ways of doing things
  - Show an adequate combination of flexibility and tenacity according to circumstances

- Stay calm and be objective
- Make decisions when required, even on partial information

The auditor must have self-control over their behavior in order to help the audit process and to be tactful. All persons involved in the internal audits program should show these attributes and qualities, but they should be more ingrained in persons having to lead an audit team.

2.8.2 Skills and Abilities

2.8.2.1 Basic Skills

Some skills are independent of the objectives and scope of audits and of the role of the auditor on the team. The most obvious is related to the capacity to:

- Read and understand documentation, on paper or in an electronic medium
- Plan and organize one's own work
- Understand and adapt to various environments
- Communicate verbally in an effective way at all levels of the organization
- Phrase simple and clear questions that are focused on the desired information
- Listen attentively and understand verbal information
- Synthesize ideas and manipulate data
- Operate effectively under moderate stress

2.8.2.2 Specific Skills

Supplementary skills are required to take the role of team leader and/or if the audit objectives include the verification of process effectiveness.

2.8.2.3 Specific Knowledge

The knowledge required by organizations for their internal auditors is often insufficient, resulting in incomplete or unreliable audit conclusions. An internal auditor has to analyze complex, interrelated processes and deal with management issues, even more so with GMP, with its emphasis on management and interrelated processes.

As with skills, some specific knowledge is required to contribute to even a simple audit. An auditor must know:

- The quality principles that form the basis of any QMS, GMP, GLP
- The general audit concepts and techniques, including the type of audits, the approaches and processes, and the subsequent activities, as they relate to internal quality audits
- The audit process actually implemented in the organization, including forms to be used
- The processes under audit, that is, the required activities, equipment, personnel qualifications, monitoring, inputs, and so on

Further, an auditor must have minimal knowledge of the industrial sector in which the organization is operating.

2.9 CRITERIA FOR THE SELECTION OF AUDITORS

An organization obtains potentially competent auditors by applying adequate criteria when initially selecting auditors, followed by appropriate training. As some personal attributes and skills are more difficult to acquire through training, the criteria should be carefully crafted and applied in order to obtain personnel with the proper attitudes and skills.

In practice, a large number of organizations screen their personnel individuals who satisfy the following criteria:

- Have all the desired personal attributes and attitudes
- Have skills related to the capacity to listen carefully to others, understand complex systems, and synthesize ideas
- Have knowledge of the industrial or service sector

Other skills and knowledge are then acquired through formal training.

2.10 TRAINING OF AUDITORS

What is the best way to train internal auditors: In the classroom or on the job? Using external specialized resources or doing it internally? The best solution for any organization depends on many variables:

1. Knowledge and skills to be acquired
2. Initial training effort or training of replacement auditors
3. The number of persons to be trained
4. Human and financial resources available

2.10.1 Initial Training

When initiating the GMP audit program, an organization usually does not have access to appropriate expertise internally.

At a minimum, the audit program manager should attend intensive GMP lead auditor training, a six-day course or equivalent, available through www.qualityacademia.com. If the organization can afford it, a second person should attend similar training. In addition to having a backup, the interaction between the two persons will reinforce the audit program.

For an organization that needs only two internal GMP auditors, the initial training is now complete. The decision of training to go external or internal will take into account:

- The geographical distance from available qualified audit specialists
- The number of auditors to be trained; a large number may justify the cost of an internal auditor training program
- The availability of an internal auditor training course in another unit of the organization
- The possibility of purchasing a commercial course to be presented internally

Creating a new course internally requires resources, the amount of which is often underestimated. For a large number of trainees, a classroom approach with practical test cases is preferable.

2.10.2 Training of Replacement Auditors

Organizations that use external resources for their initial training effort can either use the same external resources or internally train new auditors, a decision based on:

- The number of persons to be trained each year or each semester
- The geographical distance from locations where public training is available, as well as related travel costs
- The flexibility in scheduling the training
- The availability of competent trainers internally

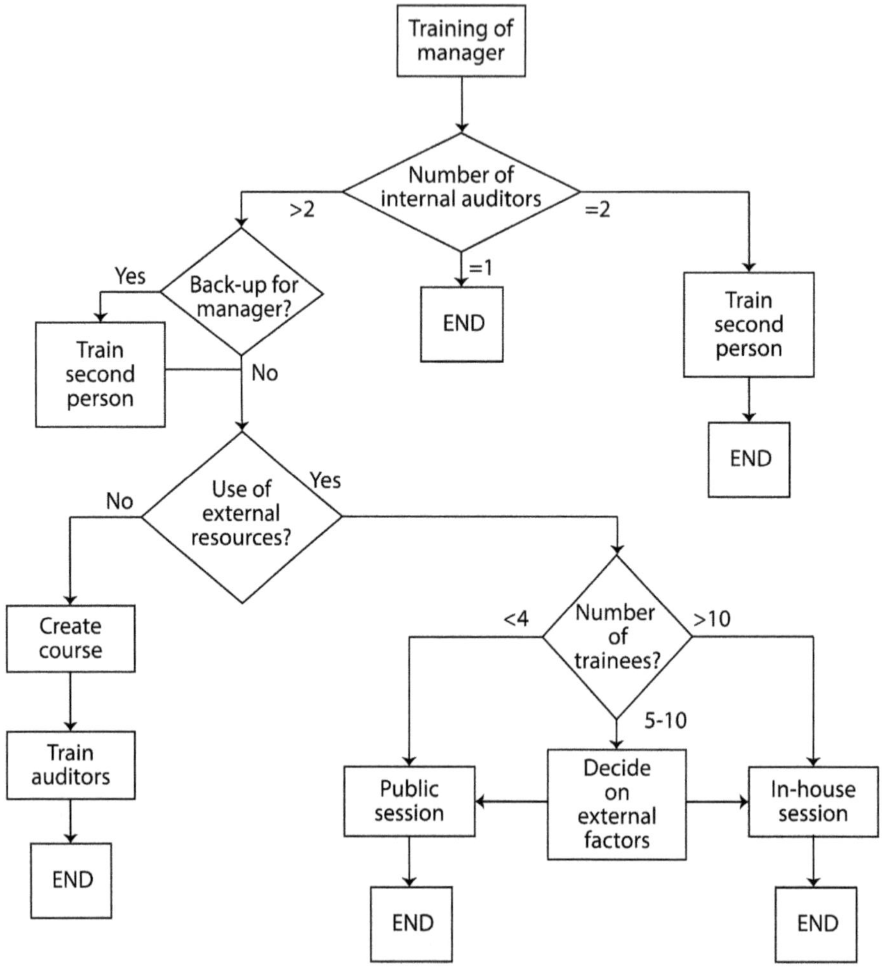

FIGURE 2.4 Decision diagram for initial training effort.

2.10.3 Assessment of Competence

Once the organization has defined the criteria for initial GMP auditor selection, as well as the skills and knowledge to be acquired through training, it is useful and necessary to specify the methods to be used for the evaluation of each candidate against the selection criteria, and for the acquisition of skills and knowledge.

2.10.4 Candidate Selection

Evaluating personnel is a complex activity, and very often specialists from human resources or the raining department will be able to help. Even when that is the case, GMP auditors are frequently selected on personal impressions or with vague criteria related to attitudes or behaviors. Tables 2.8 and 2.9 provide a sample document on how a GMP auditor should be qualified according to which criteria and how they will be certified in-house after qualification. The difficulty of the GMP auditors' selection process depends largely on the number of new auditors needed annually and on the availability of specialized HR personnel. Personnel evaluation can use fairly complex techniques.

2.10.5 Training

Evaluating the training is easier than evaluating a candidate's personal attributes. If training requirements consist of participating in one or more courses, an attendance sheet and/or a certificate of attendance constitutes a sufficient record. If each candidate must acquire knowledge or skills, then that requirement should be measured as objectively as possible.

2.10.6 Acquiring Knowledge

The classical method is to use a written exam, and most training organizations have such exams. Knowledge acquisition can also be measured through an interview, which is a more practical method for organizations that do not use external resources to train internal auditors. The interview method, however, will lack the necessary objectivity if not supported by a checklist and marking grid.

2.10.7 Acquiring Skills

The two usual methods of acquiring skills are role-playing and observing a candidate while performing the activity. Role-playing is used mostly in structured training; observing a candidate is also used in structured training that includes a real audit, as well as in on-the-job training. It is worth noting that role-playing in structured training can also be used to assess some personal attributes, such as emotional stability, capacity to work under stress, and verbal communication skills.

2.10.8 Auditor Certification Programs

Many programs are available for the certification of GMP auditors from associations, such as the Istanbul, Turkey-based company Quality Academia. Although organizations can use these programs to confirm GMP auditor competence, the requirements for obtaining and maintaining such certification are very stringent and usually are not within reach of the average internal auditor.

Some organizations are offering auditor certification adapted to internal auditors. The American Society for Quality (ASQ) has been running the Certified Quality Auditor (CQA) program for more than 10 years. Exemplar Global (exemplarglobal.org) and the International Registrar of Certified Auditors (IRCA) have also run certified GMP lead auditor courses for several years.

2.11 MAINTAINING AUDITOR COMPETENCE

2.11.1 Axes of Competence

GMP auditor certification programs also require for maintenance of certification that auditors engage in professional development activities each year, over and above performing audits. The maintenance of individual GMP auditor competence must be coordinated with the maintenance of homogeneity/consistency between auditors.

2.11.2 Audit Skills

The most effective method of maintaining audit skills is to perform GMP audits on a regular basis. Three evenly spaced one-day audits per year is a bare minimum to maintain skills at a minimal level.

Some organizations use their internal GMP auditors once per year for a one-day or two-day audit. GMP audit results are not reliable without complementary activities. In the absence of frequent audits, other activities, such as the following, must be carried out to maintain competence:

- A one-day annual training session, using practical test cases or role-playing
- When preparing for each audit, a meeting of the audit team with a qualified resource
- Participation in audit-related external events, such as conferences, congresses, and associations

2.11.3 Knowledge

The previous discussion on audit skills – that doing GMP audits on a regular basis is the best way to maintain the knowledge related to the quality management system – also applies to the audit methods and to the organization's audit procedure.

2.11.4 Homogeneity

For a given GMP audit, an auditor is asked to make a judgment on the compliance and effectiveness of one or more processes. In identical circumstances, would another GMP auditor reach the same conclusions? Would the same GMP auditor reach the same conclusions at another time, given no variation in the process? These are the two axes of homogeneity of an auditor's conclusions. This concern is twofold for each axis:

- Obtaining a reasonable quantity of factual evidence and findings about a real process
- Coming to conclusions on compliance and effectiveness from those findings

For a given process, with identical circumstances, there is homogeneity between GMP auditors if they come to identical conclusions, despite different sampling and different findings. There is homogeneity in time if the same auditor comes to identical conclusions at different times.

It is possible to evaluate the homogeneity of a group of GMP auditors by comparing the audit results of the auditors individually auditing the same process at the same time. But such a procedure would be very time-consuming, both for the auditors and auditee, and hardly feasible. Table 2.7 provides a summary of the auditing skills required of an auditor.

2.12 EVALUATION OF GMP AUDITOR COMPETENCE

The evaluation of GMP auditor competence should be planned, implemented, and documented in accordance with the audit program, including its procedures, to provide an outcome that is objective, consistent, fair, and reliable.

TABLE 2.7
Auditing Skills

✓ Evaluating and judging	✓ Drawing conclusions	✓ Observing
✓ Controlling an audit	✓ Understanding technical material	✓ Identifying problem areas
✓ Planning and preparing	✓ Presenting (reports and at meetings)	✓ Knowledge of audit principles and functions
✓ Maintaining confidentiality	✓ Verifying evidence (no assumptions)	✓ Applying knowledge
✓ Leading	✓ Discovering the truth (but not a witch hunt)	✓ Collecting data
✓ Supervising	✓ Awareness of authority level	✓ Organizing
✓ Recording	✓ Business understanding	✓ Time management
✓ Administering		✓ Reviewing
✓ Decision-making		✓ Sampling
✓ Analyzing evidence		✓ Assessing

TABLE 2.8
Auditor Qualification Form

Auditor Qualification Form
Version Nr.:
Revision Date:

1. Personal Data

Last Name:
First Name:
Title:
Address:
e-mail:
Profession:
Current Function:

2. Short CV/Professional Education, Experience

From	Until	Company-Function
YYYY	YYYY	
YYYY	YYYY	
YYYY	YYYY	
YYYY	YYYY	

3. Participation in training courses related to Quality Systems/GMP/Auditing

No	Date	Days	Title	Organization
1.	YYYY-MM-DD	0	Detailed title of the training	Performed by
2	YYYY-MM-DD	0	Detailed title of the training	Performed by
3	YYYY-MM-DD	0	Detailed title of the training	Performed by
4	YYYY-MM-DD	0	Detailed title of the training	Performed by
5	YYYY-MM-DD	0	Detailed title of the training	Performed by
6	YYYY-MM-DD	0	Detailed title of the training	Performed by

4. Participation in GMP Audits and Inspections within the past 5 years

No	Date	Days	Title	Location	Function*
1.	YYYY-MM-DD	0	**Site/Company/Scope**	**City/Country**	**A,C,P?**
2	YYYY-MM-DD	0	**Site/Company/Scope**	**City/Country**	**A,C,P?**
3	YYYY-MM-DD	0	**Site/Company/Scope**	**City/Country**	**A,C,P?**
4	YYYY-MM-DD	0	**Site/Company/Scope**	**City/Country**	**A,C,P?**
5	YYYY-MM-DD	0	**Site/Company/Scope**	**City/Country**	**A,C,P?**
6	YYYY-MM-DD	0	**Site/Company/Scope**	**City/Country**	**A,C,P?**
7	YYYY-MM-DD	0	**Site/Company/Scope**	**City/Country**	**A,C,P?**
8	YYYY-MM-DD	0	**Site/Company/Scope**	**City/Country**	**A,C,P?**
9	YYYY-MM-DD	0	**Site/Company/Scope**	**City/Country**	**A,C,P?**

5. Special Expertise /Self-Evaluation

Manufacture of API (non-sterile APIs, including starting materials and intermediates)	• Yes	• No
Manufacture of specific APIs (APIs manufactured by fermentation, biotechnology, sterile APIs)	• Yes	• No
Manufacture of excipients	• Yes	• No
Manufacture of packaging materials	• Yes	• No
Manufacture of non-sterile drug products (non-sterile)	• Yes	• No
Manufacture of sterile or aseptic drug products	• Yes	• No
Manufacture of IMPs	• Yes	• No
Manufacture of radiopharmaceuticals/PET products	• Yes	• No
Manufacture of vaccines	• Yes	• No

(Continued)

TABLE 2.8 (CONTINUED)
Auditor Qualification Form

Manufacture of medical devices/diagnostic systems	• Yes • No
Manufacture of non-drug products, e.g., cosmetics, food supplements	• Yes • No
Transport, distribution, warehouse	• Yes • No
Analytical laboratories (3rd party)	• Yes • No
Computerized systems	• Yes • No
Quality system audits	• Yes • No
Self-assessments	• Yes • No
Others	If Yes, please describe

6. Elimination and Evaluation Criteria

Characteristics	Grade	Remarks	By/Date
Integrity/honesty			
Pass/Fail			
Self-control			
Pass/Fail			
Communication			
Reading/writing			
Pass/Fail			
Verbal communication			
Pass/Fail			
Curiosity			
Pass/Fail			
Experience in pharma and biotech Industries			
1 = 3 to 5 years			
2 = 5 to 10 years			
3 = ≥10 years			
Knowledge QMS, GMPs			
1 to 3			
Knowledge of audit techniques			
1 to 3			

7. Conclusion

Based on the above data relating to

- Educational background/profession
- Expertise
- Practical auditing experience
- Training

Auditor's professional competence has been thoroughly evaluated.

The auditor is qualified and authorized to audit.

1. Insert type of manufacturers or areas the GMP auditor is qualified/authorized to audit
2. Insert type of manufacturers or areas the GMP auditor is qualified/authorized to audit
3. Insert type of manufacturers or areas the GMP auditor is qualified/authorized to audit

as a member of Acme Auditing Board.

The period of validity of this appointment is 5 years, starting with the date of approving auditor's qualification record.

***Function during the inspection or audit**

A = Lead Auditor;

C = Co-Auditor;

P = Passive participation in an authority or customer audit

The evaluation process should include four main steps, as follows:

- Determine the competence of GMP audit personnel to fulfill the needs of the audit program
- Establish the evaluation criteria
- Select the appropriate evaluation method
- Conduct the evaluation

The outcome of the evaluation process should provide a basis for the following:

- Selection of GMP audit team members
- Determining the need for improved competence (e.g., additional training)
- Ongoing performance evaluation of auditors

GMP auditors should develop, maintain, and improve their competence through continuing professional development and regular participation in audits.

Key auditing competencies include:

- Being able to implement an audit
- Maintaining communication
- Collecting and verifying audit evidence
- Composing findings and conclusions

TABLE 2.9
Auditor Certificate Example

ACME HEALTH
GMP Auditor Certificate
Name of the Auditor:
has earned this certificate for excellence in auditing and due to special expertise related to the areas indicated below;

✓ Manufacture of specific API (APIs manufactured by fermentation, biotechnology, sterile APIs)
✓ Manufacture of excipients
✓ Manufacture of packaging materials
✓ Manufacture of non-sterile drug products (non-sterile)
✓ Manufacture of sterile or aseptic drug products
✓ Manufacture of IMPs
✓ Manufacture of radiopharmaceuticals/PET products
✓ Manufacture of vaccines
✓ Manufacture of medical devices/diagnostic systems
✓ Manufacture of non-drug products, e.g., cosmetics, food supplements
✓ Transport, distribution, warehouse
✓ Analytical laboratories (3rd party)
✓ Computerized systems
✓ Quality system audits
✓ Self-assessments

(Items to be deleted as appropriate)

Date /Signature Head of ACME HEALTH-Audit Management
This certificate is valid for five years after signature

2.12.1 How to Qualify as a GMP Auditor

The three areas of qualification are:

- Objectivity
- Knowledge of the audited processes
- Ability to perform audits as a leader or member

Every person performing a GMP audit should possess a certificate of qualification. Someone has evaluated their skills and declared them to be competent.

2.12.1.1 Local Qualification

You are qualified for your own requirements. It is not necessary for everyone everywhere to use the same criteria. We have different needs and different cultures. The important thing is to set standards and meet them. One of the jobs of the GMP audit boss is to define these qualification standards and document them in a local procedure. Topics to consider include:

- Technical understanding of the processes
- Understanding of how to audit
- Ability to communicate, both in writing and verbally

The boss party that assigns the GMP audit to the auditors, for example, the top management in the company or the manager of a third-party firm, may want to qualify the auditors and reveal their strengths and weaknesses. The GMP audit team and managers should meet periodically to discuss the issues.

2.12.1.2 Professional Qualification

While not necessary, it is good to have a few staff members certified by the American Society for Quality or Exemplar Global. This adds professionalism to the audit program. It certainly makes the audit boss's job easier.

2.12.1.3 Education and Experience

GMP auditors should have a combination of work experience and education. Authorities such as the American Society for Quality, Exemplar Global, or their own organization qualify auditors. One method of proving qualification is to become an ASQ or Exemplar Global Certified Quality/GMP Auditor.

To become a certified GMP auditor, the auditor must have experience in one or more areas identified in the quality/GMP and at least a high school education. In the past, auditors usually had a general background in the quality field.

2.12.1.4 Interpersonal and Other Skills

Regardless of education or training, a GMP auditor who needs more specific interpersonal skills can conduct an effective audit. Good oral and written communication skills are as essential as excellent listening skills

2.13 AUDIT PREPARATION AND PLANNING

Health authorities state that regulated pharmaceutical companies should periodically perform GMP audits, this also paves the way for the audits of suppliers and third-party manufacturers.

As part of the audit preparations, the auditor compiles the information required for the company to be audited. Table 2.10 presents a summary of the introductory information that the firm to be audited should have during the audit preparation phase.

TABLE 2.10
Information to Be Questioned

Content of the list of companies to be audited

- Company name, address
- Addresses of the plants to be audited
- Date of the last audit
- Risk classification for the plant
- Degree of plant conformity with GMP requirements
- Audit priority

TABLE 2.11
Sample List

Preparation for an Audit

- Check the company description (Site Master File)
- Establish the persons responsible
- Compare the specified manufacturing scope with the valid manufacturing authorization
- Determine the available production equipment, and compare it with the manufacturing procedure flow charts
- Select one of several products manufactured or checked by the company for the purpose of comparing the manufacturing and test documentation with the submission file for marketing authorization
- Check reports from previous audits including reports about remedying deficiencies
- Check past complaints about drug products manufactured or distributed by the company
- Check product recalls since the last audit/inspection
- Create a list of samples to be taken
- Review GMP Guidelines and standards relevant to on-site manufacturing
- Compile an audit plan (auditors, dates of audit, work/topic apportionment)
- Inquire whether there have been any changes since the last audit date, and if so, the results.

The description of the company that will be subjected to a GMP audit can be a starting point to learn about the company. It is also examined whether the information provided by the company that has previously submitted the information is correct and whether it is credible and up to date.

GMP audit preparation not only means the preparations of the lead auditor or the auditing party but also includes determining all duties and responsibilities in constant contact with the party to be audited. Table 2.11 summarizes which documents should be examined in the company that will undergo a GMP audit.

For a first- or second-party audit, the client and/or employer of the auditing organization usually initiates the audit either by request or by approval of an audit schedule. The client and/or the employer of the auditing organization should define the purpose and scope of the audit as well as the criteria against which the audit is to be conducted.

Since GMP audits are often scheduled on a regular basis, such as semiannually or quarterly, an interoffice memo may suffice for notification. In some companies, a phone call followed by an email message or memo may be acceptable because the lead auditor and the manager of the area to be audited are likely to have greater contact prior to the actual audit. An auditing organization must carefully prepare for an audit.

2.13.1 Logistics Planning

GMP audit logistical planning is necessary for both GMP audit team coordination and establishing auditee communication links. Good planning will result in a smooth, seamless audit performance.

Poor planning will make the audit team appear less professional and might jeopardize meeting audit objectives. GMP auditors need to arrive at the right place, with the right equipment and information, at the right time, and for a suitable duration of time to get the job done right. To help the auditor properly think through what is necessary to start the audit process, it is prudent to outline a detailed logistical plan.

The logistics of a GMP audit are a crucial part of the planning process any time the auditors must travel to the audit location. The following items should be included in the plan that is typically given to the auditee in advance of the audit:

- A timeline for the GMP auditor's visit
- The opening and closing meeting location and time
- A request for a GMP audit team meeting place, such as a conference room for team debriefings and discussions of observations, equipped with telephones, computers, or any other equipment needed to conduct audit activities
- A request for facility layout maps, office locations, confidential areas, and so on, and any special requirements such as safety, environment, or health equipment like steel-toe shoes, hard hat, lab coat, hair net, head cover, grounding devices, or safety glasses
- Travel plans, possibly identifying a rendezvous prior to the opening meeting, and possibly including an extra day to meet at a nearby hotel and get familiar with the area

2.13.2 GMP Audit Plan

An audit plan identifies:

- The purpose (objectives) and scope (boundaries) of the audit
- The auditee and the organizational units to be audited
- The audit team members
- The criteria being audited against (such as a standard, contract, or procedure)
- Logistical information such as the audit date(s), place of the audit, the expected duration of the audit, interview schedule, and meetings with auditee; when applicable, confidentiality requirements, transportation requirements, required health and safety permits, or security clearances
- The lead auditor and auditors performing the GMP audit, the guide accompanying them, the duties and responsibilities of the observer
- Provision of resources considered critical to the GMP audit
- Consideration of the required sample amount within the sample plan to perform the GMP audit as intended

Table 2.12 provides an example GMP audit plan.

Now, let's prepare an audit plan.

2.13.3 Determination of Audit Purpose

The purpose sets the tone of the audit, such as the determination of compliance or identification of opportunities for improvement. The GMP auditors may verify, determine, or confirm auditee actions for the purposes of approval, certification, renewal, and so on.

Purpose statements should be concise and clear. The GMP audit conclusion should address the purpose of the audit. An example of an audit purpose may be to determine conformity to ISO 9001 (quality), ISO 14001 (environmental), and GMP (pharma and biotech and medical devices) requirements.

TABLE 2.12
GMP Audit Plan Example

Name of Auditee: ABC PharmaPlant
Date of Audit: December 24th, 2023
Purpose: Assess above company as a contracted manufacturing partner against GMPs

Type of Audit:

Corporate Audit ☐ Supplier Audit ☐ Regulatory Audit ☐ Certification Body Audit ☐
Scope of the Audit: ABC PharmaPlant manufacturing facility and support activities will be included in the audit.
Requirements: As specified in FDA, MHRA, PIC/S GMPs and existing ABC PharmaPlant policies and procedures

Applicable Documents:

- Quality Manual
- Site Master File (SMF)
- Validation Master File (VMP)
- Certificates of the Site
- Health Authority Inspection List
- SOP Index

Auditor(s): Ali Stone (Auditor), Veli Water, Mustafa Edik (Lead Auditor)

	Group 1 – ME/VW	Group 2 – AS
09:00–09:30	**Opening Meeting**	**Opening Meeting**
09:30–11:30	**Facility Tour** • Warehouse • Manufacturing	**Quality Management System** • Change Control • Deviation Management and CAPAs • Quality Risk Management • Training • Internal-External Audits
11:30–11:45	**Private Team Meeting**	**Private Team Meeting**
11:45–13:00	**Facility Tour** • Warehouse • Manufacturing	• Supplier Qualification • Pest Control • Calibration & Maintenance • Documentation and Data Integrity
13:00–13:30	**Lunch & Private Team Meeting**	**Lunch & Private Team Meeting**
13:30–15:30	**Laboratories** • R&D Laboratory • Analytical Chemistry Laboratory • Microbiology Laboratory • Packaging Materials QC Laboratory	**Stability** • Cleaning Validation • Equipment Qualification
15:30–15:45	**Private Team Meeting**	**Private Team Meeting**
15:45–16:45	• Process Validation	• Computerized System Validation
16:45–17:00	**Private Team Meeting**	**Private Team Meeting**
17:00-17:30	**Closing Meeting**	**Closing Meeting**

2.13.3.1 Examples of Internal Requirements

- Written policies, procedures, and work instructions
- Training procedures and materials
- Stated objectives in the budgets, programs, contracts, and so on
- Defined process requirements as stated by managers of areas audited
- Company (internal) product or service specifications

2.13.3.2 Examples of External Requirements

- Customer (external) specifications
- Industry handbooks, published guides
- A contract that includes references to a specific standard, such as PIC/S, European Medicines Agancy (EMA), TGA, TMMDA, MHRA, ICH, WHO, FDA standards
- ISO standards such as (9001, 13485, 15378, 22716), ANSI, ASTM

2.13.4 Lead Auditor Selection

Every audit team has a lead auditor. Large teams may have a lead auditor as well as audit team leaders who lead small auditing groups and who report to the lead auditor. When the audit team consists of only one auditor, that auditor is the lead auditor. When more than one auditor is needed for an assignment, the lead auditor, usually an individual with supervisory experience or management capabilities, is identified prior to the other team members and participates in their selection. A lead auditor whose duties include auditing, directing, and monitoring team members usually has more auditing experience than other team members and is more familiar with the standard being audited against.

2.13.5 Building a Team

When preparing an audit plan, the lead auditor (and/or client or audit program manager) should assess the complexity of the activities to be audited and select team members who possess the qualifications or expertise needed to perform the audit. Qualified auditors must be knowledgeable in the audit criteria, such as ISO 9001,13485, 15378, and 22716; the organization's quality assurance manual; and cGMPs. Auditors should be familiar with the industry of the organization they are auditing. It is common for auditors to claim industry-sector experience based on prior work experience. System audits will be more effective if auditors understand the industry jargon and common processes for that type of organization.

Personnel is selected for specific auditing assignments on the basis of experience or training that indicates qualifications commensurate with the complexity of the activities to be audited. The lead auditor is responsible for determining auditor assignments and communicating the assignments prior to the audit. The lead auditor should consider an auditor's work experience and education when making assignments.

To ensure an effective audit, audit team members must be able to report the audit results objectively and impartially. The audit team members are listed in the audit plan. The plan should include individual auditor certifications, such as Exemplar Global Approved and ASQ CQA. The plan may include individual assignments if the lead auditor thinks it is appropriate for communication purposes. If included in the audit plan, area assignments may represent specific experience or expertise, such as purchasing, engineering, testing, piloting, IT, and nondestructive testing. The audit interview schedule serves as a record of audit team assignments.

2.14 AUDIT DOCUMENTATION

Before performing an audit, an auditor must secure documentation of the policies or practices and procedures of a company so that conformity of a product, process, or system with the requirements specified in the documents may be assessed. The lead auditor should contact the auditee to determine the availability of documents related to the audit criteria and to request copies or online access. Proprietary material contained in the documents may be removed before the documents are sent to an auditor or online security programs can block access.

Historically, paper has been the conventional form of documentation, but documentation may be in any medium that can be recorded, reviewed, and retrieved. For instance, computer files and audio and video recordings are all acceptable forms of documentation. Many companies have placed their controlled documents on an intranet for use and downloading by employees. The auditor may be asked to visit the organization's website and obtain the documents independently. If the audit is a second-party or third-party audit, it may be necessary for the auditee to brief the audit team on how to access, review, and retrieve online documents.

One of the most common performance records reviewed by auditors is the prior audit report. The review may be for the auditor's professional benefit or the client may request that the auditor verify corrective actions from prior audits as part of the audit purpose.

Performance records to be reviewed may include:

- Prior audit reports
- Customer feedback
- Corrective and preventive action records
- Product returns
- Product complaints
- Product quality levels
- Process and unit performance compared to goals
- Organization and business performance
- Field reports
- Regulatory citations (if not confidential)
- Risk assessments (product, service, versus objectives)

Auditors should be especially interested in product, process, and system changes since the last audit.

2.14.1 Preparation of Audit Checklists, Guidelines, and Log Sheets

An audit checklist is the primary tool for bringing order to an audit. A well-planned and well-defined checklist helps ensure the audit will be comprehensive. Without a checklist, an auditor has a disjointed, disorganized activity and no place to record failed efforts.

A completed checklist provides:

- Objective evidence that an audit was performed
- Order and organization
- A record that all applicable aspects of the program were examined
- Historical information on program, systems, or supplier problems
- The essence of the closing meeting and audit report
- An information base for planning future audits
- A vehicle for time management
- Sampling plans and strategies

The checklist guides each member of the audit team to ensure that the full scope of the audit is adequately covered. The checklist provides space for recording facts gathered during fieldwork, with places to answer questions and numerous areas for taking notes.

A checklist is a set of notes and instructions about specific areas, with specific questions to ask and specific techniques to use during an audit. Checklists or written procedures are used to ensure continuity and comprehensive coverage of the area of interest.

A checklist may contain a standard set of information that every auditor looks for, such as the following:

- Does the person being audited have access to written procedures and applicable work instructions?
- Is the procedure up to date?
- Does the scope of this employee's training cover the observed activities?
- Is the organizational structure proper for what the employee is expected to do in the observed procedure?
- Are appropriate records being maintained?

Some checklists are yes/no questions that reference a specified requirement, others are open-ended interview questions that test the limits of a procedure, and yet others are statements providing additional instructions for the auditor.

An audit checklist is flexible; audit observations need not be restricted to what the auditor has planned. A checklist is nonlimiting in that it does not restrict what an auditor needs to do, provided the auditor stays within the scope of the audit; it is simply a way to plan so that the auditor knows what to look for and observe in the given area.

Some audit organizations issue notifications when checklists are shared, to remind the auditee that the checklist is not standard. and that areas not on the checklist may be examined as long as they are within the purpose and scope of the audit. It may benefit the auditor to review previous audits after the checklist is completed to see if anything that was a concern in the past should be added.

When used effectively, checklists can serve the purpose of standardizing controls.

Some GMP auditors do not prefer checklists because they limit themselves, there is no space to take the necessary notes, only one checkbox at a time to mark, and also because they think that the audit activity is completely out of the control of the auditor. The absence in these lists of areas to be audited can be remedied simply by leaving a field. Think about it, Annex 1 to the EU GMP guide published in August 2022 has almost four times more pages than the old version, and every subject it covers is at least as critical as the other. However, if you start creating a list of questions using the entire Annex1, you can reach almost 1000 questions and you may have to fit these questions into a checklist of at least 100 pages. Table 2.13 provides a summary of the resources to look for to create a useful checklist.

Even when you take into account the areas where the auditor can write comments or notes, trying to fill the document page by page means devoting almost the entire audit.

It is clear that different companies have different expectations, but the main focus in all health authority audits is to complete the audits without findings or at least with a few minor findings so that a patient's health is assured. For this purpose, it is very time-consuming to prepare checklists and they should be constructed in such a way as to eliminate the confusion of experienced and inexperienced auditors. Checklist preparation software prepared by Quality Academia complies with 21 CFR Part 11 and EU GMP Annex 11.

TABLE 2.13
What Are the GMP Audit Checklist Sources?

- Regulatory Documents (FDA, PIC/S, TGA, MHRA, WHO, TMMDA, EMA, etc.)
- GMP Guidelines
- Peer-viewed articles
- Non-profit organization guides (ISPE, PDA, ASQ)
- Corporate documents (Policies, Directives, SOPs)
- FDA 483s of other inspections
- Management expectations, issues

2.15 ARCHIVING

Audit reports as well as associated documents need to be archived for the purpose of evaluating the compliance history at the time of a new audit or for informing senior management about the status and developments in terms of compliance. It is thus necessary to retain such documents for a sufficient period of time. The time period of storage for these documents should be laid down in a written procedure by the company.

As there is no official GMP requirement for the storage time of auditing documentation, in general, it should be at least seven years unless otherwise justified. Third-party audit reports also should have a validity period and a maximum of three years is defined for Active Pharmaceutical Ingredients Committee (APIC) audit reports. The auditing documentation does not necessarily need to be archived on paper. All other technical options such as microfilm, CD, or electronically on computer servers are acceptable as long it is ensured that the files can be accessed over the retention period.

2.16 SAMPLING PLAN

A sampling plan must be developed for each audit. The lead auditor should determine the overall sampling policy and guidelines to avoid personal bias and ensure consistency between audits. The type and extent of sampling for each individual audit are normally the responsibility of the auditor, based on the purpose and scope of the audit. Questions regarding the adequacy of the sampling plan should be addressed by the lead auditor.

The purpose of sampling is to gain information about an entire group of units, items, things, or people based on observing and/or inspecting a segment of the group. Sampling is a technique to gain the needed information at a reduced cost (compared to 100% inspection). To be valid, the segment or sample selected should be representative of the entire group. It is important that the sampled group be recorded. Table 2.14 provides a sample checklist.

For example, an auditor may sample the shipping records for the current month or sample all corrective action records since the last audit. It is important that the auditor not let the auditee pick the samples. He or she may pick only favorable samples or pick the same ones from audit to audit. Samples include not only documents, records, and products but also areas or activities to observe and people to interview.

2.17 RECORD OF OBSERVATIONS

Recorded observations provide the audit evidence for reporting the results of the investigation. It is imperative that auditors record their observations rather than relying on memory. Observations must be traceable back to where they were made to ensure credibility. Auditors should obtain the auditee's acknowledgment (not agreement) of observations identified as noncompliance/nonconformance or negative findings. Auditee acknowledgment provides support for the auditor's work during the closing meeting. If the auditor has effectively communicated with the auditee, there should be no surprises during the audit or during the closing meeting. Observations can be positive or negative. Positive observations may be reported as a positive practice, a noteworthy achievement, or a positive finding. Negative observations may be reported as part of a noncompliance/nonconformity or finding.

Next, let's prepare a checklist.

2.18 AUDITING STRATEGY

Regardless of whether the subject of the audit is a product, process, or system, the auditor should use a general auditing strategy, or audit path, so that data are collected in a logical and methodical

TABLE 2.14
Checklist Example

Personnel (PIC/S GMP Part I, Chapter 2) Yes ☐ No ☐

2.13 Visitors or untrained personnel should, preferably, not be taken into the production and quality control areas. If this is unavoidable, they should be given information in advance, particularly about personal hygiene and the prescribed protective clothing. They should be closely supervised.

If yes, describe. If no, provide a rationale (e.g., Not applicable because…)

Deviations
Identify and describe any noted GMP deviation(s) and the rationale for the deviation, where applicable.

No.	Observations	Risk Classification

Corrective Actions
Detail the corrective action(s) taken and/or to be taken.

Attachments
Attach supporting documentation such as SOPs, and action plans with timelines for each corrective action identified above. List of attachments:

manner. Effective audit planning ensures that an auditor will have the time and the means to conduct a thorough investigation of the audited organization that is consistent with its purpose and scope. It also enables the auditor to segment the audit by element or department, or to use the discovery method or tracing techniques to gather data.

2.18.1 Tracing

Tracing, or following the chronological progress of something as it is processed, is a common and effective means of collecting objective evidence during an audit. In forward tracing, an auditor starts at the beginning or middle of a manufacturing or service process, for example, and traces forward (downstream). In backward tracing, the auditor starts at the middle or end of a process and traces backward (upstream). Backward tracing can be more revealing than forward tracing because the auditor examines the process from the perspective of seeing the results (product or service) of the preceding activity.

The mechanics of tracing are relatively simple:

- Start at the beginning, middle, or end of the process.
- Choose an action, such as mixing in the vessel.
- Gather information on the six process forces (methods, machinery, material, manpower, measurement, and environment) for that action. Be sure to record this information on your checklist. Write down what you see, the person you are talking to, where you are, when the action was done, why the action is accomplished, how it is accomplished, and how well the action was performed (results). These become your facts for later use.
- Follow the path of the transaction backward or forward through the process.

Tracing may be used in process and system audits. The tracing method is useful when a procedure is unclear, when the audit evidence to verify conformity or nonconformity is elusive, and when evaluating performance issues. Tracing is also used as a strategy for process audits and auditing process-based management systems.

When performing a supplier audit, a slight modification might be of value.

2.18.2 Process Method

The process method is an expansion of the tracing technique for use as an overall audit strategy. The process method uses a flowchart and process map as a guide during the audit. For a process audit, the scope may be a single process; for a system audit of a process-based management system, the scope may be the design function, purchasing, laboratory, document control, or operations. When auditing a process-based management system, the sequence and interaction of processes have already been defined. However, for an audit of a single activity or process, it is common for auditors to create their own flowchart.

The process method is effective for identifying weaknesses within the organization's management system, breakdowns in securing appropriate inputs, and achievement of output objectives. A disadvantage of the process method is a lack of linkage to requirements in external standards to ensure ongoing conformance/compliance. It is recommended that organizations that have a process-based management system audit by function.

2.18.3 Understand the Process

Whether you are about to perform a small process audit or a large system audit, you must know the basic process steps and the areas or groups involved in those steps. You can get this knowledge by:

- Reading the notes of previous audits in that area
- Reading the manuals and major process procedures
- Talking with a trusted friend who knows about the area
- Exploring your supplier's website
- Reading the process maps and flowcharts

Once you have a visual diagram of the processes on paper, you now know the answers to the following questions:

- Who does the job?
- What is the job?
- Where is the job done?
- When is the job done?
- Why is the job done?
- How is the job done?

These six words – who, what, where, when, why, how – will be used many times in the audit.

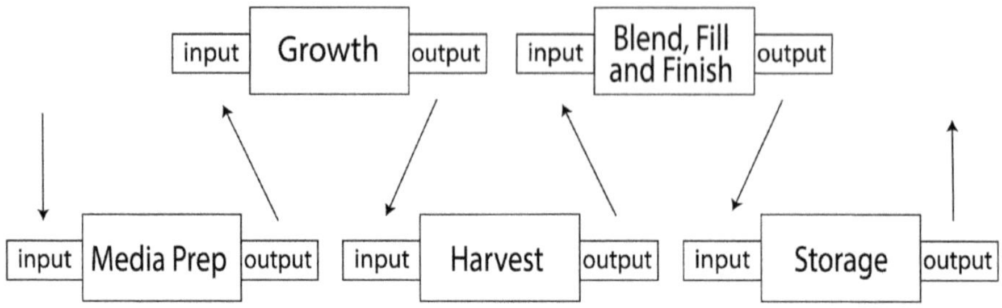

FIGURE 2.5 Process approach.

2.18.4 Process Mapping

It has been estimated that people working in organizations can waste 15%–20% of their time by redoing things that are wrong, chasing things without results, querying incomplete instructions, or doing other people's jobs.

Presumably, all companies have organizational charts. While these are useful tools for communicating reporting relationships, given the complexity of the tasks at hand, they can never fully reflect the actual relationships and interactions within a company. Organizational charts do not answer the question of what is required to deliver a given output, what contribution a department makes to the whole, or how work is accomplished. Figure 2.5 provides an example of how the process approach should be understood.

Process mapping gives a better understanding of the work to be done in the organization. Process maps are often used to show how the activity is done making the the work more visible.

Process maps aid understanding by abstracting the key elements of a process and using visual charting symbols that mask unnecessary details.

A better understanding of the important characteristics of a process generates useful data from which to derive findings, draw conclusions, and formulate recommendations. Furthermore, process maps urge us to systematically ask questions and adopt a viewpoint of business process improvement.

Process maps help to:

- Orient new employees
- Organize the work to be done by the organization
- Evaluate or establish alternative ways to organize the tasks to be done
- Clarify roles and contributions
- Identify areas of improvement
- Evaluate, establish, and/or strengthen performance measures

There are a variety of maps that can be used to visualize the work of an organization. Three different approaches are widespread and complementary when used in combination:

- Relationship maps
- Flowcharts
- Cross-functional process maps

2.19 COMMUNICATION AND DISTRIBUTION OF THE AUDIT PLAN

An audit plan is the primary product and communication tool of the preparation phase of the audit process. The audit plan tells us what will be covered in a particular system audit or sequence of process audits.

Before issuing an audit plan, the lead auditor should contact the auditee to verify the proposed date of the audit.

After an audit plan has been finalized, the date and logistics of the impending audit should be confirmed and recorded. For an internal audit, notification is done by memo or published schedule. For a second- or third-party audit, it is done by letter. The notification letter or memo should be addressed to the senior person in charge of the area to be audited or their designee, such as an auditee coordinator.

For internal audits, this person is probably the department manager or area superintendent. For external audits, it is probably the plant manager, president, or head of quality. Even if the lead auditor prepares the letter or memo, the client should sign it. This formal notice should be delivered before the visit. The timing of the notice should be consistent with organizational procedures and common courtesy, varying from one week to several months. An advance notice of 30 days is common but not required. For some compliance audits or those where there is suspicion of wrongdoing, there may be no notification.

2.19.1 Notification Letter Example

Mr. Dale Smith
1 ABC Blvd.
ABC Industries
ABC, FL XXXXX

Dear Dale

The audit team will perform a quality system audit of the ABC facility to assess conformance to ISO 15378 standard requirements. The audit's scope will be limited to the conditions stated in ISO 15378 and existing internal quality system policies and procedures. The audit process will be conducted according to recognized auditing conventions as defined for Exemplar Global Approved, ASQ Certified Quality Auditors. It would be best if you planned for the audit to require two days of on-site interviewing and verifying requirements. After the audit, a draft report will be left, and the final report will be forwarded within two business days. Jane Red and I will be the auditors. The on-site audit is scheduled for March 23 through March 24. The audit team will need a room with a table, telephone, internet access, and electrical outlets for computers and printers. The audit plan is enclosed. As agreed, please forward the management system manual to my attention to arrive two weeks before the scheduled audit. Thank you for being so cooperative in making the necessary arrangements for the planned audit. If you have any questions I can be reached at
XXX-XXX-XXXX or John.Doe@anymail.com.

Sincerely,
John Doe, ASQ CQA
Copy: Jane Red
Enclosures: 1

2.19.2 Distribution of an Audit Plan

Members of an audit team receive copies of the audit plan, as does the auditee. The auditee is responsible for distributing the audit plan within the organization to be audited.

Several parallel activities are involved in the preparation of an audit plan: scheduling the audit, notifying the auditee, and lining up audit resources. Once the audit plan is prepared, it is formally communicated to the auditee. The lead auditor should publish the audit plan; notify the client and the auditee of the impending audit; and confirm that any special needs or plans, such as travel requirements, communication devices, or equipment and security clearances, have been arranged. Once these steps have been completed, the performance stage of the audit may begin.

2.19.3 Conflict Resolution

The most effective method of resolving conflict is to take steps to prevent conflict occurrence. Within the audit team, the team leader may establish ground rules for communication, interaction, and performance of team members prior to the audit.

The audit team may reduce conflict during the audit by eliminating misunderstandings, remaining open-minded and flexible, and avoiding surprises during the audit process.

Misunderstandings are the most common cause of conflict during an audit. Sources of misunderstanding are poor communication, poor listening, and auditor or auditee bias. Auditors must always be on the lookout for signs of misunderstanding. Indicators include a less-than-cooperative attitude on the part of an interviewee, repeated questions about the audit scope and purpose, and the provision of inadequate or incorrect records for review. In these situations, the best course of action is for the auditor to repeat or clarify the information that was not fully understood before proceeding. Also, auditors must be active listeners, meaning they must not only hear what is being said but also recognize when the message is clouded by a possible contradictory meaning.

Finally, the auditor should always assume that the auditee is acting in good faith. He or she must not assume that a misunderstanding is on the part of the auditee. Instead, the auditor should rephrase the question in a different way, being careful to avoid terminology that has confusing meanings. Auditors must not assume that the auditee is familiar with specific industry jargon. However, in some cases, the auditee may not act in good faith or may be uncooperative, and the auditor should watch for those situations.

During an audit, auditors must remain open-minded and flexible. Auditors must bear in mind that when they conduct an audit, for whatever reason or from whatever source of authority, they are disrupting the daily routine of the auditee.

Unnecessary and unreasonable demands are often a source of conflict during the audit. This becomes especially true when the auditee is intimidated by the audit team and, as a result, is unwilling to say anything ahead of time. These kinds of issues can manifest themselves later in the form of uncooperativeness, avoidance, belligerence, or even sabotage.

A sure way to create conflict during an audit is to withhold important information about the audit's status or results and surprise the auditee at or after the closing meeting. It is important, therefore, that auditors share information with the auditee on a regular basis. This is usually done during informal daily meetings or during actual audit performance. In addition, auditors have an obligation to inform auditee personnel of the nature of the information they are recording, how it will be used, and so on. Being upfront with auditees not only helps address some of their concerns about the audit process but also puts them at ease, thus reducing the likelihood of conflict.

If a conflict should develop during an audit, the best course of action is for the auditor to temporarily stop the audit and allow a cool-down period before proceeding. If auditors don't take this action, they run the risk of getting overly involved and losing control of the audit process. The ultimate result of losing control is the premature termination of the audit or the auditee not allowing the auditor to proceed. Neither of these situations is acceptable to the auditor's organization or the client. Once the conflict has been defused, the auditor can meet with the auditee to determine the source of the misunderstanding. Auditors must remain open to the possibility that they may be a direct cause of the conflict. To resolve the conflict and allow the audit to proceed, auditors must remain open-minded and listen to the auditee.

An auditor does not have the option of avoiding or trying to smooth over an issue with an auditee. Nor should an auditor attack individuals or attempt to use his or her position to overpower them. Forcing a conflict escalates it and increases the chance of recurrence later in the audit with the same person. Sometimes the successful resolution of conflict is an agreement that there is disagreement.

2.19.4 How to Deal with Difficult Situations

Challenging situations include antagonism, coercion, and even time-wasting techniques employed by the auditee to slow down or stop the audit process. An auditee may use time-wasting tactics by deviating from the audit plan. In any difficult situation, the auditor should remain polite but firm, maintaining self-control and complete control of the audit.

2.19.5 Combating Time-Wasting Techniques

An experienced auditor will immediately recognize most time-wasting tactics employed by an auditee. An auditor needs to lead an audit and not allow the auditee to take control. If an auditee attempts a lengthy presentation at the opening meeting or an extended plant tour, for example, the lead auditor should halt these activities or limit the time spent on them. Often, a lead auditor can eliminate wasted time by making specific requests during the audit planning stage. For example, many audit teams prefer an on-site lunch to go off-premises for several hours. Of course, auditee management and the client should be notified when an audit team repeatedly encounters delay tactics or if the audit schedule is severely compromised by such delays.

When an audit is not moving along as planned, the auditor can say:

> "I'm really enjoying this conversation, but all of us have a lot to do; so what do you say we get started."
>
> "This tour has been very helpful, but I think it has accomplished what I needed; so let's start the audit interviews."
>
> "Since we have so many areas to cover today, can we speed things up so we don't have to stay after working hours to complete the audit?'"
>
> '"What's next on the schedule?"' (The auditee notes the next step from the provided copy of the audit agenda.) "OK. Let's move on so we don't affect the operation."

2.19.6 Problems Commonly Encountered during the Audit Planning Stage

Problems likely to be encountered during the planning stage of an audit include difficulties related to the availability of personnel and other audit resources and scheduling the audit.

2.19.6.1 Availability of Audit Team Personnel

Most audit team members are not full-time members of a quality assurance staff or of an auditing organization. They are either volunteers or people drafted from other areas of the company, such as production, accounting, sales, and personnel. It is often difficult for them to find the time that is needed to effectively plan and audit, generally four to six hours per person.

2.19.6.2 Availability of Other Resources

One problem is the lack of availability of other resources, such as funding, that can have an effect on audit planning. Another problem is that the auditee may fail to forward the necessary documents to the auditor in time to allow adequate preparation.

2.19.6.3 Scheduling Difficulties

Often other events take priority and make the audit difficult to schedule. Such events could include a major reorganization or merger. Perhaps the auditee has just received a major contract, or the audit is scheduled at the same time a regulatory agency plans a visit. An audit is an important event, but sometimes other events take higher priority.

2.19.6.4 Auditee Availability

At times, in spite of all good intentions, some auditees cannot be available due to prior commitments or personal emergencies.

2.19.7 Audit Plan Changes

The lead auditor ensures that any change in the audit plan or schedule is proactively discussed during the opening meeting and obtains consensus for changes. The lead auditor needs to be flexible enough to make audit plans and schedule changes at the auditee's request. This is critically important to ensure that the audit gets off to a good start.

The auditors are responsible for ensuring that audit plan changes do not result in sacrificing the original time allocations established to effectively evaluate conformance to requirements.

2.19.8 Confirmation of Audit Logistics

A number of logistical issues need to be settled at the opening meeting. The number of issues will depend on whether the audit is internal or external, the size of the organization, the size of the audit team, and the complexity of the area to be audited.

Issues include but are not necessarily limited to:

- Meeting rooms for audit team use and for daily auditee briefings
- Office space for auditors
- Access to necessary facilities and equipment required for verification activities
- Documents and records needed
- Special safety, environmental, or security issues
- Facilities for the safe storage of confidential records
- Access to telephones
- The distances between areas to be visited
- Housekeeping items (restrooms, lunchroom, break area)
- An escort
- Tours

An auditee should provide an escort for each subgroup of the audit team. The escort accompanies the auditor on the entire data-gathering portion of the audit and serves as a liaison between the auditing organization and auditee management. The escort performs personnel introductions and provides clarifying information as needed. For example, an escort can suggest a way of rephrasing a question, perhaps by substituting terminology more familiar to the person being interviewed. However, the escort should refrain from answering questions for the auditee unless asked to do so by the auditor.

Other duties of the escort include:

- Providing or requesting supplies and records required by the auditor
- Acting as an observer for management, and providing an overview of the auditor's observations, findings, and conclusions to auditee management
- Acting as a guide for the auditor
- Confirming or denying that a discrepancy or nonconformity has been found before moving on to the next area to be audited
- Ensuring that auditors comply with company rules, including safety rules

2.20 WHICH DOCUMENTS SHOULD BE REQUESTED FROM THE AUDITEE PRIOR TO A GMP AUDIT?

It is a time-saving and useful method in GMP audits to request some documents from the facility to be audited before the audit or, in the worst case, to have them available during the audit.

Before the audit, the GMP lead auditor may request the following documents:

- Site Master File (SMF)
- Organization chart
- Validation Master File (VMF)
- SOP Index
- Site layout
- Critical equipment list
- Quality manual
- Layout plans of critical systems (HVAC, water, gas, steam)

2.21 OPENING AND CLOSING MEETINGS

During the opening discussion, the group consisting of the lead auditor and auditors, the senior management, and the audited company personnel introduce themselves and exchange business cards. The audited company management or the person making the presentation should not use this situation only as a promotion of their company. If it is related to sales figures, market strategies, and areas to be audited, it may attract the attention of the GMP audit team. Presentation slides can be used as a tool for the auditors to prepare their reports more easily later on. The auditee should show the company presentation in confident and clear terms. While the company presentation often seems like a waste of time, it can initially offer clues about the behavioral patterns of audited party managers and staff.

2.21.1 Opening Meeting

Soon after the arrival of the audit team at the audit site, the audit team and the auditee gather for an opening meeting, sometimes called a pre-audit conference or entrance meeting. This meeting starts the data-gathering phase of the audit. An opening meeting should be held at every audit, whether on-site or remotely.

The purpose of the opening meeting is to:

- Confirm the agreement of all parties to the audit plan
- Introduce the audit team
- Ensure that all planned audit activities can be performed
- Introduce audit team members to auditee management
- Establish communication links and the role of the escort

Communicate the audit plan by:

- Reviewing the scope and purpose of the audit
- Reaffirming the reference standards being used
- Describing the methods and procedures to be used during the audit
- Inform the auditee how the results will be reported (nonconformances, findings, or other) and the methods used to classify and prioritize results
- Inform the auditee of the possible outcomes of the audit
- Confirm logistics, such as hours, escorts, tentative schedules, and needed facilities, and availability of resources
- Confirm time and date for closing meeting as well as interim auditee daily briefings

Additionally, if applicable, the opening meeting may:

- Include a review of previous audits
- Help ensure that the audit team understands the confidentiality agreement
- Verify required safety and security clearances

- Identify local language issues and concerns (different languages used in the workplace, slang, word-usage level)
- Identify work agreement requirements between management and labor that may affect the performance of the audit

Opening meetings are a good way to put both parties at ease and warm up to the audit, and are where the essentials of the audit are discussed. Before an audit starts, individuals should know the daily audit schedule and what to expect in the closing meeting, written report, corrective action requests, and possible follow-up. The structure of the opening meeting should be flexible and the level of formality will depend on several factors, including the audit scope, the purpose, the size of the audit team, and whether the audit is internal or external.

2.21.1.1 Opening Meeting Roles and Responsibilities

The auditors involved in the audit, the lead auditor, and the audited party representatives should attend the opening meeting.

The lead auditor prepares the meeting agenda and often hands out copies in advance. At the opening meeting, the lead auditor:

- Introduces audit team members and presents their credentials
- Restates the purpose and scope of the audit in a clear and diplomatic fashion
- May refer to previous audits of the facility and previously required corrective action
- May ask specific questions as a result of the document review or other information provided to the audit team
- Solicits areas of interest from the auditee or mentions areas of specific concern
- May present the audit checklist to the auditee
- Sets the detailed audit schedule
- Describes the documents used to develop the audit plan
- Ensures meeting minutes and attendance are recorded (these duties may be completed by the lead auditor, or the lead auditor may assign the duties to an audit team member)
- Verifies that the auditee's management has communicated to employees and other interested parties, such as a union, that the audit is being conducted

Table 2.15 describes the roles and responsibilities at the opening meeting.

The lead auditor should also describe the audit process and the anticipated benefits. The lead auditor should allow time in the schedule to answer any questions from auditee personnel.

The following action items may need to be included in the opening meeting:

- Introduction of participants, including audit team members, guides, and observers, and outlining their roles

TABLE 2.15
Roles and Responsibilities

Roles of the Auditors	Roles of the Company
• To introduce the auditors and the apportionment of their tasks • To show the reason, purpose and scope of the audit • To name the documents to be viewed • To make suggestions for the organizational procedure	• To introduce the persons responsible at the company • To describe the QM/QA system • To demonstrate the essential changes since the last audit • To report on the remedying of deficiencies from the last audit • To provide escorts • To allocate rooms for the auditors

- Establishment of communication links and the role of the escort
- Communicate the audit plan by:
- Reviewing and confirming the scope and purpose of the audit
- Reaffirming audit criteria and the reference standards being used
- Describing the methods and procedures to be used during the audit, including evidence based on samples taken
- Inform the auditee how the results will be reported (nonconformities, noncompliances, findings, improvement points, or other) and the methods used to classify, grade, and/or rank (prioritize) results
- Confirm that the auditee will be kept informed of the audit progress
- Inform the auditee of the possible outcomes of the audit
- Confirm logistics such as hours, escorts, tentative schedules, needed facilities, and availability of resources
- Confirm time and date of the closing meeting as well as interim auditee daily briefings
- Confirm health, safety, emergency, and security procedures
- Ensure that the audit team understands the confidentiality agreement
- Report identified audit risks to the auditee organization and how they will be monitored and/or treated; when appropriate, such as with internal audit team aspects, audit risks should be kept private and be addressed by the audit team and/or the auditing organization
- Review the findings of previous audits
- Explain conditions that could result in terminating or canceling the audit
- Verify required security clearances
- Identify local language issues and concerns (different languages used in the workplace, slang, word-usage level)
- Identify work agreement requirements between management and labor that may affect the performance of the audit
- Review information about auditee feedback regarding auditee findings or audit process issues (normally done during the closing meeting)
- Discuss how to deal with potential findings during the audit
- Schedule training or a tutorial on navigation methods for electronic documentation software, if appropriate

2.21.1.2 Attendance Sheet

The attendance sheet, which includes the names, duties, and departments of the people who attended the opening and closing meetings during the audits, must be signed by the participants. People attending the opening meeting are requested to attend the closing meeting whenever possible. Table 2.16 presents the sample form with a list of people who attended the opening and closing meetings.

2.22 WHY REVIEW PREVIOUS AUDIT FINDINGS?

Reviewing previous audit results and actions taken at the audit opening meeting is perhaps one of the most frequently preferred behavioral models in the pharmaceutical industry. Although the audited facility has issued a CAPA regarding the nonconformities identified in the previous audit and clearly stated in the audit report, it is useful to discuss the open issues at the opening meeting, as they may be actions that take a long time to complete. As I tried to express earlier, any GMP noncompliance detected in the previous audit may also have the note that it will be evaluated in the next audit.

2.23 ROLES AND RESPONSIBILITIES OF AN AUDITEE

As GMP auditors, we expect one or more representatives of the audited party to attend the opening and closing meetings whenever possible. In the GMP audits I have carried out so far, only one

TABLE 2.16
Attendance Sheet Example

	Print Name	Job Title	E-mail	Meeting Attendance	Signature
1				Opening Meeting ☐ Closing Meeting ☐	
2				Opening Meeting ☐ Closing Meeting ☐	
3				Opening Meeting ☐ Closing Meeting ☐	
4				Opening Meeting ☐ Closing Meeting ☐	
5				Opening Meeting ☐ Closing Meeting ☐	

official from the audited party attended both the opening and closing meetings. Don't you think that the fact that I left the audit with more than 50 findings explains how worthless the top management finds audits?

In some cases, the lead auditor may specifically request certain representatives or may suggest minimum attendance based on the auditor's judgment (for example, if the room is too small to hold many people, if the wrong people want to attend, or if there are culture issues). The auditee's senior management representative should be invited to introduce his or her attendees. It is expected that senior management representatives attend opening and closing meetings in third-party audits.

After the COVID-19 outbreak started, I requested that a minimum number of people attend the opening and closing meetings at the facilities I audited. In this way, I reduced the number of crowded people in the room and had the chance to interview only the fully responsible people.

There are points that the auditee should agree with the auditor regarding the audit:

- The individual who will represent the auditee on all matters during the audit
- Auditor access to areas and activities to be audited
- Facilities for the audit team's use
- Support personnel to be provided: escorts, specialists, line personnel, and so on
- Safety and regulatory requirements
- Protection of proprietary rights

As you have experienced in many GMP audits, the auditee wants to introduce the company as a presentation, and a facility tour is also one of the most anticipated audit startup models. It is important for the course of the audit that both the company introduction presentation and the facility tour are within a reasonable time frame.

The auditor should inquire in advance whether any safety, environmental, or health-risk presentations are required or suggested by the auditee so that they can be worked into the schedule.

Let us start the audit!

2.24 AUDIT DATA COLLECTION: AUDIT FINDINGS

The lead auditor and the auditors participating in the audit can use their own strategies during the data collection phase to meet at least two times a day as the data is collected and discuss what evidence the collected data is based on.

If the data gathered are unclear or contradictory, the audit team should determine whether additional data are needed that can either confirm or deny tentative conclusions.

The audit team is also responsible for ensuring that an auditee representative is made aware of potential problems as they surface. Information gathering normally takes most of the time and effort in the performance phase of the audit process. The auditor's task is to collect factual audit evidence, analyze and evaluate it against the audit criteria (requirements), draw conclusions from this comparison, and report the results to management.

As an example, a GMP auditor conducting a process audit of a gas chromatograph testing of an analysis must review the inputs and outputs (such as samples, containers, collections of samples, sample plans, output results, and the verification and use of outputs). GMP auditors are people who can establish all the actual steps of the process and their connections with other processes.

2.24.1 Information Gathering

In auditing, information or objective evidence can be gathered in several different ways. GMP auditors may examine documents and records; interview employees; physically examine samples of products, data analysis, performance reports, customer or user feedback, and websites; and observe work in progress. The GMP auditor must sort the audit evidence (observations) gathered and then summarize it in a format that will be useful to the client and/or auditee.

2.24.2 Data Collection Plan

GMP auditors must develop a plan for collecting specific evidence needed to verify conformance to requirements. The collection plan provides the auditor with space for recording the results of examinations and identifying people who are interviewed. The auditor should ask open-ended questions that result in verifiable data. The questions asked may deal with causal groupings such as methods, materials, machinery, people, measurement, and environment, which may affect a specific action. Information and data include any observation of facts and can be qualitative or quantitative. Sources of data could include records from the auditee as well as observations that an auditor makes during audit performance, such as counting the number of press cycle openings, noting the amount of liquid discharged, or observing an operator during a prescribed, defined operation.

2.24.3 Document Examination

GMP auditors must be well trained in methods for examining documents and records. A document specifies what should be done; a record specifies what has been done. Documents are written instructions or procedures that establish a practice or tell a person how to do a job. Examples of documents include written work instructions, guidelines, SOPs, diagrams, drawings, and blank forms. In contrast, records result when a particular step or process is performed as directed in the documents.

Technically, a record is a type of document whose purpose is to collect information about an output of a process. Some refer to a record as a special type of document that states results. Records are very important for GMP auditors and management review meetings, and results held at the facility during the year, results of facility audits, health authority inspections and results, test documents, outputs from equipment, and other reports are all records and can be requested by the auditor. GMP auditors thus also witness whether what is claimed in the SOPs and instructions actually happened.

2.24.5 Observation of Work Activities

GMP auditors observe work in progress to see if it meets requirements. GMP auditors gain this knowledge by monitoring a process being performed to see how the work is done. "Observation" means "to watch attentively" or "to watch or be present without participating actively." GMP auditors must be well trained in observation techniques so that they learn how to observe closely but not obtrusively. Because an auditor's presence may be distracting to people at work, GMP auditors need to minimize any interference. An auditor should also realize that disrupting a worker may be unsafe. The risk of an auditor's presence interfering with normal operations should be managed. GMP auditors should know the normal working hours of the auditee organization and should work the same hours. If the auditee starts work at 9 a.m., then the auditor should consider starting at 9 a.m. The same is true for quitting time. GMP auditors need to know when to observe work. For example, a GMP auditor who wants to observe work being done by different shifts should be aware of when the shift changes take place. GMP auditors should avoid retaining people through shift changes or during their breaks. Besides the negative effect on relations with the auditee, it could result in overtime, scheduling issues, and even union grievances if the person being held over is not next on the overtime list.

Most importantly, GMP auditors need to understand what to observe and what they have seen. For example:

- GMP auditors should establish whether the product is made according to the documented procedures by the individuals responsible for using the specified equipment. GMP auditors should record the results of the observations.
- When looking at equipment, GMP auditors should note the type, condition, use, and any identification tags or numbers. GMP auditors should also establish that the equipment has been properly maintained and calibrated, if applicable.
- GMP auditors should verify that employees are familiar with policies and procedures; that employees know their responsibilities and roles; and that employees have the necessary training, skills, experience, and authority to perform their jobs.
- When reviewing documents, GMP auditors should verify contents, type, scope, format, readability, and date. How the document has been filed and distributed and the method by which modifications have been made is also important.
- GMP auditors should look at the input and output of every process or activity to judge the effectiveness (results) of the process.

By listening and watching, GMP auditors should be able to ascertain whether what the worker is doing supports the procedures outlined in the system- or process- related documentation. It is usually best if GMP auditors observe people doing their actual work rather than creating work for them or conducting a simulation.

2.25 HOW TO INTERVIEW

Before starting interviewing a strong rapport must be built between the GMP auditor and the interviewee.

2.25.1 How to Build Rapport

Let's see how to build rapport at the beginning of an audit.

2.25.2 Before Interviewing

I try to arrive as early as possible at the facilities where I will perform audits. There is always a responsible person to meet me at the door. Before starting the audit opening meeting, I try to analyze the reflexes and excitement of people by asking small questions to everyone as an ice breaker. This is not a one-sided exchange, and I am offering clues about my tone of voice and body language while speaking. Rest assured, this tactic always works, and both you and the other party will be more willing to do work for the audit.

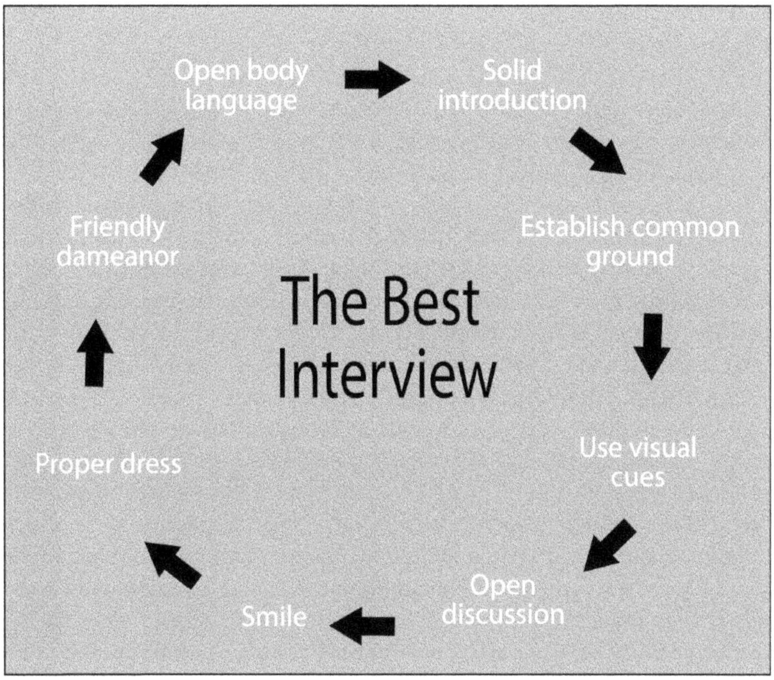

FIGURE 2.6 Successful interview.

Go, and give it a shot. Figure 2.6 summarizes a list of criteria for a successful interview.

2.25.3 Icebreaker Questions

- What is something interesting about your family?
- What was your biggest raffle/lottery ticket win?
- Where do you shop for extravagancies for yourself?
- Name one of your three wishes granted by a genie.
- Do you make time for downtime in your week?
- If you had a million dollars ….
- Do you play a musical instrument? Which one?
- Star Wars or Star Trek (neither is *not* an option)?
- What is the best restaurant in your town?
- Why is your job awesome?
- Why did you wear what you are wearing today?

2.25.4 Are You Ready for the Interview?

Interviewing, the process of obtaining information from another person in response to questions, is an important and widely used form of data collection in auditing.

Interviews are used in GMP audits to gain insights, clarify information, confirm or deny suspicions, and elicit details that may not be brought out by other audit activities.

- Put the interviewee at ease
- Explain your purpose
- Find out what they are doing
- Analyze what they are doing

- Make a tentative conclusion
- Explain your next step

When conducting an interview, the GMP auditor should begin by providing a personal introduction and putting the interviewee at ease. The purpose of the interview should be explained, followed by general, open-ended questions that get an interviewee talking about the tasks he or she performs. The GMP auditor may simply ask, "What are you doing?" or "What tells you to do that?" If the interviewee's answers confirm the documentation, the information has been corroborated. If not, the GMP auditor needs to probe by asking questions that are even more specific to ascertain whether a work problem does indeed exist or whether the interview is being hampered by miscommunication.

The GMP auditor may not understand the process the interviewee describes, or the interviewee may misinterpret or fail to understand a question. An escort can often help clarify communication problems resulting from unfamiliar terminology by rephrasing a question. An escort should not be permitted to answer for an interviewee, however. When GMP auditors make a conclusion about whether necessary controls are in place, they must communicate the preliminary conclusions to the interviewee and/or escort. For example, a GMP auditor could say, "That's good. It looks to me as though you have a thorough understanding of your job and are addressing all requirements."

The GMP auditor who suspects a potential finding could say, "It looks like we might have a problem in this area. I can't see where those controls are placed." Such comments give the interviewee an opportunity to offer more information. If the interviewee is able to allay the GMP auditor's concerns, no problem exists. In general, people respond readily to inquiries and offer helpful suggestions for improvements if they feel that the GMP audit team is sincere, appreciates their views, and has their needs and interests at heart.

The GMP auditor usually takes notes during an interview and should take a few minutes at the end of each interview to ensure that the notes are complete and will be meaningful when reviewed later in the day. If the interviewee appears apprehensive, the GMP auditor should explain that notes are taken to record information about the details of the interview. If extensive note taking is necessary, it may be best for the GMP auditor to step out of the room or into a secluded location to record observations so that the interviewee will not become unnerved by the auditor's recording large amounts of information.

In all communications with the auditee, the GMP auditor must focus on the situation, issues, documentation, activities, and/or behavior – not on the person. When communicating with auditee management, the GMP auditor should avoid naming individuals whenever possible. Interviews are an integral part of process and system audits. There may be no interviews during a product or service audit except to verify the status. For product audits, the GMP auditor simply verifies product or service characteristics.

External GMP audit interviews are more formal than internal audit interviews. Even if an external GMP auditor has audited the same organization for a number of years, the GMP auditor must not become too friendly or fraternize with the auditee personnel. Also, it is useful to have interviewees agree with records or notations of their comments, and auditee personnel verify the accuracy of the auditor's observations of activities and status of items when they are being recorded. This technique will help ensure that the auditee understands the results of the investigation.

2.25.5 Interviewing Techniques

The interview is the most challenging aspect of a GMP audit for new auditors. Yet, the interview process may be the most important element of a GMP audit. For that reason, GMP auditors must be able to apply various techniques to gather information with minimum auditee stress and anxiety. One auditee may be familiar and comfortable with the audit process, while another may be nervous and apprehensive. Eliciting information from the former may be simple, but from the latter, it may be difficult and time-consuming. The manner in which a GMP auditor initiates the interview

TABLE 2.17
Common Time-Wasting Ploys and Possible Solutions

Problem	One Solution
The requested personnel are unavailable to the audit team.	The GMP auditor could politely state that the absence of key personnel may prolong the audit or that the audit's scope may have to be modified.
The escort is repeatedly late in the morning.	The GMP auditor could ask the escort for suggestions on how to start the meeting on time and make the necessary changes to the schedule.
Auditee makes the auditor wait repeatedly for needed supplies or requested documents and records.	The GMP auditor should request needed supplies during the audit planning stage and anticipate document and record needs in advance.
Constant distractions occur during interviews (the area is noisy, constant phone ringing, or other interruptions).	The GMP auditor could suggest that they move away from area or close doors if possible; phones should be set to call forwarding or should be answered by someone else.
Interviewees state that they were not informed of the audit and are not prepared.	The GMP auditor should confirm that employees are aware that an audit is taking place and should ask auditee management about the state of readiness of the quality system to be audited.

process may well determine the success of the interview. Table 2.17 provides an understanding of tactics that could potentially delay the audit before starting the audit.

People respond readily to inquiries and offer helpful suggestions for improvements if they feel that the audit team is good, appreciates their views, and has their needs and interests at heart. Each person involved with an organization has a unique perspective, as does each member of the audit team. One person can filter information differently from another, miss an angle, or stop short of getting the full story. For these reasons, facts stated and other data collected during an audit interview must be corroborated to ensure accuracy. The information can be corroborated in one of three ways:

- Another person, preferably from another group or management level, says the same thing
- Another member of the audit team hears the same thing from the same or a different source
- An item, document, or record verifies the action

2.25.6 Steering GMP Auditors toward Specific Interviewees

Several types of problems may occur during interviews, and GMP auditors should be trained in techniques that minimize such interference. One such problem occurs when an auditee attempts to steer the GMP auditors toward specific interviewees. Auditees often prefer that GMP auditors interview certain personnel and avoid certain others. However, this person's knowledge may not be typical of others in the organization. Therefore, GMP auditors should take charge of the selection process.

GMP auditors should approach an area and look around to see who is doing what jobs. Candidates for interviews can be randomly chosen as long as it is safe to interrupt them in their jobs. GMP auditors should follow the interview protocol described on the preceding pages. Table 2.18 summarizes the dos and don'ts of behavior in an interview.

If an auditee insists that a specific person is unavailable for interviewing, GMP auditors should be considerate but persistent if the interview is vital to the success of the audit. Without being threatened, the GMP auditors can suggest that the person's unavailability may prolong the audit. A statement such as "I may have to stay over for another shift" may gain GMP auditors the desired cooperation.

TABLE 2.18
Interview Dos and Don'ts

Do	Don't
• Answer all questions honestly	• Volunteer information
• Say "I don't know" or "I'll get the answer for you"	• Be sarcastic
• Avoid phrases such as "I think" "Sometimes/often/ usually," "never," and "next time"	• Guess answers
• Avoid qualifiers, such as "typically," "normally," "generally," and "usually"	• Attempt to answer what-if or hypothetical questions
• Stop speaking once the question is answered	• Argue or disagree with any inspector statement
• Repeat every question that is asked	• Philosophize, ramble, or editorialize
• Ask for explanations or interpretations of what you do not understand	• Point out deficiencies or errors
• Control your temper; remain courteous and professional	• Apologize for problems or comments made by an inspector
• Maintain eye contact	• Feel the need to respond to every comment made
	• Become defensive or evasive
	• Look away, fidget, or look nervous
	• Enter into casual conversation during the audit

2.25.7 Answering the Interviewee

Another common problem occurs when an escort or area supervisor answers for the interviewee or intimidates them. An escort generally stays within listening distance and takes notes, but should step back slightly. An ideal arrangement is for a GMP auditor to sit or stand between the interviewee and the escort, facing the interviewee.

You might have encountered this during GMP audits. There is one person at the facility who is qualified as an escort and tries to help the auditor in all matters. This person is not really someone to comment on the content and scope of the audit and should not be involved in one-on-one interviews. For example, when the GMP auditor has questions about a machine, they can direct their gaze to the machine operator and reduce the intervention of the escort to zero because it is the operator's job to check the machine setups and machine parameters during setup. If an escort starts to answer for an auditee, the GMP auditor needs to redirect the question to the interviewee, avoiding eye contact with the escort. If an escort continues to interfere, an auditor should say, "I prefer to get the information from the staff member."

2.25.8 Asking Leading Questions

A different kind of problem occurs if the GMP auditor asks leading questions expecting to hear certain answers. The GMP auditor should not lead an interviewee with the questioning. Leading questions can be avoided if the auditor asks open-ended questions. Additionally, if a GMP auditor goes into an interview expecting to hear certain answers, the audit results likely will be biased.

2.25.9 Overcoming Language and Cultural Barriers during the Audit

GMP audit personnel must either be fluent in the language in which the audit is to be conducted or have the support of a technical expert with the necessary technical language skills. When necessary, the auditing organization should employ a skilled interpreter to assist with the audit.

Even if all primary participants in a GMP audit speak the same language, the GMP auditor may encounter language or literacy barriers when attempting to interview individual employees. These same barriers may prevent the employee from understanding or performing assigned tasks. An approved and valid procedure can provide solutions to problems, but in the worst-case scenario, when personnel cannot read and understand the procedure, the problem is unfortunately far from being solved.

If the GMP auditor understands the physical process before going into an audit and then focuses on the work, some of the literacy issues may be overcome with the aid of flowcharts and other simple diagrams.

2.25.9.1 What Can I Do If They Do Not Speak My Language?

- **Verbal behavior**
  - Clear and slow speech: Pronounce each word accurately and precisely. Be careful not to choose colloquial terms that the other person may be unfamiliar with.
  - Repetition: Repeat each important idea using different words to explain the same concept.
  - Simple sentences: Avoid compound and long sentences.
  - Active verbs: Avoid passive verbs.
- **Nonverbal behavior**
  - Visual repetitions: Use pictures, graphs, figures, diagrams, and tables from visual expressions.
  - Gestures and facial expressions: Use more gestures to reinforce the meaning of words.
  - Demonstration: Act out as many themes as possible.
  - Pauses: Pause more frequently.
  - Summaries: Hand out written summaries of your verbal presentation.
  - Proxemic distances: Respect people's spatial distances.
  - Attribution
  - Silence: When there is silence, wait. Do not attempt to fill the silent moment with words. The other person is probably just thinking more slowly in the non-native language or translating.
  - Intelligence: Do not equate poor grammar and mispronunciation with a lack of intelligence. It is usually a sign of second-language use.
  - Differences: If unsure, assume differences, not similarities.
- **Comprehension**
  - Understanding: Do not just assume that they understand; assume that they do not.
  - Checking comprehension: Have colleagues repeat their understanding of the material. Do not simply ask if they understand or not. Let them explain what they know to you.
- **Design**
  - Breaks: Take more frequent breaks. Second-language comprehension is exhausting.
  - Small modules: Divide the material into smaller modules.
  - Longer time frame: Allocate more time for each module than usual in a monolingual program.
- **Motivation**
  - Encouragement: Verbally and nonverbally encourage and reinforce speaking by non-native language participants.
  - Drawing out: Explicitly draw out marginal and passive participants.
  - Reinforcement: Do not embarrass novice speakers.

2.25.10 Using a Translator

GMP audits should always be conducted in the language of the auditee. Unfortunately, this is not always possible and the services of a translator must be utilized. If at all possible, the translator should be independent of the operation being audited as well as the firm conducting the audit in order to ensure the credibility of the GMP audit. Thanks to the inclusion of a translator in the audit, the time spent on collecting the necessary evidence and understanding the documents more clearly by conducting one-on-one interviews is considerably reduced. However, in addition to the advantages of using a translator, the lead auditor and the representative of the audited facility must also consider the time to be allocated to this person within the audit period.

The GMP auditor's questions must be short and to the point for the translator to convey the proper meaning. Otherwise, questions will have to be interrupted for the translator to communicate a portion of the question to the auditee, allow the auditor to complete the question, and then complete the translation of the question to the auditee. While simultaneous translation may be a little faster, the GMP auditor will not be able to read body language as easily since the auditee will be attending to the translator's voice rather than the auditor's voice. Only appropriate planning with a skilled translator will result in an effective audit.

2.25.11 Conversational Process

Interviews should be relaxed and conversational. GMP auditors should not simply dive into the checklist.

From that point, the GMP audit is underway, with the auditor linking questions to other topics of interest and documenting evidence, as appropriate.

As a GMP auditor, one of the methods we use to gather objective evidence is to interview the person doing the work. In this case, I suggest keeping in mind that the area we focus on, rather than the checklist, is the person we are talking to. The GMP auditor's focus should be to understand the process and the person's role in that process. Once the focus shifts from the process to the checklist, the conversation and the information GMP auditor needs in order to verify effectiveness will lag.

The use of open-ended questions is necessary to gather information during the interview process. GMP auditors need to ask who, what, when, where, how, and why questions, as these types of questions elicit information from the interviewee. These types of questions will actually allow the person you are talking to explain in detail about the process, thus opening the way for you to ask more questions as a GMP auditor.

When the GMP auditor relies only on questions beginning with phrases such as "Does the organization ..." or "Is there a ...," the auditor will be talking more than the auditee and, as a result, will learn very little.

As you know, we may prefer checklists from time to time, as they make it easier to take notes and compare the questions and the requirements, and most question lists contain yes/no answers. We can refer to many sources such as a procedures, standards, corporate documents, GMP guides, etc. to create the questions. Isn't the main task of the checklist, in short, the conversion of the statements written in these sources into questions?

For example, a document may state, "Process settings must be recorded on the work order." The GMP auditor will construct checklist questions from this document, such as "Are process settings recorded on the work order?" Yet the questions during the interview may be very different. An interview with an operator may go something like this:

Auditor: How do you know what the process settings should be?
Operator: I look at the SOP before I start the job.
Auditor: Could I see the SOP for the job you are working on?
Operator: Yes, here it is.
Auditor: Thank you.

The auditor is able to verify that the operator has access to the SOP and that the settings are recorded in the logbook.

Auditor: Could you show me where you make the setting adjustments?

Then the auditor verifies that the settings on the SOP match the settings on the machine. Good follow-up questions for the auditor to ask may be:

- What do you do if the settings are not on the SOP?
- What do you do if you think the settings are wrong?
- Do you ever change the settings?

2.25.12 Knowing How to Ask Questions

Much has been written about the importance and benefits of a solid auditing program. Table 2.19 provides an example of how a GMP auditor might detail questions at five levels.

Most of us know the feeling of an auditor coming within one small question of finding significant issues. A good auditor is a "platinum." This means the auditor questions systems down to at least five levels. Let's look at these levels:

- Bronze: This first level of questions merely asks for a description of systems in place to address GMP requirements. This level is the most fundamental area and relies heavily on the word of the audited party.
- Silver: This level usually involves asking to review a copy of the SOP or procedures in place that describe a system or operation. In most cases, the auditor will compare the operation described at the silver level with the SOP provided for review at this level.
- Gold: The third level of auditing involves reviewing examples of the system under review. The auditor, at this point, is examining the real-life output of the system described at the silver level and written at the gold level (only a very good auditor routinely advances beyond this level).
- Diamond: In the fourth level, the auditor focuses on a problem or non-routine example of the system or activity under review. Now the auditor has focused on an area that likely will truly reveal how the system operates or is controlled.
- Platinum: At this level, the auditor is verifying that the problem was properly handled, the system actually functions as required, and that those auxiliary systems close to the system reviewed are controlled and functioning.

It is difficult to visualize the levels of auditing depth by merely reading the aforementioned descriptions. So, let's look at some real auditing examples, shown in the table below. Can you see the progression in questioning from the auditor in Table 2.19. The questions have progressed from an

TABLE 2.19
Five-Level Question Examples

	Audit of Out of Trend Specification Results
Level	How to Ask Question
Bronze	Would you please explain your way of handling Out of Trend results?
Silver	Could you show me your OOT SOP?
Gold	Could I see a list of all OOT investigations for this product covering the last 12 months?
Diamond	In reviewing this OOT investigation for a failed friability test result, how did you decide that the root cause for the failure was an improper sampling step?
Platinum	As you stated that the root cause of this OOT was attributed to sampling error, can you show me the documentation of retraining that will prevent a recurrence? Were all other laboratory staff also trained? Can you show me their training documentation? Has this specific staff been involved in any additional OOT investigation?
	Could you please describe your laboratory personnel training program?

overview of the OOT handling procedures to a specific example. Then, the details of an example are reviewed, leading to a more in-depth inquisition about the cause of the failure. Finally, you can begin the cycle again with the question, "Could you please describe your laboratory personnel training program?" We have progressed through all five levels of OOT investigations to a new topic and a bronze question. An approach that drills deep into a subject, then moves to a related one is a sure sign of an experienced and accomplished auditor.

Again, you can see the progression to the platinum level where the auditor is digging rather deeply into the details. The real assessment of GMP compliance is that assessment that occurs when platinum questions are asked. The very best, most accomplished auditors quickly move into platinum questions, and challenge the proper function and quality impact of the system under review. Auditing is a key prevention tool in the GMP toolbox. When done correctly, and with the ultimate goal of improvement, auditing can yield progressive and sustainable improvement. To make the most of auditing, a progression to those tough questions – platinum questions – will provide the most significant opportunity for revealing issues that need resolution.

Open-ended questions – seek specific information

- Give me an example of ....
- How does this work?
- Any other comment?

Advantages
- Clarifies
- Verifies the meaning of words and statements
- Avoids misunderstandings

Closed-ended questions – short answers
- Were you involved in this process?

Advantages
- Allows the auditor to verify the information and avoid misinterpretations and misunderstandings.
- Good for timid and communicative auditees.

Biased questions – influence the auditee
- I suppose you already know this violates basic GMP norms?
- Shouldn't this be under quarantine?

Disadvantages
- These questions suggest the answer.
- The information gathered is strongly distorted or influenced.

Aggressive questions
- Were you the incapable person who did this?
- Couldn't you have placed this in the right place?

Disadvantages
- Places the auditee on the defensive, stressed, and out of control.
- Interferes with the audit process.
- Information gathered may not be reliable.

Multiple questions
- When did you start in the company, how were you trained for the task you are performing, and do you believe you are following the right procedure?

Disadvantages

- They confuse the auditee.
- The induce the auditee to deviate from important points.

Table 2.20 is composed of sample questions so that the reader can understand more clearly which question is of what type.

CASE STUDY

Identify the questions in Table 2.20 as open, closed, biased, or aggressive.

TABLE 2.20
Sample Question Types

Question	Type
Of course, for you, it is obvious this is the wrong valve, correct?	
Would you please tell me more about the addition of an antifoam agent?	
After having been working all this time, why didn't you read the cleaning procedure?	
Did you record the fermentation processes?	
What do you mean by inactivating the bulk?	

2.25.13 Active Listening

1. *Face the speaker and make eye contact* – Eye contact is the most valuable form of behavior when speaking face to face. But trying to make excessive eye contact can do more harm than good, so you need to tailor the act of eye contact to the situation. It will help if you try to break eye contact about every five seconds or look at one eye for five seconds, then the other for the next five seconds. Another approach is to look at the person's mouth after five seconds of eye contact. Control your posture and maintain communication with open, collaborative body language. You should avoid crossing your arms or legs and being overly defensive. Leaning slightly forward from a sitting position indicates that your listener is most attentive, so take care to preserve this moment.
2. *"Listen" to nonverbal cues* – Pay attention to what the other person is saying with their body language. As important as your body language is, what the other person is trying to convey with their body language is also essential. Gestures and facial expressions are as effective as your tone of voice. You can analyze the body language of the person in front of you with a bit of attention; for example, you should try to understand what message their arms and legs are conveying. By rubbing his/her eyes as if she/he is tired and sleepless, so much so that it is possible to detect more or less the body language of a person speaking.
3. *Don't interrupt* – People generally never want to be interrupted while speaking. By interrupting someone you imply that you are going to say more important things; it isn't comforting. The person may even think you do not have time to spare when you interrupt them while speaking. However, if you are a person who tends to speak quickly, you should look for ways to slow yourself down. The most important thing to remember is that a short pause does not require immediate intervention. Be sure to let the other person finish speaking.

4. *Listen without judging or drawing conclusions* – When you start to react emotionally to what the other person says, it will cause you not to understand any words that will be said later. Whatever happens, focus on listening. Never try to make a judgment by guessing the person's following words.
5. *Don't start planning what to say next* – You can't listen and prepare at the same time.
6. *Show that you're listening* – To demonstrate that you are listening, nod slowly, try to smile, and say short phrases like "yes" and "of course." Only look at your watch if you have to, don't constantly change your sitting position, and don't play with your hair or nails.
7. *Don't impose your opinions or solutions* – It may not always be possible to cope with this situation, but showing that you listen to the other person can be more effective than telling them what to do. For example, someone sick may want to talk about an issue that is on their mind instead of asking you for your opinion; in such cases, try not to give advice. Often, people try to come up with their own solutions and if you have a solution that you think is great, you should find out if the other party is willing to listen before you pass it on.
8. *Stay focused* – If you are having difficulty concentrating on what the other person is saying, try repeating their words in your head so you have the opportunity to reinforce what has been said. Try to eliminate any distractions during the conversations.
9. *Ask questions* – One of the easiest ways to show that you are listening to the person speaking is to ask questions about the discussed topic. If you think you have understood correctly, wait for the speaker to finish and say "I think you meant that …" or "I really think I understood what you meant by what you said…" You can even ask questions to encourage the speaker to speak more such as "So what happened next? How did you react to this situation?"
10. *Paraphrase and summarize.*

2.25.14 Why Is Body Language So Important?

You need to be able to read body language as you interview people, as well as be aware of what your own body language is telling them. Following are seven categories of body language. Pick each category and think of two examples that would fit into that category. Determine if the examples would be positive or negative examples of body language.

- Body posture
- Hand or arm movements
- Gestures
- Facial expressions
- Dress or attire
- Distance between the GMP auditor and auditee
- Eye contact

Body language is normally a combination of each of these. Also, bear in mind individual and cultural differences in body language.

Clues to build a strong rapport:

- Open body language
- Solid introduction
- Establish common ground
- Use visual cues
- Smile
- Proper dress
- Friendly demeanor
- Interview

How to start to ask questions at the beginning of an interview with the, let's say, operator.
Ask good questions and draw out good answers.

- What have you been working on lately?
- Tell me about what you do here each day.
- What kick-starts your day/this process, etc.?
- How did you get started here?

There are many ways auditors can become valuable tools for continuous improvement. I have broken these down into six subcategories to demonstrate the various roles of the auditor and how these roles support the organization's continuous improvement efforts.

- Ears
- Eyes
- Mouth
- Hands
- Feet
- Back

Let's explore each of these in greater detail.

Ears: I put ears first because they are the most important. It is with our ears that we hear of opportunities for improvement. We often hear workers grumbling and complaining. Some, it seems, do little but complain. But have you ever really considered their gripes? In many cases you can learn a lot from really listening to them. They may overexaggerate, embellish, or even arrive at the wrong conclusions, but often there are some real opportunities in there.

During our interviews, the auditee often will give us little bits of information that may not be critical to the audit but can yield a wealth of knowledge for continuous improvements. In some cases the employee may not be aware of the impact of the statements. In other cases, they may provide information about a different area. The key is to attune yourself to listen for information that might make a dramatic difference in the operation of the business.

Eyes: The auditor sees many things while auditing. Some are good, and others are not so pleasant. When viewing the quality management system through the eyes of the GMP auditor, the organization gets a fresh look at itself. In continuous improvement, what the GMP auditor sees may be important. But it is not nearly as important as what the GMP auditor understands. The GMP auditor can see the organization as a whole, perceiving how everything fits together. The internal sees the entire organization because they audit the complete system. The GMP auditor can see both strengths and weaknesses. What the GMP auditor does with that information is just as critical.

Mouth: The mouth is an interesting device. It can either bless or curse. The mouth of a GMP auditor is used for two purposes. First, it is used to communicate. The GMP auditor communicates continuous improvement opportunities to management. It also communicates with the plant to help identify opportunities, and communicates management policies and directives concerning continuous improvement.

The second function of the mouth is as a sales tool. The GMP auditor communicates the need for and awareness of continuous improvement. Through the GMP auditor, the organization can show its commitment to a program of continuous improvement.

Hands: Through the GMP auditor, continuous improvement activities are documented. Hands put to work the thoughts of the mind, and the GMP auditor puts to work the thoughts of management. The GMP auditor helps mold the form, figure, and future of the continuous improvement program.

Feet: The GMP auditor's movement throughout the organization becomes the feet of the continuous improvement system. The GMP auditor helps carry the organization toward its continuous improvement goal. The GMP auditor helps determine the direction of the organization. Whether the organization goes forward or back, left or right, may well hinge on the GMP auditor.

Back: The GMP auditor supports the organization's continuous improvement activities. This internal support is the backbone of organizational growth. With growth, there is sometimes pain. This too is often laid on the back of the GMP auditor. In many organizations, the GMP auditor endures overwork, underappreciation and conflicting schedules. Perhaps through continuous improvement, we can turn that around as well.

PSYCHOLOGICAL ASPECTS OF THE AUDIT SITUATION

Observation versus Cooperation

- A complete and water-tight surveillance to identify areas of risk is not possible.
- An audit can only cover a fraction of the material.
- Auditees who want to hide something will usually achieve this (at least partly).
- Auditees who look at the auditor as an unwanted person have no acceptance of this person's role.
- Auditees who do not accept an auditor will not cooperate with him/her. They will try to hide information as much as possible.
- It is difficult to get information from people when they intentionally withhold information from you.
- It is a lot easier to be a consultant for people who want to and are willing to cooperate.
- Consultants who are accepted within an organization get a lot more information concerning mistakes and weak points.
- Consultants should not be overfriendly but also not too tough.

Figure 2.7 provides a graphical example of which areas of auditing and audited parties are better when it comes to GMP audits.

Different question types and behavior patterns are summarized in Tables 2.21 and 2.22, respectively.

2.26 CLASSIFICATION CRITERIA FOR OBSERVATIONS

2.26.1 Objective Evidence

The purpose and scope of the GMP audit, as you know, bring together as much evidence as possible on each subject. Evidence can include procedures, work instructions, evidence of how and by whom, when they were applied, evidence of education, training and experience, qualification, and validation results required for personnel to perform these jobs.

Evidence needed to support the auditor's findings may be physical, testimonial, documentary, or analytical.

Evidence derived from records is likewise more reliable than oral evidence. Evidence must be sufficient, complete, and relevant to be useful as a sound basis for audit findings and recommendations.

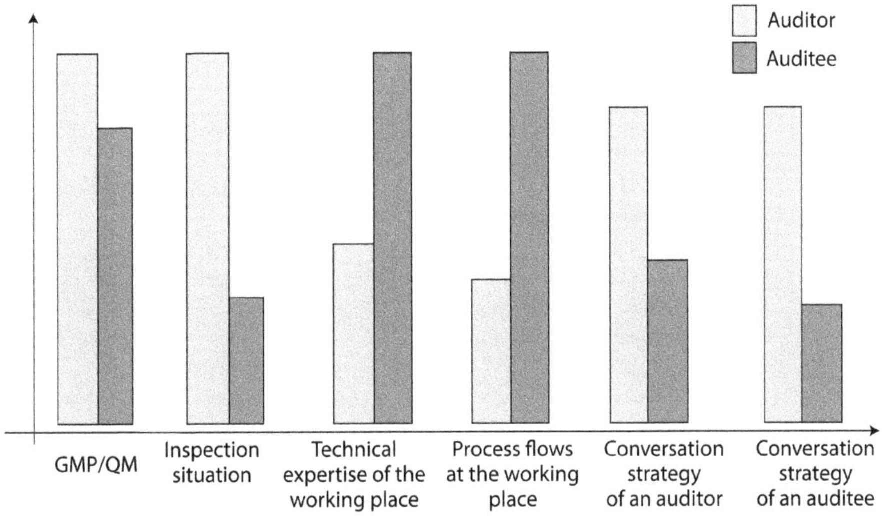

FIGURE 2.7 Auditor and auditee comparison.

TABLE 2.21
Question Examples

	Examples	Criteria	Effect
Closed questions	• Did you forward the notice? • Do you have an SOP for this? • Is the process validated?	• Start with a verb • Only a few possibilities to answer; usually only YES or NO	• Only minor yield of information • Forced to a distinct statement • Recognized as having strong control • Help to collect facts
Open-ended questions	• How do you perform…? • What has happened? • Why do you…? • How do assess…? • What happens if…?	• Start with a question word (e.g., 5 Ws) • Cannot be answered with YES or NO • Allow a lot of possibilities regarding the formulation and content of the answer	• Extensive yield of information • Experienced as a partnership • Recognized as low control
Maneuvering questions	• I agree with you Mr./Mrs. But shouldn't we get back to the real issues? • Should we first talk about this subject? • On what points of the problem do you agree with me?	• Can start with a questioning word as well as a verb	• Help the conversation to stick to and focus on the principal subject matter
Reflective questions	• Is it your opinion that…? • If I understand you correctly, you mean…? • You also think it's possible that…? • What you really want to say is that…?	• Usually don't begin with a questioning word • Reflect the contents of the last statements made by the answering party	• Signal interest • Confirm mutual understanding • Prevent each party from talking about a different subject • Provide a positive atmosphere of mutual cooperation • Reduce the possibilities for free interpretation

TABLE 2.22
Behavior Examples

How to Behave	Auditor	Auditee
Good	1. Selects a topic and elaborate on a minor finding including applicable evidence. 3. Acknowledges compliance level.	2. Responds properly, and shows appropriate evidence.
Arguing between auditees	1. Selects a topic and elaborates on a major or critical finding including applicable evidence. 3. Manages the situation; communicates the potential deviations.	2. QA and Production argue between themselves with contradictory statements and they give details that were not asked by the auditor which represent new GMP violations. 4. Get very nervous and reveals a lack of upper management support including a lack of personnel.
Biased	1. Selects a topic and elaborates minor finding(s) including applicable evidence. 3. Reacts in a biased manner (e.g., "This is a nonconformity because I experienced a similar problem where I used to work, and I have read and heard about this before"). 5. Insists in a biased position.	2. Elaborates proper response/attitude. 4. Prepares a proper response/attitude.
Confronting / Aggressive	1.Selects a topic and prepares a finding including applicable evidence. 3a. "You are entitled not to agree, but the observation is based on this objective evidence. XXX shows noncompliance of point XXX of guidance XXX." 3b. "I am the auditor, and what I say is the truth and I do not need to give you any explanation." 5. Leaves without collecting evidence.	2. Simulate confronting physical and body language response (e.g., "I do not agree with your observation which is completely wrong and is showing your lack of knowledge"). 4. Escalates reaction.
Consulting	1. Selects a topic and elaborate on a finding including applicable evidence. 3. Gives consultancy on GMP fundamentals and exactly how to do things (e.g., "You need to hire two more people in this area who should report to you only, create a new position, replace your existing disinfectants with this new very good and inexpensive one…"). 5. Keeps on giving recommendations without auditing.	2. Prepares an answer acknowledging the finding. 4. Acknowledges but tries to show current disinfectants are adequate.
Lack of data integrity	1. Selects a topic and elaborates on a finding including applicable specific evidence (e.g., no double verification on calculations of release testing, backdating, inaccurate data, test-to-pass, wrong cross-outs, false entries). 3. Concludes/assesses compliance level.	2. Auditee says, "Production manager minimizes finding showing lack of knowledge." QA remains silent showing surprise and a lack of oversight.
Lack of GMP knowledge	1. Selects a topic and elaborates finding including applicable evidence. 3. Shows evidence that what is currently done by production is not what was validated.	2. Shows evidence of compliance but shows he does not understand the main aspects of the topic, was not involved in the study, and does not believe in the GMP or the associated requirement (e.g., "Qualification studies were done by the vendor, but are of no real use and are just a bureaucratic exercise to please regulators").

(Continued)

TABLE 2.22 (CONTINUED)
Behavior Examples

How to Behave	Auditor	Auditee
Jumping to early conclusions	1. Selects a topic and elaborates on a finding including applicable evidence. 3. "This is a disaster, you are doing everything wrong! This for sure will be a nonconformity, you will see. I am going to talk to your boss about this. Who trained you? Did you have any experience before coming here?" Also, gets very close to auditee and finger-points.	2. Starts to explain a proper response, but auditor interrupts and jumps to conclusions and disqualifies auditee. 4. Nervous, out of control, silent, depressed. QA manager intervenes and deescalates the situation.
Lack of objective evidence	1. In the closing meeting, communicates to auditee a finding with incorrect or no objective evidence (e.g., "Temperature in one of the freezers is not recorded correctly, difficult to read, incomplete. I saw it myself").	2. Politely disagrees and shows evidence of compliance.

2.26.2 Observation of Work Activities

The GMP auditor observes work in progress to see if it meets requirements. The GMP auditor gains this knowledge by monitoring a process being performed to see how the work is being done.

Observation means "to watch attentively" or "to watch or be present without participating actively." Hence, auditors must be well trained in observation techniques so that they learn how to observe closely but not obtrusively. Because an auditor's presence may be distracting to people at work, the auditor needs to minimize any interference resulting from it. The GMP auditor also should realize that disrupting a worker may be unsafe.

The GMP auditor needs to know when to observe work. For example, if GMP auditors desire to observe work being done by different shifts, they should be aware of when shift changes will be taking place. GMP auditors should avoid retaining people through shift changes or during their breaks.

2.26.3 GMP Audit Findings Classifications

The GMP auditor uses several techniques to classify and prioritize observations before the closing meeting. In any audit, the risk is associated with reporting a finding on the basis of available data. The GMP auditor may fail to identify a significant problem or may identify a problem that is not significant. Evaluation is often based on a relatively small sample and, therefore, involves a high sampling risk. A problem area may be classified as nonconformity, noncompliance, or finding. The GMP auditor should also group observations under similar findings or nonconformances.

In some cases, observations may be chronic or marked by long continuation or frequent recurrence. The chronic nature of observation may mean risks are higher, and management may want to avoid the risk or treat it. Observations may also be systemic.

2.26.3.1 Finding

A finding is an audit conclusion that identifies a condition having a significant adverse effect on the quality of the activity under review. A finding has the following characteristics:

- It is negative. Something is amiss.
- It is a violation of a requirement. A promise was made and it was not kept.

- It is significant. It is a big deal. It relates to business values and is affecting those values in an adverse manner.

A finding may be systemic, but either the probability or the consequences may be insignificant. This is an important concept.

2.26.3.2 Critical Deficiency

This is any deficiency that produces a product harmful to human or animal patients, makes residues detrimental to health in an animal used in food production, or could output at risk. We can also speak of a critical deficiency if the pharmaceutical product manufacturer makes fraudulent, misrepresented, or falsified data on its products or the production of products. A critical deficiency may occur alone, or the coexistence of multiple primary deficiencies may suggest acritical deficiency.

2.26.3.3 Major Deficiency

It is a major deficiency that is not a critical deficiency, but has a degree of severit. For example:

- If the manufacturer produces a product that does not comply with the marketing authorization, clinical research permit, product specifications, pharmacopoeia requirements, or registration dossier expectations.
- If the product manufacturer cannot take the necessary GMP control measures before, during, and after production.
- Experiencing a significant deviation from production permit requirements.
- If the qualified person (QP) has not fulfilled its duty, the procedures required to release drug product batches have not been fulfilled.
- In these cases we can talk about a major deficiency. If more than one defect in the "other" category is detected, the sum of them can also be described as a "major" deficiency.

2.26.3.4 "Other" Deficiency

If the detected deficiency is not classified as either a critical or major deficiency, then the "other" category is used as a definition. This deficiency still characterizes a significant deviation from GMP guidelines and regulations. This deficiency is named as such because it falls into neither the critical nor the major category.

2.26.3.5 Critical Deficiency Examples

A critical deficiency is a very serious situation that could result in health authority sanctions.

- Inadequate sterilization validation in all sterile-filled drug products
- Lack of measures to be taken in case the amount of residue in the product from the methods used in cleaning exceeds the limit values
- Uncontrolled pest infestation in the company
- Deliberate falsification of analytical and microbiology lab results or destruction of records
- Failure to ensure the quality and identity of API, excipient, and packaging materials
- No master batch record is available
- Inability to access production and packaging and labelling records, or deliberate falsification or destruction of said records
- Failure to validate the water for injection (WFI) system used in the production of sterile products
- HVAC system is an unvalidated sterile product production environment
- Unmaintained, dirty facilities that may lead to contamination and cross-contamination
- Actions that are not suitable for recall processes

2.26.3.6 Major Deficiency Examples

- Inadequate validation of processes that are considered critical (these deficiencies can be regarded as critical when the subject is biotechnology and sterile products)
- The absence or inadequate positioning of air filters that should be used in the environment (we can talk about a critical deficiency when there is a situation like the above)
- Constructional deficiencies in areas where the product is produced, such as holes/cracks in walls, floors, and ceilings and unpainted surfaces)
- Structural defects that will prevent adequate and effective cleaning of production areas
- Production areas that can cause confusion
- Absence of sampling sections in warehouse areas
- Inadequate protection of equipment that may be affected by environmental contamination
- Lack of education, training, and experience of warehouse, quality control, production, and quality assurance personnel
- Insufficient initial training and follow-up training and lack of qualifications of trainers
- Reduction of sampling frequency and the number of tests before the supplier's qualifications are complete
- Incomplete testing of API, excipient, and packaging materials
- Lack of validation and verification of test methods
- Failure to plan in-company audits, insufficient qualifications of auditors
- Absence of a procedure for the management of complaints and returned products

2.26.4 What Are the Most Common Approaches When Finding a Critical Nonconformity?

Nonconformances play a very important role in the day-to-day running of a pharmaceutical company. Their impact on the final quality of products brings with it other negative consequences, such as efficiency, productivity, recalls, warning letters, and company reputation loss.

Extract from records made during an audit (raw documentation) are shown in Table 2.23.

2.27 HOW TO KEEP NOTES

It is very important to take readable notes during the course of the audit. A good GMP auditor is a person who can easily read their notes when evaluating the findings and answers of auditees and when preparing an audit report. So well-written notes can help auditors to be good auditors.

Good communication skills mean understanding processes, performing tests, and assessing risks and related controls, which can result in good grades.

Well-trained GMP auditors are expected to be proficient in taking notes, interpreting, reviewing, and correcting their notes. Some GMP auditors are skilled at taking highly detailed notes, while others only take topical or superficial notes. Note-taking has no GMP guide or regulation. You can take notes in detail or superficially, but the main thing is maintaining the auditee while taking notes. Qualified GMP auditors must present their messages thoroughly and clearly at the closing meeting and in the reports they prepare. Otherwise, they will likely need clarification during the preparation of the report.

2.27.1 On My Own

The GMP auditor may wish to conduct the interviews on their own. This style can be called "on my own." Sometimes, it may be difficult for the auditor to speak, ask questions, listen, and take notes simultaneously.

TABLE 22.23
Observation Record Example

Audit of the example company on 12–14.12.2022 Page 8

Auditor: M. Edik Escort: A. Stone, B Water

Company tour on 12.12.2022

Area	Observation	Evaluation
Semi-solid manufacture Room 301	(1) Stirring vessels 1 and 2 are maintained in an established rhythm (log books 12+13).	NAD
	(2) Pipes and extraction plug valves in room 18 are not indicated.	Violation of 3.42* major; Violation of 4.7* minor
	(3) In the manufacturing record for cream XY, Ch.B. 23455, on 12.12.2022, there are many changes without initials and date.	SOP 1234 is suitable, NAD
	(4) Personnel do not wear PPE in the working clothing in accordance with SOP 1234.	Violation of 2.18* major
Cream manufacture Room 301	(5) An approval sticker on container CNT 22XX33YY44ZZ has been heavily soiled and is not legible.	Violation of 5.13* minor
	(6) The maximum storage time (standing time) for the bulk has not been determined.	Violation of A9.9* major
Instrument washing room Room 309	(1) Equipment and tools are cleaned with purified water.	NAD
	(2) The containers used to contain numerous blind angles and are therefore difficult to clean.	Violation of A9.2* major
	(3) The cleaning process for the containers used in penicillin production, and also used for other products, is not validated.	Violation of 5.19* critical

* Refer to chapter or appendix of the EU GMP guidelines

2.27.2 Team Approach

Although it is not easy to establish a team of two or three people for a GMP audit, the "team approach" can be one of the most effective methods when an audit is performed. For example, one team member can ask questions during the interview, while the other can take notes. In this approach, which can be performed alternately, the questioner can take the task of note-taking while others ask questions. In this way, both GMP auditors can check each other's grades. A minor disadvantage is that more than one examiner could try to ask questions simultaneously.

2.27.3 Good Guy/Bad Guy

If you remember detective movies, there is a good cop and a bad cop who interrogate criminals. These people would examine the suspect who was thought to be guilty. However, in GMP audits, I have seen few auditors who take this style seriously apart from FDA inspectors. They ask most of the questions and after making an inference from the answer they receive, they ask if a new question is needed and take a note.

2.27.4 The Outline

Before the interview, the GMP auditor takes due care and creates an outline to be filed in during the interview. In another form, the GMP auditor can outline the format and fill in the form outline during the interview. This style is very effective.

2.28 TEN KEY POINTS FOR OBSERVATION AND NOTE TAKING

1. Select what you want to see for yourself; do not just look at what may be presented to you.
2. Make sure you see what is actually happening, not just what you expect to see or are told should happen.
3. Obtain an overall picture of the surroundings and be alert to activities taking place around the audit area; do not be too focused.
4. Observe the auditees – their body language and the things they look at.
5. Be alert for any areas the auditees might wish to move you through with undue haste; they might be interesting.
6. Ensure you have your facts right; confirm your observations and facts with the auditees.
7. Make sure your notes are legible.
8. Make sure your notes have sufficient details for you to provide a specific observation, e.g., document titles and numbers, interviewees' names, and actual words.
9. Diagrams or schematics (either drawn by yourself on the basis of responses to questions or taken from procedures) can be a useful basis for making notes and determining what to look at next.
10. Develop a mechanism for highlighting in your notes those points that are issues or that you wish to follow up on at a later stage. This can aid in rapid review during the audit of points still to be covered or at the end when the time comes to prepare for the closing meeting.

2.29 TEAM MEETINGS

You should make time prior to lunch and at the end of each day for your team to meet. About 20–30 minutes for each of these meetings should be enough. These caucus sessions should be informal discussions among only the audit team members. Ask your guide to take a break. You accomplish three things during these short team meetings:

1. Share facts, concerns, and problems.
2. Adjust your data-gathering activities. This is sort of a repeat of the preparation phase.
3. Develop the report by proposing possible conclusions.

The valuable information obtained by the GMP auditors during the audit and the mutual sharing of their knowledge benefit the audit and relieve the auditors. With such an approach, the question of whether to go deeper is asked, and the most appropriate answers can be found by exchanging ideas. We can predict positive or negative results thanks to the information we obtain during the GMP audit. This method also allows for different control tactics.

The result of this sharing could be a redirection of the audit. Keep in mind, however, that you are obligated to stick to the original purpose and scope. Based on the information now available, the following issues can be addressed:

- Are the results of the interviews and other data gathering sufficient to reach a conclusion?
- Should there be additional interviews, additional checklist questions, or additional records reviewed? Do you need more data?
- Are there administrative problems to be resolved with the group being audited or your audit team?
- Does the audit seem to be accomplishing its objectives?

It is worth re-emphasizing one issue. The interviews you have conducted with those working in the production or laboratory are extremely helpful in data collection and provide clues about the GMP performance of the auditee. Whatever you do in the interviews, express them with very accurate and understandable sentences so that these impressions, which were drafted in the team meeting, can support the main report.

Before the closing meeting, you should meet with your audit team for the last time and create an opportunity to evaluate the findings and observations.

This last meeting should be separated from the previous ones in terms of content and duration, and there should be no exposed parties. Your goal is always to further develop the findings and observations, good or bad, to make them more understandable.

First, collect and discuss all of the good and bad facts uncovered during the course of the audit. All the team members should review their individual field notes. The audit team leader acts as a scribe to make a master list of these good and bad facts.

An amazing thing happens. You will always have one or two piles that are bigger than the rest. Focus on these big piles. Set aside the problems with only one or two bad facts. These are minor issues and can be dealt with later.

Table 2.24 summarizes the tactics that the auditee may apply in the audit and the auditor's responses.

2.29.1 Daily Briefings

You can see that effective team meetings are important to success. Likewise, daily briefings with the auditee will enhance the quality of your GMP audit. If a goal of improved performance is to be attained, it is important that there be no surprises at all levels. This communication can be strengthened by a short and informal briefing of about ten minutes at the end of each day with a representative from the group, normally your guide or shadow. Topics to be discussed include:

- Checklist areas completed
- Checklist areas to be examined (or revisited) tomorrow
- Any areas of concern
- Any problems experienced

If you've uncovered something out of whack, it is certainly not necessary (or desirable) to present polished conclusions yet. Rather, explain to the representative that these are potential problem areas or areas of concern at this stage. At the end of the meeting in GMP audits, the person or people you talk to share this information with the manager who is not present. It's almost as if all facility employees go to great lengths to improve findings immediately, and these initiatives will result in the compilation of information that either confirms or ignores your concerns. Let's say you couldn't

TABLE 2.24
Tactics Chart

Auditee Tactic	Auditor Response
Amnesia	Verify the availability of needed documents and then provide a reminder.
Bribes	Refuse and report them. Advise client firm to deal with others.
Cook's tour	Cut it short – stick to the audit plan.
Extended lunch	Auditors can eat alone. Ask for a substitute escort.
Gobbledygook	"Interesting, but the audit is my primary concern."
Insincerity	Ignore it. Proceed with the audit.
Language barrier	Prepare in advance. Provide a translator.
No-show	Go to head auditee contact and ask for a substitute escort.
Sympathy	Ignore it. Proceed with the audit.

reach a consensus with the other party on a subject, you should re-examine this situation until it is fully clear and decide whether to include it in the main report.

Either way, both parties win.

2.30 CLOSING MEETING

The closing meeting is sometimes referred to as the exit meeting and the final meeting. At the closing meeting, as the GMP auditor, you can either conduct the audit alone or with a team, and share all the findings that you think are clarified at the end of the day with the audited party as a draft report. The point I would like to reiterate here is that you should never reach new findings other than those agreed upon at the closing meeting, and you should not overshadow the content of the audit report with these new inferences. In addition, you should definitely take into account that the audited party has something to say, and you should try to resolve any concerns, if any, so that there are no last-minute surprises.

The art of the closing meeting lies in persuading the auditee (and your audit boss) that your conclusions represent the true state of things. If problems were uncovered, then the consequences are serious and something can and should be done about them. You have to deal with the problems in a timely manner and look for solutions; otherwise it will be very annoying to collect things after the report is prepared and published, and it can hurt your reputation. The more encouraging you are, the more the auditee will have the opportunity to clear all the problems in his mind at the last minute.

Table 2.25 contains an example of a GMP auditor asking the reader for the outcome of a conversation with the auditee about the time constraint.

2.30.1 Presentation of Audit Results

The closing meeting is a presentation of the GMP audit results to the auditee organization and, in some cases, the client. The purpose of the closing meeting is to present audit observations to top management to ensure that the audit results are clearly understood. The closing meeting normally takes place after the conclusion of the interviews and the final team meeting.

2.30.2 Discussion of Follow-Up Actions

The GMP lead auditor is expected to clarify all follow-up actions with the auditee and the client and to arrive at a consensus. The GMP lead auditor will have an open discussion with the auditee organization at the closing meeting to determine an adequate time for corrective action plans and/or corrective action implementation. The following are some guidelines for discussion with the auditee:

TABLE 2.25
Allotted Time

Scope and Allotted Time for GMP Audit

- Auditor requests the auditee extra time to complete the Internal GMP audit since the examination of various procedures and staff interviews took up a lot of time.
- Auditee responds with a no – indicating the auditor had a plan but strayed from it and spent excessive amounts of time interviewing staff. An extension request cannot be accommodated.
- Auditor acquiesces and closes the audit even though it was not completed and states in the final report that there was not enough time to complete the items in the scope.

Question: What personality type was the auditor? What difficult behavior do you see from the auditee?

- Clear definition of nonconformance/noncompliance/finding
- Process for effective corrective action determination/implementation
- Emphasis on timely corrective action plans with accurate root-cause determination
- Follow-up audit dates or follow-up corrective action review
- Areas and departments to be audited in follow-up audit

2.30.3 Keeping a Record

Reporting what was observed is what the closing meeting is all about, but there are three housekeeping items that should not be overlooked:

- An agenda for the closing meeting
- A record of who attended the meeting
- Minutes of any discussions or agreements made at the meeting

Recordkeeping does not need to be a big deal and can be handled nicely with a one-page form that includes logistical information, agenda, minutes, and attendance. For remote audits or eAudits, the GMP auditor can either record the closing meeting using video equipment or give participants keyboard and mouse controls to enter their name, and the auditor can use the same form to record the minutes of the meeting.

The agenda should include items typically covered at every closing meeting, plus space for specific topics to be added. Prior to the closing meeting, the auditor can review with the auditee contact what will be discussed and add any items that the GMP auditor and the auditee representative agree to. People can sign in by printing their name on the form as they arrive at the closing meeting, and then the same form can be used to record any important items discussed during the meeting. The objective is to get everything on one piece of paper that can be filed or scanned into the records database.

The closing meeting agenda may include:

- Audit purpose (objectives) and scope
- Background information
- Categorization and prioritization of results
- Audit sample limitations
- Summary of overall effectiveness
- Detailed findings
- Expected follow-up actions
- Timing of final report

2.31 PROBLEMS COMMONLY ENCOUNTERED DURING THE GMP AUDIT REPORTING AND CLOSURE STAGE

The following are examples of the types of problems commonly encountered during the audit reporting and closure stage.

2.31.1 Lack of Response or Inadequate Response

Sometimes an auditee will fail to take the action specified in the corrective action plan. The auditee may fail to notify the auditing organization when action is taken or the action taken may not be sufficient to warrant closing out the finding. The auditee may lose sight of the original intent of the corrective action request and address a different issue. At times, an auditee may not understand what a standard says or means and may argue the issue's importance or relevance.

2.31.2 On-the-Spot Corrective Actions

Some organizations may ask for a list of findings prior to the closing meeting so that corrective action can be implemented before the auditors leave the building. The immediate management response should be recognized by the audit team, but in most cases, the underlying cause of the problem is not eliminated.

2.31.3 Continual Requests for Extensions

An interesting note from a reputable auditor:

> The mutually agreed flow of the GMP audit was a quick tour of the facility from raw material entry to finished product shipment. The GMP auditor stated that he wanted to see the area that is not in the plan or even out of scope during the audit. He said he knew when he left the area that it was not on the agenda, but that he needed to ask a few questions. Fortunately, even though we thought that this field tour would not last long and he stated that he did not want to disrupt the main audit, he insisted on staying in that area for about a full day, and the flow of the 2-day audit was naturally interrupted.

2.32 ROLE OF AN AUDITOR

The closing meeting is conducted by GMP lead auditor, who should:

1. Thank the auditee for their cooperation
2. Present the summary
3. Read the results of the audit without interruption from the auditee
4. Discuss audit details
5. Allow individual auditors to clarify statements or respond to specific questions about the areas they audited
6. Allow the auditee to clarify any misunderstandings that may change or remove a nonconformity
7. Indicate how the audit results are categorized or graded and prioritized
8. Explain the required follow-up and expected corrective action response
9. Obtain acknowledgment from the auditee representative
10. Ensure that minutes and an attendance record are kept
11. Affirm confidentiality of the report

Table 2.26 provides information on the results of a unique GMP auditor and multiple auditors taking part in audits.

At the closing meeting, auditee representatives have the responsibility to listen attentively. There should be no extensive discussions unless the auditee needs clarification on items in the audit report.

Now, let's prepare a GMP audit report.

TABLE 2.26
Unique versus Multiple Auditors

Unique Auditor	Large Number of Auditors
The cost of initial training and of an eventual replacement is low	Cost for initial training, maintenance, and replacement of auditors is substantially higher.
Auditor becomes very efficient.	Audit experience being limited; each auditor maintains a minimal ability.
Auditor requires no support, and only negligible supervision.	Better support and supervision is required, as well as more detailed audit methods.
Audit scheduling is very simple.	The scheduling of audits and auditors is much more complex.
Loss of an auditor has a high negative impact.	Loss of one of the auditors has negligible impact; loss of a program manager will not disrupt audit schedule.
Auditor may be perceived as the QMS "police."	This is a negligible risk since auditors come from many departments.
	This is an excellent way to promote the quality concepts and the QMS; auditors gain knowledge of other functions of the organization.
	Auditors are more likely to perceive the organization as a system of interrelated processes.
	It promotes cross-pollination of good practices within the organization.

2.33 THE REPORT IS YOUR PRODUCT

The GMP audit report is your final product. All of the sights, sounds, smells, observations, scraps of paper, tensions, and anxieties are finally reduced into something for others to read. When everything is closed out, the only evidence of your presence is the report! It is your means of communicating information to others. As such, it should have certain characteristics in order to be successful.

2.34 REPORT CHARACTERISTICS

Reports should have accuracy, conciseness, clarity, timeliness, and tone.

They must be concise so that superfluous words do not block the reception of the message. Clear reports put your thoughts into the mind of the reader. Timely reports examine topics while they are of interest. Results are published before they are forgotten. Finally, the tone of the report must be courteous and professional. It must sound like the voice of management.

In addition to these traits, which are common to all reports, your report needs to possess the additional characteristics of relevance, consistency, and comparability. Unless your report is relevant to the business needs of your customers, it will be largely ignored. Consistency of reports over time will show trends and provide for greater comprehension by the reader. After they are familiar with the format and contents, they know where to look for desired information. Finally, reports from a number of auditors should be comparable to allow for maximum efficiency and fairness.

Your reports should be verifiable. The reader may not always be able to personally verify them, since we cannot track down the evidence for every piece of history known. But if you use generally accepted names for things (for example, foot, yard, milling machine, and batch preparation files), there is less danger of your message being misunderstood. When you refer to specific items or locations at the audit site (but not individual names!), the perception of verifiability is enhanced. Of course, one of the main purposes of a structured checklist is to record supporting information (facts). The reader tends to become tangled in the checklist details and may miss the overall message of the report. Your completed checklists should be kept in the office file as background data. These are field notes and should not be included in the report.

2.35 GMP AUDIT REPORTING

The purpose of the GMP audit report is to communicate the results of the investigation. The report should provide correct and precise information that can be valuable for managing critical organizational issues. The GMP audit process may end when the lead auditor issues the report or after follow-up actions are completed. After the GMP audit results have been reported at the closing meeting, the lead auditor formally communicates the audit results in a written audit report. The GMP audit team may have prepared a handwritten or typed report for the auditee at the closing meeting. The closing meeting should discuss all significant findings so that the auditee is not surprised by the information in the final audit report.

The final GMP audit report formally communicates the audit results to the client and the auditee. The GMP audit report is prepared, signed, and dated by the lead auditor. It should be reviewed and approved by the management of the auditing organization as well as by the client before it is sent to the auditee.

The final GMP audit report, if not already delivered at the closing meeting, should be sent to the auditee in accordance with a mutually understood time frame. Short timelines between performing and reporting on the audit are highly desirable. Table 2.27 provides an example of a GMP report.

GMP audit report serves the following functions:

TABLE 2.27
GMP Audit Report Example

Content of an Audit Report	
1 General details	**2 Carrying out the audit**
1.1 Name and address of the audited company	2.1 Cause for the audit
1.2 Scope of manufacturing authorization	2.2 Audit time period
1.3 Persons responsible	2.3 Auditors involved and task assignment
1.4 Current version of site master file	2.4 Company contact person
1.5 Date of the last audit	2.5 Audit procedure, areas visited
1.6 Essential changes since the last audit	
1.7 List of the last regulatory authority audit	
3 Determinations and observations	**4 Miscellaneous**
3.1 Quality assurance	4.1 Sampling data
3.2 Personnel	4.2 Comments about the site master file
3.3 Buildings, Rooms, Facilities, Equipment	4.3 Distribution of audit report
3.4 Documentation	
3.5 Production	
3.6 Quality control	
3.7 Contract manufacture and analysis	
3.8 Complaints and recall	
3.9 Self-inspections	
3.10.Related GMP Annexes	
5 Enclosures	**6 List of errors and deficiencies**
5.1 Plans	6.1 Critical
5.2 Copies of company documentation	6.2 Major
	6.3 Minor
7 Recommendations and notes	**8 Summary and conclusions**
7.1 Information for third parties	
7.2 Details of conditions and deadlines	
9 Date and signature of auditors	

- It supplies information that verifies adherence to requirements or that initiates corrective action and system improvement.
- It guides management in subsequent decisions and activities.
- It establishes a record of the investigation and conclusions.

A report should be sent even if there are no deficiencies. Organizations may compare the results from year to year to identify positive and negative trends. This same type of analysis is valuable to the auditor, the auditing organization, and the auditee (internal department, supplier, or customer).

2.35.1 Audit Report Details

While the format of the written GMP audit report can vary widely from company to company or industry to industry, certain information should be included. Most GMP audit reports contain an introduction, a summary, a listing of the findings/nonconformities, and requests for corrective action.

The GMP audit report should not name individual employees of the auditee in connection with specific findings, observations, or nonconformances; a report should refer to positions, titles, or systems when identifying problems. However, an auditor may encounter a situation where naming an individual is appropriate and essential for effective corrective action. For example, if the individual has special knowledge of a situation, product, or process, the individual's name may be used in the GMP audit report.

There is no standard length for GMP audit reports. However, reports should be concise and should contain the prescribed information.

Reports may or may not contain attachments. Attachments (exhibits or examples) may include minutes from the opening and closing meetings. The GMP audit report starts with an introduction or background. Much of the introductory material included in an audit report can be completed using the audit plan before the audit is performed.

2.36 DISTRIBUTE THE REPORT

The distribution list for a GMP audit report should be noted on the report. An internal GMP audit report is normally distributed to the supervisor of the audited area as well as to someone in higher management. An external GMP audit report typically goes to a division manager or CEO. A second-party external audit report may be distributed to a site production or quality manager.

The GMP lead auditor should not automatically send a report directly to the auditee unless approved to do so by the client. The client is given the option of receiving the audit report first and attaching a cover letter before the report is distributed to the auditee. This procedure promotes accountability of the auditing function. Responses by the auditee to requests for corrective action should be sent to the same people.

As a GMP auditor, your job is complete when the audit boss accepts your report. Sure, you will probably get involved in any correction activities after the report goes out. Or follow-up tasks may be done by others. The auditor's job is to report. The follow-up work is more of a consulting role. Follow-up also provides you with valuable feedback on how you did. You learn from your mistakes in logic or communication and become better. Be careful not to confuse auditing and consulting.

2.36.1 Communication Technology

GMP audit report distribution was much easier before the days of the internet, intranets, wide area networks, virtual private networks, and the like. The report with finding sheets, the corrective action report (CAR) sheets, and the cover letter were copied as a package and sent out. The process today is much different. The days of paper-only distribution are forever gone. We know that very little information is private once it is released to the network.

An email message, once sent, cannot be unsent. It is difficult, if not impossible, to destroy virtual data with today's technologies. With the communication technology we have today, audit reports can be distributed within 24 hours of the closing meeting. Electronic communication is fast and convenient. However, there are some issues beyond internet or intranet security that should be considered when using electronic distribution systems.

At the end of the audit I carried out in a software company abroad, during the closing meeting, while I was conveying my findings to the audited party, I allowed them to take notes before the official report. The technical operations director asked the following question with particular emphasis: "Besides the findings you are currently counting, will there be any other findings in the report?"

I asked, "Why would you have such a surprise?"

"We go through several audits each year and the auditor who came to the last audit only counted three to four minor findings at the closing meeting, but the report that came after 15 days had more than 20 findings, five of which were major. It really shocked us", he said.

Be very cautious, and do not add new findings to the report other than the ones you have verbally expressed.

2.37 AUDIT FOLLOW-UP

Follow-up GMP audits may be required in some situations, but due to cost considerations, many organizations do not require them. Instead, follow-up activities are included with the next regularly scheduled audit. When follow-up GMP audits are performed as a separate activity, the GMP audit schedule is determined by the findings opened during the original audit. These GMP audits can also provide independent verification of corrective action effectiveness.

For example, a "for cause" audit was performed of raw material supplier XYZ because a contaminated product was sent to customer ABC. The audit resulted in critical observations regarding the lack of cleaning processes at XYZ. ABC halted orders of raw materials until corrective actions were implemented at XYZ and a follow-up audit could be performed to confirm their effectiveness.

When a GMP auditor accepts an audit assignment, the scope may include verification of corrective actions from previous audits. In most cases, the verification activity can be easily integrated into the audit schedule by looking at the auditor assignments and areas visited during that audit. Sometimes, there may be too many corrective actions or they may be too complex to verify in the amount of time allotted for the audit. If corrective action verification jeopardizes the objectives of the current audit process, the GMP lead auditor should notify the audit program manager or client.

2.38 AUDIT CLOSURE

After verifying corrective action implementation, the auditing organization may prepare a follow-up report and distribute it in a manner similar to the original audit report. Closure requires verification that corrective actions have been implemented as planned. The GMP auditor notifies the auditee that the audit is closed. The GMP auditor may then discard certain papers accumulated during the audit but is generally required to retain defined records for specific time periods.

2.39 PROBLEMS COMMONLY ENCOUNTERED DURING THE AUDIT REPORTING AND CLOSURE STAGE

2.39.1 Lack of Response or Inadequate Response

Sometimes an auditee will fail to take the action specified in the corrective action plan. The auditee may fail to notify the auditing organization when action is taken or the action taken may not be sufficient to warrant closing out the finding. The auditee may lose sight of the original intent of the corrective action request and address a different issue. At times, an auditee may not understand what a standard says or means and may argue the issue's importance or relevance. As a result, an auditee may try to fix a problem haphazardly without understanding the need for action or training. Sometimes the person involved in fixing a problem was not involved in the audit and may not have all the needed information. An auditee may try to pit one audit team against another and argue that the other audit team had no findings in a certain area.

2.39.2 On-the-Spot Corrective Actions

Some organizations may ask for a list of findings prior to the closing meeting so that corrective action can be implemented before GMP auditors leave the building. The immediate management response should be recognized by the audit team, but in most cases, the underlying cause of the problem is not eliminated. GMP auditors should use their best judgment, but true root-cause analysis, leading to effective corrective actions, is difficult at best when the auditee proposes to solve a problem quickly.

2.39.3 Continual Requests for Extensions

Sometimes a GMP auditor and an auditee may not agree on the significance of a problem. When vested interest is not the same, agreement on the importance or priority of solving a problem may be difficult to achieve. An auditee may request continual extensions for fixing the problem. Excuses can range from "We're working on a big order and don't have time to fix it right now" to "We forgot."

2.40 CORRECTIVE AND PREVENTIVE ACTIONS (CAPAS)

The principle of corrective action is that conditions adverse to quality are identified and corrected. The cause of identified problems is determined. Steps are taken to preclude repetition. These steps include fixing the underlying cause and making sure the fix actually worked. True corrective action is probably the most difficult part of a quality management system to implement. The real underlying or root causes of problems are seldom easy to identify. But without an effective and objective corrective action program, the quality of all operations will suffer greatly.

The corrective action system has four linked processes (actions), which all work together to keep the problem from recurring:

- Identify problems worthy of correcting
- Understand each problem's underlying (root) cause
- Fix the underlying cause conditions
- Make sure the changes work

2.40.1 Identify Problems for Correcting

Corrective action takes a great deal of organizational energy. It is possibly one of the most difficult of all the quality systems explored over the last several decades. Organizations have a limited amount of this energy. It must be spent wisely. If you and your suppliers attempt to correct each and every identified problem and nonconformance, you will all waste that limited store of energy. You will burn out. So, you need to set a minimum threshold of pain before a problem is worthy of corrective action attention. As stressed earlier, the pain factors should be associated with business values. These are always cost, production, and risk.

Try not to confuse corrective action with remedial action. All identified nonconforming items need remedial action. Only problems that exceed the local pain threshold require corrective action.

Your pain threshold should be high enough to filter out the background noise, but low enough to achieve results before it is too late. If you wait until a fire destroys the warehouse, it is too late to work on the sprinkler system maintenance issues. There's nothing left to fix.

Through the GMP audit, you have made this decision for the auditee. Because it is a finding, it is (by definition) worthy of corrective action. You even helped by entering each problem (from the finding sheets) into the corrective action tracking system. You attached a CAR sheet to the audit report package. You have declared, "This is a problem in need of fixing."

2.40.2 Determine the Underlying Cause

By listing the supporting bad facts under the finding, you have helped. Now the auditee must apply all the problem-solving tools to find the underlying (or root) cause. Teams, Pareto charts, nominal group techniques, and affinity diagrams are all ways to do this. This is very difficult to do. It generally requires more than one person. It usually takes more than a day.

2.40.3 Fixing the Underlying Cause

Fixing the underlying cause requires change at the system level. New tools, new methods, increased or different skill levels, new capital equipment, and all manner of resources are but a few of the options. This takes time and money. Rarely can this level of change be done in less than a month.

2.40.4 Verify the Fix Really Worked

Even with the best minds in the company trying to solve a problem, sometimes we just don't get it right. The problem comes back. We didn't really understand it or failed some place in implementing changes.

The last step in corrective action is to take a backward look to see if the problem is recurring. In some cases, constant changes in markets, resources, or technologies will introduce enough variables to guarantee that the problem recurs. We need to analyze data to show that the problem remains fixed. We usually need a year's worth of data.

The GMP audit program, along with normal management review and supervision, will address the first step of identifying problems needing solutions. But as was discussed in the previous section, you will make judgments that may result in different interpretations of the severity of the reported problem. This is why the identification of pain is so important in the way findings are presented in your audit report. If you show the disease, rather than just the symptom, the cause becomes obvious. It is the underlying reason for the pain. It is the cause of cause-and-effect analysis.

2.40.5 Strategies When Corrective Action Is Not Implemented or Is Not Effective

GMP auditors must be objective when determining which strategy to use when corrective action is not implemented or ineffective. Change the audit function and the auditee's organization may be necessary. GMP auditors must leverage their communication and negotiation skills take appropriate action. GMP auditors should evaluate each application's shortcomings or ineffective corrective action on its own merits.

Rather than assume that the auditee is at fault, the GMP auditor should reevaluate the corrective action and follow-up process with an open mind. Perhaps an incorrect conclusion was reached due to a lack of time to verify evidence. Available evidence may indicate an error in judgment on the part of the corrective action team or there may be inadequate resources. The lack of resources may be due to a change in business conditions beyond the auditee's control. A lack of success when attempting to implement corrective action does not always indicate a lack of management commitment.

When the corrective action is not implemented, the GMP auditor should take time to evaluate the situation thoroughly before making any attempt at escalation (i.e., getting higher-ups involved). Jumping into the role of enforcer prematurely puts the auditor in a negative position. It is in the auditor's best interest to be viewed as one who is interested in assisting the auditee to achieve the desired results. As a last resort, the auditor may need to go up the chain of command to the next level of management.

When the corrective action is not effective, the GMP auditor should either recycle the finding or issue a new finding (referencing the original) to start the corrective action process again. Failure of the corrective action to be effective is serious and should be considered for a form of rework. If this is a frequent recurrence, the corrective action process may be suspect.

2.40.6 Rating of the Individual Quality Elements Guideline

1 = In compliance (no observations)
2 = Individual improvements needed

- No major observation based on a system failure
- Not more than three individual major observations

3 = Major improvements needed

- No critical observation
- More than three individual major observations
- One or more major observation(s) based on a system failure

4 = Major improvements needed immediately

- One or more critical observations

2.41 EVALUATION OF GMP AUDIT PROGRAM EFFECTIVENESS

Periodic evaluations of all GMP audit programs and audit teams cause improvement opportunities to surface. GMP audit programs are evaluated through periodic management reviews. Management can review records, analyze performance, or appoint an independent audit team to review/fill audit results. In the case of a multisite company, auditors from another location can come in annually to audit the audit team.

This evaluation could include:

- Observing GMP audit teams perform an audit
- Examining GMP auditor training records and audit schedules
- Looking at sample GMP audit results, corrective actions, and follow-up activities to see if the program is working as intended

Additional data can be collected by asking an auditee to rate audit team members in the areas of:

- Interviewing skills
- Interaction with personnel
- Reporting results

Critical reviews by peers, subordinates, and superiors provide an exceptional opportunity for growth and maturity. While it may be slightly unpleasant to those being reviewed, it also helps keep things in perspective.

2.42 RISK-BASED AUDITING

In order to carry out a risk-based GMP audit, risk-management tools such as ISO 31000,ISO 14971, or ICH Q9 can be used.

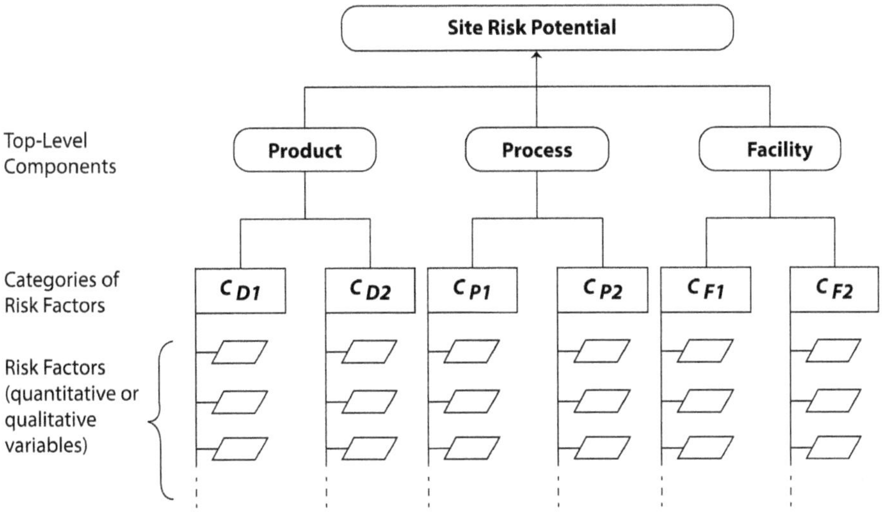

FIGURE 2.8 Site risk potential.

2.42.1 What Is Risk Management?

A risk management program is a series of interrelated processes and tools designed to identify, describe, assess, mitigate, and track risk, thereby managing organizational risk. The industry has various types of tools to both identify and assess risk, such as failure mode and effects analysis (FMEA), fault tree analysis (FTA), and hazard analysis of critical control points (HACCP). Another commonly used tool is the SWOT (strengths, weaknesses, opportunities, threats) analysis. Figure 2.8 shows how to organize site risks.

2.42.2 Principles of Quality Risk Management

To understand and apply quality risk management successfully, paying attention to two principles is necessary:

- Any risk that will affect the product quality is to be considered; it must have a scientific basis, and the patient's health, being the end user, must be taken into account.
- Documentation of risks with the effort made in the management process should be proportionate to the existing or thought to be a risk.

Table 2.28 provides an overview of product, process, and facility risks.

Table 2.29 summarizes the dos and don'ts when preparing for a GMP audit.

In Table 2.30 is a list of what a GMP auditor should react to when faced.

2.43 MHRA AUDITS

GMP is the minimum standard that a medicines manufacturer must meet in its production processes.

Pharmaceutical products should:

- Always be of high quality
- Serve the patient's intended use
- Be able to respond to the requirements of the market's authorization and product specifications

TABLE 2.28
Risks

Factor Category	Description	Example(s)
Product	Factors pertaining to the intrinsic properties of drug products such that quality deficiencies could potentially and adversely impact public health.	Dosage form; intrinsic chemical properties
Facility	Factors relating to characteristics of a manufacturing site are believed to be predictive of potential quality risks, such as the lack of effective quality systems.	Poor cGMP compliance history
Process	Factors pertaining to aspects of drug manufacturing operations that may predict potential difficulties with process control and/or vulnerability to various forms of contamination.	Measuring; mixing; compression; filling

TABLE 2.29
Dos and Don'ts of Audits/Inspections

Do	Don't
Motivate employees	Scare or threaten employees
Highlight strengths and weaknesses in a process or system	Emphasize deficiencies
Analyze the entire process or system	Narrow the evaluation of noncompliance
Support global competitiveness	Appease regulators
Increase quality (Lean Six Sigma, OPEX)	Disqualify quality
Promote best practices	Point a finger at deficient practice
Clearly define quality standards	Use obsolete standards
Establish measurable quality and compliance outputs	Rely upon FDA inspection outcomes
Offer a rating system	Use FDA 483 as a measuring stick
Determine the priorities for correcting deficiencies	Determine priorities by regulatory authorities and resulting actions
Identify areas for improvement	Assign blame for the failure
Provide a process to identify the genesis of deficiencies and lack of or need to improve the quality system	Stick to unsustainable improvements
Foster continuous improvement, total quality management, and forward-quality thinking	Focus on deficiencies only
Identify, install, and maintain quality standards exceeding regulatory expectations	Focus on only meeting regulatory expectations
Provide executive management with a clear picture of the overall compliance status of each system or procedure	Identify only key deficiencies of the systems and procedures but lack specifics
Provide a process gap analysis rating system for a quantifiable assessment	Provide only qualitative indicators
Present corrective actions to bring the system back to a 100% compliance status	Adopt a band-aid approach rather than a comprehensive, far-reaching approach
Identify and correct deficiencies by analyzing root causes	Miss to identify root causes from symptoms

On the other hand, GDP, or good distribution practice, requires the consistent storage, transportation and processing of drugs under appropriate temperature and humidity without compromising product specifications throughout the licensed supply chain.

TABLE 2.30
Possible Criteria to Categorize Nonconformances

Nonconformances Deficiencies	Isolated/Systemic/Major/Minor
Accidental rip of glove in class A	
Aseptic gowning SOP does not instruct to check gloves	
Wrong expiration date on a reagent bottle label	
BR poorly written; instructions may be confusing	
Operator repeatedly spills samples	
Management does not reinforce the requirement to record information properly (Data Integrity)	
A preventive maintenance task was delayed	
Preventive maintenance does not include filling machine	

2.43.1 Inspections under the Risk-Based Compliance Program

As you know, each medicinal product license holder carries a risk rating or score, and the Medicines and Healthcare products Regulatory Agency (MHRA) prioritizes inspections for those with the highest ratings or scores. Auditees will be told about these inspections in advance, although under the short-notice inspection program, the MHRA may send little or no notification. At the inspection, the GMP and GDP inspectors examine the systems used to manufacture or distribute medicines.

Please note:

- Your GMP rating is based on your compliance report and internal information about previous inspection history and organizational changes.
- You can't appeal against your rating.
- An increase in risk will be peer-reviewed by a GMP operations manager, a compliance management team (CMT) member, or a GMP expert inspector before being finalized.
- Each manufacturer takes a a full copy of the reasons behind their risks at the end of the inspection.
- When it comes to GDP inspections, the distributor risk score is based on on-site activities and non-conformities. It allows the distributor to know the date of the next inspection.

Time-to-time, inspections can be performed with other inspections, such as GCP (good clinical practice) and GPVP (good pharmacovigilance practice).

2.43.2 Product-Related GMP Inspections

When MHRA evaluates the marketing authorization application; it performs a GMP inspection regarding the product. This inspection takes into account the manufacturer's compliance with GMP. The MHRA may inform the facility to be inspected in advance.

2.43.3 Triggered Inspections

The MHRA may inspect a manufacturer if it is informed about possible GMP or GDP breaches by:

- A whistleblower
- Other MHRA departments
- Another regulatory authority

The following points should be taken into account in the GMP inspection:

- Interview with relevant personnel
- Examination of relevant documents
- Facility tour

Facility tour visits can be any facility or process related to the production, purchase, and distribution of pharmaceuticals, including:

- Production areas including weighing facilities
- Analytical and microbiology laboratories
- Stock and inventory management
- Storage areas
- Returned, rejected, and quarantined products
- Temperature and humidity monitoring
- Purchasing and selling
- Transportation activities

The MHRA Inspection team may request additional documents, records, diagrams, graphics, pictures, and additional samples if needed during the inspection. If it is concluded that there is a serious non-compliance during the GMP inspection, the flow of the inspection can be rescheduled. In the closing meeting of the inspection, as in other GMP inspections, the inspector should provide feedback on all aspects and share the deficiencies with the inspected party and convey their ideas about the timing of corrective and preventive actions and the institution's approach.

2.43.4 Grading of Inspection Findings

Deficiencies found during inspections are graded at three levels. The following definitions are summaries. For further information, please see "Compilation of Union Procedures on Inspections and Exchange of Information" 20 April 2022 EMA/224865/2022 Rev 18 Corr.

Grumblings I hear about FDA, MHRA, EMA, PIC/S, and WHO inspections are often that health authority inspections frequently fail. The issue that I always emphasize in the consultancy or audit preparation plans I present to pharmaceutical and biotechnology companies is that taking the event into consideration from the perspective of the health authority plays a key role. If you do not read the guides and directives published by the authority, even if you do not understand them, it is not possible to look at them from the authority's perspective. For example, in the guidelines, the health authority plans the inspections as risk-based, and the frequency of inspections and how many days an inspection will take is determined according to the risks of each facility and even the product. However, the practice I have seen so far is that the inspected party lags far behind this approach.

For instance, while the health authority reserves more than ten days for the inspections of aseptic product manufacturing facilities, how many days do you, as pharmaceutical manufacturers, reserve for internal audits for your aseptic manufacturing operations, or how many days do you reserve for the audit of a contracted manufacturer?

Before you say it, let me guess, maybe one to two days for internal audits, and two to three days maximum for contracted companies, right? Comments please?

2.43.4.1 Critical Deficiency

A critical deficiency is a deficiency that has produced or significantly risks creating a product that is harms to humans or veterinary patients or could result in a harmful residue in a food-producing animal.

Any departure from good distribution practice results in a significant risk to patients. This situation consists of an activity that broadens the risk of counterfeiting medicinal products.

2.43.4.2 Major Deficiency

A major deficiency is a noncritical deficiency that:

- May be producing or likely to produce a medicinal product that does not comply with the marketing authorization
- Although this represents a deviation from GMP and GDP, it may also mean that the conditions of the manufacturer or wholesale license are not met
- Indicates that party release conditions are not fully met or that a QP is not fulfilling its legal duties within the EU
- A combination of several "other" deficiencies that on their own may not be major but together may represent a major deficiency and should be explained and reported as such

2.45.4.3 Other

An other deficiency is a deficiency that cannot be classified as either critical or major or there is not enough information to classify it as critical or major but indicates a departure from good manufacturing and distribution practice.

2.43.5 Actions after the Inspection

After the inspection closing meeting, the auditee will receive a post-inspection letter confirming any deficiencies found.

The auditee is expected to send their action plans by e-mail to let the inspectors know the proposed action completion dates and deadline. The inspector will review the auditee's response. If they accept it, the auditee will receive a GMP or GDP certificate with their inspection report.

If the medicinal product manufacturer is being inspected for GMP compliance, they should complete an interim assessment if the site has changed since the first inspection.

All information on the daily fees, costs, and in which cases additional charges may be required for inspections can be found on the MHRA's website.

2.44 WHO INSPECTIONS

As you know, GMP inspections are one of the most important control points of pharmaceutical quality assurance. The main purpose of inspecting pharmaceutical product production facilities is to control whether GMP compliance is ensured and at the same time, it is a necessity to obtain sales and marketing permits of the pharmaceutical product. Inspections are also an effort to minimize the entry and mobility of counterfeit drugs into the supply chain, which has become more common in recent years.

2.45 TYPICAL PIC/S INSPECTION OF A MEDICINAL PRODUCT MANUFACTURER

2.45.1 Before the Inspection

- Lead inspector assigned
- Inspection team selected
- Technical specialist sometimes included on team
- Company notified
- Company requested to provide Site Master File (SMF)
- Inspection team reviews documentation
- SMF, complaints, recalls, testing failures, marketing authorizations

- Lead inspector prepares inspection plan and sends it to the company
- Inspection conducted

2.45.2 After the Inspection

- Caucus of the inspection team
- Interim inspection report prepared (deficiencies only)
- Exit interview with company
- Attendance sheet completed
- Interim inspection report provided (discussion encouraged)
- Written response requested within four weeks
- Objective evidence assessed by the lead inspector
- If the response is judged OK, inspection is closed out
- Final inspection report sent to company
- If response is not OK, refer to independent committee for appropriate action

2.45.3 PIC/S Inspection Report

- Identical to the EU inspection report format
- Format used by PIC/S and EU inspectorates to prepare GMP inspection reports
- Uniform system of classifying GMP deficiencies: critical, major, other

2.46 EMA AUDITS

The guiding principles of the European Medicines Agency (EMA) state:

> Any medicine manufacturer, no matter where it is located, must comply with GMP if the products manufactured are supplied to the EU. Any medicine manufacturer located in the EU must comply with GMP no matter where the medicines are supplied to
>
> - Medicines for human use (and their active substances)
> - Medicines for veterinary use (and their active substances)
> - IMPs [investigational medicinal products] (used in clinical trials)

2.46.1 Manufacturing/Importation Authorization (MIA) Only Granted after Inspection

Manufacturers and importers of finished (drug) product manufacturers and importers in the EU need to be authorized to carry out their activities. This applies to IMP manufacturers. All finished products manufacturers and importers are regularly inspected by an EU authority, unless a mutual recognition agreement is in place:

- Irrespective if the site is in or outside of the EU
- Frequency of inspection is based on risk

2.46.2 GMP Certificates and EudraGMDP

EudraGMDP is a database that contains (nonexhaustive list) public information on:

- Manufacturing/importation authorizations (MIAs)
- GMP certificates
- GMP statements of noncompliance (SNCs)

MIA and GMP certificates are uploaded directly by EU National Competent Authorities (NCAs)

EU NCAs can be contacted for clarifications.

Allows verification of the GMP status of manufacturing sites.

2.46.3 The Role of EMA

The agency has a coordinating role for GMP inspections of manufacturing sites for centrally authorized products (CAPs) or as part of a referral procedure.

Key roles:

- Ensuring common interpretation of EU GMP requirements and related technical issues
- Developing EU-wide procedures on GMP inspections and related activities
- Facilitating cooperation between member states for inspections of manufacturers in third countries
- Developing and maintaining the EudraGMDP database
- Sampling and testing planning for CAPs
- Coordination of the actions at EU level in case of quality defects

International collaboration

- Maximize inspection resources
- Broaden inspection coverage
- Minimize duplicates
- Focus on sites of highest risk

2.46.4 EDQM Inspection Program

- The European Directorate for the Quality of Medicines & HealthCare (EDQM) establishes an annual inspection program:
  - On-site inspections carried out with official inspectors from European Member States
  - Use of information obtained from inspectorates (e.g., inspection reports)
  - Distant assessment (paper evaluation in the frame of the reinspection program)
- Inspection resources worldwide are limited and API manufacturers having CEPs (Certificates of Suitability) supply generally the global markets. Optimization of use of inspection resources by exchange of information:
  - International API Inspection Program
  - PIC Scheme (PIC/S)
  - Confidentiality agreements

2.46.5 How Does an Inspection Procedure Work?

- Inspection performed by a team composed of one EDQM inspector and one inspector from an EU/EEA/MRA authority (joint inspections may also be performed, e.g., with WHO, TGA, USFDA, PMDA).
- Inspection lasts generally three days.
- Inspectors may take samples for analysis by official medicines control laboratories (OMCLs).
- An (initial) inspection report is issued within six weeks after the inspection.
- Immediate actions regarding the validity of the CEPs are taken in case of major or critical deficiencies.

- The company should reply to the deficiencies found with a CAPA (to be fully documented and reflect actual measures in place).
- Discrepancies with the CEP dossier should be specifically addressed and managed by the revision process at EDQM.
- The CAPA is reviewed and the final report is issued.

2.46.6 What Are the Inspection Outcomes?

- According to the inspection results, the company is quoted as GMP compliant, "borderline," or noncompliant.
- Borderline status is only provisional: after assessment of the CAPA plan, the outcome is upgraded to compliant or downgraded to noncompliant.

2.46.7 Positive Outcome

- In case of a positive conclusion of the inspection combined with a satisfactory evaluation of the submitted CAPA (and the CEP dossier is up to date), an inspection attestation is delivered by EDQM, stating compliance with the CEP and with GMP.
- A GMP certificate is issued by the EU/EEA/MRA participating inspectorate via the EudraGMDP database (public information): https://eudragmdp.ema.europa.eu/inspections/displayWelcome.do;jsessionid=nHmaiSR1SUVOMkezNoFrmbGT-gqOQAPTExHx-pXoKoOqXZD2eWwf!-684463328
- The company becomes a candidate for reinspection. It may be reinspected/reevaluated within two to five years depending on the numbers and classification of deficiencies found.

2.46.8 Negative Outcome

- In case of critical/major deficiencies to GMP and/or the CEP dossier:
- Actions are taken by EDQM on the CEPs/CEP applications:
  - CEP(s) of the site is suspended or withdrawn.
  - Site removed from the list if more than one involved in a CEP dossier.
  - Ongoing CEP application(s) rejected.
- A Statement of Noncompliance is issued by the EU/EEA/MRA participating Inspectorate via the EudraGMDP database (public information).
- Decisions to take action on CEPs/CEP applications:
- Recommended by the inspectors.
- Discussed within the EDQM Certification Department.
- Decided by the EDQM Ad Hoc Committee.
- Holder and manufacturer are notified of the decision.
- Public information is available on the EDQM website (https://www.edqm.eu/en/) and on the CEP database, as well as information to international partners.

2.46.9 How Are CEPs Suspended?

- CEPs are suspended for a period of two years.
- The company is requested to apply within this time frame for a reinspection.
- Based on a valid justification, the company may ask for an extension of this period.
- Lifting the suspension can only be done after an inspection demonstrating GMP and CEP compliance, as well as full implementation of the CAPA.

2.46.10 Suspension versus Withdrawal of CEPs

- Suspension: A temporary cancellation CEP can be restored.
- Withdrawal: A definite cancellation, decided:

- When no corrective actions are deemed possible
- For extensive cases of falsification of data
- After repeated noncompliance

If the company is still interested in having a CEP a new dossier must be submitted and a successful reinspection performed.

2.46.11 Manufacturers and Importers of Active (Drug) Substance

Manufacturers and Importers in the EU need to be registered with the National Competent Authority of the Member State where they are located. This applies to IMP manufacturers.

For inspection of active substance manufacturers in case of suspicion of noncompliance, the responsibility for using active substances manufactured according to GMP is with the manufacturing authorization holder (finished product manufacturer or importer). The qualified person declaration is needed in marketing authorization dossiers (based on audit).

2.47 FDA INSPECTIONS

The FDA inspects manufacturers or processors of FDA-regulated products to verify that they comply with relevant regulations.

Those inspected include:

- Vaccine and drug manufacturers
- Blood banks
- Food processing facilities
- Dairy farms
- Animal feed processors

2.47.1 When Does the FDA Inspect a Company?

The FDA inspects manufacturers or processors of FDA-regulated products after an application is submitted to the FDA for a new product; when a facility is due for a "routine" inspection or as a follow-up to a previous inspection (for example, typically once every two years); or to investigate a specific problem that came to the FDA's attention. Inspections include foreign manufacturing or processing sites for FDA-regulated products that are sold in the United States.

2.47.2 What Is an FDA Form 483?

The FDA 483 form is most unusual, and it is the most undesirable form to be submitted with a post-inspection finding. Its history can be traced back to 1953. After the FDA inspector or inspectors inspect a facility such as a drug production facility, warehouse, or laboratory, they fill this document with the findings they have obtained before leaving the building and deliver it to the responsible manager of the enterprise. The inspector must provide a copy of the report to the secretariat if, in his opinion, it is determined that any food, medicine, medical device, or tobacco product produced

TABLE 2.31
Form 483 example

DEPARTMENT OF HEALTH AND HUMAN SERVICES FOOD AND DRUG ADMINISTRATION	
DISTRICT OFFICE ADDRESS AND PHONE NUMBER Minneapolis District 250 Marquette Ave. South, Suite 600 Minneapolis, MN 55401 Industry information: www.fda.gov/oc/industry	DATE(S) OF INSPECTION MM/DD/YYYY FEI NUMBER:
NAME AND TITLE OF INDIVIDUAL TO WHOM REPORT IS ISSUED	
TO: Mr. Apostrophe Vice President, Production	
FIRM NAME ACMEHealth Pharmaceuticals	STREET ADDRESS xxx street No.: xxxx
CITY, STATE AND ZIP CODE xxxx, MN 55326	TYPE OF ESTABLISHMENT INSPECTED Tablet Repacker
THIS DOCUMENT LISTS OBSERVATIONS MADE BY THE FDA REPRESENTATIVE(S) DURING THE INSPECTION OF YOUR FACILITY. THEY ARE INSPECTIONAL OBSERVATIONS; AND DO NOT REPRESENT A FINAL AGENCY DETERMINATION REGARDING YOUR COMPLIANCE. IF YOU HAVE AN OBJECTION REGARDING AN OBSERVATION, OR HAVE IMPLEMENTED, OR PLAN TO IMPLEMENT CORRECTIVE ACTION IN RESPONSE TO AN OBSERVATION, YOU MAY DISCUSS THE OBJECTION OR ACTION WITH THE FDA REPRESENTATIVE(S) DURING THE INSPECTION OR SUBMIT THIS INFORMATION TO FDA AT THE ADDRESS ABOVE. IF YOU HAVE ANY QUESTIONS, PLEASE CONTACT FDA AT THE PHONE NUMBER AND ADDRESS ABOVE. DURING AN INSPECTION OF YOUR FIRM (I) (WE) OBSERVED: List your significant observations ranked in order of significance.	

SEE REVERSE OF THIS PAGE	EMPLOYEE(S) SIGNATURE Jane Hypen	EMPLOYEE(S) NAME AND TITLE (Print or Type) Jane Hypen, Investigator	DATE ISSUED MM/DD/YYYY

or kept in the facility is contaminated, odorous, or decomposed for some reason, and is kept under these unsanitary conditions.

2.47.3 Are FDA Observations during a 483 Necessarily Violations?

No. The 483 documents and communicates concerns regarding objectionable conditions observed during inspections. The form states explicitly that it "lists observations made by the FDA representative(s) during the inspection of your facility. They are inspectional observations, and do not represent a final Agency determination regarding your compliance." The agency considers the 483, together with a report called an Establishment Inspection Report (EIR), all evidence collected on-site, and any responses from the corporation to determine what further action, if any, is suitable. In Table 2.31 is the content of a 483 form cited by the FDA.

2.47.4 When Will My Firm Receive a 483?

If an inspection results in observations of objectionable conditions or practices that indicate violations, a 483 is issued at the end of the inspection, often in a close-out meeting at which the FDA investigators may either read or summarize the observations to responsible representatives of the inspected establishment.

2.47.5 What Is a Warning Letter?

Warning letters are letters from the FDA that serve a variety of purposes. First, a warning letter "allows pharmaceutical companies to take voluntary action that allows individuals to recuperate before an enforcement action by the FDA." Second, warning letters are "issued to ensure voluntary compliance and provide advance notice." This approach is "actually based on the belief companies will willingly comply with the law" and also serves to conserve the agency's scarce resources. Under FDA policy, warning letters should only be issued for "regulatory violations" or "violations that, if not promptly and adequately corrected, could result in sanctions." Such a letter indicates that the FDA's concerns go beyond the observations of the particular investigator and are shared by district or center-level compliance staff. Although a warning letter should be taken very seriously and poses a threat of sanctions, from the FDA's legal perspective, it is "unofficial and advisory." In other words, a warning letter "communicates the agency's position on an issue but does not commit the FDA to take enforcement action."

2.47.6 Typical Inspection Techniques

After introductions and obtaining the background information on the facility, most inspections include a walkthrough. This preliminary tour allows the investigator to become familiar with the operation and refine the inspection plan/strategy. It also allows for determining the depth of the inspection, gaining knowledge about the products and processes, assessing the source of manufacturing records, and identifying potential areas of concern. A variety of considerations are made in establishing the depth of the investigation, including the current compliance program, the nature of the inspection assignment, general knowledge of the industry and its problems, the history of the firm, and the conditions found at the establishment as the inspection progresses. Some other features considered are the size of the facility, number of employees, employee practices, environmental conditions at the site, raw materials used, whether the processes are manual or automated, potential sources of contamination, the flow of manufacturing, and methods used for data collection. During the tour, the investigator looks for potential problem areas.

2.47.7 Fact-Finding Mission

The investigator is on a fact-finding mission during the inspection. Their job is to report facts in such a way that they are accurate, informative, and detailed. The facts cover a variety of observations such as the condition of equipment, facilities, products, labels, and labeling. Additionally, the investigator observes the behavior of employees and the statements that the employees make during the interview and inquiry process. This fact-finding mission is accomplished in a variety of ways. Some investigators start collecting information during the preliminary facility tour. For example, they may collect equipment identification numbers, lot numbers of raw materials or components, information from calibration stickers, and so forth. Questions asked or answered during the tour may drive the direction and level of detail that occurs in subsequent activities. Another method used is to follow written FDA inspection guides for the specific systems reviewed. Still, other investigators follow a plan that they have generated and that works best for them. One approach is to have the tour start with the receipt of materials and follow through to finished product production. The auditing interviews and techniques follow the same approach. In most cases, investigators will use an approach that works best for them to gain the information they seek.

2.47.8 Sample Collection

There are a variety of reasons that the FDA may collect samples during a site inspection, including samples collected as part of surveillance programs, samples collected during inspections as evidence of violations, samples representative of lot production, in-process samples, aseptic samples, or collecting post-inspection samples.

2.47.9 Interview Techniques

FDA investigators are trained in a variety of methods to be used when interviewing personnel at a facility. These include the use of direct, indirect, and leading questions. During this process, the same question may be asked of different employees to determine whether an answer is truthful and accurate, or to confirm the method. Investigators are very experienced in interview techniques. During the interview process, the investigator is aware that interviewees may be nervous. It is also common knowledge that firms prepare their employees for inspections, and they specifically do not want to divulge adverse information. Part of the FDA's training includes developing an ability to detect deception while being aware that true deception is rare.

2.47.10 Wrong Dialogue

Investigator: Are calibrations of HPLCs performed periodically?
Lab Analyst: Yes, they are. As a matter of fact, most of the equipment is calibrated regularly, except for this one over here. Would you like to see all SOPs and calibration logs?
Investigator: Sure, let me see them. Also, could you find out which products were released using the one over there?
Lab Analyst: Our current calibration program is in really good shape. But you should have seen our situation when a former manager was on duty last year.
Investigator: That's very interesting. Could I see last year's calibration records as well?

The preceding conversation conducted between an FDA investigator and a lab analyst will result in that laboratory being subjected to the kind of scrutiny that could seriously damage its reputation and credibility. The analyst tried to detail all the questions, even gave information voluntarily, and gave the opportunity to go into more detail based on the information they provided.

Although the FDA inspector only asked a simple question about HPLC calibrations, he got a lot of information about the entire calibration program and its shortcomings. There was no need for all these details. In many audits, such a clear point would perhaps not have emerged. Although this analyst tried to cooperate out of goodwill, he made statements that were not good for the lab.

2.47.11 Alternative Conversation

Investigator: Are the HPLCs calibrated on a regular basis?
Lab Analyst: Yes.
Investigator: Do you have any procedures governing the calibrations and a document where the calibrations are recorded?
Lab Analyst: Yes.
Investigator: May I see them?
Lab Analyst: May I know which SOP and logbook you want to look at?
Investigator: Just your HPLC calibration procedure and HPLC logbook will suffice.
Lab Analyst: Is there an entry in the HPLC logbook that you particularly want to see?
Investigator: No. I want to look at the whole logbook.
Lab Analyst: The logbook is pretty thick; I can help you if you want.

Investigator: Thanks, no I want to take a look myself.
Lab Analyst: Glad to bring the HPLC calibration SOP and logbook for you right away.

Note that this time the analyst kept his answers as short as possible and focused on what the inspector specifically wanted to see. No information was given voluntarily during this dialogue and the answers did not need to be elaborated. When the dialogue continues in this way, the inspection will not go beyond the information requested by the inspector, and thus, even if a 483 is received, its scope and content will be narrower. In fact, the trick lies in knowing how to talk to an inspector. The presence of a subject expert who is well-versed in the subjects in the laboratories as well as in the production and warehouse areas during FDA inspections will help to get through the inspection more smoothly.

2.47.12 Practice: Simulating Inspector Interviews

Subject matter experts (SMEs) are expected to participate in at least two audit simulations where they can develop their skills and knowledge before participating in a real-time FDA audit.

Who should conduct this simulated inspection? The best practice is to use personnel who are independent. If your company has different production facilities, these people may be experts from within your own company. You may be able to bring in somebody who is independent within your company. You can also use external, experienced consultants who have experience with FDA inspections— they may be former inspectors—but also have industry experience to really understand the business side of things. The person should also be very experienced with quality systems.

So what methods are used? During a simulated inspection, the mock inspector will ask questions that would be anticipated during an FDA inspection. Your subject matter experts will face issues and questions as in a real-time inspection, and experience all the stress they might experience in a real-life inspection.

2.47.12.1 Just the Facts

Inspector: Do you have a procedure for complaint handling?
SME: Yes, we have a procedure for complaint handling. It's been recently updated because of an internal audit observation.
Inspector: I see. Could I see a copy of the updated procedure, the associated change record, the internal audit report, and the CAPA associated with this observation?
SME: Sure, we'll get you that information.
Inspector: While we are waiting for the documents I requested, can you tell me what the observation was?
SME: Sure. The observation was that the procedure didn't require that we document the evaluation we perform to determine whether or not a further investigation is required.
Inspector: What are the actions that you have taken?
SME: One of the actions is overdue. We updated the procedure to require that the evaluation is now documented, a rationale for why we're not going to perform further investigation, if that's the case, and the signature of the individual who performed the evaluation. We're making a change to the software to accommodate this.

2.47.12.2 Feedback

As part of these simulated inspections, the person acting as the inspector can provide coaching and feedback. In this scenario, they note that the SME made some mistakes that should be avoided in a real inspection:

- *Providing extra information* – In this case, the SME provided extra information that was not required to answer the question. When asked about the procedure, the SME said, "We changed it because of an internal audit observation." The inspector didn't ask if the procedure had been updated, only if there was a procedure. The SME also stated that corrective actions were overdue. And then the SME included the details of that internal audit finding, which the FDA is not entitled to see.

- *Offering documents the inspector is not entitled to receive* – The SME agreed to provide an internal audit report. The FDA is not entitled to see internal audit reports, although inspectors can request the resulting CAPA report. This mistake opened up issues for the inspector to probe even further. In fact, this led the inspector to ask what the observation was when really none of this information should have been entered into the inspection.

The SME again offered up extra information when she stated that the actions were overdue. The inspector had asked only what the actions were, not whether or not they had been completed on time.

2.47.12.3 Recommended Follow-Up

The external expert recommends the site develop a procedure that outlines the company policy for providing documentation to regulatory agencies. The procedure would specify items that the company would not normally provide in an inspection, such as internal audits, pictures, and video recordings, and personnel or financial information. By outlining this information in a policy, it's easier to prepare your SMEs on how to answer if an inspector asks for these items. This also gives SMEs some backup if they have to deny an inspector's request.

So it's helpful to have such a policy and be sure everyone is trained on it. When issues arise, you can manage that risk and maintain a good, positive relationship with the inspector, which is very important to success.

2.47.13 Audit Readiness (Being Audited)

FDA inspectors are knocking at your door: Are you ready? Your first thoughts may be of logistics – meeting space, document availability, condition of your facility – but what about personnel? Are your employees prepared to face FDA inspectors who are trained to dig out the kind of information you may not want them to have?

The employees who know the most about your front-line operations are usually the ones that have the least experience with inspections and therefore are more likely to slip up.

That's why training SMEs deserves at least as much attention in your inspection readiness plan as more tangible aspects like documentation and equipment function. It is important to know how FDA inspectors get started, as well as understand which systems, procedures, and processes to review. What techniques do they use, what do they hope to gain from the interviewee, how do they dig deep and capture the details, how do they prepare for inspections, what might make them always curious, and how are they motivated?

On the first day on the job, FDA inspectors take an oath of allegiance to the US government and swear to uphold the Federal Food, Drug and Cosmetics Act. Right from the beginning, they are officers of the government, just like an FBI agent or a US marshal, and they take their job just as seriously.

An inspector's goal is to find violations, and they are good at it. They are highly educated and trained not only in the tenets of good manufacturing practice but also in techniques for eliciting information from individuals. And their advancement within the agency is tied to their results.

Putting an unprepared SME in the same room with these professional interrogators is a clear invitation to disaster. No matter how smart, hardworking, and educated your employees are, and how robust your processes are, FDA investigators will always find an open door and they can break through. FDA investigators can draw inferences from what subject matter experts say and don't say.

The best way to guarantee those conclusions reflect well on your business is to prepare your staff for their moment in the spotlight and make sure they practice, practice, practice.

The following sections offer guidance on developing an SME training plan, including selecting the right kind of people to participate, assessing vulnerabilities and strengths, understanding common FDA interrogation techniques and testing readiness with simulated inspection interviews.

2.47.14 Preparation: Selecting and Training SMEs

Surviving an FDA inspection is about managing risk: understanding your weaknesses, anticipating how inspectors perceive them, and preparing your staff to respond. The human factor may present the most significant risk. But beware, you may not be able to fully control what your employees say or not at FDA inspections, but constant practice will reduce the risks.

- *Evaluating your SMEs* – The first step is to decide who will fare best in an inspection and who presents too much of a risk to put in front of inspectors. Several factors should be considered in this analysis.
- *Competency in answering an inspector's questions* – Subject matter experts must be prepared for questions to be asked in designated risk areas. They also need to be ready to provide clear explanations of your system, your procedures and your associated records.
- *Performance under stress* – Consider how individual SMEs react to stress. Will they keep calm or will they panic? Inspectors always ask for challenging clarifications and questions, but do your staff skip such questions or target senior management as the first to blame? I think it is necessary to remind this, and I recommend that employees with such an approach be kept out of sight during the inspection period.
- *Demeanor and attitude* – Think about the kind of impression an individual may make on inspectors. Does he or she appear professional, appropriately dressed, and confident? Is his or her attitude one of openness and cooperation or hostility and fear? Can the SME make eye contact and speak clearly and with authority?
- *Adaptability* – Can the SME handle unexpected requests and changes in tone or direction?
- *New personnel* – The key personnel, who are the subject experts you will determine, may have just started to work and may need to work for a while with help. It will take time for them to learn how to work with FDA inspections and how to talk to investigators. At the end of this period, you will start the training process of your subject experts who reinforce the subject.

The training of subject matter experts should not be included in any inspection preparations and should be planned much earlier. In this way, you will avoid the risks of last-minute trainings by using the time more efficiently and without haste. The topics of the training should focus on basic GMP requirements, SOPs, and FDA regulations, and should not include unnecessary information. The critical issues that participants need to learn will explain how the general course of inspections should go, how to avoid disconnection with inspectors, the types of questions that can be asked, and how to organize answers.

2.47.15 Practical Tips for FDA Inspection of a QC Laboratory

You should know for what purpose the US FDA health authority conducts inspections. The FDA conducts inspections to determine whether a pharmaceutical or biotechnology company, whether in the country or abroad, is producing in accordance with GMP regulations and FDA guidelines and directives.

Usually, the inspectors stay abroad for three weeks and perform inspections at three to five different organizations.

2.47.15.1 Materials, Reagents, Reference Standards

- Delivery notes of materials received shall be provided on request.
- Goods receipts are documented (paper and/or electronic version).
- All reference substances are tested for validity, certificates are complete, and invalid substances are removed.
- Material safety data sheets (MSDSs) are reachable for all references.
- Mixed solutions, reagents, and mobile phases are documented; all used test equipment, such as balances, pipettes, refrigerators, and freezers are identified and calibrated.
- Reagents, solvents, and mobile phases are controlled regarding expiration date and replaced if necessary, including laboratory water and other solvents.
- Samples, reagents, solvents, and mobile phases on laboratory benches and shelves are labeled clearly and per (legal) requirements.
- For example, samples and specimens in refrigerators and freezers are checked for the expiration date and clear labeling.
- Labels on solvent and reagent flasks are as specified, not manually overwritten.
- The US Pharmacopoeia (USP) reference standards must be used, even if there are other alternatives.

2.47.15.2 Facilities and Equipment

- Access to the building and other areas is enabled.
- A laboratory coat is worn; safety glasses are carried in the jacket pocket and worn where required.
- Visitor coats and glasses are kept ready for the inspectors.
- Hygiene requirements (e.g., cleanroom locks and disinfection) also apply to the inspectors.
- All equipment binders/logbooks are checked for good condition and maintained up to date.
- A validation master plan that contains concepts of qualification, requalification, risk assessment, validation, maintenance, periodic inspections, and system retirement is kept.
- All laboratory equipment is included in an inventory list with at least the following attributes: device ID, location, the person responsible for the equipment, purpose of use, and reference to SOPs.
- According to USP <1058>, lab equipment is categorizedinto three classes (A, B, C).
- The qualification status of the relevant equipment is checked.
- The relevant equipment's calibration and maintenance status (SOP) is checked, including refrigerators, freezers, drying ovens, incubators, balances, dilutors, pipettes, and so on.
- For computerized systems, risk-based validations are available that cover access controls (data security), audit trails, viewing the data formats, recovery, handling in case of failure, backup and archiving, and data integrity.
- Calibration documents of reference equipment are maintained.
- The person responsible for the equipment can present clear audit trail functions and records for a past time period on the screen.
- The screen lock is active, or a password-protected screen is set automatically after a defined time (e.g., company logo as a screensaver).
- Equipment that is expired or no longer in use is labeled as "Rejected" or qualified before the inspection.
- It should be possible to retrace and explain recent and former changes (change control).
- The persons responsible for equipment are trained to present critical laboratory equipment (e.g., HPLC, balances, pipettes, pH meters, refrigerators).
- Wastepaper baskets are emptied to avoid discussions about raw data disposal.

The persons responsible for equipment have, in this part of the inspection, the important role of demonstrating to the inspectors that all significant aspects of GMP are under control. The persons responsible for equipment have to fulfill the following essential tasks:

1. Documented briefing and training by the supplier ("in-house expert")
2. Issue of an SOP for operation, function testing, calibration, maintenance, and cleaning of the equipment
3. Issuing and maintenance of an equipment logbook to document relevant events during the life cycle
4. Responsibility for periodic functional tests and calibrations
5. In-house training and documentation of additional equipment users
6. Updating of qualification documentation after modifications and repairs
7. Definition and training of a representative
8. Closure of equipment in the case of a violation of acceptance limits
9. Ordering and stocking of important consumables and spare parts
10. Presentation of the equipment during audits and inspections

In practice, it is essential that the responsible persons can perform these topics using work instructions and that the operation is compliant with established procedures (SOPs). It is common for FDA inspectors to interview two persons about the same issue (investigative interview) and observe actual laboratory operations on-site.

2.47.15.3 Laboratory Controls

- Specifications for the tests to be performed are recorded in the method of the device used to measure and can be easily distinguished by an ID.
- Instrument methods must be identical to paper-based versions.
- The current test procedure is located near the workplace.
- All samples in the autosampler of the HPLC are listed precisely in the order of the analysis sequence and include system suitability tests (SSTs) and calibration samples.
- No longer needed (elapsed) samples in the autosampler are removed.
- The validation documents for the relevant methods are available in written form, are signed, and comprise a validation plan and report.
- The corresponding raw data undergo a meticulous review, during the preparation phase.
- Critical elements of validation should be documented by raw data, for example, robustness, the mobile phase composition, the shelf life of reference samples, and the column temperature.
- If the method indicates stability, this should be demonstrated by stress-test data, measured during method development.
- The essential data for complete traceability (who, when, what, how, why, where) are documented on the printouts of the analytical results.
- Work instructions for all aspects of stability testing are available (e.g., planning, logistics, implementation, evaluation, and qualification of stability storage cabinets, and respective storage rooms).
- Written stability testing protocols are available for all relevant tests. They are dated and signed.
- If manual integrations are performed in automated data analyses, they have to be conducted according to a current procedure. If only manual integrations will be applied, this has to be described using illustrated and comprehensive examples.
- Ensure that a second person reviews all essential calculations to ensure accuracy, completeness, and compliance with the applicable regulations and that this procedure is regulated and documented.

2.47.15.4 Personnel

- A training plan for the employees is available and is updated.
- Each employee should review his/her training records, regarding, for example, vocational education and training, GMP training, and update as appropriate.
- Job descriptions for all key personnel are in an updated form and reflect the actual responsibilities of each employee.
- For groups of employees, function-related collective job descriptions may be defined.
- Annual GMP refresher training is conducted.
- Specific training is available for all testing activities (test procedures, equipment training).
- Training of employees who were absent for a longer time, for example, maternity leave, have to be refreshed upon comeback.
- Each employee catches up on their current job profile before the inspection and can also reflect on it.

2.47.15.5 Quality Management (QM)

- There is a list of all SOPs.
- The SOPs governing the activities of the QM department are updated, valid, and trained.
- All SOPs on the topics "change control" (Change Control), "deviations," "results out of specification" (OOS), "corrective and preventive actions" (CAPAs), as well as "training" are updated, valid, and trained.
- Self-inspections are planned and conducted as scheduled.
- Supplier inspections are planned and conducted according to plan.
- Management reviews are performed periodically and are documented.
- The job descriptions of all QM staff are available and updated.
- Lists of complaints, results OOS, deviations, and changes for the product in the inspection focus are available, covering the past 24 months.
- For deviations from the past that have not been satisfactorily addressed, review documents with conclusive explanations and consequences available.

2.47.15.6 Documents and Records

- All SOPs, specifications, manuals, and instructions are available in the specified binder in the laboratory; check with table of contents, and that all documents are up to date.
- Documents related to deviations, changes, OOS results, complaints, and CAPAs are stored and managed centrally by the QM.
- All laboratory notebooks and worksheets are checked by a second employee with initials and date (accuracy, completeness, compliance), according to 21 CFR 211.194 (Lab Records).
- No self-adhesive labels (Post-its) are used in the laboratory, even for private information.
- Blotting pads are blank, no calculations on calendar sheets.
- Wastepaper baskets are emptied; no raw data, chromatograms, or reports (even drafts) are in the bins.
- No weighing printouts lying around on the balance printer or nearby.
- All data storage media such as floppy disks, CDs, DVDs, and memory cards are placed inside drawers, because usually they are not labeled sufficiently.
- Folder labels are uniform and comply with the regulations (e.g., colors).
- Older versions of documents are removed from the laboratory.
- No handwritten entries in specification documents (SOPs or similar).
- Handwritten entries, for example, in verification documents only with indelible pens (DIN ISO 12757-2).
- No use of thermal paper and inkjet printing for archival purposes.

- An archive is established and meets all the criteria of protection against fire, flood, and other catastrophic data loss.

2.47.15.7 Company Presentation

- Overview of the company's name, affiliated group, headquarters, and location(s).
- Overview of the economic indicators, number of employees, and so on.
- Overview of the main products and markets, related to the US market.
- Overview of the product focused by the current inspection, within formation on history, processes, techniques, locations, subprocesses, key suppliers, or subcontractors, for example, sterilization process.
- Organizational chart of the group, the location, and the departments that are focused on by the FDA, for example, QM, R&D, production, packaging, and quality control laboratory.
- Location plan; inspection of relevant buildings should be marked.
- Material, personnel, airlocks, and lay-out of essential facilities such as production, packaging, storage, and quality control laboratory.
- Presentation of the audit core team (persons in the audit room are direct contact persons during the inspection).
- Presentation of the quality systems established on-site (e.g.,ISO17025,cGMP 21 CFR 211), with a short historical overview.
- Information on audits of authorities (brief history and current inspection status, accreditations, and certifications).
- Information about any previous FDA inspection (on-site).
- Information on the integration of QM into the value chain (opportunity to present the QM as an integral part of the company processes).
- A brief overview of the documentation systems of the company, for example, the QM manual, process instruction (PI) sheets, test procedures and associated protocols, and operating procedures (SOPs).

Immediately after the company presentation, the inspectors should be invited to ask questions.

2.47.16 During the Inspection

The "audit room" is the room in which the inspectors and the core team stay during the inspection and where the documents and records are reviewed by the inspectors. This is also where the opening meeting and the closeout meeting take place. The management or management representatives must be present both at the beginning and at the end of the inspection.

The following requirements apply to the audit room:

- Room with good indoor climate (air-conditioning in summer, heating in winter, natural light if possible and windows)
- Sufficiently large workspace on the tables, so that documents can be well spread
- A sufficient number of seats for the participants
- Ideally, for the opening and the closeout of the audit, a meeting room for more participants should be provided.

Who should be in the audit room?

- Inspectors (typically, one to two persons)
- Management (only for the welcome and farewell of inspectors)
- Head of quality management as the main contact person
- Representatives of the regulatory affairs department (optional)

- Note-taker and moderator, usually one person
- Interpreter(s), if this option is selected
- Two vacant seats for persons who have to present results or documents (e.g., manager and employee).

Lists in the audit room:

- Audit room protocol (paper or better: intranet-based, online connected with back office)
- List of participants (registration of all participants in order of occurrence after the beginning of the inspection, including name, position, and topics discussed)
- Audit room tracking list (document for tracking the documents that come from the back office into the audit room, including comments on the status of the documents and whether the documents have been reviewed or copied and taken by the FDA inspectors)

The "back office" is the communication center in the background of an FDA inspection. At this point, the information requested by the inspectors performing the inspection and all the information regarding the inspection needs is collected. Since the inspectors have a high probability of requesting all kinds of information, such as documents, records, images, and drawings, all of this information is brought together in the back office and quickly presented to the attention of the inspectors if requested. Since their knowledge and experience will be benefited from in the back office, subject matter experts will also wait in the back office to be called. These activities should be recorded on the back office follow-up list by the note-taker.

The following requirements apply to the back office:

- Room situated within walking distance of the audit room (not more than one to two minutes away)
- Sufficiently large work surfaces on tables, so that documents can be well spread
- A sufficient number of seats for the team and persons waiting to be called to the audit room
- Space for the binders needed for the audit

The technical equipment should enable network connections to access important documents (e.g., SOPs, process instructions, and manufacturing and test documents), records, journals, guidelines, or pharmacopoeias (e.g., USP). Internal roles and responsibilities as well as contact persons should be defined individually for each FDA inspection, as this may change due to the inspected issues. The back office administrates the "document tracking list," which contains the status of all requested documents.

2.47.16.1 Behavior in Inspections

It is expected that some personal characteristics will be at the forefront of people who will meet face-to-face with FDA inspectors. These people should have balanced moods, the ability to discuss without getting excited, the ability to think logically, and the ability to resist stress are among other features, at least from experts in the subject.

For the dress code, there is only one rule: You should feel comfortable in your clothes. Not too formal, not too casual, but comfortable and customary clothes that do not overly exceed the level of the inspectors' clothes. If you generally do not wear a necktie, you should not wear it in the inspection. Women should avoid shoes with high heels and use decorative cosmetics, accessories, and jewelry with restraint.

Your first impression of the FDA inspectors is very important, and if there is not a very tense environment at the beginning of the inspection and the inspectors do not act very formally, the inspectors are addressed by their names, but perceiving this as an expression of sincerity, one should not try to touch the soft belly of the inspector from this point. Although inspectors speak in a relaxed and friendly manner, it should be kept in mind that inspection is a very formal activity.

2.47.16.2 Lab Walkthrough

As already mentioned, a laboratory walkthrough is usually conducted immediately after the opening meeting. FDA investigators may plan the route on the facility tour with you and may want to follow your suggested route. This point should not be hung up on too much because the main task of the inspectors is the main buildings and critical processes, and they are well trained in this regard.

If there are two inspectors, it must be expected that the group splits, which should be considered an option when planning the escort of the inspectors. Some inspectors inspect according to the material flow and track the full path of the testing procedure, but there are also some who take a seemingly arbitrary way and change from one topic to the next during their lab walkthrough. The employees in the different areas of the company should continue to work normally, as if it were an ordinary working day, even if the FDA inspectors approach. Only on request, people should speak with the inspectors, otherwise they should continue to work routinely. Table 2.32 tests typical behavioral patterns in an FDA audit.

2.48 QUESTIONS TO AUDITORS

- Read the 20 questions in Table 2.32 carefully and tick the questions that should or should not be said in FDA audits.
- You are auditing a pharmaceutical company for compliance with GMP. On arrival, you are met by one of the company's vice presidents. The first thing they want to do is introduce the company to you in detail but you think that this will waste your time. You also learn

TABLE 2.32
Don'ts versus Dos

Behavior	Don't	Do
1. Answer all questions honestly		
2. Say "I don't know" or "I'll get the answer for you"		
3. Avoid such phrases as "I think," "Sometimes/often/usually," "never," and "next time"		
4. Avoid qualifiers, such as "typically," "normally," "generally," and "usually"		
5. Stop speaking once the question is answered		
6. Repeat every question that is asked		
7. Ask for explanations or interpretations of what you do not understand		
8. Control your temper; remain courteous and professional		
9. Maintain eye contact		
10. Volunteer information		
11. Be sarcastic		
12. Guess answer		
13. Attempt to answer what-if or hypothetical questions		
14. Argue or disagree with any inspector statement		
15. Philosophize, ramble, or editorialize		
16. Point out deficiencies or errors		
17. Apologize for problems or comments made by inspector		
18. Feel the need to respond to every comment made		
19. Become defensive or evasive		
20. Enter into casual conversation during the audit		

that they had made reservations for lunch at a nearby restaurant and that there will be no key people of the company available.

a. What do you think is wrong with this facility?
b. How could you have avoided this situation?

Use examples from your own experience to illustrate how you can handle situations when the manufacturer wants to prevent you from doing a thorough audit.

- During an internal audit, you are interviewing a machine operator to determine if he knows the procedures, and if he follows them. The operator's answers indicate that he does in fact know and follow all the procedures that apply to him. Just as you are starting to thank him for his participation, he responds about an exception to the operational procedures. There is one part that is handled differently from all others. The setup is entirely different, as the actual operation and the inspection criteria are not the same. It requires a special form, which he shows you a copy of. A review of the procedures indicates that this particular product is not governed by any procedure. When questioned about this, the production manager responds that this part is rarely run, and the specifications are ever changing. Procedures would tie the department down and remove the required flexibility.
What should you do?
- On DD/MM/YYYY, XXXXX Pharmaceuticals received a warning letter because of conditions at its XXXXX plant. The letter, dated MM. XX, followed an inspection at the facility initiated in MM YYYY. The most important part of the warning letter was the quality control lab's inability to investigate incidents at the facility, contrary to the company's approved standard operating procedures (SOP). The FDA considered some of the observations to be repeat observations. A finding by the FDA states: "The persons performing and double-checking the cleaning and maintenance are not dating and signing or initialing the equipment cleaning and use log." The company responded by saying: "Company SOP ___ requires that all personnel performing the cleaning and inspection of equipment sign the record concurrent with the operation. Personnel have been reinstructed in this procedure. All personnel during this training were reminded that appropriate actions up to and including termination will be taken for infractions or deviations from our procedures." Do you think that the response from the company contains GMP problems?
- In a recent audit, the auditor said he had to review an issue with his supervisor to decide if a Form 483 was necessary. Are FDA auditors now being required to verify 483s with their supervisors?
- Is the submission of a response within 15 days of receiving a 483 likely to decrease the probability of a warning letter? Or, does it depend on the magnitude of the 483?

BIBLIOGRAPHY

Clifford L. Nilsen, *Managing the Analytical Laboratory Plain and Simple*, Interpharm Press, 2002
David M. Bliesner, *Establishing A CGMP Laboratory Audit System*, Wiley, 2006
Dennis R. Arter, Charles A. Cianfrani, John E. West, *How to Audit the Process-Based QMS*, Second Edition, Quality Press, 2013
FDA Inspection Readiness: A Guide to Preparing Subject Matter Experts. FDANEWS.
Jeanne Moldenhauer, *System-Based Inspections for Pharmaceutical Manufacturers Regulatory Basis for FDA Inspections*, Parenteral Drug Association, 2006
J.P. Russell, *Terry Regel After the Quality Audit Closing the Loop on the Audit Process ASQ*, 1997
J.P. Russell, *The Internal Auditing Pocket Guide Preparing, Performing, Reporting, and Follow-Up*, Second Edition, ASQ Quality Press, 2007
J.P. Russell, *The Quality Audit Handbook*, Second Edition, ASQ Quality Audit Division, 2000

P.A. Carson, N. Dent, *Good Clinical, Laboratory and Manufacturing Practices: Techniques for the QA Professional*, RSC Publishing, 2007

Sandy Weinberg, *Cost-Contained Regulatory Compliance: For the Pharmaceutical, Biologics, and Medical Device Industries*, Wiley, 2011

Stavros Kromidas, *The HPLC Expert, Possibilities and Limitations of Modern High Performance Liquid Chromatography*, Wiley-VCH, 2016 Guideline for good clinical practice E6(R2)

3 GMP Auditing in Various Areas

 DOI: 10.1201/9781003284970-3

3.1 AUDITING IN MANUFACTURING ENVIRONMENTS

Whether it is a sterile or nonsterile production facility, a supplier, or a contract manufacturing facility, I would like to share with you how I prepare as an auditor. Good manufacturing practice (GMP) guides and directives can be downloaded from web pages that are easily accessible. When it comes to active pharmaceutical ingredient (API), excipient, or packaging material audits, the sources I will review include:

- EU, PIC/S GMP and GDP guides and annexes
- Active Pharmaceutical Ingredient Committee (APIC) guides and supplementary documents
- International Pharmaceutical Excipients Council (IPEC) guidelines and supplementary documents
- EXCiPACT guides
- PS 9000
- ISO 14644
- Pharmacopoeias
- World Health Organization (WHO) guides and supplements
- FDA 210, 211
- International Council of Harmonisation (ICH)

I also look at documents of nonprofit organizations such as the Parenteral Drug Association (PDA) and International Society for Pharmaceutical Engineering (ISPE).

For example, let's say I'm auditing an oral solid dosage facility.

I start with the documents at hand.

I visualize the audit flow and the production processes of these drugs in my head. I try to adapt the clauses of EU, PIC/S GMP guides to that area, especially at every step or in every area I visit. So, how?

Although what I will convey are short explanations, it is aimed to provide the reader with a new audit perspective.

This GMP consists of nine items and annexes. I'm starting the audit at the warehouse and I'm starting to question these nine clauses and supplements of the GMP in this area.

1. *Pharmaceutical quality system.*

How do the Pharmaceutical Quality Management systems work in the warehouse, how are the change control, deviations and risks, which are the important components of the QMS, handled? Are the indispensable documents of the quality management system available in this area? Starting from here, I'm digging deeper.

2. *Personnel.*

Under the title of personnel, I question how many people work in the warehouse and how many of these personnel work temporarily or seasonally. I always evaluate the lists of the trainings personnel has received or planned, the results, and the effectiveness of the trainings so that personnel can do it correctly and completely. I control whether the health checks of the personnel are carried out regularly. When the personnel is absent from the facility for reasons such as leave or illness, after questioning who is acting as the delegate, I look at the job descriptions of the one or two key personnel I have chosen and try to ask level 5 questions with those people as explained in the following sections.

3. *Premise and equipment.*

Does the warehouse have enough space to carry out the activities expected from it in terms of premises and equipment? I compare the layout with the physical appearance and ask the relevant questions to the responsible persons. I inquire about the entrance and exit to the facilities; the security measures; the equipment used in the warehouse; and the calibration, qualification, and validation status of the equipment, if there is software connected to the equipment.

4. *Documentation.*

When I come to documentation, I access the master list of standard operating procedures (SOPs) used in the warehouse areas and verify that there are valid documents at each station. I would like to see proof documents that the personnel's understanding of documentation and data integrity is not just about correcting and initialing typographical errors. I'm questioning who has access to what level of documents.

5. *Production (operations).*

The next item is about the production part of GMP, but I treat it as warehouse operations and I haven't had any problems so far.

6. *Quality control.*

Clause 6 of the GMP is quality control and I am trying to apply this clause to warehouse operations. For example, there is no one who does not know that sampling operations are usually carried out in warehouse areas, so I immediately go to the sampling area and ask questions about the activities there, and I interview the working personnel face-to-face with the help of a supplementary document such as an SOP or instructions.

7. *Outsourced activities.*

First, I question which activities are outsourced in the warehouse, for example, cleaning and pest control. When the company does not have sufficient transportation vehicles, I question how the transportation companies are selected, qualified, inspected, and how they are included in the approved supplier list. I try to consolidate my findings by reading the source documents and interviewing personnel.

I leave it up to you to adapt the annexes related to clauses 8 and 9, or GDP guides and annexes.

This assessment is an approach that can be applied throughout the book and will give you many different perspectives.

3.1.1 Plant Tour

One of the first significant and critical activities conducted in an audit is the plant tour. The plant tour usually occurs after the presentation of credentials, and introductions of site and audit personnel. A short opening meeting describing the specific objectives of the audit (e.g., GMP, pre-approval inspection, complaint investigation), scheduling, and so on then follows. The plant tour provides a broad overview of the site operations, both in technical manufacturing and regulatory compliance. The opening presentation associated with the tour can be used to help "set the stage" for the audit team. In addition to preparing them for what they will see on the tour, it can also be used to familiarize the audit team with the company, the specific manufacturing site, and organizational structure.

There is usually a time and a place that one can pinpoint the exact moment the regulatory inspection takes a turn for the worse. At any cost, this must not happen on the opening plant tour. The entire tone of the inspection process is greatly influenced by what the inspector sees while on tour. An experienced auditor will make important judgments about the site based on the tour experience. If the plant tour is done well, these judgments will be positive. The plant tour is critical and must be considered to be of utmost importance. Planning, evaluation, and practice of the plant tour should be conducted long before a tour is needed in an actual audit.

If bad things happen on the plant tour on day one, what can be expected for the rest of the audit? The plant tour is definitely not a perfunctory or superficial exercise.

3.1.2 The Plant Tour Represents the Entire Plant

This discussion focuses on the plant tour. It describes potential problems and an approach to assuring a successful plant tour. However, do not infer that areas in the plant tour are special or should be different from nontour areas in the plant.

The "audit readiness" mantra is typically lost on the specific tour route. This approach thoroughly cleans the corridors and rooms on the tour route, making sure all of the "sins" are sufficiently hidden away. Only the best and brightest employees will be working when the tour is conducted. The tour route is freshly painted and literally shines. But what happens when the auditor wants to go into an area that is not on the planned tour route? Or when the auditor looks in a closet off the tour route?

The plant tour should demonstrate the function and compliance strategy in the entire plant. The plant must have a culture of audit readiness in all areas at all times. The following are two axioms that are particularly applicable to the entire pharmaceutical or medical device plant:

- "A place for everything and everything in its place."
- "Cleanliness is next to godliness."

These axioms must be demonstrated not only in the plant tour but also in all areas of the plant. Do not neglect the readiness of the entire plant – every day and all the time. The plant tour represents the entire plant.

3.1.3 What Can Go Wrong?

There are many often overlooked items on the plant tour that can negatively affect the entire audit. Even when everything is thought to be in order and generally clean, there are items taken for granted or assumed to be of no consequence to auditors. Experience has shown that seemingly trivial items can lead to serious audit consequences and observations. A sampling of these with unintended consequences suggested by several experienced auditors follows.

3.1.4 Ceiling Tiles and Air Ducts

The floor may be clean and the walls painted, but have you looked up at the ceiling tiles and the entry ducts? Fresh stains look different than old stains, so stain appearance can tell a story – for sure, that housekeeping is not doing its job and, perhaps, that the HVAC filters are leaking or are behind their scheduled replacement. Stains from leaky pipes are an invitation for piping drawing reviews. Any incident that causes a review of site drawings is high risk. The drawing may not be current or not consistent with recent facility changes. The site engineering department may be well behind, even years behind, in updating drawings. How long has it been since piping has been "walked down." What about dead legs in those systems?

3.1.5 Surfaces

All horizontal surfaces have been cleaned, so we should be in good shape, right? International regulators are known to actually perform the dreaded white-glove test on the top edge of doors and easily accessed equipment. Also, what about the undersides of horizontal surfaces, such as in the drug dispensing area? What is the dust on the walls in the drug dispensing area? What about the rust on the bottoms of the stainless steel chairs in the aseptic area?

3.1.6 Pest Control: Flying Insects

Pest control is always a hot topic. Making sure that any bug zappers that kill flying insects have operating bulbs is always a plus. Burned-out bulbs create a terrible impression of the organization and of all those working in an area. Why didn't workers in the area report the outage? If they did, why didn't they follow-up on the correction? How long have the bulbs been burned out?

3.1.7 Pest Control: Rodents

If there are rodent traps in warehouses and along external walls, make sure they are routinely inspected. Who is supposed to do this and at what frequency? A trap full of dead mice is not a pretty sight for the inspector.

3.1.8 Aseptic Operations

If there is an aseptic area, make sure there is no operator who is chasing down a fly because, oh wait, the bug zapper wasn't working. Do you really want to have to show the inspector the last two years of environmental data because he saw the operator squash that fly? Housekeeping procedures and records requests can be avoided if the plant presents itself well.

3.1.9 Training

Whoever is working in an area is vulnerable to having their training records for what they are doing requested by the auditor. It is not easy trying to explain why an operator is not trained in preparing the cleaning solutions that he has been using for the last three years. Supervisor training records may be a problem. Supervisors sign off on operator performance, but often supervisor training is delinquent – not a good example to operators and an obvious embarrassment for the organization. The operators are trained, but the supervisors are delinquent.

3.1.10 Crowded and Sloppy Rooms

Avoid excessively crowded conditions (i.e., equipment, materials, and other miscellaneous supplies). When hiding the "sins" was previously mentioned, it did not suggest shoving them all in a closet hoping they are never to be seen. There should not be areas in the plant akin to your home attic, basement, or garage. Cleanliness in the plant is not like asking your child to clean their room, only to eventually find what they did with the mess.

3.1.11 Wood Pallets

Be sure wood pallets are all heat treated (Pfizer and J&J learned about this). Better yet, use only plastic pallets in areas where products and packages are exposed within the manufacturing area.

3.1.12 Cardboard

No cardboard boxes should be in the production area near open products or where commodities are prepared. If this is observed by the auditor, the nonviable particulate data for the last two years may be requested.

3.1.13 Status Labeling

This may sound easy. If an inspector ever has to ask what the condition of the product tanks is, you have failed miserably. You show them the equipment using a log. Why isn't the equipment use log bound? How are the pages tracked? No quality assurance review? No review for the last two weeks? A simple status label avoids these questions because they don't ask to see the logs. There is nothing worse than an inspector picking up a bottle of excipient and finding that the material expired three days ago. SOPs in force on how materials are controlled will be requested. Inventory logs will also be reviewed.

3.1.14 Paper Systems

If your site uses paper copies of procedures in various locations, all copies of procedures at the various locations must be current. This is very difficult to control when the facility has multiple locations with hardcopy procedures, and even more difficult for a large facility or for multiple locations. Electronic systems enable procedures to be immediately updated and always be current.

3.1.15 Access to Procedures

Your operators are all trained in their job functions. An inspector asks an operator what would seem to be a simple question. The operator does what he was trained to do. He replies that he is not sure and will have to check the SOP. Just where is the SOP? It is down two flights of stairs in the department office. Do you really think the operators will make the trip to the office to consult procedures when questions arise during processing?

3.1.16 Equipment Identification

Equipment must be clearly identified. Equipment identification numbers must not be painted over so that traceability is lost. Equipment identification is necessary for linkage to qualification documentation. If equipment qualification cannot be verified by means of documentation, you cannot use the equipment.

3.1.17 Equipment Calibration

Equipment calibration status tags are always checked in audits. There must be no equipment with tags that indicate calibration is past due. Further, calibration status tags must be readable. Ink must not be faded so that calibration status cannot be determined. It is reasonable to conclude that equipment with tags that are so badly faded and unreadable is probably out of calibration.

3.1.18 Unusable Equipment

Equipment that is not working, awaiting repair, or otherwise unusable should be tagged "out of service." If portable equipment, it should be removed from the work area. If it is not clearly tagged, how can you be sure it will not be used?

3.1.19 "Band-Aids" on Equipment

Operators do what is needed to keep equipment running. This includes taping parts in place, catching dripping oil in a bucket, shimming wobbly equipment, and so on. When these "Band-Aids" work, they become acceptable parts of the equipment – they become invisible to those working in the area. However, when fresh eyes (i.e., the auditors) look at equipment, they easily see the Band-Aids, which, of course, is an obvious failure to properly maintain equipment.

3.1.20 Hoses

Hoses, tubing, and transfer lines should be hanging to facilitate drainage. Often it is the last thing to be done when the room is cleaned. But sometimes it is not done. One auditor reported that hoses were stored on the floor in a southwestern US manufacturing plant. A snake crawled into the hose. When the solution was added through the hose, the snake was expelled into the product solution. The operator was bitten by the snake and was sick for several days. This incident would have been interesting reading for an auditor.

3.1.21 Site Intercom Systems

Be sure to turn off or, even better, disable any intercom system that may be used. You would not believe what can be inadvertently broadcast throughout the plant by an unaware employee while the auditor tour is in progress. Site intercom systems must not be a vehicle for personal humor.

3.1.22 Doing it Right

Another two axioms are very applicable to the plant tour:

- "You never get a second chance to make a first impression."
- "You pass or fail an inspection in the first 30 minutes."

As mentioned earlier, the plant tour usually occurs on the first day of the audit. How do you think the rest of the audit is going to go if the plant tour on day one is poorly conducted or demonstrates multiple problems?

The plant tour is a critical activity. Like any critical activity, planning, evaluation, and practice are necessary to ensure success. These activities should be done far in advance of an actual audit. The site should be well prepared for an audit. The plant tour being conducted with the auditors must not be the first time it has ever been done.

3.1.23 Planning the Tour

The route of the plant tour should be carefully chosen. The route must not be a haphazard walk-through of the hallways of the plant. The tour should be helpful in understanding the plant activities. In a typical opening presentation, one of the first questions asked by an inspector is about the process flow. In keeping with the planned agenda, the plant tour should start at the beginning of the process flow to best demonstrate the plant's function. Start with the dock where incoming materials are received. Thereafter, the passage of materials to the quarantine storage area can be toured. The drug dispensing area can be shown at this time. An inspector will most likely want to see any reject or quality-hold areas during this part of the tour. Be ready to show these often overlooked areas. Thereafter, the tour should demonstrate all other major plant functions. The tour ends with the approved product being stored in an area awaiting shipping.

The tour should demonstrate that "form follows function" (i.e., the plant design and structure are supportive of the logical process flow of activities in the plant).

3.1.24 Evaluating the Tour Plan

After designing the planned tour, the route should be carefully evaluated. Evaluation should be done by a team of people with different attitudes and perspectives. High-level evaluation by management is not usually sufficient. High-level people don't pay attention to details. Personnel who are known for attention to cleanliness, detail, fastidiousness, and other applicable characteristics are valuable participants in an evaluation. They are able to see things that others do not see, for example:

- Do hanging mops smell musty?
- Are the undersides of tables dirty with formulation dust?
- Are the tops of doors clean?
- Are corners of rooms – floor and ceiling – clean?
- Are coveralls, lab coats, booties, and other protective clothing available for auditors to enter a process room?
- What training is needed for proper gowning and entry?
- Areas that do not show well (yes, they do exist) should be avoided if not absolutely necessary.
- Make sure all stairwells, elevators, etc. are uncluttered, clean, and safe.

3.1.25 Tour Narrative

After the tour has been finalized, a tour narrative of key points should be written. This will be spoken by the tour leader during the tour with the auditors.

3.1.26 Tour Presentation

A tour presentation (PowerPoint or equivalent) should be prepared to introduce the audit team to the organization and site. Although usually presented immediately prior to the tour, it can be provided to the auditor before visiting the site in cases when advance notice of the audit is given by the agency.

3.1.27 Practicing the Tour

After the plan has been evaluated and problems mitigated, practice runs of the tour route should be conducted. The tour leader is the key person on the tour. Actual walking tours should be practiced with the tour leader and associated personnel as often as needed so that there are no unknowns or anxiety associated with the actual tour. The narrative of the tour should be practiced. Sitting in the conference room is not the way to practice the tour. The objective of multiple practice runs is to ensure an excellent tour when auditors are in the plant problem-free and without anxiety for the tour leader and participants.

3.1.28 Tour Leader

The selection of an appropriate tour leader is important. This person should speak the native language of the auditor if possible, have good communication skills, speak clearly, be able to think on their feet, and be knowledgeable of all plant operations. The plant manager, who may be a financial person with expertise in finance but minimal GMP knowledge, may not be suitable to lead the plant tour. Designated people may be selected to lead certain areas to provide the best possible tour. Technical competence is important but is not the sole criterion for leading the tour.

3.1.29 What Else?

The site should now be ready to conduct a successful tour. The following are additional considerations for the actual tour.

3.1.30 Before Starting the Tour

An overview of the tour using the aforementioned tour presentation before the actual tour is recommended. The tour leader should describe the presentation. Tell the auditor what functions the plant does, and then show that the tour will cover all relevant areas. Follow a basic rule of presentation skills. Tell the auditor what they will see, show them (the actual tour), and summarize what they have seen (after the tour is completed).

3.1.31 Communication with the Auditor

The auditor should be asked if they have a special interest in any particular area. For example, the auditor may have an interest in the prevalence of wood pallets in the plant as a result of the recent Tylenol recall. Asking the auditor about special interests enables the plant to openly and prospectively address the auditor's objectives. This also gives the plant the opportunity to assemble site experts as needed for the auditor.

This is a subject that should not be taken lightly. If you ask, the auditor may show interest in something you are not initially prepared to discuss. It is much better to ask and get the right people to answer questions than to struggle with finding information or people under time constraints or with the auditor waiting for an answer.

3.1.32 Clothing Requirements

Many processing areas have particular clothing requirements. You should make special reminders about the facility to the people participating in the tour. For example, make-up, jewelry, food/beverage restrictions, and gowning requirements should be explained. Booties don't work with high heels. This topic should be discussed in the conference room before the tour starts. Don't overlook this topic. It is much more easily discussed in the conference room before the tour starts. It will be uncomfortable to refuse an auditor at the point of entry into a restricted area.

3.1.33 Restricted Areas

The tour leader must be prepared to explain why the inspector cannot go into a restricted area such as the aseptic processing area. There is extensive training in aseptic technique and gowning that must be completed to enter these areas. On the other hand, do not attempt to mislead an inspector on the amount of training just to keep him out. If an area can be seen from windows, suggest this as an alternative. If the auditor is adamant about entering a restricted area, training can be set up, protective clothing provided, etc. as needed so that the auditor can enter the area after all training and clothing requirements are met.

Site procedure must not be violated for the convenience of the auditor. If you violate a procedure to allow an auditor to enter a restricted area (for example, safety shoes are required and no high heels or sandals in manufacturing areas), you may (and should) get an observation for violating the procedure.

3.1.34 Deviating from the Tour Route

The auditor may ask to go to areas that were not part of the planned tour. For example, the auditor may ask to see cold rooms, walk through warehouse aisles, and other areas not originally planned. As long as these are not restricted areas, the tour must be modified to accommodate these requests. You should expect this. When things look too good or are too well planned, what is being hidden? All areas of the plant must be ready for inspection by the auditor at all times – not just the planned plant tour route.

3.2 CONTRACT MANUFACTURERS

As you know, there is news about the merger of small production facilities and increasing their power from time to time in the written and visual media pharmaceutical and biotechnology industry, including drug development, production, and analysis, are transferred to organizations with more opportunities in these areas, at least with sufficient capacity. The purpose and scope of the contract signed by the two parties may change occasionally. If the contract buyer party undertakes all activities, including, for example, the supply of raw materials, this service procurement is considered a full-service contract.Therefore, since the controls of the API and excipient supplier and manufacturer are completely outsourced, they may pass into the responsibility of the contract manufacturer. Sometimes, when some substeps of the main production steps are with the contract manufacturer, other activities may be with the license-holder company.

Let us consider the application of the coating process on tablets for film-coated tablets. The subcontractor delivers the pills it has produced to the manufacturer, and the manufacturer only performs the film coating process and then returns them. Therefore, before determining the scope of a service procurement or production contract, the party giving the contract must also consider any risks if the contracting party outsources the activities that must be carried out within the firm. This may include production, development, analysis, procurement, storage, and distribution.

3.2.1 Determination of the Number of Contract Manufacturers

According to the approach of the contracting party, it is important whether an agreement will be made with only one contract manufacturer or with two contract manufacturers. The main idea of the agreement signed bilaterally is the desire to easily switch to the other in case of any problem with one of the parties signing the contract

- *Example 1*: If a long-term equipment failure in a manufacturer results in the subcontractor being unable to produce for at least a few weeks, the contracting party may pass product production to the alternative manufacturer.

- *Example 2*: The contracted company may not respond immediately to its customer in case of a new market or capacity increase due to its own capacity constraints, in this case, the alternative contractor may respond more quickly and thus the supply constraint can be eliminated.
- *Example 3*: The contract manufacturer encounters regulatory problems with a government agency such as the European Medicines Agency (EMA) or US Food and Drug Administration (FDA) and has to cease operations. Having an alternative contract manufacturer can help ensure continued supply of either an API or preparation.

3.2.2 Audits

The contract giver may perform a GMP audit of the contract acceptor's premises before the contract giver accepts the contract acceptor's proposal and the parties formally sign the agreement. In this way, an opportunity may arise between the parties giving and receiving the contract in the early period and an environment where both parties can express their mutual views on the GMP capacity.

3.2.3 Approval of the Contract Manufacturer

An audit of a contract manufacturer should usually provide information on which contracted company will work with in the future.A detailed audit report is prepared after each GMP audit is completed. In this report, the company information of the audited party, the contracted manufacturer, and the GMP deficiencies observed in detail and corresponding to each defect should be referred to the national and international GMP guidelines.As the GMP-audit report is part of the firm's quality assurance system, it should also be considered that it will be questioned in official health authority inspections.It should be understood that in the USA, the FDA will generally not ask to see documentation about internal audits conducted by a firm. They will usually accept documentation of audits being performed and that follow-up actions are being performed. However, under certain circumstances, the FDA could seek to obtain copies of audits when significant legal concerns exist. Under such conditions, it is suggested that legal counsel should be consulted to discuss the local situation being encountered.

3.2.4 Scope of the Contract Assigned

Let's say that the party giving the contract accepted the contract acceptor party after some activities and transactions; in this case, the the duties and responsibilities of the parties should be determined and put into use quickly. Duties and responsibilities can be determined for the following topics, but not limited to the following:

- Who will produce the development and pilot batches?
- Who will supply the primary and secondary packaging materials?
- Who will supply the raw materials, APIs, and excipients?
- Who will approve the starting materials?
- How will the transfer of analytical methods take place?
- How will validation batches be produced?
- Which ways will be used as production technology?
- What resources will be available for packaging and labeling technology?
- How will sampling be done during both validation and routine production?
- How will physical, analytical, and microbiological testing be performed?
- How will intermediates and finished dosage forms be released?

3.2.5 Agreement Signed Between the Giver and the Acceptor of the Contract

The scope and volume of production to be ordered by the contract giver to the contract acceptor is the basis for defining mutual responsibilities. The duties and responsibilities of the contract manufacturer and the license holder giving the job should be determined as I have stated before, and should also lay the groundwork for the contract. By the way, it is useful to remind you that the relevant agreement does not only refer to the duties and responsibilities of the parties, but also directly or indirectly to the outsourcing, which is clearly stated in the GMP guidelines.

3.2.6 Contract Manufacturer Agreements for Audits

The data and deep experience obtained as a result of numerous audits and health authority inspections in the production facilities of the party accepting the contract bring along the need for updating between the parties that signed the contract. When the knowledge, skills, and experiences of the contracting acceptor, namely the contracted manufacturer, are examined in accordance with the guidelines and directives in GMP audits, changes in some activities may come to the fore. In the inspections of the health authority, the responsibilities of the companies that are party to the contract and the chronological continuity of the duties are taken into account. In addition, whether the responsible persons whose names and signatures are in the contract are still working in their companies is also among the subjects under review. Since such contracts are not updated periodically, the contracted manufacturer may be mistaken in many cases. For this reason, it will be in the interest of both parties to review the contract periodically. In the USA, contracted manufacturers are expected to be periodically inspected by the official health authority, named the FDA.

The EU GMP Guide requires that a contract giver carry out audits of a contract acceptor. This same expectation exists in the United States. According to the EU Guide, the contract giver must audit the contract acceptor's facility. When a contract is made, the giving party is expected to audit the acceptor party. This is one of the most important ways for the contract giver to understand whether the contract acceptor's knowledge, skills, infrastructure and technology are sufficient. This requirement is explicitly demanded in both EU GMP guidelines and FDA guidelines. So, how and with how many people will the audits be carried out? Audits should usually be done with two or three GMP auditors and should take a minimum audit of one to three days.

Performing GMP audits regularly ensures that the auditor and the auditee are in constant communication, for example, the contract acceptor can transfer facility and personnel changes to the other party. However, as an expert GMP auditor who has been auditing for years, I have witnessed that not all contract-giving parties fulfill this obligation. We can count four groups regarding the time period in which the audits of the party accepting the contract can be carried out.

- Group 1: Yearly
- Group 2: Two years
- Group 3: Three to five years
- Group 4: Sporadically

Organizations that are in the fourth place in the ranking, that is, those who give contracts, are those who conduct audits while the relevant health authority is expected. Such companies have an interesting structure. It is worth noting here that this type of behavior model is against the drug law. At this point, the party accepting the contract, unfortunately, has to explain the situation to the party giving the contract with a consultant manner. Therefore, there are some situations that make the audits of the party accepting the contract difficult.

As is known, many licensed companies can work with many contracted companies. One reason may be the organization's philosophy and the fact that the company has different dosage forms produced by other contract manufacturers. Therefore, the contract manufacturer accepts these audits.

As it is known, the personnel performing the GMP audits are department personnel such as production, purchasing, laboratory, and quality assurance, and these people have limited knowledge and skills in every field, and even their duties may be left unfinished during the audits.Another issue is the increase in the number of complaints and deviations related to one or more products occurring in the contract manufacturer. In these cases, it may be inevitable to carry out GMP audits again. I can even say that the fact that the contract manufacturer has not experienced any problems or complaints for a while can be a factor in deciding the frequency of GMP audits.To summarize, the situations that restrict the regular performance of GMP audits;

- Working simultaneously with more than one contract manufacturer
- Knowledge, skills, training, and capacity of auditors
- The contracted manufacturer does not cause any problems, complaints, or deviations do not occur

3.2.7 Types of Audits

As part of the preparation of an audit, the focus of the audit should be defined. Generally three audit types can be differentiated:

- System audit
- Procedure or process audit
- Product audit

While performing a system audit, the contract manufacturer's quality management system and quality assurance approach are verified and questioned first. In fact, the main reason and focus of the audit is to try to determine how much the company complies with the legal requirements. It is the evaluation of the stages of the drug product from the starting materials to the finished drug product in a standard system audit. Evaluation can be extended to post-production storage and distribution. Depending on the size of the operation, these audits will take 1 to 3 days. In addition, 1 to 3 auditors can successfully complete these audits, depending on the circumstances.

The firm's operational concerns may exist in this type of GMP audit. For example, it may be desirable to follow the production flow of a predetermined drug product. In such an audit we can say that the entire documentation system, including the logbooks that the auditors pursue, the records of environmental conditions, the records of the parties, the test results performed, all kinds of records, and qualification and validation documents related to cleaning.

The purpose of the product audit is to evaluate whether the contract manufacturer fully complies with the terms of the contract. Before these audits, the auditors of the licesnce company, that is, the contracting party, are expected to make a detailed preparation.

3.2.8 Audit Priorities

When we look at the GMP audit preparations, we see that the contracting organization, that is, the contracting party, tries to collect as much information as possible about the contract manufacturer or supplier company by sending detailed question lists that try to cover every area to prepare for the GMP audits.

Answering these types of questions is the responsibility of the contract manufacturer, and if they are answered, it is unreasonable for the contracting party to have any doubts. If such a list of questions is not sent, the chance of obtaining information about the contracted manufacturer will decrease, and perhaps this information will be provided through a consultant. As I tried to express before, the questionnaires consist of question titles related to GMP audits. Some license-holder companies may also want to ask specific questions highlighting the process or system they want to examine in the contracted manufacturer.

3.2.9 Result of an Audit

The contract giver prepares a comprehensive report at the end of the GMP audit. The audit report can usually consist of a completed questionnaire or a list of questions. However, some companies may expect the audit report to be written in full and in detail. The report includes the observations and findings of the contract giver about the contract acceptor. Detected findings and observations are evaluated in different categories, such as high-risk, medium risk, and low-risk production facilities. The report prepared after the GMP audits is expected to include not only the findings, but also sentences that address the shortcomings of the contract manufacturer.Thus, the contracted party can read the report accurately and completely and therefore can complete what is expected from it more accurately and quickly.After the GMP audit report is sent, the other party is expected to submit their responses, including the actions they have taken or will take, within 4–8 weeks. Thanks to the audit report, the contract giver can reach one of the following conclusions about the audited party.

- *Not approved* – The audit of a contract acceptor was carried out and revealed that the contract acceptor cannot be approved.
- *Qualified* – The audit of a contract acceptor was carried out and revealed rectifiable deficiencies.
- *Certified* – The audit of a contract acceptor was carried out and revealed no deficiencies.

3.2.10 How Does a Contract Acceptor Prepare for an Audit?

If you are manufacturing contract medicinal products for many companies and are also undergoing inspections by many health authorities, you should take a professional approach in GMP audits/inspections as a contract manufacturer and make a preparation plan that includes observations obtained from your previous audit/inspection experiences. However, as it is known, not all contract drug manufacturers can perform all GMP observations in a short time. The point I want to make is the more GMP audits you experience, the more regulations you will master. After the GMP audit report is received, the current status of the actions regarding the findings in the report can also be shared with the contract giver.

3.2.11 Carrying Out Follow-Up Audits

As readers are well aware, it is common practice nowadays for the contract giver to allow a period of time after the GMP audit for the contract acceptor to evaluate the observations and determine the actions. As I have mentioned before, if the contract giver does not receive a completed audit action plan from the audited party after the 4-8 week period has expired, it may conduct a re-audit at the facility. In addition, if the contract acceptor declares that he has completed all the actions, but does not include sufficient evidence documents, such as pictures, images, and documents in the action plan, it may bring up an audit again. In fact, an audit can be made to see if the images sent belong to the actions taken in reality or not. All these audits are called follow-up audits. What needs to be considered here is whether the follow-up audits are related to only the previous audit findings or whether a brand new audit is attempted from the beginning.

3.2.12 Positive Spin-Offs of Audits

As it is known, GMP audits are activities with high costs for both the contract giver and the contract acceptor. Regardless, they are still valuable. When we consider the GMP audits from the perspective of the contract giver, it allows him to compare the companies he audited with other contracted companies and to understand who has superior qualifications. In addition, obtaining detailed information about how to do business with the contract giver can broaden the perspective of the contract acceptor. In this way, the contract acceptor receives consultancy services for free, in a way, with the different GMP approaches of other customers.

3.3 DUE DILIGENCE AUDITS

Due diligence auditing is a crucial activity that should be undertaken when contemplating activities such as mergers and acquisitions.

The areas covered in the audits and inspections include:

- *Facility design and services* – Exploring operational and GMP compliance, compliance of process, material, and personnel flows with regulatory requirements
- *Quality systems* – The workings of quality systems and regulatory requirements
- *Customer complaints* – Nature of system and use for information, training, and improvement
- *Fabric and technical infrastructure* – Nature, condition, maintenance schedules, and plans
- *Engineering and maintenance* – Nature of system of planned preventive maintenance and a calibration program, and extent of fulfillment
- *Management team and structure* – Suitability, strengths, weaknesses, capability, stability, succession planning
- *Health and safety* – Company policy, track record, recordkeeping, and potential future liabilities
- *Relationship with regulatory authorities* – Nature and extent
- *Work in progress* – Validation of quantity and time frames and value
- *Training* – Adequacy, frequency, nature of records
- *Products and processes* – Compliance with specifications and regulatory requirements
- *Quality control laboratories* – Level of equipment, staffing, and operation relative to regulatory standards
- *Plant and equipment* – Adequacy, condition, validatability, validation period
- *IT systems* – Nature, extent, validation, replacement plans
- *Security* – Physical security of plant, plans and programs for theft and sabotage prevention

In addition to the aforementioned areas, during the due diligence process, other areas covered include:

- Legal
- Financial
- Commercial
- Environmental
- Tax
- Technology
- Employment

The acquisition of a brand or industrial site operating in the healthcare sector requires, in addition to the classic acquisition phases, an assessment of the target's ability to meet the regulatory requirements specific to the products concerned. Table 3.1 summarizes the primary differences between due diligence and GMP audits.

Why is pharmaceutical due diligence important?
You want to put a state-of-the-art TV unit in the living room of your newly purchased home. Your greatest expectation is that the TV unit is of the best quality. Just as you expect a TV unit to meet a number of criteria, the evaluation of a pharmaceutical production facility should be done with this sensitivity. All components of the TV unit should not cause you any problems after the purchase. The best way to do this is to do regulatory due diligence. Well, who will do this? It should be done by an expert party who can ensure that nothing goes unnoticed.

TABLE 3.1
Primary Difference between Due Diligence and Audits

Element	Audit	Due Diligence
Objective	Performance assessment	Potential partner
Scope	cGMP compliance	Multiple business evaluations
Deliverable	Observations and CAPAs	Recommendation to do business and remediation
History	Established	Very limited
Audit strategy	Risk based (data)	Risk based (processes)
Cost/Time	Fairly standard	More variable

3.4 MOCK INSPECTIONS

Mock FDA inspections allow the identification of probles that may affect the quality of the drug product or medical device produced to create a compliance problem with the FDA and GMP. Mock inspections help to determine audit readiness by testing knowledge of procedures and creating a process that simulates the pressure of an actual inspection. One of the most effective methods to prepare subject matter experts (SMEs) is to conduct mock inspections.

Mock audits are simulations of actual FDA audits conducted to assess Good Clinical Practices, Good Laboratory Practices, and Good Manufacturing Practices

3.4.1 Who Performs Mock inspections

- Internal team using FDA procedures
- External team of ex-FDA employees

An internal team consists of qualified professionals with experience in regulatory compliance. If an external team is used, it generally consists of members of an outside consulting firm with FDA work experience.

3.4.2 Purpose of Mock Inspection

- Determination of what inspectors are looking for during an actual inspection
- Preparation for an actual FDA inspection
- Identification of common issues of compliance and how to rectify them

In preparation for inspections, it is useful to have an external organization perform a mock inspection. A mock inspection should completely imitate a real-time inspection, and it would be to the benefit of the mock inspection if the inspection report exactly imitates an authority inspection report.

Before leaving a German pharmaceutical company, in 2012, at a time when the Turkish Ministry of Health Medicines and Medical Device Agency was not yet established and no concrete steps were taken regarding the PIC/S membership, I was a GMP lead auditor who had proven himself in the field of audits, so the factory manager called me. I realized how big of a responsibility I was actually going to undertake when FIMEA said that there would be an inspection soon and that the whole factory should go through a preparation process so that the opportunity to export to Finland could be obtained.

FIMEA is the Finnish health authority and all our dermatological product exports were subject to this inspection. More precisely, the entire fate of the factory depended on me and the mock inspection team I would set up. I immediately rolled up my sleeves and published an announcement about the inspection to the entire factory. For the mock inspection, I included another experienced friend of mine from Germany in this preparation process. We worked day and night nonstop. I gave all the necessary training to all the personnel in the factory, I announced that there would be a real-time mock inspection to prepare the people for the inspection, that everyone should be as natural as possible, and that staff working at the bottom and a senior manager would be included in the mock inspection.

I planned the mock inspection for three days. Like the FIMEA's inspectors, I made all correspondence in an official language and wanted the factory to be ready for this special day. I shared nearly 50 findings I found in the mock inspection with all stakeholders, including senior management. I held separate meetings with each responsible department and personally participated and led the improvement activities. Days later, the FIMEA team was at the facility. All the issues that I and my auditor friend focused on during its stay were questioned. However, we had already completed all the improvement activities.

The FIMEA inspection was completed with a few minor findings and the German pharmaceutical company in the Istanbul factory had the chance to export to Finland, perhaps for the first time, to a PIC/S member country.

The lessons we learned were that the better a facility always prepares itself for a health authority inspection or client audit, the less pain it will suffer in a real-time audit and the less costly returns are to be feared. If I hadn't done a mock inspection for the nonconformities I found during those three days, maybe the FIMEA's daredevil inspectors would have detected it, and maybe it would have taken years to close the corrective and preventive actions (CAPAs) and greatly reduced our chances of exporting.

3.4.3 Outcomes of Mock Inspections

- Obtain evidence that management is overseeing operations effectively
- Review the compliance of contract service providers
- Verify that quality assurance and regulatory affairs departments are operating appropriately
- Evaluate process controls, records, and document controls
- Determine whether the mechanisms are compliant

A mock audit is usually conducted over four to five days and utilizes FDA inspection procedures. At the end of each day, the auditor provides a recap of findings at a closing meeting and identifies personnel to interview.

- Day One: The mock inspector meets with people from the company, tours facilities, and reviews internal quality files.
- Day Two: The mock inspector reviews the policy for retaining records &and product controls at facilities.
- Day Three: The mock inspector reviews procedures for corrective action plans, handling of complaints, and oversight of suppliers.
- Day Four: The mock inspector reviews any remaining compliance programs and interviews any personnel that he or she still needs to interview.
- Day Five (if needed): Additional areas of the quality systems not previously covered will be reviewed.

At the end of the fourth or fifth day, the mock inspector conducts a closing meeting with the company. During this meeting, the mock inspector provides a review of major and minor findings and suggests actions to improve the issues found.

The mock inspection moves around three factors: people, preparation, and process.

- *People* – Put the right people in the right roles.
- *Preparation* – Appropriately prepare team and documentation.
- *Process* – Follow the industry's standard inspection process.

Mock FDA inspections are the most important tool that can detect possible gaps in quality systems and GMP compliance in pharmaceutical, biotechnology, and medical device productions. The mock inspection can minimize the risks that may put the company in a difficult situation if these gaps are caught by any health authority, and detect noncompliances on time and in advance.

For this reason, all license-holder companies are expected to know what mock inspection means and the importance of performing a standard mock inspection in the most similar way to actual authority inspections.

3.5 SUPPLIER AUDITS

Conducting GMP audits of suppliers by GMP guidelines and directives is an expanding practice with increasing expectations. While supplier audits determine whether a supplier's systems are advanced or have a rather old structure, they are also an opportunity to reduce or even eliminate the risks that the end user consuming the drug product, the patient, may face.

3.5.1 When Should You Audit a Supplier?

The responsibility of the pharmaceutical company "to approve and monitor suppliers of materials" is clearly defined in EU GMP rules. However, the decision as to whether such monitoring should extend to auditing a supplier is an important one that should be taken after careful consideration of a range of criteria that can have relevance to the specific situation faced by the organization. An organization may be acting in a preventive mode or a reactive mode and both have validity within the overall management of the quality of supplied pharmaceutical materials or services. While a well-developed audit program will be biased toward preventive auditing, time must always be found to be able to react to unexpected problems or commercial imperatives.

However, in order to avoid the situation of always being in the reactive mode of perpetual fire-fighting, which can be dispiriting and commercially expensive, it is essential that a well-developed preventive plan be developed. This plan should be based on a cycle of auditing that should stretch over several years and be based on a policy that has been debated and accepted by key stakeholders within the organization. Primary stakeholders who should be included in the development of this policy are representatives of Engineering, Production, Purchasing, and Quality Assurance (QA). This list is not exhaustive and may extend to other groups such as Development units, where new products are involved or the computing department where computerized systems are used in quality critical processes. Figure 3.1 describes the qualification and disqualification process of a supplier.

It is also important for large organizations to attempt to integrate the interests of different parts of their organizations to avoid the potentially embarrassing scenario where the supplier points out that someone from the same company conducted an audit recently. Such embarrassment would of course be exacerbated if the supplier politely pointed out that different audit standards were being applied. Avoiding such incidents will require a degree of internal mutual trust and cooperation that may take some time and effort to develop. It is probable that this trust will only become sufficiently developed when auditors from different parts of the same organization experience each other's audit processes.

The audit policy that is developed will necessarily be specific to each organization, but a number of generic factors should be considered. The expectations of regulatory authorities can have a major impact, depending on the particular sensitivity and experience of the authority involved. There is

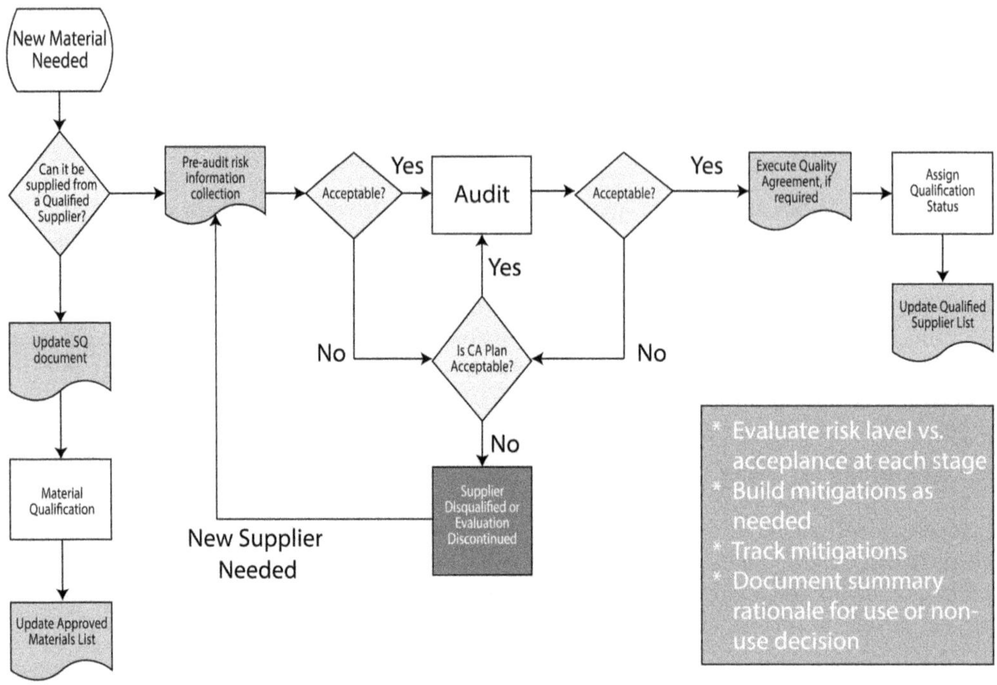

FIGURE 3.1 Supplier qualification versus disqualification.

no substitute for seeking to understand what these expectations are, and although an introduction to this subject is provided later in the chapter, the audit plan owner should stay aware of current GMP trends on this subject. The trend over the last 20 years has been for such expectations to gradually increase.

Where an audit is not mandated by a regulatory authority, a helpful tool used by many organizations is to perform a risk assessment for each supplier/material combination. This assessment should consider the risks to the business and patients of a failure to detect a potentially defective batch of material either within the supplier's quality control (QC) system or the pharmaceutical company's QC system. It is not enough to only consider each supplier, as alternative types of materials manufactured by the same supplier could have widely differing uses.

For example, a supplier may supply printed cartons, patient information leaflets (PILs), and labels. With leaflets, it is very important that each side is printed and there are no blank sides so that every patient understands how they should use their medicine. This problem would not occur with either labels or cartons as they are not printed on two sides, but such blanks can occur on leaflets if two sheets of paper pass through a print station simultaneously. It is highly unlikely that they would be detected through convenience sampling and QC sampling techniques, so assurance of the very high performance required must rely on automatic prevention or detection systems installed on the supplier's printing lines. Such assurance is usually achieved by understanding the critical risks and auditing the systems designed to protect against them occurring.

A rationale for making the decisions required must be based on good data. This requires that metrics should be in place to assess "confidence in the supplier." Typically, this will include an assessment of the extent of development of the supplier's quality system, combined with a measure of delivered performance for each material. This important measure should include measures of incoming defective batches, feedback from production performance, and any customer complaints that are attributable to supplied materials. The measure should be normalized on the volume of material received, using either batches or parts per million, to enable comparisons with other suppliers and measure progress over time without distortion due to changes in received volumes. Monitoring moving annual averages or CUSUMs can be a powerful aid for viewing emerging trends and can be useful in providing trigger points for action.

The question of how often audits should be scheduled is one that should evolve from the risk assessment process. Typical frequencies employed are for suppliers of critical materials to be audited at least annually, whereas suppliers of low-risk materials might be audited every three years.

The amount of time typically "allowed" by a supplier for a second-party audit is one day, but this could be extended for critical/high-volume supply situations. This time is usually considerably less than the times recommended by the International Accreditation Forum for third-party certification audits. Pharmaceutical companies should always maintain the capability to respond to urgent situations that require evaluation and dialogue with the supplier.

3.5.2 What Is the Appropriate Standard for Auditing?

Selecting the appropriate standard that will be used for an audit is paramount. There is a wide range of standards available that have been published and these are continually being revised. Most are based on either the EU GMP or ISO 9001 frameworks. The auditing organization may alternatively have developed its own audit standard, which may expand on an existing standard. The key principles that should be borne in mind here relate to the appropriateness and openness of the standard. For example, if an organization wanted to assess the customer focus of a potential supplier, it would probably want to ensure that elements of the new ISO 9001 standard (which is much better than EU GMP in this respect) were employed. The standard selected should be recognizably appropriate for both the customer and the supplier, and it should be available to the supplier prior to the audit.

For example, a pharmaceutical company may wish to audit the supply chain for a starting material excipient being sourced from China. Appropriate standards to consider are IPECs that cover manufacturing and agents and the WHO good trade and distribution practices guideline that covers distribution. The supplier should always be advised well in advance of the standard and provided with a copy if needed.

3.5.3 Who Should Audit the Supplier?

This is a very important question on which the effectiveness of the audit process will depend enormously. It is very rare, if not unknown, to find an auditor who has the competence to audit all types of suppliers or audit to a range of standards. Consequently, the representative(s) selected for each type of audit should have skills and knowledge appropriate to the product or service provided.

Although a member of QA will usually initiate the audit process, it is very important that the decision as to who the auditor(s) should be involves colleagues in other appropriate departments such as Development, Microbiology, Production, Purchasing, and Engineering. The question of how many auditors should be involved can usually be answered after careful consideration of the audit plan. Writing this plan down well in advance is the key, as it requires a clear understanding of who will be auditing which clauses and of which standard.

A second key question to ask is "What do we need to understand better?" For example, the organization may want to understand whether it can be confident that a prospective contract testing laboratory will deliver reliable results for a product under development. In this case, the auditors might be representatives of the Development Unit's laboratory team and a QA audit specialist, who jointly understand the requirements for the performance of the analysis, communication, and appropriate elements of GMP. They should define clearly how they will achieve this early in the process. Again looking at the clauses carefully in advance is invaluable. No clause should be audited unless it will provide useful information.

Another common audit situation is where a company needs to monitor the quality system of an established supplier of materials used regularly in manufacture. An example might be printed labels. In this case, the audit could be performed by a single auditor skilled in the appropriate supplier GMP standard (e.g., PS 9000) and familiar with the supply sector.

Another scenario might involve a planned change of manufacturing location by a supplier of a critical excipient made using specific process to the customer. This complex issue raises questions concerning the risks to supply from the current location, which purchasing colleagues would seek continued reassurance on. This situation also brings to mind the problems of the quality of the systems used to control the production facility. This would prompt the involvement of an auditor skilled in PS 9100. Finally, the nature of the process might start a validation specialist to audit the proposed process to understand the implications of revalidation.

These examples serve to illustrate the range of questions that can arise and usefully show how the number and skills required by auditors will vary with each scenario.

3.5.4 How Do You Conduct Audits?

The subject of learning how audits should be conducted is an enormous one that is normally addressed by a combination of reading, in-house training, and attendance at external courses. The manager considering which type of training to select has a number of different options to consider, ranging from accompanying the current auditor to attendance at a weeklong internationally recognized course such as a pharmaceutical lead auditor course. Each option can deliver training of value. The primary options to consider are as follows.

3.5.5 Accompany the Current Auditor

This method is very common and often the entry point for people new to auditing. The novice will quickly experience the wide range of skills they will need to become competent, ranging from listening to time management. Learning from an experienced auditor and given opportunities to practice, supported by constructive feedback, the novice will learn fast. This process can be strengthened by providing guidance on auditing that might take the form of company standards, policies, and guidance or published guides such as the Pharmaceutical Quality Group's (PQG) monograph. However, it can be difficult to provide a recognized competency assessment using this route.

3.5.6 Run an Internal Training Course for New Auditors

This method will take longer to organize but has the advantage over the first option in that audit practice performed by the novices can be done in a risk-free environment where it does not matter if they make what might otherwise be potentially embarrassing mistakes. It also has the advantage that discussion of difficult, confidential, or other company issues can be facilitated throughout the training. Role-play can be a valuable part of this process, which can be combined with training for auditees on how to react during an audit. The power of such role-play can be accentuated if video footage is taken and subsequently reviewed with the auditor. An internally run course could also be designed to assess the competency of trainee auditors, with the help of the ISO guidelines for quality auditing.

3.5.7 Attend an External Accredited Course

Attendance at a recognized course offers some significant advantages that merit consideration. Courses such as a pharmaceutical lead auditor course are usually based on PS 9000 and PS 9100, which integrate GMP appropriate to suppliers of pharmaceutical materials into the internationally recognized ISO 9001 framework. ISO 9001 is the de facto quality system standard for a wide range of suppliers, and current experience shows that over 90% of suppliers to the European pharmaceutical industry are certified by it. Consequently, the vast majority of the quality manuals that an auditor will see will be based on this standard and he or she will need to understand the language of ISO 9001. An excellent way to achieve this would be to attend a lead assessor course. A further significant benefit of these courses is that the discipline required to be demonstrated by the auditor during the course should improve their confidence and hence their performance in what can be a demanding role.

An additional benefit can come from the exposure the attendee gets to auditors from different organizations. Such interaction can open minds to new and better ways of working, which may enable returning attendees to influence their organization for the better.

It is recommended that new auditors attend such a course but only after they have learned the basics of their company approach from colleagues, have shown some basic competence, and are determined to develop their skills. All of the accredited courses have a final assessment, so a "pass" is an externally recognized attainment of competence.

3.5.8 Auditing to Good Manufacturing Practice

This is the most difficult scenario to manage and requires that the auditor selected should have been exposed to GMP standards for a considerable period. As the subject is so large, it will usually often be necessary to select auditors who are specialists in the appropriate subject, e.g., good laboratory control practice, cleanroom standards, transmissible spongiform encephalopathies (TSE), or electronic signatures.

Of paramount importance, regardless of the type of audit, is the need for each auditor to brief all stakeholders on the scenario of the outcome, the implications, and the follow-up process. The understanding achieved must be shared to release their maximum value. This step is crucial to help convince the various stakeholders that they are getting good value for their insurance premium – the cost of audits.

3.5.9 Benefits of Auditing Suppliers

Initially, supplier audits are usually arranged to meet the GMP compliance expectations of regulatory authorities, but this changes with increasing experience. It is a common experience of auditors (and their managers) that as the mutual understanding grows between the supplier and the pharmaceutical industry customer, the quality of the supply performance and the supplier's quality system tend to improve. This would be expected to show up as an improving trend in the defect rate of supplied materials measured, which should be measured for all key materials. This experience is reflected in the valuable research performed in the United States by Raju and Cooney. The careful measurement of each supplier/material history enables the pharmaceutical company to manage its supply base with more clearly delineated strategies. Table 3.2 provides an example of critical mistakes made in supplier audits.

TABLE 3.2
Critical Mistakes in Supplier Audits

#1: Poor risk identification
#2: Poor evaluation and prioritization
#3: Broad audit scope
#4: Lack of expertise in the audit team
#5: Poor audit preparation
#6: Not understanding the full supply chain
#7: Failure to prepare supplier
#8: Poor reporting
#9: Failure to communicate results
#10: Not tracking to closure

For example, a key decision might be whether the organization should continue to purchase materials from a particular supplier. The creation of a policy that defines minimum acceptable levels of performance for both the quality of the supplier's system and a measure of the quality of supplied materials can be very helpful. Such measures, which if evaluated in addition to commercial considerations such as responsiveness, price, and reliability of delivery, contribute to a holistic appreciation of the overall competence of the supplier, which can then be compared with the alternative sources of supply. A very useful approach is to convert each set of audit findings into a semiquantitative overall score for the system. This can then be plotted as a histogram for each supplier in the supply base.

Historically, there have been many examples of critical defects and health risks caused by supplier failures to meet common GMP expectations. While not necessarily directly linked to regulatory requirements meant for drugs, supplier performance or practices have been associated with illnesses and even fatalities due to their poor or even fraudulent practices. This fact alone provides enough concern to suggest that drug companies need to have adequate systems and procedures to assure that suppliers are reliable and trustworthy in providing materials, equipment, and even services that will not undermine the quality of drugs being produced. Suppliers to drug companies play a significant and very important role in safeguarding the reliability and safety of drug products. Suppliers that provide goods or services to the drug product or the API manufacturer are key to the continued success of these firms.

Table 3.3 provides a list of supplier types.

3.5.10 Supplier Selection

There are numerous ways to select suppliers. All depend upon the importance of the service, equipment, or product being purchased. It is always beneficial to select, whenever possible, multiple suppliers. Although there are commercial benefits to working with a single dedicated producer, there are also risks to supply lines under such arrangements.

Single-supplier agreements in the drug industry may actually impose greater risks to a firm than in other less regulated industries. If an approved single supplier runs into a significant regulatory problem, a drug manufacturer may become a victim of the previously undetected issue and be unable to continue his own manufacturing without another approved supplier. For this reason alone, drug producers need to cautiously consider if single-supplier versus multiple-supplier sourcing is worth an increased risk or added costs. These types of business decisions need to be evaluated along with the potential impact of a regulatory closure of a manufacturing, testing, warehousing, or distribution facility should a government regulatory agency take aggressive action against a drug firm due to an inappropriate activity by a supplier to the drug firm. Programs should be established to monitor and watch critical suppliers over extended periods of time and to ensure that all quality programs stay consistent with agreements and quality conformance requirements. Drug and drug product manufacturers need to establish clear contracts and quality agreements with important suppliers. Practices and procedures should be established that protect the purchasing firm and provide it with the authority to examine and review both facilities and records associated with the manufacturing of supplies purchased from the supplying companies.For basic commodities, the simplest purchasing activity is the acquisition of material/equipment as a direct purchase or under a commercial agreement based upon the description agreed to by the purchaser and the supplier. In this case, the supplier may be the original manufacturer, a sales organization, or a distributor. Such materials are commonly purchased without special quality requirements and are supplied via normal or standard supply chain systems. These materials and services are general in nature, made for or available to the general public, nonspecific commercial uses, or even services of a general nature such as cleaning services in nonmanufacturing or control areas.

Table 3.4 provides information on the supplier quality system.

TABLE 3.3
Suppliers to Drug Manufacturers

Type of Supplier	What They Typically Supply
Manufacturing equipment – API	• Weighing equipment and scales • Vessels of various construction • Reactors • Mixers • Motors and pumps • Vacuum equipment and systems • Distillation columns and units • Dryers • Mills and particle size classification systems • Purifying equipment • Conveying systems and equipment • Piping • Computer control systems including distributed control systems • Packaging systems • Labeling systems/equipment • Filter systems for liquids, air, and gases; these can include both sterile and nonsterile systems • High-pressure reactors
Manufacturing equipment –Drug product	• Weighing and dispensing equipment and systems • Mixing equipment • Milling and classification systems • Dissolving tanks • Blender and weighing systems • Drying equipment and systems • Compression systems including tablet presses • Packaging systems including bottlers, unit dose packers, liquid packaging units for vials, ampoules, encapsulation systems • Sterile environment equipment including sterilizers, HEPA air systems • Lyophilizers (freeze dryers) • Appropriate filters and systems
Laboratory equipment and supplies	• Laboratory reagents • Solvents • Glassware • Laboratory instruments including physical, analytical, and biological systems • General laboratory equipment such as centrifuges, ovens, baths, mixers • Balances and scales • Laboratory installations
Intermediate suppliers to API manufacturers	• Compounds produced by chemical or biological procedures • Compounds isolated from natural products for further processing into different compounds • All materials that are used to produce an intermediate that contributes a significant structural element to the final API
Raw materials for API manufacturing	• All materials that are used in chemical, biological, or physical processing during the early stages of production to manufacture intermediate materials on the way to producing the final API
Raw material and excipient suppliers for drug products	• All Inactive ingredients (excipients) used directly in a dosage form and are not reacted with something else to form another compound • Raw materials that are consumed during the manufacturing of a dosage form
Active pharmaceutical ingredient for drug product	• The active ingredient placed into the drug product that provides the pharmacological activity of the drug product

(*Continued*)

TABLE 3.3 (CONTINUED)
Suppliers to Drug Manufacturers

Type of Supplier	What They Typically Supply
Facility equipment or services	• Design or engineering • Water systems • Gas systems • HVAC systems
Service suppliers	• Pest control • Calibration • General maintenance • Cleaning • Contract services such as testing labs
Packaging materials and equipment	• Drums • Bottles • Primary contact materials such as plastics, glass, metals • Secondary packaging materials such as shipping cartons • Cardboard, adhesives, labels

3.5.11 Suppliers of Critical Equipment, Services, or Materials

Important equipment, services, or materials will always be purchased against official specifications and acceptance criteria. Such specifications and acceptance criteria should be in writing and agreeable to both parties. Equipment should be fully described as to materials of construction, design specifications, appropriate Installation Qualification (IQ) and Operational Qualification (OQ) documents, as well as, when appropriate, validation documentation and design and installation drawings. Important services such as calibrations or pest control services should be performed according to written preapproved processes and procedures, and documented by the supplier as performed.

Commercially available laboratory equipment may not need the same level of documentation as manufacturing equipment but should include all appropriate records to support its intended use by a drug manufacturer. Literature that supports the scientific and technical validity of the laboratory equipment should be retained by the drug firm together with all appropriate qualification and/or validation documents of laboratory equipment used to control and release materials purchased for use in an API or final drug product. This same requirement should be satisfied for equipment used to test and release any drugs themselves. All critical equipment used to produce drugs should be properly installed and qualified and, where appropriate, validated. Agreements with suppliers should establish responsibilities for the supplier to document and record all important activities needed to ensure that proper GMP-related steps are performed by suppliers. Where appropriate, all records associated with these actions should be provided to the purchasing firm and held by the purchaser for future reference should they be required or requested by any regulatory authorities or firms purchasing products in the future.

3.5.12 Review and Approval of Suppliers

Quality systems and/or procedures should exist to establish the appropriate selection and approval of suppliers of important equipment, systems, and materials. Every firm should have written procedures for the selection and approval process used for suppliers. Those procedures or programs should identify who within a firm is responsible for the selection and approval of all key suppliers as well as the programs or systems that are used to monitor and evaluate suppliers. Suppliers of critical

TABLE 3.4
Supplier Quality Systems

Supplier Classification	Type of Applicable Quality System to Consider
Basic suppliers	• Commodity materials should be appropriately identified and meet normal business documentation • Purchase orders and appropriate shipping and receiving documentation are consistent with the material or service purchased • Purchase through normal local or international commercial sources • Identification of sourcing • Where appropriate, identification of materials for construction • Purchasing firms would normally not perform special reviews or the examination of supplier facilities or records for such materials
Critical or key suppliers	• Purchasing firm should establish a written procedure identifying its internal programs used to identify, select, and oversee such suppliers • The minimum requirements for these suppliers should be written records of all operations and a documented review and approval system that supports the quality of the product provided to the purchasing company • Commercial contracts should exist and include GMP compliance requirements as appropriate with local and international requirements • Quality agreements, which supplement any commercial contracts, should clearly identify the quality-related responsibilities of a supplier and the purchasing firm • Generally, the authority to visit and review operational site(s) by the purchasing firm should be part of applicable quality agreements • Certificate of analysis (COA) documents should be provided to the purchaser and reviewed for compliance with established acceptance criteria • Change control systems should be established in writing and agreeable to both purchaser and supplier
Final product suppliers	• All of the identified quality systems noted above plus: • Authority for facility visitation and review of supplier facilities should be required • Such suppliers should be audited on a predefined basis as identified and according to a documented written quality procedure; if audits are not required, internal procedures should identify this and identify alternative applicable procedures used by the purchasing company • Such suppliers should have all materials tested and released according to programs that are within established and approved quality systems and in conformance with local and/or international GMP expectations • Quality systems that meet GMP are expected; other quality systems, such as ISO and internal programs, may supplement GMP, but the primary quality system is always GMP • Changes performed by suppliers should require prior approval of the purchaser before changes are implemented by suppliers

materials or services should be monitored under an "umbrella" control system that will maintain a review and evaluation program to ensure that reliable materials or services will be available over extended periods of time. Drug firm procedures should clearly identify the processes used to evaluate suppliers and which drug firm departments or functions are responsible for controls, reviews, and approvals of critical suppliers.

Table 3.5 provides information on critical supplier selection.

3.5.13 Quality Agreements

Suppliers of any critical service, material, or equipment should be covered by an appropriate quality agreement. A quality agreement is a detailed agreement between firms to ensure clear identification

TABLE 3.5
Critical Suppliers

Action	Type of Activity or Control
Selection of critical suppliers	• Usually selected by either R&D, manufacturing or purchasing functions.
Approval of critical suppliers	• These suppliers are initially screened by the Purchasing department or selecting group within the drug company. The quality function then uses a written survey to determine what manufacturing and quality systems or procedures are used by the supplier. • After review of the written survey, an on-site audit of the supplier is scheduled and performed by the Quality Unit. In some cases, it may be appropriate for a team to perform the review. Teams can be formed across various groups such as Purchasing, R&D, Manufacturing and even laboratory personnel. Any team should always include the Quality Unit. • The on-site audit verifies capacities and capabilities to produce and control the material being supplied.
Audit Scope	• Warehousing and distribution facilities. • Manufacturing facilities and personnel. • Laboratory facilities and personnel. • Quality functions and controls. • Change control systems. • Releasing system and procedures.
Incoming supply	• Sample all materials delivered to drug firm. • Analyze all lots. • Accept materials based on approved specifications and test procedures. • Release all lots meeting acceptance criteria. • Reject and return any lots failing acceptance criteria. • Analyze all data annually to ensure acceptable trends and quality. • Determine if test results that are obtained by the drug company agree with those reported on the Certificate of Analysis (COA) from the supplier. • Examine long-term trends and quality history to determine if a reduced testing protocol can be considered and established in the future.
Establish reduced testing protocol if appropriate	• If allowed by local and international rules, directives, laws, or legal requirements, establish a protocol that will allow material from this supplier to be accepted according to the defined protocol requirements. • Any such protocol should require at a minimum: • Physical examination of the material, packaging, and labeling upon receipt by the drug firm. • An identification test should be used to identify the material received. • Appropriate retention samples should be kept. • Appropriate lot trace identification records should be established.
Establish long-term acceptance criteria protocols or procedures for critical suppliers	• Test a minimum of the first ten deliveries of the material. • If all ten meet acceptance criteria, show good agreement with Certificates of Analysis provided by the supplier, and all other quality agreement conditions are met, then adoption of a reduced testing protocol for this material and from this supplier can be proposed for future implementation. • The umbrella control system should mandate, at a minimum that all deliveries be subjected to ID testing, physical examination of deliveries, and a prior review of COA from the supplier. • If all of the above are found acceptable, the material delivered can be accepted and released for use. • This reduced testing program can remain in use provided that its reliability is re-verified on a periodic basis using scheduled audits and full testing programs. • Firms should be reaudited on a reasonable basis such as every two years or another agreed-upon schedule. • Full testing by the receiving drug firm of a predefined number of lots should be conducted to ensure continued compliance with the COA data being reported by the supplier. The exact number of lots and the frequency would depend upon the number of lots delivered on a predefined basis such as annually or quarterly. • The exact number of lots (e.g. 5 or 10) should be prospectively determined and documented within approved procedures established by the drug firm. • The supplier should not be made aware of the exact testing frequency that will be used by the drug firm. This is to keep the supplier "honest" in the way it provides materials and COA information to a drug company.
Disqualification of a supplier	• A failure discovered by the drug company either by audit, document review, or testing should result in the disqualification of the supplier under any reduced testing protocol. From that point, every delivery must be fully tested and reviewed until such time as the supplier can be requalified for consideration under such protocols.

TABLE 3.6
Contents of a Quality Agreement

Topics to Consider for Inclusion in a Quality Agreement	Requirement (Examples as Appropriate)
Parties	• Identify purchasing drug firm and location • Identify supplier or contractor and location(s)
Identify all activities covered by the quality agreement	• Records to be maintained • Testing to be performed • Auditing requirements • Reporting of auditing • Records to be provided and by whom • Who will perform auditing • Who will perform testing • Who will release materials • Change controls and reporting requirements • Responsible approving authority • Reporting authority and required distributions
Define appropriate acceptance criteria	• Acceptance testing • Packaging • Labeling • Material of construction
Define auditing authority	• Who can perform on-site audits • Frequency of audits that can be performed • Notice requirements
Required change systems	• Notification system • Prior approval requirements

of individuals and groups responsible for specific quality-related responsibilities for each party. In its most basic form, such an agreement may be part of a contract or other legally binding document. Nevertheless, it can also be a stand-alone agreement that is binding according to the contract signed by both parties. If local or national requirements further designate specific necessities for quality agreements, they should always be followed by drug firms. It can have a quality impact on a drug manufacturer, and therefore, suppliers need to clearly understand their obligation to the purchasing drug firm.

The quality agreement helps ensure that both the supplier and the purchaser have agreed to both parties' responsibilities, and what each party will perform and document under the terms of the agreement. Such agreements should always exist between a drug firm, its major suppliers, and all of its contractors. Aside from legal protection that may be afforded to the drug firm, such agreements make routine operations and practices easier to maintain and track. These agreements make oversight of suppliers and contractors more effective and enhance documentation. They also allow for more effective implementation of testing programs and reduced auditing requirements, and a mechanism for establishing auditing activities that may be needed over the span of time.

Table 3.5 summarizes the topics that should be included in quality agreements. Such documents should always identify the parties; the locations of operations; and the responsible representatives from the suppliers, contractors, and purchasing drug firms.

3.5.14 Critical Suppliers to Drug Firms

Figure 3.2 illustrates the potential relationship between critical suppliers and drug manufacturers as it applies to lot traceability. While lot traceability starting within the supplier's manufacturing

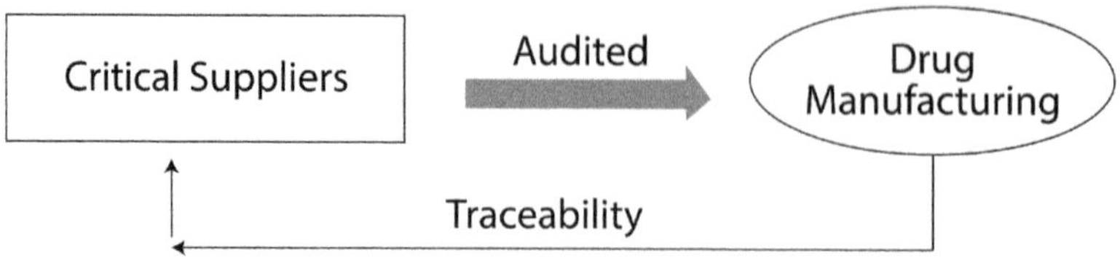

FIGURE 3.2 Critical supplier audits.

operation is not always critical for the drug manufacturer, there are indeed times when what is done by a supplier of critical materials can seriously impact a drug manufacturer's quality or even safety in a very important way. Historically, there have been changes made by supply firms that affected their own supply chain and their sources. These changes were eventually found to be the source of some very serious quality deficiencies that actually impacted the end drug product's safety. For this reason, including traceability and even sourcing concerns as part of an audit program needs to be seriously considered by drug firms.

3.6 SUPPLIER AUDITS

Reviews of new and existing suppliers are appropriate and in many cases necessary to safeguard the quality of the drug supply. What will usually vary are the frequency of reviews, the type of reviews to be performed, and the depth of such reviews. Will such reviews be performed before establishing a supply agreement and will these reviews be performed annually or more frequently? All initial work performed on suppliers should be used to evaluate and define the quality system and procedures that would be appropriate for each supplier. Not all suppliers will be addressed in the same way. Initial analytical testing of materials and how this data compares with the Certificate of Analysis (COA) received from suppliers will need to be evaluated and combined with audits performed at key suppliers. The combination of all this information will play a vital role in the monitoring and control of suppliers. The next section will provide the type of things that should be considered both before and during an audit.

3.6.1 Audit Programs

While auditing of suppliers is not mandatory in all areas of the world, implementing such programs can be valuable and an important safeguard for drug companies. When purchasing important materials or equipment, site visits to a supplier's manufacturing site can offer invaluable insight into the true ability of a supply firm to deliver a reliable product that is made under good manufacturing and control programs. These visits also allow drug producers to ensure that a supplier has the facilities, staff, and organization to supply a critical material and not just buy and resell a material from an unknown or unidentified source without the knowledge of the purchasing (drug) firm. Before initiating an audit program of a supplier, drug firms need to evaluate the quality risks associated with any particular supplier. Where risks are minor or relatively low, on-site auditing may be inappropriate and wasteful for both the drug firm and the suppliers. However, when risk assessment or even history would indicate an increased risk from a particular material or supplier, then it is recommended for the drug firm to seriously consider establishing an appropriate audit program to monitor suppliers. Early review of suppliers should identify these risks, determine if on-site audits are appropriate, and what additional information may satisfy any concerns with supplier quality impact. Sometimes additional documentation from a supply firm can provide everything necessary to alleviate any concerns or at least help define what should be examined by any on-site programs and what frequency should be considered acceptable in the future. Using the preliminary work just described, an official audit program can be justified and designed to achieve the goals established as part of a sound quality review program. Independent verification by using an on-site audit provides insurance for the drug company. However, always remember that such audits should be allowed by supply firms over time with some reasonable element of surprise. This randomness will add reliability to the findings obtained by actual lot testing and protocol reviews. All of these activities should be a part of critical supplier quality reviews.

3.6.2 Special Considerations

Audit programs can play a vital part in protecting drug companies. Aside from routine auditing of suppliers, there are times when special reviews and audits should be considered and performed.

3.6.3 Supplier Quality Auditing

A quality audit determines whether a new or existing supplier is suitable for supplying components to the specification required by the customer or whether an existing supplier is continuing to meet the specification required. Oversights at this stage can have serious implications, both from a quality and financial standpoint. It is therefore essential that continuous quality audits be performed as part of the supplier qualification program.

3.6.4 Quality Auditor

Since there is considerable complexity in quality auditing, various types of industries involved, and a limited time period for performing the audit, the quality auditor must have the appropriate requirements, training background, and experience to perform the audit. Some required attributes include:

- The auditor must be a realistic, practical person capable of quickly understanding process details and practical problems encountered by the supplier. This will ensure that unrealistic demands are not placed upon the supplier by the auditor.
- The auditor must be capable of communicating with staff at all levels, from the production supervisor to the laboratory personnel. This ability is necessary for the auditor to fully access the quality system of the supplier.
- The auditor must be very observant, be prepared to ask questions that provide background information efficiently, and be able to look behind areas and bypass equipment during the plant tour.
- The auditor should have extensive experience with the intended use of the components or packaging materials, and in particular, potential quality problems and standards that may be encountered during use.
- The auditor should have full awareness of the GMP requirements for component manufacture and usage. The auditor should also be aware of the regulatory requirements and the particular country requirements in which the final product will be sold using the sourced component or packaging material.
- The auditor should have experience with the component or packaging material manufacturing process prior to conducting the supplier quality audit.
- The auditor can usually acquire knowledge of component and packaging material processing and quality requirements by touring companies involved in manufacturing similar materials. By doing this, the auditor obtains experience for the required GMP standard for that industry. Not having this experience can often result in an unrealistic standard being requested of the supplier.

3.6.5 Auditing

All suppliers should be quality audited at least once prior to initial use, and then on a reasonable periodic basis as required under the supplier qualification program. The quality auditor must prepare in advance of the audit. This usually involves some background information on the company to determine the main business the company is involved in. It is also important to determine whether the company normally supplies components, packaging materials, or other materials to the pharmaceutical industry. A company that has never supplied a customer with strict GMP standards may require reeducating from management down to the operations level. Unless there is an alternative supplier, this is usually a monumental task and requires extra resources from the customer. It is also important to determine if nonstandard production is being requested. In other words, the supplier is providing the customer with very small quantities of the material when they usually produce very large quantities using dedicated lines. This can sometimes highlight several problems for the quality auditor to follow-up on during the audit. This includes:

- Line changeover (clearance) procedures will require special attention, particularly with respect to clean-down and reconciliation.
- A forecast production system may be operating to produce the materials for the customer. In other words, six months' or a year's predicted process output would be produced at a time to minimize costs. For example, for the production of molded bottles using special glass, this system would require the supplier to store stocks for a considerable amount of time. Under these circumstances, the packaging and warehouse would need special attention because of the possibility of pest contamination or material deterioration during prolonged storage. Based on this, auditor oversight can sometimes occur if a set auditing sequence (e.g., the checklist) is not followed, leading to important areas being missed. It is usually best to start at the beginning, i.e., raw material storage area, and follow through the process in the manufacturing sequence to the final dispatch to the customer. In this way, all aspects of GMP requirements at each stage can be reviewed.

During the heparin crisis in 2008, it was really chilling when a swine flu epidemic broke out in those years and most pigs perished. So, heparin could not be obtained from pigs as the starting material, so a fraudulent manufacturer imitated the raw material exactly and sold it to Baxter as if nothing had happened.

Consequently, unexpected reactions had been reported during the administration of a heparin sodium injection. Unfortunately, hundreds of people died in the USA.

In the investigation, it was determined that heparin was contaminated with excessively sulfated chondroitin sulfate (OSCS) and caused an adverse drug reaction. The API was manufactured by Scientific Protein Laboratories (Changzhou, China). Table 3.7 summarizes drug-induced fatalities up to 2009.

This event is proof of how a contaminated API can seriously threaten patient health and cause death.

A critical component such as API need not threaten the health of the patient, even if it is contaminated with a foreign particle, in order to activate the health authorities. A similar situation occurred in an API exported to the USA, and even with this contamination alone, the entry of the product into the country was prevented.

In 2022, approximately 300 children in Gambia, Uzbekistan, and Indonesia were first referred to hospitals with acute kidney failure due to the cough syrup they used and then lost their lives. Due to the increase in cases, the WHO stated that the contamination source might be diethylene glycol (DEG). Also well recalled in the industry is the contamination of sartan APIs in 2018.

The fact that APIs account for almost 80% of the total cost of a drug product sometimes necessitates the need to procure cheaper APIs, especially from Asian countries such as China and India.

TABLE 3.7
788 Deaths Due to DEG Contamination of Drug Products, 1937–2009

Year	Country	Product	Number of Deaths
1937	USA	Sulfanilamide	107
1969	South Africa	Sedative	7
1986	India	Medicinal glycerin	14
1990	Nigeria	Acetaminophen syrup	47
1990–1992	Bangladesh	Acetaminophen syrup	339
1995–1996	Haiti	Cough medicine	85
1998	India	Cough medicine	33
2006	Panama	Cough and anti-allergy syrup	46
2008	Nigeria	Teething formula	84
2009	Bangladesh	Cough medicine	26

Here, we cannot expect the license-holder pharmaceutical company to compromise on drug quality and effectiveness while trying to reduce costs.

3.6.6 DEG Contamination in Glycerin and Propylene Glycol

3.6.6.1 Why Does This Continue to Happen?

- Similar organoleptic properties between similar materials
- A long and questionable supply chain
- Alteration/adulteration of documents throughout the supply chain
- Lack of analytical testing/verification of contents

3.6.6.2 Which Pharmaceutical Ingredients Are at Risk?

- Problems of contaminated ingredients not limited to glycerin or propylene glycol
- Other instances of contamination
    - 2008, melamine in human foods
    - 2008, Heparin, OSCS contaminant
    - 2007, DEG contamination in toothpaste
    - 2007, Melamine contamination in pet foods
    - 1999, gentamicin pyrogenic reactions

3.7 SUPPLIER AUDITS EXAMPLE

Supplier audit is the most important criterion for selecting the relevant supplier or manufacturer. Although supplier qualification is not the main topic of this book, one of the most important inputs of qualification is undoubtedly audits. Figure 3.3 shows the flow of accepting a sample API supplier is not that easy at all. Figure 3.4 shows which supplies pose a risk to a pharmaceutical company. The API, which is at the top of the pyramid, can be said to be the most critical among a pharmaceutical company's suppliers.

3.7.1 Auditing and Resources

It is often the case that it would appear that there are far more audits required than resources available. There are finite resources available so just adding more isn't an option. A different solution is necessary. Spending less time on each audit is one option but that could result in you missing some

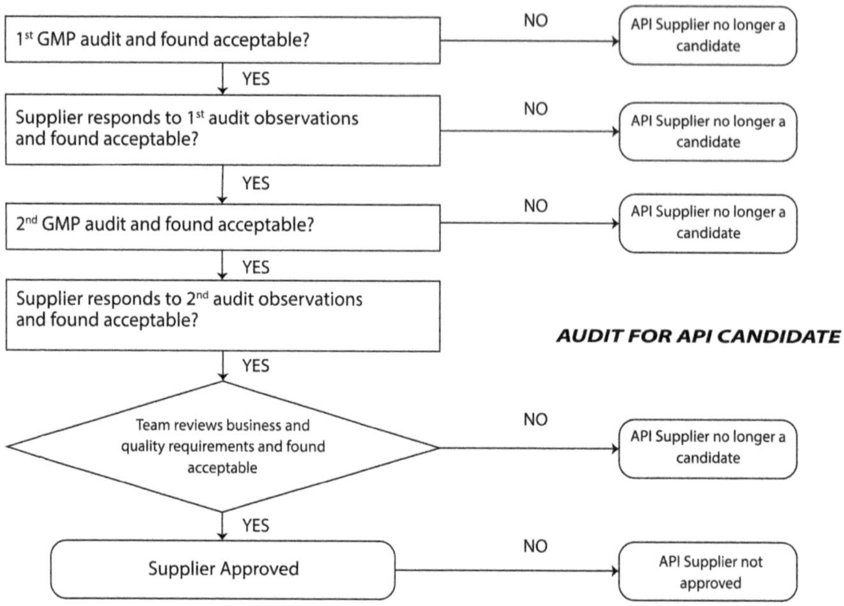

FIGURE 3.3 Supplier approval flow.

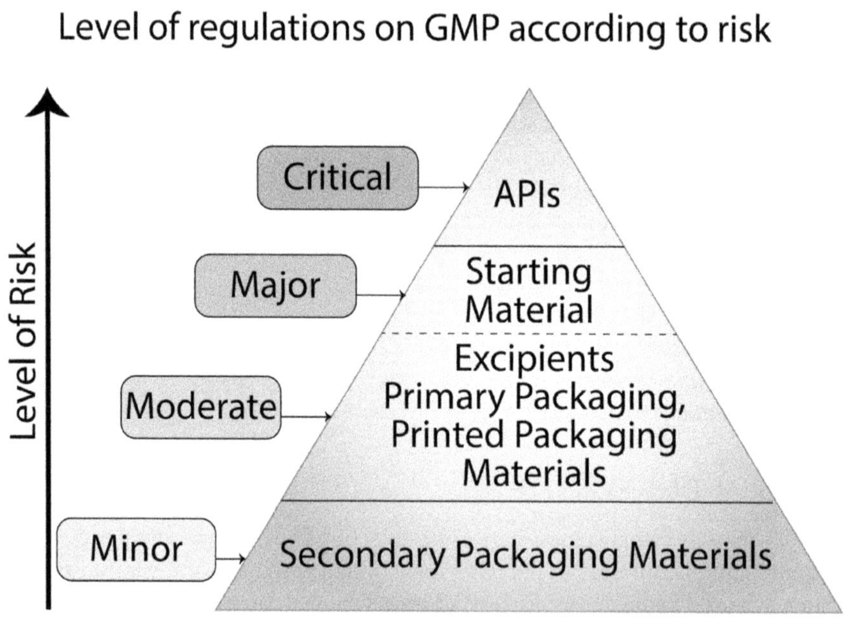

FIGURE 3.4 Supplier risk classification.

critical findings. A process that allows you to concentrate on the more important audits first and leave the less important ones till a later date is required. The key is to use a tool that aids the user to decide which ones are the most important to audit and at what frequency you should reaudit. In short, what are the risks?

The EU Guide to GMP states in Chapter 9, "Self-inspections should be conducted in order to monitor the implementation and compliance with Good Manufacturing Practice principles and to propose necessary corrective actions."

In September 2004 the FDA issued a paper titled "Risk-Based Method for Prioritizing cGMPs Inspections of Pharmaceutical Manufacturing Sites – A Pilot Risk Ranking Model." The US agency, along with all the other regulatory authorities around the world, and the pharmaceutical industry itself, were all trying to get to grips with a common problem. This problem was the ever-increasing need to carry out audits to ensure that the appropriate levels of current good manufacturing practice (cGMP) compliance were being followed. This was compounded by the ever-increasing number of vendors, laboratories, distributors, contract manufacturers, packaging operations, and, more recently, API suppliers, which were required to be audited in even wider geographical locations throughout the world than there had been previously. To ensure that auditing resources are used to their maximum efficiency there must be a way of assessing all the various auditable units and assigning a risk factor that can be used to rank each unit against a predetermined set of criteria.

3.7.2 Assessing the Risk

There are many different ways to assess risk. ICH Q9 states there should be "a harmonized pharmaceutical quality system applicable across the life cycle of the product emphasizing an integrated approach to risk management and science." The key is to understand what the risk is and put the appropriate controls in place. It should always be remembered that in the case of pharmaceuticals, it is the risk to the patient that is of primary importance. In this instance, it will be the level and frequency of audits required to ensure any risk will be identified before it can become a significant issue that should be assessed. Risk is generally associated with the type of process being used, for example:

- An orally delivered product would generally exhibit lower risk than a parenteral product.
- A robust process that can tolerate a degree of change without affecting the product, is in control, is capable, and remains compliant would exhibit a lower risk than a highly variable process. Hence, a chemical process would usually have less variability than a biotech process and would therefore have a lower risk.
- A dedicated facility will be at a lower risk than a multiproduct facility with respect to cross-contamination.

Other factors that could influence the assigned risk factor could include such things as the use of inline monitoring to pick up issues immediately (Process Analytical Technology (PAT)), or the fact that the manufacturing process is controlled using well-defined science and good quality systems.

3.7.3 Criteria for the Audit Strategy

To establish an audit frequency for each individual unit, there should be a linkage between the science, the risk, and the degree of regulatory scrutiny.

To identify the level of risk there are a number of questions that need to be addressed.

- *Systems-based approach* – What is the state of control of each system? The focus must be on the critical control points/parameters and look for systemic issues, not one-off minor occurrences.
- *Product quality* – What are the properties and routes of administration of the product?
- *Process* – How easily can you maintain a state of control? What is the rate of reprocessing or rejection? What is the risk of contamination?
- *Facility* – What is the state of compliance currently and what is the compliance history?
- *Organization* – Are there sufficient resources and are they appropriately qualified and trained?
- *What type of unit is it?* – Is it a manufacturing operation? If so, an early intermediate step would carry less risk than a final formulation step. A laboratory may be a higher risk than

a distributor, while a packaging operation could be high risk if the packaging was a complicated clinical trial blinded study.

3.7.4 Audit Program Strategy: Timings

All of the aforementioned questions need to be assessed to determine what the audit strategy should be. This, together with an agreed audit-frequency baseline, can then be used to establish a risk-based audit schedule. An example is shown in Table 3.8.

Once the baseline auditing frequency has been agreed, then the "risk" associated with that activity can be used to either maintain or modify that baseline frequency as described in Table 3.8. The baseline established needs to be in line with regulatory expectations, but in general, a typical set of frequencies could be as shown in Table 3.9. The principle of using a system-based inspection, first advocated by the FDA in 2002, will be used. This requires the mandatory inspection of at least the quality system at every audit plus one or more of the following: facilities and equipment, laboratory, packaging and labeling, production (manufacturing), and materials systems.

3.7.5 Audit Program Strategy: Baseline Frequency

Table 3.9 gives an approach to the baseline timing with regard to the facilities and systems to be audited.

3.7.6 Quantifying the Risks

What we now need is a tool to help quantify the risks associated with each auditable unit so they can be compared to one another and hence provide a ranked audit schedule for the next three years. The further away in the time it is, the more likely it will be required to modify the program as more information becomes available.

The criteria against which the units will be assessed need to be established. By asking the questions previously proposed against six different categories, an assessment of the risk associated with each auditable unit can be calculated. More or fewer criteria and/or questions can be used depending on how much information is required to be included in the risk assessment. The use of the six question categories and six criteria is for example only.

3.7.7 Criteria Used

- Probability of regulatory action will depend on the experience of the regulatory authorities and the rigor of their inspection program.
- Audit history, which should include evaluation of previous findings and whether they have all been satisfactorily closed out within the agreed timelines.
- Quality processes and management to include an assessment of the design and implementation of all the quality systems and the role of management in the quality process.
- Site characteristics to include a review of the fitness of use of all facilities and equipment.

The type of product produced at the site should be established; parenteral and highly potent products being a higher risk compared to standard topical/oral products.

- People should be assessed against their qualifications, training, and competence to carry out their functions at all levels of the organization. The extent and effectiveness of the training and education should be reviewed.
- Customer concerns as defined by adverse events and complaint reports, significant deviations, and recalls must also be taken into account.

Every auditable unit can then be assessed against these criteria and a score assigned to each between 1 (little risk) and 5 (high risk).
These can then be entered on a scorecard where they can be compared and ranked. An example of this is shown in Table 3.10.

3.7.8 Ranking the Risk

An example of a scorecard is shown in Table 3.10.

In the table, which has been populated with fictitious figures, the last column shows the compliance risk percentage. This was obtained by adding the figures in the six criteria columns and expressing them as a percentage of the maximum score possible (in this case $6 \times 5 = 30$).

For example, let's look at auditable unit F, a microbiological laboratory. It was deemed to have a very high risk for regulatory actions and site characteristics. It had an average risk based on audit history, quality processes and management, people, and customer concerns. This equated to a score of $5 + 3 + 3 + 5 + 3 + 3 = 22$, expressed as a percentage of the possible total (30) = 73%, or high risk as defined in Table 3.11.

3.7.9 Classification of the Auditable Units

The compliance risk percentage is then color coded, based on Table 3.11, so that the actual scores are no longer used. This is important to make the process more user-friendly, otherwise, it can become very unwieldy as people tend to put too much emphasis on individual scores when in reality all that is needed is a basic ranking, which in this case is just three classifications, which is sufficient to establish a risk-based audit schedule. The choice of high/medium/low or red/yellow/green is up to individual preferences.

This would result In the auditable units being classified as follows:

A, C, E, and H are classified as low risk (green).
G and J are classified as medium risk (yellow).
B, D, F, and I are classified as high risk (red).

Sites could also be classified as red if there had been an upward shift in the final compliance risk score of more than 5% from the previous assessment. The reason for this upward trend should be subjected to a management review and appropriate actions should be agreed upon.

3.7.10 Compliance Risk Definitions

It is important to agree with a set of definitions associated with each of the compliance risk categories. The following definitions have been used in this example:

- *Green (low risk)* – The audit reveals that the state of control is adequate. Processes/procedures/practices are generally in place and are sufficient to meet quality, compliance, and/or business objectives. Conditions exist that demonstrate the implementation of systems as described in local procedures is adequate in controlling related quality processes.
- *Yellow (medium risk)* – The audit reveals that processes/procedures/practices are in place, but the effectiveness of these systems needs to be enhanced. There are important quality, compliance, and/or business risk issues that justify management action and/or resource allocation. The identified deficiencies contribute to gaps in the state of control.
- *Red (high risk)* – The processes/procedures/practices are insufficient to give reasonable assurance of meeting quality, compliance, and/or business needs. The audit reveals that control systems are either not in place, are in direct violation of company or regulatory

requirements, or are not being executed consistently such that the required state of control of the operations is at risk. Direct and immediate management action and/or resources are required.

The data shown in Table 3.10 can now be used to construct a three-year rolling audit plan based on the perceived risk. The baseline audit frequency in this example has been set at three years. The assigned risk factor would modify the frequency accordingly.

3.7.11 Constructing a Three-Year Audit Schedule

Thus instead of trying to audit all ten units in 2023, only five would require auditing that year, three in 2024, and two in 2025. The schedule should be reviewed every three months and modified as new information becomes available.

This type of data can then be used by the audit function management to justify the number of auditors required. By having this data, the number of auditors can be justified on a year-by-year basis thus providing important input into the resource planning function. For example, if each audit required three auditors for one week each and also required one-week preparation and one-week follow-up and documentation, then each audit would require 3 × (1 prep + 1 audit + 1 follow-up) = 9 weeks person time/audit. Therefore in 2023, five audits would require 5 × 9 = 45 person weeks resources.

3.7.12 Proceduralizing the Process

The head of QA together with the head of the audit group should meet on a regular basis, at least every three months, together with members of the production, regulatory, IT, QC, purchasing, distribution, marketing, and engineering departments (as required) to review all available information and assign the risk factors described in Table 3.11 to each auditable unit. A company SOP should be written to describe this process and how the data is then used to construct a risk-based audit schedule. This can be shown to any inspector to support the fact that the various units are not necessarily audited with the same frequencies, but the baseline frequency is adjusted depending on the assigned risk.

Supplier audits one of the most essential parts of GMP audits, can be performed to verify on-site whether the supplier can supply the ordered raw material, API, excipient, and packaging material.

EU and PIC/S GMP guidelines state that starting materials should only be purchased from "approved" suppliers, if possible, from manufacturers. This expectation is valid for primary and secondary packaging materials. Naturally, the pharmaceutical manufacturer guarantees that the

TABLE 3.8
Audit Program Strategy

Components of Strategy	Timing Impact
• Base factors	Baseline timing
• Ownership of an auditable unit	
• Audit type	
• System evaluation	
• Previous "critical" audit findings	Increase frequency
• External Inspection expected	Increase frequency
• Identify compliance risk parameters	Modify frequency
• What regulatory commitments have been made?	Adjust as needed to ensure compliance

TABLE 3.9
Baseline Timing with Regard to Facilities and Systems Audited

Facility/System	Baseline Timing	System Audited
Manufacturing sites and laboratories	Annual Biennial Biennial	Quality All applicable systems Independent auditors
3rd party manufacturers	Annual-biennial depending on the process	All systems applicable to the site
Warehouse and distributors	Triennial	Quality materials and computer systems
Marketing affiliates	Triennial	Quality, materials, and computers
Vendors, including API and excipient suppliers, computer and equipment suppliers, third-party service providers	As requested	All applicable systems
Due diligence laboratories	As requested	Quality, laboratory, computers
For cause	As requested	Quality and other applicable systems
Subsystems evaluations	As requested	Verify implementation and compliance of such subsystems as deviations, change controls, stability, annual product reviews, training, etc.
Preapproval	As required when new products/ processes are implemented	Quality plus all applicable systems

drug product it produces is stored and distributed in appropriate packages. In this way, audits mean deciding whether the supplier can be approved or whether it will continue to have this status if it has been approved before. The EU, PIC/S GMP guidelines are in strict expectation of the control of starting materials and mentions a control within specifications. If no test can be done, it is recommended that at least the identity test be done..

For this reason, the pharmaceutical manufacturer or the contract giver will decide to what extent the supplier should rely on the certificates of starting materials and packaging materials, thanks to these audits. Therefore, due to a supplier's audit under GMP, it would be appropriate to discuss another decision beyond whether the relevant supplier can provide raw or packaging materials. Auditing will lead to mutual understanding and trust, as everyone will agree. Thus, both parties are at the beginning of the problems that may arise. It also prevents future disasters. So, although the pharmaceutical manufacturer takes into account the cheapest source that the purchasing department is trying to identify, it needs to understand that this is much more than just ordering cheap goods. The contracting company's ability, that is the licensee, to produce pharmaceutical products with the right and expected quality every time is closely related to the sensitive audit approach and the choices it will make. This approach makes it mandatory for equipment, device, machine, apparatus manufacturers, and software companies to be audited on the axis of design competence. Regardless of the type and scope, the only goal of the pharmaceutical manufacturer when performing audits is whether their products meet the requirements they have determined.

TABLE 3.10
Compliance Risk Assessment Scorecard

Auditable Unit Identifier	Type of Manufacturing Site, Laboratory, Vendor, Warehouse, Etc.	Probability of Regulatory Action	Audit History	Quality Process and Management	Site Characteristics	People	Customer Concerns	Maximum Score Possible	Total Scored	% Compliance Risk
A	API, for oral	1	3	1	1	3	1	30	10	33
B	API, for sterile	3	5	3	5	3	3	30	22	73
C	Fill/finish, oral	3	1	3	1	3	3	30	14	47
D	Fill/finish sterile	5	3	5	5	5	3	30	26	87
E	Warehouse	3	1	1	1	3	3	30	12	40
F	Laboratory, microbiological	5	3	3	5	3	3	30	22	73
G	Laboratory, chemical	3	3	3	3	3	3	30	18	60
H	Vendor, excipients	1	1	1	3	1	1	30	8	27
I	Vendor, API	3	3	1	3	3	3	30	16	53
J	Labeling	3	5	3	3	3	3	30	20	67

TABLE 3.11
Risk Factors and Color Codes

% Compliance Risk Score	Level of Risk	Assigned Color
0%–49%	Low	Green
50%–70%	Medium	Yellow
71–100%	High	Red

3.8 CARRYING OUT THE SUPPLIER AUDIT

Before the pharmaceutical manufacturer audits a new supplier under GMP, it should obtain at least basic information about its likely future partner company. The easiest solution for this requirement is to prepare a list of questions prepared by the drug manufacturer and prepared by the supplier with the correct information. Table 3.12 includes the issues that must be reviewed in a supplier audit.

While the audit is carried out according to the relevant question list, the drug manufacturer should clearly explain the parts in the question list that should be answered as "mandatory" and "optional."

A preferred model in GMP audits is all customers' joint audits of packaging and raw material suppliers. These audits are a method that considerably reduces the costs of both customer companies and suppliers. However, when this type of audit is chosen, the parties should agree on the expectations. For example, a firm not included in the audit may give up on its own GMP audit based on the audit report. In this case, creating a checklist and classifying the deficiencies is essential.

Sampling of starting or packaging materials should be determined before the inspection. If this has been agreed with the supplier in advance, taking samples during the audit can provide additional information that will be very useful.The drug manufacturer should compare the results of the tests with the supplier's tests, which will form the basis for the customer to reduce the number and frequency of subsequent sampling and analytical testing.

The pharmaceutical manufacturer cannot be equipped to control the supplier properly. This is especially clitical for biotechnologically produced starting materials and software. The most plausible thing is to work with an experienced consultant who is quite knowledgeable about the subject rather than making a "blind" purchase by the drug manufacturer customer.

3.8.1 Supplier Audits: API

In November 2000, the ICH Q7 guideline was published for the first time. This document is the API GMP guide, which contains the minimum expectations required for API production. Later, a similar document was included in Eudralex Vol 4 by the European Commission. This guide regulates the production of sterile APIs up to the point just before sterilization. It is intended for API production by using processes such as chemical synthesis, extraction, fermentation, natural resource use or combination. Excluded are vaccines,, blood and plasma derivatives, and APIs used for gene therapy. Pharmaceutical manufacturers can benefit greatly from this guide as they prepare for API audits with qualified GMP auditors. Another source document that can be used to audit API suppliers is APIC manuals.

3.8.1.1 Preparation of the Audit

The API manufacturer's product specifications are often submitted by the pharmaceutical manufacturer's development or purchasing department to the quality assurance department responsible for supplier qualification activity. In such cases, unfortunately, information other than the name and address of the manufacturer does not reach the quality assurance department. The listed questions

TABLE 3.12
Areas to be Checked during a Supplier Audit

1 Supplier quality assurance system
1.1 Is there certification in accordance with ISO 9000, 13485, 15378?
1.2 Is a quality assurance instruction manual available, and is it kept up to date?
1.3 Is quality defined on all levels of the system?
1.4 Is the responsibility of the upper management stipulated?
1.5 Are an organizational structure and job descriptions available?
1.6 Are change control and OOS procedures established?
1.7 Does the supplier contract work out to third parties? If it does, are there contractual regulations?
1.8 Are appropriate approval procedures established?
2 Personnel
2.1 Are the number and qualification of personnel in manufacture and quality control satisfactory?
2.2 Is there a company training scheme?
2.3 How are personnel clothing and health monitoring controlled?
3 Rooms and facilities
3.1 Is the standard of hygiene of the rooms and facilities defined and is it monitored?
3.2 Is the maintenance/calibration of the machines controlled and carried out?
3.3 Are measures in place to prevent cross-contamination and misidentification?
4 Manufacturing procedure
4.1 Are the substances/materials used specified?
4.2 Are written process instructions in force?
4.3 Is the suitability (reproducibility) of the procedure verified?
4.4 Are in-process controls being carried out?
4.5 How is the batch size established?
4.6 Is documentation batch related?
4.7 How are deviations dealt with?
5 Quality control
5.1 Is suitable analytical apparatus available?
5.2 Is the analytical apparatus maintained and calibrated?
5.3 Do written analytical procedures exist?
5.4 Does a sampling plan exist?
5.5 What is the procedure for issuing certificates?
5.6 Are control tests of third parties transferred to the certificate and declared accordingly?
6 Sales
6.1 Who is responsible for dispatch from the supplier to the customer?
6.2 Are the supporting sales documents relating to the customer and the batch retained?
6.3 How is the accuracy of the delivery and adherence to delivery dates guaranteed?
7 Complaints
7.1 Is there an established procedure for handling complaints?
7.2 Are complaints recorded and evaluated in relation to the product?
7.3 How is the customer involved in dealing with complaints?

should be evaluated to make preparation for GMP audits more efficient. Table 3.13 includes questions that may come to mind during the preparation phases.

3.8.1.2 Types of Audits

To establish the scope of the audit and the time required for this, it is helpful to define at the early planning stages which kind of audit has to be carried out.

- *System-based audit* – Qualification of areas and procedures relevant to the system's functionality. In this case, the procedures and organizations at the company are checked.

TABLE 3.13
Organizational Preparations

- Can information on the active pharmaceutical ingredient manufacturer be obtained from the Internet?
- Can company brochures be ordered?
- Are the names of contact persons at the company already known?
- Does the pharmaceutical manufacturer provide a questionnaire that can be completed by the active pharmaceutical ingredient manufacturer prior to the inspection?
- Does the company currently have information on specific products in a Drug Master File?
- Is the active pharmaceutical ingredient manufacturer represented by a broker or trader? If necessary, would it be possible to carry out the inspection jointly with the brokers or traders?
- From the point of view of the API manufacturer and the pharmaceutical manufacturer, who will be taking part in the inspection?
- Do alternative dates/times for the inspection exist?
- Does the API manufacturer produce other starting materials?
- Are other starting materials – particularly active pharmaceutical ingredients – manufactured at the company site?

- *Product-based audit* – In this case, the development of the active pharmaceutical ingredient is audited. This includes the delivery of the starting materials, the manufacturing process, and the laboratory controls.
- *Procedure-based audit* – Is the API to be used produced as specified in the production test instructions? Are the requirements in the instructions clearly expressed?
- *Area-based audit* – This audit is carried out to cover the areas where the API manufacturer makes production.

It is important to know whether the audit to be carried out will be the first audit of the API manufacturer or whether the audit will be carried out for some lack of quality experienced in the API manufacturer.

If the pharmaceutical manufacturer is auditing the API supplier/manufacturer for the first time, it is recommended that this audit be system-based. Through this audit, procedures and systems form a basis for learning about the company.

Lets say that if only one API will be audited separately from all organizations in the API manufacturer, then the audit will be product-based.

In case of any deficiencies in the API, procedure-based audits will include control of the extent to which the procedures for production and testing procedure are followed. In addition, if any deficiencies are seen in the production and testing this will be a critical area of interest in field-based inspection.

An API audit is an audit of the buyer site to the suppliers. The goal of an API audit should be to verify that the sites comply with the principles of GMP for APIs. In order to certify/qualify deliveries, the manufacturing and testing facilities used by the manufacturer need to be audited in order to evaluate the possibilities of accepting material based on a Certificate of Analysis rather than performing full in-house testing. API audits are part of the certification program and should be performed regularly to an approved schedule.

3.8.1.3 Preparing for an Audit

Regardless of whether the audit is conducted internally or by a third party, the auditor and/or the audit team should answer the following standard questions in preparation for API audits:

- How is the audit process defined in the quality management system?
- What information or regulations are available to the auditor with reference to the planned audit (e.g., API product quality review [PQR], risk analysis/risk management plan, qualification of auditors, forms for the audit documents, etc.)?

- What are the relevant auditing criteria (i.e., GMP, GDP, ISO, or other regulatory frameworks)?
- What areas (audit scope) are to be included in the audit (general system and/or product audit, for-cause audit, etc.)?
- Has the company to be audited (auditee) been informed about these areas and are they acceptable to them?
- What is the schedule for carrying out and concluding the audit?
- What is the necessary or expected time allotment for conducting the on-site audit?
- What possible or anticipated problems could appear during the audit, for example, with reference to:
- Unexpected line stops
- National holidays
- Special weather conditions
- Visa requirements
- Vaccinations or special health requirements
- Availability of manpower at the auditee
- The auditability of production of the APIs during manufacturing
- What will be the language of the audit? Are competent translators needed for the audit?
- How much experience has the auditor had with different cultures?

As soon as these questions have been answered and the auditee has confirmed the auditing date and criteria, the next stage of preparation for the audit can begin. The necessary travel arrangements for API audits can be very time-consuming and complex and will not be dealt with in greater detail at this time. A mutual nondisclosure agreement should always be concluded prior to API audits (if none is already in place).

In addition, a preaudit questionnaire, a proposal for an audit agenda and a reply request should be sent to the auditee well before the date of the audit (at least approximately 14 days before the start). This ensures that while preparing for the audit, in addition to internal information relating to the audit, the auditor and/or audit team will have adequate specific information from the auditee as well as a mutually approved audit agenda before the audit begins.

3.8.1.4 Conducting the Audit

It is advisable to subdivide API audits into the following parts:

- Opening discussion
- On-site audit (tour of operations)
- Review of documentation
- Closing discussion

The opening discussion and the closing discussion together make up the formal framework within which the on-site audit takes place. Their significance is often underestimated.

The opening discussion should include an initial statement and introduce the participating companies on both sides (auditee, auditors). It should also define responsibilities and accountabilities, functions, and expectations. In addition, the audit protocol should be addressed and adapted as needed to any new information. Finally, any open questions the auditee or auditors may have can be settled during this discussion, if deemed appropriate.

The auditor should not begin the on-site auditing process without adequate information gathered from the audit preparations or at the latest from documentation submitted during the audit. This particularly includes details of the manufacturing process and the location of the areas to be audited, which are often widely scattered over the company site. It is advisable to conduct an on-site audit in accordance with the process flow, either "forward from start to finish" or vice versa.

The latter possibility is especially recommended if the time allotted for the audit is short and the routes of synthesis of the APIs are long. This makes it possible to set appropriate risk-based priorities during the audit. "Open" questions should be directed to the participants while the on-site audit is in progress. This is the best way to implement the frequently encountered tripartite structure of (GMP) questioning techniques for audits.

The amount of time allotted for the review of documentation in audits should be roughly equal to the amount scheduled for the on-site part. Risk-based priorities in the review of documentation are useful for API audits. As a rule, the following documents should be reviewed:

- Manufacturing authorization
- GMP certificate
- Certificate of Analysis
- Site Master File
- History of official inspections, including results (compliance, noncompliance, warning letters or the like)
- Other certificates (such as ISO)
- Organization charts (company, production, quality, etc.)

SOPs for core processes, including proof of implementation through appropriate records, such as

- Deviation/nonconformities
- Out of specification (OOS)/out of trend (OOT) results
- Change control
- Reuse of solvents
- Ongoing stability
- Supplier qualification/assessment /quality agreement
    - Self-inspection
    - Training
    - Pest control
    - Customer complaints
    - Recalls
    - Reprocessing/reworking
    - Rejects/returns
    - Blending
    - External contractors
    - Transportation

The following documents should be taken into consideration in the special (specific to APIs) part of the review of documentation:

- Proof of conformity to marketing authorization (Certification of Suitability [CEP], open part of Drug Master File, Active Substance Master File, or the like)
- Product quality review (PQR)
- Randomly selected, complete batch documentation of a released batch (production, quality control, packaging and labeling, QA release)
- Process validation (SOP)

Protocols and reports on process validation (manufacturing/synthesis, drying, mixing, grinding/finishing, if applicable micronization, sterile filtration, or the like)

- Cleaning validation (scope depending on the risk profile of the substances and the presence of a multipurpose or dedicated system)

- Qualification of systems and utilities (SOP)
- Qualification data on utilities (water, compressed air, nitrogen/argon, ventilation systems/cleanroom etc.)

Additional documentation must be reviewed during the audit in accordance with the individual risk profile of the API to be audited (such as sterile APIs, and APIs of human or animal origin).

The organization and personnel, including appropriate qualifications and training, employed in any given system, will be evaluated as part of that system's operation. Production, control, or distribution records are required to maintain cGMPs and those selected for review should be included for inspection audit within the context of each of the above systems. Inspection of contract companies should be within the system for which the intermediate or API, or service is contracted and also include evaluation of their quality system.

3.8.1.5 GMP and API

An active pharmaceutical ingredient, or API, is the therapeutic material found in a drug product. It can be derived from chemical or biological sources. Along with excipients, it is the starting or raw material for a drug product. It is also defined as a drug substance in the ICH Q7. This GMP guide is a truly harmonized guide developed through the ICH process involving both industry and regulators across the world. The guide is implemented in most parts of the world.

3.8.1.6 Differences in API Facilities

API facilities may range from large vessels of thousands of liters for chemical synthesis or fermentation to small-scale laboratory glassware. Some equipment may be located outdoors but must provide adequate protection to the material. Controls are most critical when API systems are open. Controls for APIs that could be sensitizing, toxic, or require microbial control will be different. The level of control may be different for the final dosage due to quantities and concentrations of materials.

3.8.1.7 GMP in Relation to API Operations

- GMP is applied in the receipt of materials, production, packaging and repackaging, labeling and relabeling, quality control and release, storage, and distribution. The GMP levels increase throughout the production process from starting materials through intermediates to APIs, and it is important for the manufacturer to define where GMP starts. In a cell culture/fermentation process, ICH Q7 applies to the following:
- Maintaining the working cell bank
- Cell culture and/or fermentation
- Isolation and fermentation
- Physical processing and packaging

3.8.1.8 Overview of API Manufacturing

The chemical synthesis process begins when raw materials and solvents are brought together and charged into vessels (reactors) with subsequent chemical reactions. Typical chemical reactions in API manufacturing include Grignard reactions (addition of a reagent to a ketone or aldehyde to form a secondary or tertiary alcohol), hydrolysis, halogenation, hydrogenation, and oxidations. Frequently a catalyst is added to the raw materials to initiate a chemical reaction. The catalyst will not be consumed during the chemical reaction. Multiple chemical reactions typically occur to form the "raw" API active molecule. This molecule is not pure and must undergo purification to eliminate process impurities, which may include unreacted raw materials, residual solvents, degradation products, and by-products.

The purification may include:

- Distillation
- Extraction
- Filtration

The final molecule is usually in liquid form and has to be isolated through crystallization. Crystallization may be achieved by cooling the reaction mixture or adding a solvent that causes solid crystals to form within the liquid. This results in a mixture called a slurry. Often the crystallization may be induced by a seed material. The seed material's role is to induce the desired crystal form of the API. A centrifuge filter pad is used to separate crystals, also known as "cake" crystals, from the liquid and then discharge the mother liquor. Another liquor may contain unreacted materials, intermediates, levels of the API, and/or impurities.

The diagrams for a synthetic chemical, a single-reaction step process and a typical single product are taken.

When we look at the two flow charts, it is possible to say that both have the same number of unit operations and the process startswith weighing. The main difference is that there are sterilization, inactivation, and storage unit processes in the fermentation process.

The purpose of critical unit processes to be monitored and optimized is to include reaction and fermentation steps to increase API yield while reducing the remaining impurity profile. The main functions of the unit that are important to the end user are precipitation or crystallization, grinding, sizing and purification, which affect the physical properties.

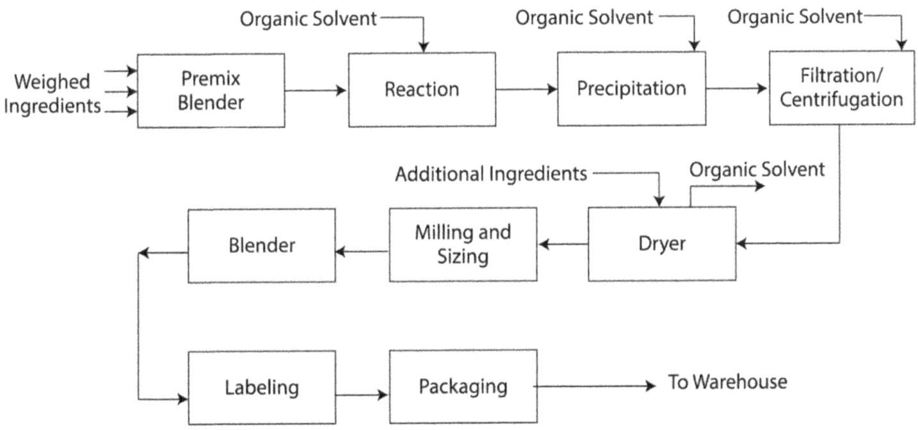

FIGURE 3.5 API process.

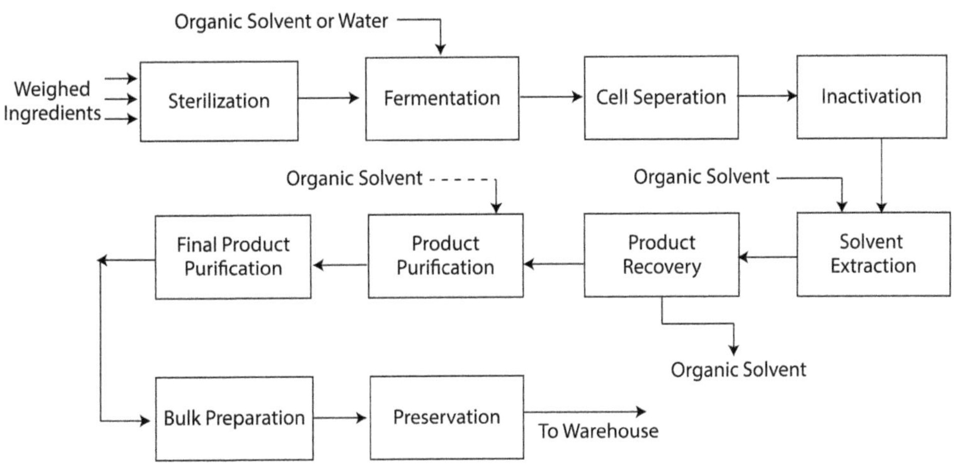

FIGURE 3.6 Typical fermentation API process.

In fact, Current Good Manufacturing Practice (cGMP) requirements are witnessed in every unit operation performed at the facility and compliance with validation documentation is observed. This applies not only to the final API, but also to the basic intermediates used in the preparation of the final intermediates up to the starting materials accepted into the plant. As it is known and clearly stated in the relevant GMP guidelines, the level and criticality of all control and validation activities that will be required increase as a multi-step process approaches the end. As a result, if the critical and final intermediates are outsourced, the acceptance of these substances is ensured not only by certificate of analysis, but also by detailed audits under the leadership of the quality assurance department of the pharmaceutical company.

Figure 3.5 explains the API production process by chemical synthesis, while Figure 3.6 summarizes the API production process obtained by fermentation.

As can be clearly seen from the process flow charts of APIs produced by chemical and biological processes, respectively, in Figures 3.7 and 3.8, as a GMP auditor I can clearly express how such flow charts can change the course of audits. Although mere checklists, GMP guides or corporate documents of the company open doors of opportunity for the GMP auditor. I can say that I have always liked the tidy process flow charts and I have seen a lot of benefits.

3.8.1.9 Definitions of "API"

- Eudralex, Volume 4, Part II
- ICH Q7
- FDA Q7A

A substance or substance mixtures intended for manufacturing a pharmaceutical (medical) product and which becomes an active ingredient when used to manufacture a drug. Such substances provide pharmacological activity or other direct effects in diagnosis, treatment, alleviation, medical intervention, and to prevent disease or to affect body functions and structure. Such substances provide pharmacological activity or other direct effects in diagnosis, treatment, alleviation, medical intervention, and to prevent disease or to affect body functions and structure. The presence of such substances in drug products is intended to provide pharmacological activity in diagnosing, treating, alleviating, and preventing diseases. It is also essential to provide other direct effects that influence the structure and function of the body.

The API's structure, origin, or manufacturing method can be decisive for this. The following API types are distinguished in the aforementioned regulatory works or in works subordinate to them in the regulatory system:

- Chemical
- Herbal
- Microbial
- Biotechnological (including genetically engineered)
- Human origin
- Animal origin

It would certainly be possible to continue the outline, branching off into many subgroups or a combination of different types of origin. However, the given differentiation is adequate for the purpose at hand to clearly identify the regulatory structures to be applied to the particular active substance.

3.8.1.10 API Scandals: Heparin

Table 3.14 describes the most notorious API scandals.

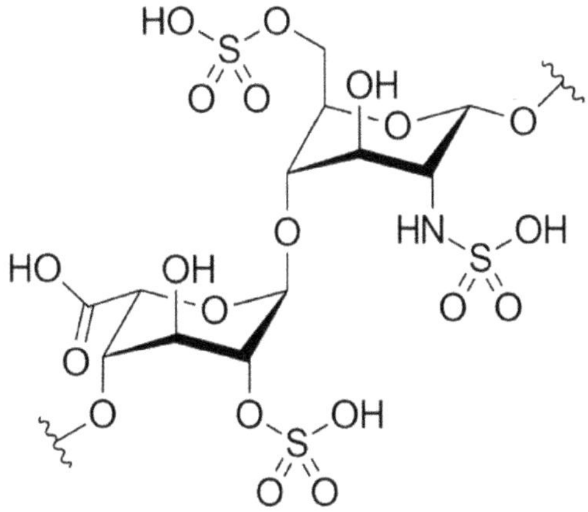

Heparin structure. From Wikipedia, https://en.wikipedia.org/wiki/Heparin#/media/File:Heparin.svg

TABLE 3.14
APIs

- Heparin
  - Baxter – 2008, more than 80 deaths in USA
  - Investigated – FDA (GMP – sourcing of starting material, lack of control)
- Nelfinavir, Lopinavir/Ritonavir
  - Roche – 2008, a global recall of batches
  - Genotoxic substance (GMP, cleaning of tanks)
  - Morphic forms

One of the most tragic events in the pharmaceutical industry is undoubtedly the case of heparin. Before explaining how this event happened, we need to get to know heparin.

- What is heparin?
- What does it do?
- Who is producing it?
- How and where and under what conditions is it produced?
- How and under what conditions is it supplied?

Heparin is a very effective anticoagulant. Anticoagulants are also known as blood thinners. Products such as heparin are given by intravenous or subcutaneous injection. It is frequently used in surgeries and dialysis. Its main source is pig gut and has been known since the 1930s. After heparin is taken from the pig gut, it goes through a series of drying processes. After these stages, crude heparin is refined and sold by companies producing APIs.

Heparin is actually an API. There are a number of directives and rules that companies related to API production must comply with, including ICH Q7, EU, PIC/S GMP, and FDA GMP. Although heparin used for pharmaceutical purposes is obtained from cattle and pigs, some methods have been tried to obtain it outside of animals. Baxter is one of the pioneers of companies marketing heparin as a pharmaceutical product. The first address in the supply chain actually goes back to more than ten crude heparin plants established in China.

After heparin was sold as a commercial product by Baxter, complaints about critically ill patients began to be received from many hospitals. In response to this situation, Baxter initiated a large-scale recall, and the FDA decided to investigate the case thoroughly. As a result of the research, it was determined that a substance called chondroitin sulfate, which chemically imitates heparin and is much cheaper to produce, was responsible and was among the banned chemicals. It has been understood that this substance, which is not found in nature and is produced only with chemical modifications, is intentionally included in the supply chain. In detailed laboratory tests, it was determined that approximately 2%–50% of the lots were contaminated with this chemical. Although the FDA is investigating in-depth, the main reason is that Baxter stepped into this facility without prior approval. What is even more tragic is that the Chinese government has not audited the company that produces this chemical on the grounds that it does not serve the pharmaceutical industry.

What questions should you ask as a GMP Auditor?

1. Why did Baxter purchase a product from a facility that the FDA had not previously approved?
2. Why did Baxter wait until 2007 to perform an on-site inspection of the facility?
3. How did the FDA come to the conclusion that it is a safe facility without going there and without pre-approval inspection?
4. How did the FDA confuse SPL-CZ (manufacturer) with another facility in its database?
5. Why did Baxter want to buy this product from China, which is experiencing constant contamination cases?
6. Would this event have happened if the FDA had carried out this inspection on time and determined that the facility did not comply with the GMP criteria?
7. Do we need a comprehensive understanding of heparin and glycosaminoglycan structure?
8. Should we take a closer look and update our analytical testing methods?
9. How can we control and qualify our suppliers more tightly?
10. Should we return to a non-animal heparin solution, and if so, what should be the criteria?
11. How should the FDA increase its control over high-risk facilities overseas?
12. How does the Chinese government distinguish chemical factories from pharmaceutical-grade API facilities?

Which parameters should be considered in APIs?

- Polymorphism
- Solubility in water
- Route of synthesis
- Solvents used
- Impurities
- Sterile API
- Fermentation
- Toxicity
- Activity/potency
- Particle size
- Other properties to be considered
- Site compliance information (WHO/EDQM/Other)

EXAMPLES OF OBSERVATIONS OF NONCOMPLIANCE

Batch records and related SOPs

- Before continuing the step-by-step… a complete centrifugation before the actual operation
- Step is in progress in real time but BMR is incomplete
- There is no evidence of when the cooling started ... temperature log is not available … although it is written in the BMR and equipment logbook, there is no entry that must be made every 30 minutes …
- There is no information in the relevant SOP that the centrifuge bag should be cleaned … … Evidence for the following is not available:
  - Bags defined as dedicated
  - Labeling of drums used in storage
- Also risks associated with fluid bed bags and cross-contamination

Examples of major noncompliances

- QMS
  - No evidence of knowledge and experience
  - No report for deviations
  - No completed change control
- Documentation
  - Lack of reporting and reports full of errors
  - Process validation is not complete
- Management of materials
  - Sampling procedure (sampling methods, date of release is before the manufacturing)
  - Material storage and use is not appropriate (FIFO)
- Buildings and facilities
  - Poor design of water and HVAC systems
  - Cleaning of equipment – equipment still carries residue after the cleaning process; cleaning of the equipment in another area

3.8.1.11 GMP: From Starting Material (SM) via Intermediate (IM) to API

- Definition of Starting Material
- Definition of Critical Parameters
- Definition of Critical Process Steps
- (= identification of key intermediates)

Note: Definition for "critical" parameters is the direct influence on the quality of APIs and not on this specific process (performance).

Figure 3.7 describes the GMP requirements in an API produced with multiple steps, while Figure 3.8 summarizes the increasing GMP requirements.

As can be clearly seen in Figure 3.8, while a simple finished drug product is produced and packaged, for example, with a three-stage production process, it is not possible to tell how many complex stages API products go through from starting materials to the finished API, and each of these stages is more GMP critical than the previous step. Therefore, auditing of a finished drug product production process has always seemed to me an easier and shallower approach.

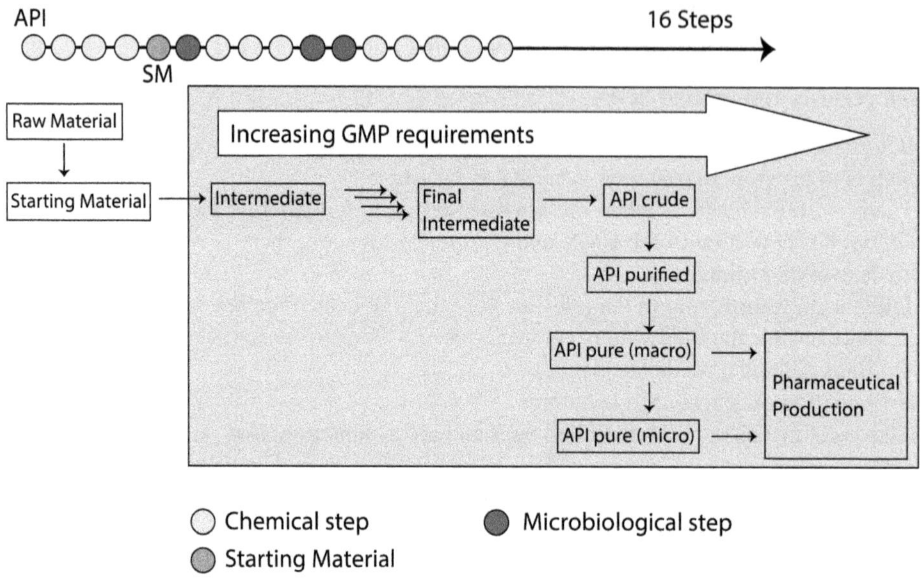

FIGURE 3.7 Multiple steps.

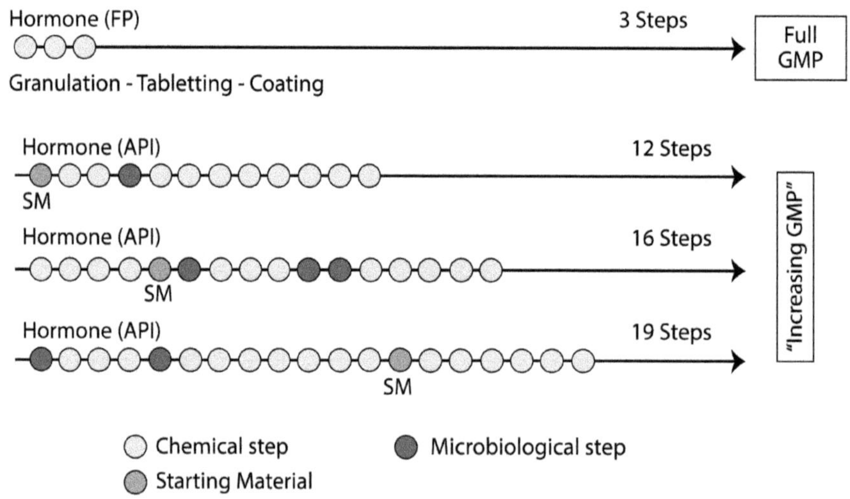

FIGURE 3.8 GMP requirements.

Auditors question cycle:

- Are starting materials, critical processes, and critical parameters/key intermediates defined?
- What are the rationales behind (risk assessment, development report, etc.)?
- What is defined in the submission? Check for compliance as is ⇔ submission.

3.8.1.12 Dedication Strategies for APIs

Dedicated – Equipment and facility are dedicated to one API and its intermediates all the time.

Multiuse – The facility has a mix of different APIs and its intermediates, but they are manufactured in dedicated trains of equipment, generally concurrent operations

Multipurpose – The facility has a mix of different APIs and its intermediates, trains of equipment have no specific API assignments, both concurrent and campaign operations.

3.8.1.13 Batch Release

One of the FDA's biggest concerns is whether the released API product has been fully tested. The API release process is usually based on the valid test results written in a final Certificate of Analysis, and the main expectation of the FDA is that the person making the release sees all the tests related to the relevant lot and makes a decision. The main expectation is that the batch will be released as a result of the independent evaluation of the production and analytical tests. The procedures used at the API facility are expected to address an OOS/OOT situation, if any, to the root cause during production. The release task should not be performed by the person performing the analytical tests. Generally speaking, the release must be done by QA personnel independent of the QC and production departments.

3.8.1.14 Deviations and Failure Investigation

Deviations encountered during production in API production facilities are usually not reported, so it is not possible for production and quality assurance management to be aware of the situation.

If deviations and failure investigations opened, as a result, are not closed in a timely manner, it raises great suspicion. Also, failure to detect the root cause or causes of defect investigation is a separate source of concern. Another major concern is that no future measures are taken to avoid similar deviations and that other processes and batches affected by the deviation are not examined.

Timely documenting and reporting of deviations from approved procedures in periodic trainings given to people at API facilities should be addressed, and proof that they understand the subject clearly should be secured with documents. The GMP may accept human error as the root cause if all deviations are examined to the root cause and if possible, other factors affecting the root cause can be excluded. The deviation and failure investigation SOP used in API production facilities should explain in clear and understandable language what steps should be taken when deviation occurs. QA personnel are expected to manage such deviation or failure investigation activity. There are those who argue that deviations should be handled as planned and unplanned. If a deviation is to be evaluated as planned, I think it would be more appropriate to manage it with change control instead of deviation. Therefore, the deviation can be thought of as going beyond the predetermined procedures completely. Deviations are one of the most questioned topics not only in the FDA but also in all health authority audits. For this reason, the FDA may take the term "planned deviation" seriously and reveal some findings that may upset you. Your task as an auditor is to start reviewing lists of rejected lots, malfunctions, and deviation reports. GMP auditors do not care too much about how and by whom routine events are handled, and the steps that the API facility will take or take steps in the face of unexpected events, deviations, and malfunctions that will make them think are much more critical.

3.8.1.15 Change Control System

It can be expressed as a systematic approach in which any change is proposed, evaluated, approved, implemented an reviewed throughout the fim (ICH Q10).

- The scope of change management is much broader than change control, which is typically applied to one change at a time.
- Change management includes oversight and management of the entire portfolio of changes and the change process, including all the change control components.

- In a pharmaceutical quality system (PQS) developed according to Q10, change management applies across the entire product lifecycle.

Unfortunately, change control in API facilities is not fully understood and is not applied correctly because it is only used for equipment and processes. Whereas, in facilities, equipment, instruments, processes, procedures, personnel, cleaning, suppliers, areas, analytical methods, specifications, etc., in many cases, a change control process can be initiated. Regardless of whether changes are major or minor, they need to be accounted for and evaluated. As GMP auditors, you should know that just because an API facility has a change control procedure does not mean that the company is doing its job right. You should find that changes are allowed to be considered unimpeded, even in the face of worst-case scenarios, for example, urgent changes. Seeing that changes that do not directly affect the quality of the product are rejected does not mean that you should skip that issue, but rather you should pursue how the change was decided as a minor change. As with deviations, QA personnel are expected to manage the process in this procedure. Another anomaly that I noticed in API facilities is that maintenance and engineering departments that are directly or indirectly affected by the change control process are not included in the process. It is a GMP auditor's job to handle changes as critical, major, and minor, and to obtain evidence that risk and impact assessments have been made.

You should see that the changes are handled within the framework of the quality assurance system, that each change is assigned an automatic or manual number, that each change is closed at the specified time, and you should also find evidence that the effectiveness of these changes has been evaluated. As a GMP auditor, you should expect validation, qualification, calibration, stability studies, and training to be provided to you with proof documents after each change.

Impurities that start with the active ingredient valsartan left deep traces in the entire industry. I consider this finding invaluable, as it provides evidence of exactly how the change control procedure was violated, which I got from a warning letter posted on the FDA web page:

> *Failure to evaluate the potential effect that changes in the manufacturing process may have on the quality of your API.*
>
> In November 2011 you approved a valsartan API process change (PCRC - 11025) that included the use of the solvent **(b)(4)**. Your intention was to improve the manufacturing process, increase product yield, and lower production costs. However, you failed to adequately assess the potential formation of mutagenic impurities when you implemented the new process. Specifically, you did not consider the potential for mutagenic or other toxic impurities to form from **(b)(4)** degradants, including the primary **(b)(4)** degradant, **(b)(4)**. According to your ongoing investigation, **(b)(4)** is required for the probable human carcinogen NDMA to form during the valsartan API manufacturing process. NDMA was identified in valsartan API manufactured at your facility.
>
> You also failed to evaluate the need for additional analytical methods to ensure that unanticipated impurities were appropriately detected and controlled in your Valsartan API before you approved the process change. You are responsible for developing and using suitable methods to detect impurities when developing, and making changes to, your manufacturing processes. If new or higher levels of impurities are detected, you should fully evaluate the impurities and take action to ensure the drug is safe for patients.
>
> Your response states that predicting NDMA formation during the Valsartan manufacturing process required an extra dimension over current industry practice, and that your process development study was adequate. We disagree. We remind you that common industry practice may not always be consistent with cGMP requirements and that you are responsible for the quality of drugs you produce.

There must be written procedures addressing all changes in the following areas:

- Facilities
- Production equipment
- Production processes

- Utilities
- Raw materials
- Specifications
- Analytical methods
- Analytical Instruments
- SOPs
- Computer-related systems

3.8.1.16 Annual Product Review/Product Quality Review

As a GMP auditor, one of the most important nonconformities that may be encountered in an API facility is the incomplete or incorrect execution of annual product review (APR)/product quality review (PQR) activities. Unfortunately, what many API facilities understand from APR/PQR and what the GMP auditor understands can be very different. The facility's understanding is that it claims to have completed the APR simply by performing the analytical tests and demonstrating that it is within specification. In fact, what we need to look for as a GMP auditor is whether this activity is covered in the relevant SOP.

3.8.1.17 Warehouse Operations

In a situation that we are accustomed to seeing in API facilities, it is an environment where warehouse areas are dirty, messy, and poorly lit. In the warehouse areas, labeling that expresses the status of the materials is not done completely, or it is possible to come across unlabeled containers and barrels. Another annoying sight I encounter during the audits is that no precautions are taken to prevent the materials kept in the warehouse garden from being affected by the weather and environmental conditions. Warehouse areas are the starting point of the facility tour in a GMP audit. The fact that the materials are taken to the warehouse areas means that you are at the very beginning of the production process of an API product. Therefore, your first impression of the warehouse areas will give you an idea about the overall facility. Warehouse areas are also where you will have the chance to interview an employee for the first time as a GMP auditor. An interview here can give you an insight into how, for example, training activities or job descriptions are managed throughout the facility.

3.8.1.18 Qualification/Validation

Equipment and utility qualifications, as well as process and cleaning validations, are critical to an API facility. It is not uncommon for a health authority to want to see all validations complete before an API product is released. Completing certain validations and qualifications, especially before proceeding to process validation, is critical from a GMP perspective. For example, it would be more accurate to talk about process validation after the validation of analytical methods, completion of cleaning validations, qualifications of equipment used in tests and production, and computerized system validations of the software to which these equipment are connected. A test or a step of the production process carried out with equipment that is not qualified, that is, does not complete the DQ, IQ, OQ, and PQ phases, should be easily caught by a careful GMP auditor. It is preferable to have a policy for validation and qualification activities, but its absence is often not considered a finding. However, a well-functioning SOP that explains each step of qualification or validation in detailed and understandable language is very helpful. It is also expected that there will be a validation master plan throughout the facility and that all validation and qualification activities will be defined under this plan. The purpose of the validation or qualification activities and who they cover in which departments should also be clearly stated. Some of the deficiencies I have encountered in the audits are as follows:

- Equipment is put into use before the qualifications are completed.
- Risk assessments that will affect the operation of the equipment in the change of location or part are not made.

- The software built into the equipment is not validated.
- Qualifications of HVAC, water, steam, and gas systems are incorrect or incomplete.
- No questioning of the competencies of companies performing calibration activities.
- Process validations do not include critical process parameters, critical quality attributes, or critical material attribute information.
- Taking into account one or two variables in the worst-case scenario study in cleaning validation.

Critical quality attributes (CQAs) may include:

- Chemical purity
- Impurity profiles
- Particle size
- Density
- Moisture and solvent content
- Homogeneity
- Microbial quality

Critical process parameters that affect critical quality attributes must also be identified and ranges established. Different API processes may have different critical process parameters.

Typical critical process parameters are:

- pH
- Pressure
- Temperature
- Mixing speed

These critical process parameters should be routinely recorded with specified ranges during the manufacturing stages and verified as acceptable at batch release. Time limits for the completion of significant manufacturing steps should be established and documented.

For intermediates that are isolated and stored before further processing, storage conditions, hold times, and container types must be specified. If a process change is mentioned in APIs, it is expected that this change will not change the impurity profile and API quality properties.

Printing devices should be monitored to ensure all imprinting conforms to established specifications. If the intermediate or active substance is to go out of the manufacturer's material management system, then information such as retesting and expiration dates should be on the label. The label should contain special shipping or storage conditions for an API or intermediate. Packaging and labeling facilities should be inspected before use, as a precaution for mislabeling. This should be documented. APIs shipped outside of the manufacturer's control must have tamper-evident seals attached/securing the container. A system must exist to ensure that the distribution of each lot can be determined to facilitate recall procedures.

3.8.1.19 Equipment

Common problems observed regarding API production equipment include the following:

- Lack of status labeling (e.g., clean, dirty, out of service)
- Lack of labeling to show identification of the batch in-process

- Lack of unique equipment number identification, especially for ancillary/portable equipment (pumps, filters, mobile carts, and mobile vessels)
- Lack of labeling to show calibration of instruments (e.g., temperature and pressure gauges, flowmeters)
- Failure to have logs of equipment use, cleaning, and maintenance
- Failure to have an effective preventive maintenance program
- Poor management of material transfer hoses
- Difficulty in distinguishing "clean" hoses from "to be cleaned"
- Lack of permanent identification
- Lack of cleaning records

As auditors, hoses in production environments should always remain a point of attention. The ends of the hoses may be open or they may not be where they should be. In such cases, the cleanliness of the inside of the hose is often questioned. In fact, it is a situation that can be seen in API facilities, and as a GMP inspector, you should look for labels that clearly show the maintenance, cleaning, malfunctioning, calibration, and qualification status of any equipment.

Labeling should not be left to the disposal of any department or personnel. There should be a company-wide SOP where all labels are managed, and it should be stated that the labels are managed in such a way that they contain the correct information and that they are printed in the correct number each time. Labels must be legibly placed on materials such as reactors, tanks, vessels, pumps, drums, and bottles in API facilities. Another problem that may be encountered in API facilities is the monitoring of material transfer hoses. These hoses are often in contact with the product and often their status is not tracked and recorded. It is important to prove that the inner surfaces of the hoses are cleaned effectively and without leaving any residue. Hoses labeled as cleaned and dirty should be stored in separate locations to avoid mixing.

3.8.1.20 Solvent Recovery

Recovery of solvents in API production is a common and necessary industry practice. The important issue with recovered solvents is the potential carryover of impurities into steps or products for which there is no design in the production process to remove such impurities. There are three sources of solvents for recovery:

1. Recovery from and reuse within the same step
2. Recovery from and reuse within the same product at different steps
3. Recovery from multiple streams and reuse in common distribution

3.8.1.21 Personnel Training

Problems observed regarding training programs include the following:

- Indicators of ineffective training
- Training documentation that does not demonstrate the training received by individuals
- Training documentation that does not demonstrate who has been trained on specific procedures
- Documentation not available to demonstrate training has occurred following changes in processes or procedures
- Training program and responsibilities not clearly defined in written procedures
- Lack of training effectiveness assessment
- Lack of GMP training documentation for top management
- Lack of training documentation for temporary and contract employees who perform GMP-related functions

CASE STUDY

As a GMP auditor, prepare observations and audit findings based on the pictures in Figures 3.9 and 3.10.

Write an audit finding based on the following gathered information:

- These drums were stored in the backyard of a site that manufactures API intermediates.
- All drums were empty, two were half-full (content unknown, strong odor of organic solvents noticed).
- The site also handles hazardous materials.

Write an audit finding based on the following information:

- The site manufactures intermediates and active ingredients.
- The water seen in the picture is the outcome of the washings of the empty reaction vessels and reaction room.
- This is the general practice at the facility.
- This was repeatedly observed in a follow-up audit.

Standards vary.
Equipment conditions vary, too…

General

Much less automation in the processes/equipment.
Manpower reduction is not a topic.

FIGURE 3.9 Two drums.

FIGURE 3.10 Water seen.

Example (India):
The building and equipment have extremely high cleanliness standards. The whole production looks more like a Solida production than API production…

Example (China):
Equipment was partly in very bad condition, no proper maintenance was done.

For one vessel the axle was not centered, oil was leaking from the gasket, and a fiber cloth was bound around the axle (in the vessel) to catch the drips … Additionally, two of the bottom attachment were rusted through.

Much less automation for heavy work too.

Example (India):

- Carrying dozens of 25 kg bags from the warehouse over the courtyard and a staircase on the first floor of the manufacturing building to fill the vessel
- Filling of the tumble dryer with the finished product (starting material for parenteral API) is done by a co-worker with a shovel out of drums (no controlled environment, the body of the co-worker "hangs" above the tumble dryer). This takes more than two hours → no surprise there were deviations due to particles in the solution.

Transportation: Road Condition/Size

- Accessibility of the company for trucks
- Road conditions during monsoon

Example (India):

- Sandy road, one-car wide
- No asphalted or gravel road

Energy/Water
Power and water supply availability; Power failure/restriction; water supply during dry and monsoon season

Example:
Restrictions for power supply by the authority during the hot season:

Every Wednesday & Thursday power consumption was restricted to 50 kW = just enough to maintain storage conditions in the warehouse, no air conditioning or any other activity in the manufacturing area as possible

Experiences … Education/Language

China
Language: Audit is without a translator impossible (documents), even then questionable points: "nice try" or misunderstanding?

Mostly one or two of the management have basic skills in English (e.g., sales function)

Education: Chinese education, almost literate

India
Language: Easier according to the language: management is fluent in English, staff partly: some fluent, some little basics

Education: Upper management has sometimes overseas education, overseas experiences or consultant, management who worked for global pharmaceutical companies; illiterate people employed

The questionnaire filled out by QA: Synthesis performed at the site in dedicated equipment; batch size 500–1500 kg (COA for 1000 kg was delivered with samples)

Reality: Crude material is bought, equipment is not dedicated, mixer for final mixing has a Volume of max. 100 kg.

Example (India):

- Opening big bags for qualification purposes on our site revealed overshoes and stones in the product.
  Explanation: A supervisor was downgraded and this was his revenge… (he is fired by now.)
- Authorities visited a company with a good GMP level, however later it came out that there was a federation of companies that built/upgraded one company for visitors, and the product was produced in the other ones…

Experiences … Intellectual Property

Example (China):
(Introductory presentation)

- One advantage of the location of the company was that there are no other (pharma) chemical plants in this area.

Explanation: Because new synthesis/products will not be directly copied by neighbor companies like in other industrial areas.

- No parameters are described in the batch record due to the fact that they do not want to reveal the parameters even to their co-workers (they had already bad experiences…)

Environmental, Health and Safety Aspects

Environmental protection

Not much information, not the audit focus.

China: it is interesting to note that almost seven of the ten most polluted cities in the world are in China, and 70% of all rivers are polluted.

Example (China):

- 100 tons of benzol were flooded into the river after an explosion in a petrochemical company. The production site was quite new and had been recently initiated
- Mother liquor containing chromium oxides is recycled to split the chromium oxide components (which are sold afterward), then biological wastewater treatment of the solution

Example India:

First country of the world who put environmental topics in the constitution (1976) BUT: union states are responsible for control; due to systematic corruption in the official and private sector no real implementation of environmental protection took place

Health and safety aspects

- Not comparable to European standard

Example (China):

- The vessel filled up with acid solution manhole was open, and the stirrer was running.
- Laboratory people are wearing sandals, no socks.
- The production area air condition was not separated properly to labs and offices.
- Part of the staff had exceeded hormone blood levels and women had bleeding in between.

Example (India):

- Wiring to measurement devices was diagonal through the manufacturing area and not fixed properly.

TABLE 3.15
API Audit Report Example

Audit report number	Use the audit report number which is used for the detailed audit report (e.g., ACME HEALTH report number) to reference this report
Company audited	Name and full address of the audited company (manufacturing site and Headquarters if appropriate): Company name Address City Country Telephone/Fax Email/Homepage
Activities performed	Manufacture of active pharmaceutical ingredients ☐ Manufacture of finished medicinal products ☐ Packaging ☐ Import ☐ Laboratory control ☐ Control and release of batches ☐ *List of APIs
Audit date(s)	Date(s), month, year
Auditor(s)	Lead auditor/subject matter expert/support: name, function department, site
References	Reference to marketing/manufacturing authorization numbers, Drug Master File (DMF), Certificate of Suitability (CEP No.), GMP certificates, etc.
Introduction	• Short description of the company and the general activities of the company (history, location, activities, products, employees) • Overview of current status and future development of cooperation with ACME HEALTH (products and processes) • Previous audits (official, with date(s), names of auditors) • Short description of production processes covered by the audit
Short report of the audit:	Type of audit: Routine ACME HEALTH surveillance audit
Scope of audit	The GMP systems and processes related to the manufacturing control and final disposition of the above-listed APIs for the ACME HEALTH Divisions have been inspected; audit covered all key processes including quality assurance and quality control activities involved in the manufacturing of APIs In the course of the audit the production and quality control facilities for [API] were toured and various representative GMP-related documents, data and corresponding information (e.g., organizational charts, management reviews, training documentation, standard operating procedures, validation and qualification protocols/reports/master plans, batch records, product quality reviews, the calibration and maintenance system and records, analytical records) were thoroughly reviewed and evaluated regarding compliance with the applicable GMP requirements
Areas audited	Each audited area should be specified, e.g.: • Quality control • Quality assurance • Warehouse(s) • Production buildings • Water plants, etc.
Persons involved in the audit	The names and titles of key personnel met should be specified

(*Continued*)

TABLE 3.15 (CONTINUED)
API Audit Report Example

Overview of the audit	The audit covered the following; Quality system Management responsibilities Personnel qualification and training Documentation systems Controlled environment Regulatory compliance management Change management Deviation management Self-inspections Sites, facilities, and equipment Qualification and validation Materials handling Production and process control Reworking and reprocessing Packaging and labeling Maintenance and calibration Sampling, testing Stability testing Supplier management Computerized systems Technology transfer
List of major and critical deficiencies	No critical (and major) deficiencies have been identified. Major deficiencies: Major improvements are needed in the following areas (all major observations listed in the report, if any, have to be included; quote the respective chapter and shortly describe the necessary improvements): • Documentation systems: improvements needed regarding • Qualification and validation: improvements needed regarding
Recommendations	The company has been asked to identify the underlying root causes for all observations and to provide an action plan with corrective and preventive measures including appropriate timelines for completion. The progress made regarding the specified corrective/preventive actions will be continuously followed up, tracked and verified by ACME HEALTH Compliance Audits & Support (or another responsible audit organization) until close out of the inspection.
Summary and conclusion	[Company at the site name] has established all necessary systems and controls and is generally in compliance with the appliable EU GMP requirements for APIs as established within part II of the EU GMP (for other categories of APIs, e.g. APIs for Ectoparasiticides, a-typical APIs, or sterile APIs, refer alternatively to other applicable regulations, e.g.: EU GMP Requirements for sterile APIs as established within part I, Annex 1 of the EU GMP). Identified deficiencies have been acknowledged by the responsible site management including the Quality Unit and will be addressed and corrected according to the established action plan Therefore, [company] is being considered acceptable for the production, release and distribution of [products: list all APIs here] for the ACME HEALTH
Lead auditor (Signature, Date)	
Responsible person of the auditing organization (signature, date)	
Responsible person of the Quality Unit of the manufacturing site/qualified person	

TABLE 3.16
Key for Classification of Findings

Level	Classification Rating
Critical	A deficiency that has produced, or leads to a significant risk of producing, an active pharmaceutical ingredient that could be harmful to the human or veterinary patient. The condition violates essential cGMP rules and/or essential quality assurance practices.
Major	A noncritical deficiency that has produced or may produce a product, which does not comply with its marketing authorization or which indicates a "major" deviation from cGMP, or a combination of several "other" deficiencies, none of which on their own may be major, but which may together represent a major deficiency and should be explained and reported as such.
Other	A deficiency, which cannot be classified as either critical or major, but which indicates a departure from cGMP. (A deficiency may be "other" either because it is judged as "minor," or because there is insufficient information to classify it as a major or critical).

An API audit report example is shared in Table 3.15.
Table 3.16 provides examples of the classification of API audit findings.

3.8.2 Supplier Audits: Excipients

Let's look back at our memory a little and go back to 1998, and the WHO, in its written statement, should sadly remind that more than 500 people lost their lives due to fake excipients in the last 60 years.

1995 – Haiti, glycerin poisoning by diethylene glycol
2007 – Panama, diethylene glycol contaminations
2008 – Heparin with over-sulfated chondroitin sulfate (OSCS)
2008 – Milk and derivatives with melamine
2009 – Nigeria, diethylene glycol contaminations
2010 – Melamine in milk-containing products
2011 – Breast implants made with industrial-grade silicon
2012 – Sodium nitrite marked and used as sorbitol

3.8.2.1 Assessing the Excipient Supplier

Questionnaire

- Advantages
  - Efficient
  - Low cost
  - GMP compliant?
- Disadvantages
  - Trust supplier responses
  - "One size fits all"
  - No assessment of production risks
  - No assessment of laboratory operations

3.8.2.2 What Is a Pharmaceutical Excipient?

According to the IPEC: "Pharmaceutical excipients are substances other than the API, which have been appropriately evaluated for safety and are intentionally included in a drug delivery system." (Note that this definition would exclude medical devices.)

The IPEC definition has several key properties:

- First, the phrase "other than the API" indicates that excipients have no therapeutic effect. That is, they should be pharmacologically inert.
- Second, there is an expectation that the excipient has been appropriately evaluated for safety. This indicates that they should not harm the patient who is being administered the drug product and that at some time there has been an evaluation to demonstrate that they are indeed safe.
- Finally, the word "intentionally" indicates that the excipients are included in the drug product by design.

In terms of patient risks, it is best to add as little as possible to the drug product that is not absolutely necessary. But in reality, a dosage form comprised of only the API would not be simple to administer especially for those that are very potent. Often it is the route of administration that determines the other ingredients that aid delivery of the API. For example, a surfactant in the preparation of a cream for topical applications, a binder in a tablet, and even water for injection in a parenteral product, all of these additional ingredients are excipients.

There are other definitions of excipients, most notably the European Pharmacopoeia definition and the definition given in the Falsified Medicines Directive.

The European Pharmacopoeia provides the following definition:

> Excipient (auxiliary substance): Any constituent of a medicinal product that is not an active substance. Adjuvants, stabilizers, antimicrobial preservatives, diluents, antioxidants, for example, are excipients.

The Falsified Medicines Directive (2011/62/EU) provides the following definition:

> Any constituent of a medicinal product other than the API and the packaging material.

Some excipients may have more than one function making them especially valuable. It is the function of an excipient that is its most important attribute. Often functionality is a result of the fact that an excipient is a complex mixture of molecular entities, which is in sharp contrast to APIs.

Table 3.17 presents the different functions of different excipients as an example.

As long as excipients are pharmacologically inert, safe for the patient, and have the desired functionality, then there is no restriction on what they may actually be. In that regard, excipients can be gases, liquids, or solids. Their origins also encompass a broad spectrum including materials derived from animals, minerals, vegetables, synthetic entities, and all mixtures of these in almost any combination.

A survey of UK marketing Authorizations revealed that over 1200 excipients were in use, excluding flavors and colors. Similarly, a survey of the FDA Inactive Ingredients Database indicated that a similar number of excipients were listed. If these numbers are repeated worldwide, then it is likely that the number of different excipients in use in the industry is very large indeed.

- Another defining feature of excipients is that they are rarely designed for use in pharmaceutical applications. Typically they are entities that have been created for other purposes, such as cosmetic ingredients and food additives.
- Although excipients have a great deal in common with APIs in that they are both components of drug products and so come into direct contact with the patient, an overriding difference is that the excipient is pharmacologically inert and this can justify a different and less rigorous application of GMP in the manufacture and distribution of excipients.

Whereas an API is only used in a limited number of drug products, an excipient can be used in a very large number of drug products. Thus one supplier could provide an excipient to many users

TABLE 3.17
Possible Functions of Excipients

- Binder
- Compression aid
- Disintegrant
- Color/flavor
- Filler
- Sweetener
- Lubricant
- Preservative
- Glidant
- Suspending/dispersing agent
- Humectant
- Buffer
- Coating/film former
- Release modifier
- Adhesive
- Penetrant
- Vehicle
- Thickener
- Solvation

and therefore have an impact on a large number of drug products if the excipient was found to be improper in its quality. This distinction sets excipients firmly apart from APIs where the API is most likely used by a smaller number of users and in a more limited set of dosage forms. It follows that the risks to patients posed by excipients are potentially more far-reaching than those related to APIs and it is all the more surprising that there remains limited legislation that requires them to be manufactured and supplied in compliance with GMP and GDP.

Before looking at the international GMP guidelines, it is possible to say that many pharmaceutical excipient manufacturers or suppliers are dealing with ISO 9001. With the ISO 9001 quality management system, a favourable benefit has been achieved in establishing appropriate systems for the production of pharmaceutical excipients.

Excipients significantly affect the finished product quality, in some cases making up almost the entire formulation. Many excipients are used in industries, such as chemistry, food, cosmetics, and dye rather than the pharmaceutical industry. As a GMP auditor, you may not often require a supplier or manufacturer with such a large customer portfolio to have as stringent rules as pharmaceuticals. However, it should not be acceptable for this excipient supplier or manufacturer to base all its processes solely on ISO 9001 or a similar standard.

Therefore, a program must be in place that will monitor these excipients and provide the necessary assurance that they meet the quality parameters for pharmaceutical manufacturing processes.

Excipients are purchased and used in this way without going through further purification and refining processes. If there are some impurities in the excipient that are not detected by the manufacturer or are hidden even though they are detected, and because the drug manufacturer does not have advanced equipment and laboratory equipment, these impurities will pass into the drug product. Before EU GMP guidelines added detailed risk assessment in excipients (both excipient manufacturer and drug manufacturer would do it) as a condition, when we questioned the drug manufacturers, the answer we usually received was that their own resources were limited and they were controlled by performing the analysis certificate or representative tests of the excipient. Therefore, the excipient manufacturer has more control over the excipient in terms of all other controls. As an excipient manufacturer, we expect the processes to be periodically monitored for performance trends and the pharmaceutical manufacturer to have information on this subject. In addition, the majority of equipment used in excipient production and environmental conditions reflect the chemical industry rather than a pharmaceutical production facility.

In the manufacture of excipients, the environmental conditions, equipment, and operational techniques employed reflect the chemical industry rather than the finished drug manufacturing industry. In some processes, chemical and biochemical mechanisms have not been fully characterized; therefore, the methods and procedures for materials accountability will often differ from those applicable to the manufacture of finished dosage forms. Many chemical processes are performed in closed systems that tend to provide protection against contamination, even when the reaction vessels are not enclosed in buildings. However, this does not preclude the introduction of contaminants from equipment, materials used to protect equipment, corrosion, cleaning, and personnel.

Some excipient manufacturing processes may require observance of GMP applicable to finished drug products or bulk active ingredients because of the excipient's intended use. However, such observance is neither feasible nor necessary in many processes, particularly during the early processing steps. The requirements increase as the process progresses.

An excipient manufacturer should be able to identify critical or key points in the process where selective intermediate sampling and testing are necessary in order to monitor process performance. Toward the end of the process, the records should be increasingly thorough.

Significant processing steps, required to produce an excipient that meets the established physical and chemical criteria, should be identified by the excipient manufacturer. These steps can involve a number of unit operations or unit processes. Unit operations include physical processing steps involving energy transfer where there is no chemical change in the molecule. Unit processes are those processing steps where the molecule undergoes a chemical change.

Significant processing steps include but are not limited to the following:

- Phase changes involving either the desired molecule, a solvent, an inert carrier or a vehicle (e.g., dissolution, crystallization, evaporation, drying, sublimation, distillation, or absorption)
- Phase separation (e.g., filtration or centrifugation)
- Chemical changes involving the desired molecule (e.g., removal or addition of water of hydration, acetylation, formation of a salt)
- Adjustments of the solution containing the molecule (e.g., adjustment of pH)
- Precision measurement of added excipient components, in-process solutions, and recycled materials (e.g., weighing, volumetric measuring)
- Mixing of multiple components
- Changes that occur in surface area, particle size, or batch uniformity (e.g., milling, agglomeration, blending)

An audit team consisting of appropriate personnel (e.g., auditors, engineers, laboratory analysts, purchasing agents, and computer experts) should participate in audits. The operational limitations and validation of the critical processing steps of a production process should be examined to make sure that the manufacturer is taking adequate steps to check that the process works consistently.

It should be clarified at what stage and how the excipient will be used. Knowing whether the excipient should be included in the drug product first, directly or indirectly, will guide the next steps and questions. For example, whether the excipient will be used in a sterile dosage form, it must be known whether the excipient will contain pyrogens or endotoxins. If the drug license holder claims in the registration application file that the endotoxin level is zero in the excipient used in the drug product, this responsibility lies with the excipient manufacturer and this question may be asked during the audit.

A good starting point for an excipient plant audit is a review of the following areas:

- Nonconformance, such as the rejection of a batch not complying with specifications, return of a product by a customer, or recall of a product. The cause of nonconformance should have been determined by the manufacturer, a report of the investigation prepared, and subsequent corrective action initiated and documented. Records and documents should be reviewed to ensure that such nonconformance is not the result of a poorly developed or inconsistent process.
- Complaint files. Customers may report some aspects of product attributes that are not entirely suitable for their use. These may be caused by impurities or inconsistencies in the excipient manufacturing process.
- Change control documentation.
- Master formula and batch production records. Frequent revisions may reveal problems in the production process.
- Specifications for the presence of unreacted intermediates and solvent residues in the finished excipient.
- Storage areas for rejected products.
- Type of system (e.g., open or closed). "Closed" systems in chemical plants are often not closed when they are being charged and/or when the final product is being removed. Also, the same reaction vessels are sometimes used for different reactions.
- Form of the material (e.g., wet or dry).
- Criticality of the processing stage. Stage of processing and use of the equipment and/or area (e.g., multipurpose or dedicated).
- Degree of exposure of the material to adverse environmental conditions.
- Relative ease and thoroughness of clean-up.
- Sterile versus nonsterile operations.
- Multiuse of reactors (are the same reactors used for different reactions or multipurpose, or dedicated reactors).
- Continuous versus batch processing.
- Potential for cross-contamination.
- The potential risk in terms of raw materials, lubricants, and manufacturing process in regard to Transmissible Spongiform Encephalopathies and compliance to current legislative guidance, such as the "Committee for Proprietary Medicinal Products (CPMP) – Note for Guidance on Minimizing Risk of Transmitting Animal Spongiform Encephalopathies via Human and Veterinary Products," should be evaluated.

Other considerations of potential raw material contamination impacts, e.g., avian flu, should be evaluated during the audit.

3.8.2.3 Who Supplies Excipients?

The "excipient industry" does not exist as a defined entity, and it certainly has no common identity. The IPEC is a trade association that has both suppliers and users as members. But even the IPEC would acknowledge that its members comprise only a limited part of the organizations that supply excipients to users. The main reason for this is that few excipients are specially made for pharmaceutical use.

The processes used in production and the surrounding systems may belong to the chemical sector, as pharmaceutical excipients have applications such as food additives, cosmetics, cleaning, and industrial products other than pharmaceuticals and biotechnology. Excipients may be manufactured, for example, through a significant chemical change, physical modification, blending, or purification, which causes many of the other components present in the starting materials to be removed or reduced.

3.8.2.4 Key GMP Principles to Consider during an Audit

Usually, the customer does not perform further chemical processing or purification on excipients, and the material is used as purchased. If an impurity is present in the excipient, it will probably be present in the finished drug product.

The manufacturing environment should be evaluated. External contamination can arise from the manufacturing environment when the product is exposed if adequate controls are not in place. In excipient manufacture, chemical processes are often performed in closed systems that offer protection against environmental contamination.

3.8.2.5 Change Control

Since the excipient is used without further processing, the GMP site requires consistent excipient quality – both chemical and physical parameters. It is the responsibility of the excipient manufacturer to control excipient manufacturing processes to ensure consistent conformance to excipient specifications.

Changes in excipient manufacturing processes may result in changes in physical or chemical properties that only become evident in the finished dosage form. This is important to a GMP site from a finished dosage form manufacturability as well as from the context of bioequivalence comparisons. Changes made to the excipient should not impact the quality and performance of the finished dosage form.

Changes to test methods and specifications should also be controlled.

3.8.2.6 Traceability

The traceability of excipient batches is a key GMP requirement. This includes raw material suppliers, manufacturing documentation, batch numbers, retest or expiry date, test data, Certificate of Analysis, and transportation and distribution records.

3.8.2.7 Testing of Excipients

It is important that each lot of excipient is tested in accordance with established methods to ensure compliance with agreed-upon specifications. Testing controls should include equipment qualification and calibration, training of personnel, change control, out of specification result investigations, and retention of all raw data.

3.8.2.8 Things to Look For

- Determine which excipients the supplier manufactures or, if considering a new supplier, what potential excipients the new supplier will provide.
- Determine how the excipient(s) will be used and in what finished drug product. If the excipient will be used as a direct component of a drug product, used in the preparation of a sterile drug product, or is represented by the manufacturer as being pyrogen-free, this should make a difference in the audit.
- Review the status of the corrective actions from previous audits, if applicable.
- Determine if there have been any issues with excipients received since the last audit of the supplier.
- Collate data about the supplier from all related receiving sites and review that data for trends or issues.
- Contact Procurement and determine if there have been any issues regarding the supply of the excipients.
- Obtain a copy of the site's current specifications for the excipients.
- Obtain a copy of a recent COA received by the site and compare it with raw data at the excipient supplier.
- Review the quality agreement between the site and the supplier.

3.8.2.9 What Types of Problems May Rise before an Excipient Manufacturer Audit?

In almost all cases, excipient suppliers are not subject to GMP regulation. This makes it difficult to conduct an audit to prove the supplier's qualification. Well-known and common processes per GMP specifications are only available to a limited extent. Ultimately, it requires a high level of decision-making ability based on results to give the supplier a satisfactory score.

Companies that manufacture excipients do not basically comply with GMP rules in most cases. Having a good knowledge of pharmaceutical regulations, the GMP auditor should have a high degree of flexibility to evaluate the results obtained for supplier qualification during the audit. All well-known processes based on GMP are rarely expected. Finally, the GMP auditor must have a broad knowledge and decision-making ability to rate the excipient manufacturer based on the information received during the audit.

If we consider that 80%–90% of the weight of a tablet consists of excipients, you can estimate the amounts taken by the patient during administration.

The required registration of an on-site audit with the excipient material supplier represents the first hurdle even before the audit begins, as the authors have repeatedly cited various reasons from their own experience to avoid an audit.

The supplier to be audited frequently tries to circumvent an audit or shows little interest in the outcome, for example, by the following question(s)/arguments:

- Is the audit required by law?
- We supply many food companies and we have not received any complaints to date.
- Your purchase quantity is so small that it is not worth accepting an audit.
- If you want to come, you pay 10,000 euros per day.
- We do not allow on-site audits at our facilities.
- When conducting an audit, we acknowledge/ignore any inconsistencies and may consider correcting them if it makes sense to us.
- We are not dealing with an audit report.

Other creative rejections or combinations from the preceding list can certainly be cited; however, you will eventually find that it is relatively difficult to get an audit appointment from an excipient supplier.

Under certain circumstances, why would it make sense, at least initially, to conduct an audit and convince the supplier of the excipients that it makes sense?

There are three main arguments for this:

- Medicines are used on sick people; Manufacturer does not want to cause further possible damage due to any quality defect in the applied materials.
- Excipients should also be clinically tested as part of the drug.
- The amount of excipient administered is many times the amount of the API in most cases and can be significant in patients who need to take several different drugs over an extended period of time.

This does not mean that every excipient supplier should be audited.

Finally, it should be taken into account that the supplier of excipient materials may have a monopoly position so if the audit result is not satisfactory, the decision to change supplier is invalid in any case. In such cases, the desired product quality needs to be checked with your own additional inbound excipients.

3.8.3 Supplier Audits/Packaging Materials

What purpose do packaging materials serve?

- Physical protection
- Barrier protection

- Containment or agglomeration
- Security
- Convenience
- Portion control

The issues with packaging materials include:

- Most recalls and rejects are packaging related.
- A wide variety of materials from many different suppliers.
- Suppliers serve many different industries.
- Packaging materials are delivered in very large quantities.
- Individual packaging items are often not labeled.
- Printed materials often look very similar.
- Printed materials may have foreign language text.
- Problems and changes that take place at the suppliers are very difficult to identify.

Many different types of packaging materials serve a wide range of purposes.

Primary packaging materials contain the product:

- Containers such as vials, ampoules, bottles, tubes, cans, jars, foil, blisters, syringes
- Closures such as bungs, tops, stoppers, caps, valves, patch materials, plungers
- May contain printed information

Secondary packaging materials protect the primary pack and/or contain printed information:

- Such as folding cartons, labels, and leaflets

PS 9000 for packaging materials (full title "PS 9000: 2016 Pharmaceutical packaging materials for medicinal products") is the application standard for packaging materials manufacturers:

- Contains ISO 9001 plus appropriate GMP requirements added to certain clauses.
    - Mainly concerned with mix-up prevention, contamination control, and change control.

The application of ISO 9001:2015 and ISO 9004 to pharmaceutical packaging materials is widely used in the UK and Ireland by pharmaceutical printed materials suppliers.

The Pharmaceutical Quality Group (PQG) produced PS 9000 standard for suppliers of pharmaceutical packaging materials in 2001. Like PS 9100 for excipients, it is based on the ISO 9001 framework, with GMP added at certain clauses. As with PS 9100, the main focus of the standard is to include control of the critical areas of control of change, contamination control, and mix-up prevention. PS 9000 also contains the requirements of ISO 9004, making it somewhat complicated to follow.

The PQG has also produced a publication that is aimed at assisting packaging material suppliers understand why the additional GMP clauses have been added to PS 9000. This publication is titled "PS 9004: A guide to the GMP requirements of PS 9001:2001 Pharmaceutical packaging materials." PS 9004 has many flow charts and case studies that can help explain the rationale behind PS 9001 to individuals and companies unfamiliar with the standard requirements of GMP.

3.8.4 Primary Packaging Materials

The different types of primary packaging include:

- Ampoules
- Vials

- Containers
- Strip packages
- Blister packaging
- Syringes
- Sachet packaging

ISO 15378:2015 specifies requirements for a quality management system for manufacturers of pharmaceutical and medical device primary packaging materials. These organizations need to demonstrate their ability to consistently meet customer requirements, including regulatory requirements and international standards applicable to primary packaging materials for such medicinal products.

The ISO 15378 standard enables you to comply with legal requirements for pharmaceutical and medical device primary packaging materials. ISO 15378 integrates the requirements of ISO 9001 as well as GMP, a regulatory requirement for the pharmaceutical and medical device industries as per all international regulations. The standard also helps to reduce the risks of safety hazards and product contamination, and ensures product efficacy and shelf life.

3.8.5 How to Audit Packaging Material Manufacturers

The pharmaceutical and biotechnology industry relies on primary and secondary packaging materials. These materials are supplied by approved and qualified suppliers. Packaging material manufacturers must have at least ISO 9001 or equivalent certification. In the case of secondary packaging materials, the manufacturer must comply with PS 9000 and PS 9004 requirements. Primary packaging material manufacturers must comply with ISO 15378 in order to get approval from their clients.

Packaging material suppliers must have an established quality management system in place. This system must include a quality policy, quality manual, and quality objectives. The quality manual must state PS 9000 and ISO 15378 requirements. There should be a document management system. This system must state how to prepare, issue, distribute, and withdraw documents in a timely manner.

The supplier must have critical documents related to:

- Change control
- Deviations
- Complaints
- CAPAs
- Recalls
- OOS/OOT
- Quality control activities
- Management of artworks
- Contamination
- Validation, qualification, calibration, maintenance
- Quality risk management
- Cleaning
- Sterilization
- Training
- Internal audit
- Supplier management
- Batch release
- Storage and distribution
- Identification and traceability
- Personnel and facility hygiene and sanitation

- Pest control
- Utilities
- Cleanroom management
- Manufacturing operations
- In-process controls
- GMP requirements

Key GMP principles to consider during the audit are:

- Specification
- Identification and traceability
- On-line and segregation controls
- Change control
- Line clearance
- Contamination control
- Validation, qualification, or capability documentation
- Sampling
- Documentation
- Print security
- Special considerations

3.8.5.1 Specification

There should be a documented, controlled, and approved specification for the components being manufactured. The specification normally consists of:

- Physical attributes with ranges and limits
- Quality attributes and defect classifications
- Technical drawings
- Storage conditions and precautions
- Approved material grades
- Performance characteristics

3.8.5.2 Change Control

A good pharmaceutical manufacturing site requires consistent packaging quality. It is the responsibility of the packaging component manufacturer to control manufacturing processes to ensure consistent conformance to the packaging component specifications.

Changes in component manufacturing processes may result in changes in the physical properties of components or final pharmaceutical products that may not be evident on visible inspection of the component. It is important to a manufacturing site that any changes made to the component should not impact the quality and performance of the package component. Also, any regulatory implications are assessed as part of the change process.

3.8.5.3 Line Clearance

Line clearance is an essential element in product mix-up prevention and needs to focus on:

- Input materials on the line from the previous batch
- Samples and waste from the previous batch
- Documents on the line from the previous batch
- Verification that any electronic data is wiped from consoles etc.
- Clearance after maintenance activity or major interruptions as appropriate

- Line clearance activities need to be documented and cover all areas, not just the machine or line:
- The whole machine, including hoppers, conveyors, reject stations, etc.
- The floor around the line
- Benches, cupboards, and shelves around the line
- Pallets

The line clearance should be documented and signed and an independent check completed and signed before the area is released for use.

3.8.5.4 Print Security

Special attention should be given to how assurances are made for print security and media management. The processes should be assessed during the audit to ensure no corruption of print media occurs during the transfer or storage of electronic artwork files.

There are agreed processes of artwork approval and good practices for changing artwork including customer approval, version history, change history and archiving of old versions of artwork. There should be set-up and maintenance procedures to include initial printed material verification against an approved source such as artwork, confirmation of press set-up, and documented procedure and schedule for cleaning.

Printing plates should be uniquely identified and access controlled. Verification of the print impression should be completed against a controlled master version as part of the make-ready and approval to run the job. If reference standards, such as color-matching books are used, they should be appropriately controlled.

Gang printing is not a preferred practice due to the potential risk of mix-ups. However, if performed, stringent controls should be in place. To reduce the risk of unprinted material in the finished component, processes should be employed in the printing process. The processes include double sheet detectors for sheet-fed printing machines and sensors that alarm or highlight the depression of printing plates from the print medium. In addition, systems should be evident to ensure material that is outside of print registration or images with ink starvation are securely flagged for segregation and removed.

For manufacturers/suppliers who supply labels on reels, in the case of missing labels, these should not be replaced and the number of splices allowed per reel should be defined. Where digital printing is utilized, processes should be controlled to ensure the correct electronic artwork files for the job are in the digital printing press and there are security measures for printed products.

3.8.5.5 Braille

Where Braille has been incorporated into a printed package component there shall be a system for ensuring symbols meet appropriate regulatory and market requirements, and comply to specifications for size and embossed height.

3.8.5.6 Counterfeit

It is important that basic security at the suppliers is followed to ensure that all print media including waste is stored securely and rendered unusable prior to secure disposal. The preferred mode for the controlled destruction of materials with proprietary information is supervised incineration. This control helps in the fight against counterfeit material. In some markets, a variety of anticounterfeiting measures have been put into place, e.g., the use of holograms.

3.8.5.7 Intellectual Property

Intellectual property is the intangible property that is the result of creativity (such as patents, trademarks, or copyrights). It is important that components are reviewed in relation to the registration of trademarks, copyrights, or patents, as this could have an impact on potential future changes to specification and registration details. In the case of rubber stoppers, the exact details of the formulation may not be disclosed by the supplier (if intellectual property), and minor changes in formulation may have a direct impact on the final pharmaceutical product stability. Expectations for notification/approval authority of changes by the site should be spelled out in business and/or quality agreements, respectively.

3.8.5.8 Printing

The processes for the production of printing plates should be controlled. Master specifications should be internally maintained and available for inspection. The buyer site should retain the right to inspect as necessary to confirm the current status, make modifications as necessary, be notified, and possibly approve any changes. The expectations should be defined in the quality agreement.

3.8.5.9 Prior to the Audit

- Develop an understanding of the vendor manufacturing process specific to company requirements.
- Obtain a list of company components that are manufactured at the site.
- Review recent rejections, complaints, and issues of the receiving site(s) and the respective statuses.
- Review any quality agreements and relevant registration requirements.
- Review the compliance status of the site by checking for service history, recalls associated with the site, recent regulatory inspections (if applicable), and outcomes.
- Review previous audit reports and actions.

3.8.5.10 During the Audit

- Perform a walk-through of the manufacturing area.
- Ensure the production areas are clean and tidy.
- Ensure the fabric is in good condition and appropriate design for control of the process.
- Verify each line is physically separated or segregated from each other.
- Verify the manufacturing process and dress discipline are suitable for the prevention of contamination.
- Verify containers used for filling are clean before use.
- Verify line clearance procedures have been thoroughly followed before a new order/batch is introduced.
- Verify nonconforming raw materials, not suitable for use, have been identified and removed from the area and are secure.
- Verify samples are not returned to the line.
- Verify the correct order is on the line, and if approved changes have been made, they are clearly indicated.
- Verify all inline controls are being conducted.
- Verify unprinted material, defective material, and material not meeting the specification are securely segregated throughout the process to avoid mix-ups.
- Verify product waiting to be packed is clearly marked and segregated from the product that is being packaged. For example, printed and nonprinted material labels.
- Verify that before raw materials are delivered to the line, they have been identified and released to the approved specification.
- Verify the materials and personnel flow is adequate to prevent a mix-up.
- Inspect cleared packaging component areas and verify that all materials have been removed from the area, including refuse.

Perform a walk-through of the area where raw materials are received, issued, and stored.

- Ensure current specifications are accessible, agreed upon, approved, and utilized.
- Ensure environmental controls are in place to maintain temperature and relative humidity at acceptable levels where materials sensitive to temperature/relative humidity are stored.

- Ensure the area is appropriately secure and materials are stored in segregated containers or locations.
- Verify that access to the storage area is limited to authorized personnel.
- Ensure sufficient secure control of material of nonconforming products.
- Ensure all raw material is adequately labeled as to their identity and there is an unambiguous system for material status.
- Ensure that there is a system so materials are not used outside the recommended shelf life.
- Ensure that there is an approved acceptance procedure for all raw materials.
- Verify that written approved procedures exist and are current.
- Verify incoming materials have a unique reference number for identification.
- Verify incoming materials are tested/inspected and undergo an approval process.
- Verify materials that fail to meet acceptance specifications are rejected and quarantined.
- Ensure documentation exists for the receipt of each shipment of materials which includes:
  - Receipt.
  - Examination or testing status (either acceptable or rejected).
- Ensure that all personnel are trained and qualified to perform their function.
- Review training records for selected individuals to ensure that they have received adequate GMP training and job skills training.
- Ensure that there is a written procedure describing the training program and the steps needed to qualify for the job.
- Ensure a control system is in place for implementing changes in manufacturing.
- Verify that there is an approved, written change control procedure in place.
- For master documents:
  - Ensure that the order is an accurate version of the master packaging component instructions that were previously company approved.
  - Verify working copies of a master document are only released after a signature has been obtained.
  - Ensure that changes to the master component orders are reviewed, authorized, and issued by the appropriate department following the approved change control procedure.
  - Verify only qualified individuals may make changes to a document.
  - Verify that these individuals have signed the document verifying its accuracy.

Ensure confirmation of control systems for in-process visual or electronic verification devices.

- Verify electronic code readers, vision inspection systems, label counters, or other similar devices are calibrated and part of a preventive maintenance program.
- Verify electronic code readers, vision inspection systems, label counters, or other similar devices are challenged prior to start-up and periodically during packaging to ensure proper operation.
- Verify results of these challenges are documented.
- Verify computer systems used to control inspection equipment are secure and designed to prevent unauthorized changes by mechanics and operators.
- Verify that electronic verification devices, such as pharma code readers, are set up utilizing a controlled approved specification.
- Verify that components are checked for:
  - Sufficient quantity.
  - Identification.
  - Conformance to packaging specifications.
- Verify that in-process printing is checked frequently for accuracy and legibility and that the periodic check is recorded.
- Ensure that there is a system for the destruction of waste printed packaging component material.

- Verify that the firm has an approved procedure for destruction of printed components which includes:
  - Method of the destruction.
  - Documentation for the destruction.
  - Verify if these responsibilities are outsourced they are controlled and maintained by the firm.
- Print security
  - Ensure the system for artwork/proof approval is considered secure.
  - Ensure the system to control plates and origination material is considered secure.
  - Ensure procedures cover plate generation and control (security), plate unique numbering, plate and artwork secure disposal, and plate issue and return.
  - Ensure reference standards are current and controlled.

CASE STUDY

After examining Figures 3.11 and 3.12 and Table 3.18 together, it is aimed for the GMP auditor to answer the following questions:

Date:	28/04/2010	Sample:	T=10
Product Code	FLS 001		Top
Sample Weight	0.9976 g	A	
pH	4.23		
Date:	28/04/2010	Sample:	T=10
Product Code	FLS 001		Bottom
Sample Weight	2.5599 g	B	
pH	4.23		

FIGURE 3.11 Excel spreadsheet used to calculate assay.

T = 10	2.5813 g	A
M = 10	2.5911 g	
B = 10	2.5405 g	B

FIGURE 3.12 Corresponding lab book entries for sample weights.

TABLE 3.18
Acquisition Time

Sample Name	Acquisition Time	File Name
Acet.@250 REP 1	17:13:19	090811-003.rst
Acet.@250 REP 2	17:17:10	090811-004.rst
Acet.@250 REP 5	17:28:19	090811-007.rst
Acet.@250 REP 5	17:34:07	090811-007-20110809-173718.rst
Acet.@250 REP 6	17:37:58	090811-008.rst
Acet.@250 inj acc	17:41:58	090811-009.rst

- Where are REP 3 and REP 4? (We have an 11-minute gap and the .005 and .006 datafiles are missing.)
- Why has REP 5 been reinjected?
- Why does the 6th injection have a different sample name?

3.8.6 Computerized System Audits

Computerized systems are critical to patient safety and robust validation procedures, and they prove that the system is always operating within its predetermined specifications is a requirement. Documents such as 21 CFR Part 11, EU, PIC/S GMP Annex 11, and GAMP5 (second edition) have various expectations about computerized system validations.

Before performing computerized system audits, it is necessary to answer why this audit is needed. The company may examine the computerized systems in its own internal audits, or it may want to audit the computerized systems of a supplier or a company from which it receives contracted production or testing services. In addition, the health authority may want to put computerized systems on the inspection agenda as one of the main subjects of their inspections.

Regarding computerized systems, the following questions should become clear:

- Which computerized systems are used in which departments (Warehouse, Production, Quality Control, Quality Assurance, Licensing)?
- Why were these computerized systems needed?
- Were the systems purchased from a company or developed in-house?
- How were the vendors contacted for the first time?
- Has a questionnaire been sent to vendors?
- Do vendor companies have a business history with pharmaceutical and biotechnology companies?
- Did pharmaceutical company staff prepare a solid user requirement specification (URS) before purchasing the system?
- Who is on the URS preparation team?
- Does the system have security vulnerabilities?
- Is there a change control system to be operated in case of modifications to the system in certain periods?
- Has a quality agreement been signed with the vendor?
- Why do computerized systems have to be validated?
- Which approach should be used for validation?
- In which category are system software and hardware included?

- Have the roles and responsibilities of the members of the validation team been determined?
- How will risks be managed for validation?
- Which methods have been chosen to identify system risks?
- How is traceability of system validation ensured?
- How is information security ensured in computerized systems?
- How is the relationship between data integrity and system validation managed?
- How is the software product tested?
- How is the first data entry provided to the system?
- How are the personnel trained regarding the functionality of the system?
- How are deviations managed in computerized systems?
- How are system backup and data storage managed?
- How are disaster recovery and business continuity managed?
- How is the periodic quality review of the system done?
- How is the sustainability of the system guaranteed?
- Is the to-be-purchased/purchased system developed for a specific user?
- Has any system been purchased from the vendor before?
- Does the system have interaction and interfaces with other systems?
- Have possible effects on other systems been evaluated?
- What is the recognition of the system in the pharmaceutical and biotechnology market?

If you set a weighting factor for each criterion, this will help determine the vendor's audit risk. An example is presented in Table 3.19. Generally speaking, assuming that the pharmaceutical company has past cooperation with the vendor, the pharmaceutical company knows that the customer is satisfied with the vendor's products and will get an immediate response when they have any problems. If the computer system is used for the release of semifinished products, raw materials, packaging materials, and finished products or for data storage, the weight factor will be high. Let's imagine a software, even if this software is not presented to any user in any laboratory, in this case the weight factor will still be high. Assuming that the system is not connected to any network, the weight factor will be medium this time. As GMP auditors, we expect such a matrix to be developed for all software and all vendors. By considering the criteria one by one and calculating the total risks, it can be revealed how often the vendor will be audited.

3.8.7 Data Integrity Audits

- *Raw data* – Original records and documentation, retained in the format in which they were originally generated (paper or electronic).
- *Data* – All original records, including source data, metadata, subsequent transformations and reports, generated or recorded at the time of the activity. Allows full and complete reconstruction and evaluation of the facility.

TABLE 3.19
Risk Criteria for Vendor Audits

#	Criterion	Weight Factor
1	Experience with vendor	1
2	Business impact of system	5
3	Maturity of system	5
4	Influence of system on other systems	0
5	Complexity	3
Total Risk Factor		14

Factors 0 to 5; 5 = Highest Risk

- *Original record* – Data as the file or format in which it was originally generated, preserving the integrity (accuracy, completeness, content, and meaning) of the record.
  - Original paper record of manual observation
  - Electronic raw data file from a computerized system.

3.8.7.1 General Data Integrity Concepts

- *Data integrity* – The extent to which all data are complete, consistent, and accurate throughout the data life cycle.
- *Data life cycle* – The data life cycle covers data generation, processing, reporting, archival, retrieval, and destruction. Figure 3.13 shows the data life cycle.

Many health authorities have published guidelines on data integrity. For example, health authorities including MHRA, FDA, WHO, and PIC/S have guidelines with examples on the subject.

Likewise, PDA, ISPE, APIC, IPEC, Rx360, and OECD have very useful documents about data integrity. It is possible to provide examples, but the task of the GMP auditor here is to find out how data integrity is embedded in the audited organization and to seek answers to a number of related questions.

Table 3.20 provides examples of mistakes, bad practices, and fraud.

ALCOA+

ALCOA, with additional emphasis on data being:

- Complete
- Consistent
- Enduring
- Available

Table 3.21 provides examples of ALCOA principles.

3.8.7.2 New Approach to Audit

Focus on the potential for fraudulent activity within your quality systems.

Assumptions

- You will assume fraudulent activity is taking place if they identify weaknesses in your quality systems.

FIGURE 3.13 Data life cycle.

TABLE 3.20
Mistake versus Bad Practice versus Fraud

Mistake	Bad Practice	Intention or Fraud
Error is made; acknowledge error so it can be addressed	Action is taken that is not compliant with cGMP to save time, for instance	Procedures and/or cGMPs are bypassed knowingly

TABLE 3.21
ALCOA

Good Document Practices Principle	Paper	Electronic
Attributable	• Initials/signature by the person who performed the activity	• Electronic signature (username and password)
Legible	• Indelible ink • No erasure or correction fluid/tape • Changes are corrected properly	• Data is visible in fields
Contemporaneous	• Recorded at the time of activity • No pre- or postdating	• Record is saved when data is entered • Time date stamps • Synchronized time (for accuracy)
Original	• Paper form • No Post-its or temporary method of recording then transcribing data	• Data generated by a computerized system • Includes metadata • Not able to modify original data
Accurate	• Data recorded is accurate • Calculations are checked • Second-person verification	• Data transfers from the original data system accurately (e.g., to LIMS, archival) • Calculations correct • Second-person verification

- "Guilty until proven innocent" approach to auditing!
- "Data too good to be true!"

There is a much higher probability of identifying fraudulent activity within an organization if metadata is reviewed.

Hard copy (flat data – printed, pdf, photocopy) is no longer considered to be acceptable by regulatory authorities, as this data is not complete and not original.

If a company states that paper is its original raw data in its internal procedures, this should alert an auditor that the company is probably not managing and reviewing electronic (meta) data.

What is Metadata?

- Data that describe attributes of other data, and provide context and meaning (data about data).
- Describe the structure, data elements, interrelationships and other characteristics of data.
- Permits data to be attributable to an individual.
- Can be any attribute that expands information related to the data.

Electronic data (metadata) is the preferred choice for regulatory authorities, as this is the original ("official") data:

DATES:14 Mar 23
TIME:07:34
UNITS:g/L
ID #:BR-0123

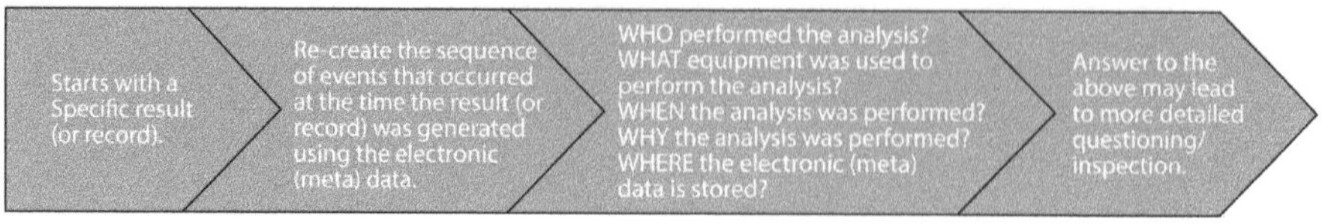

FIGURE 3.14 Data integrity audit preparation.

Five key data integrity (DI) questions you must ask during an audit are:

1. Is electronic data available?
2. Is electronic data reviewed?
3. Is metadata (audit trails) reviewed regularly?
4. Is there a clear segregation of duties?
5. Has the system been validated for its intended use?

The answers to these questions will determine whether companies are in compliance with 21 CFR part 11 (Electronic records and signatures). Leave the original metadata in the chromatography data system (CDS) and review/approval electronically to avoid increased data integrity risk (the paperless lab).

Figure 3.14 describes the stages of data integrity audit preparation.

Risk Assessment

- Do you have metadata for each system?
- Implement short- and long-term CAPAs.

Can become a high risk if older, stand-alone systems are in use.

Procedures/SOPs
Examples of typical SOPs include:

- IT policies
- System administration (CDS access, roles, and privileges)
- Data management and storage
- Data acquisition and processing
- Data review and approval
- Date archiving and backup
- Anti-fraud monitoring

IT Infrastructure
As a GMP auditor, you will need to see the server room. Verify:

- The room is secure.
- IT access only.
- Tidy and in good working order.
- Has backup and disaster recovery procedures in place.
- Date/time functionality of servers is correct.

As a GMP auditor you will select a number of instrument-controlling PCs within the lab and check:

- Date/time functionality is correct.
- Date/time cannot be changed by the lab personnel.

As a GMP auditor you must confirm that date/time functionality on all PCs within the lab is locked down and can only be changed by IT personnel with administration privileges.

Administration

As a GMP auditor, you will want to understand how access to the CDS is authorized and controlled. Also, you will need to justify the access levels within the CDS and the user privileges at each level.

- Specific user profiles and passwords are required to access instrument software and provide audit trail traceability.
- Administration control should be independent of analytical function to eliminate conflict of interest.
- Clear segregation of duties with no overlap of privileges.

As a GMP auditor, you will need to see the company's password policy or related SOPs. In these documents, you must confirm that:

- Passwords are not shared.
- Passwords are changed on a regular basis to protect user profiles.
- Passwords must have mix of alphanumeric characters and have high strength.
- User needs to log off the CDS immediately after use to avoid the profile potentially being used by other personnel to acquire, process, or manipulate data.
- User profiles are set to auto-lock after a period of inactivity to protect the user profile and data within the CDS.

As a GMP auditor, you will want to confirm that the audit trail functionality is switched ON within the CDS.

Figure 3.15 describes the various user levels.

In chromatography data systems, you must ask for administration reports:

- Active users
- User privileges
- Administration audit trail report

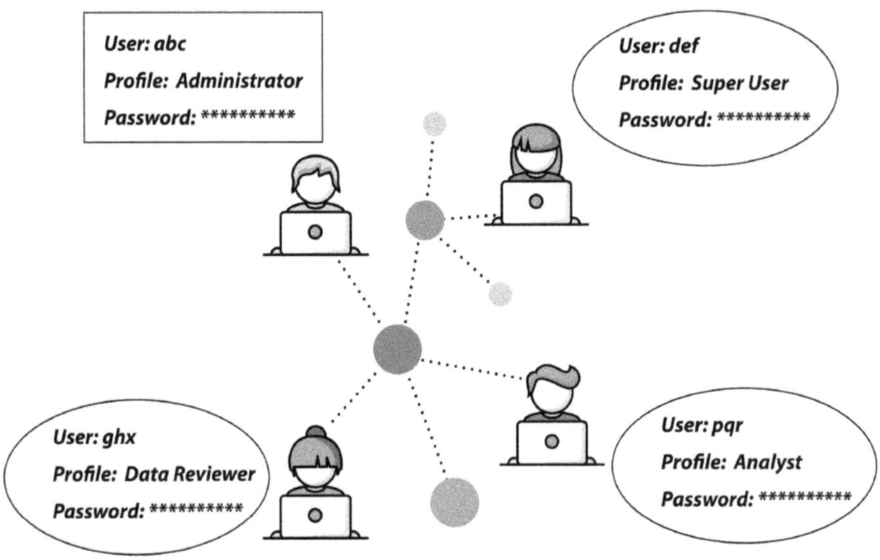

FIGURE 3.15 Various user levels.

What are the specific privileges within the user profile?

- They will want assurance that data cannot be deleted by a user once acquired.
- They will want to know if data can be moved to a different folder to potentially "hide" it (e.g., trial injections).
- They will want to see that electronic data that has been processed must be saved before it can be submitted for review (or printed to hard copy). Make sure you understand the privileges applied to each user profile and be prepared to justify them to the regulatory auditor.

Data Management

- Periodic GMP data archiving. Make sure that data archiving is defined in the company's procedure and performed regularly.
- This approach minimizes the amount of "live" data that can be accessed by users and potentially reprocessed to change previously reported results.
- The users should not have access to an archive folder(s), which adds an additional layer of protection to the electronic data.

What parameters need to be checked in a QC Laboratory?

- Analysis performed as per the monograph
- Sequence information correct
- Chromatography is typical
- System Suitability Test (SST) acceptance criteria achieved
- No "conditioning" or "test" injections using the sample (use a standard or control sample if specified by your procedures and monograph)
- Correct integration (pay attention to manual integration)
- Chromatography appropriately scaled

Data Review

- Individual results duplicate and meet specifications.
- Check the sequence and individual injection audit trail. Any atypical/suspect activity?
- Data processing:
  - Do the audit trail comments provide traceability?
  - Can the reprocessing be justified?
- Check electronic results within the CDS match results reported on hard copy chromatography or in LIMS/SAP systems.

External (Auditor) Data Review

Auditor checklist:

- Administration control.
- Individual user profiles and passwords.
- Clear segregation of duties within user profiles.
- Restricted privileges for users (can't delete/overwrite/move).
- Audit trail functionality switched ON.
- Date/time functionality locked by IT.
- Lab demo – User log-on (multiple), date/time locked, can't delete data.
- Data recall – Electronic sequence/data file recall in the lab using staff members. Data recall needs to be fast and efficient.

- Data review – Chromatography scaling, integration, and electronic results.
- Audit trail review – Look for suspicious activity and justification of processing.
- Training – Assess staff competency with CDS in the lab. Make sure staff are trained to interact with the auditor. Have a CDS superuser present during the lab inspection.
- Query search – Assurance that batch hasn't been analyzed multiple times as part of an investigation.

3.8.7.3 How to Prepare for a Data Integrity Audit

If you are expecting a GMP audit at your facility, one of the main topics will undoubtedly be data integrity, which is the trendy topic of recent days. You should try to collect as much information as possible before the audit and analyze this information in a systematic way.

So what should you consider in your research?

- Research the background of the auditor(s) to gain knowledge of their experience and areas of expertise. Use available resources, existing industry contacts, internet searches, and LinkedIn.
- Risk assess all lab areas prior to the audit to identify equipment that produces electronic data files.
- Categorize the equipment according to USP<1058> and GAMP5.
- Auditors will focus on instrumentation that falls under:
- USP<1058> categories B and C.
- GAMP5 categories 3, 4, and 5.
- Perform an internal data Integrity audit on medium- and high-risk equipment.
- Does the equipment meet the requirements of 21 CFR Part 11?
- Check that electronic data can only be accessed through the instrument software and not via the operating system.
- Identify gaps and implement short-term corrective action before the audit (if possible).
- Discuss longer-term corrective actions with the management team.

What are the focuses of Regulatory Inspections?

Warning Letters (Data Integrity)

- Lack of control over access to computerized systems
- Noncontemporaneous recordkeeping
- Deletion, falsification, alteration, or other manipulation

Two kinds of data integrity problems are widespread:

- Top-down, pervasive, dictated by management
- Bottom-up, isolated, one or a few operators or analysts

If the FDA finds a data integrity issue in one place:

- Assume data integrity is an issue everywhere unless a credible audit determines otherwise.
- When something doesn't look right, take a tangent and zero in on it.

The following list, as well as Table 3.22, provides examples of the FDA's data integrity concerns:

- Test results for one batch used to release other batches.
- Test file has the same date and time as other tests.

TABLE 3.22
FDA Concerns with Data Integrity

Shared login accounts	Cannot identify a unique individual
System administrator role not independent	Assigned to personnel responsible for record content
cGMP record does not complete	All electronic data generated to satisfy cGMP requirements is not included
Testing into compliance	Actual samples were used during "the system suitability" test or equilibration runs

- Too many tests run in the time available.
- Audit trail disabled until passing test results were obtained.
- Failing and passing batches of active pharmaceutical ingredients blended to produce passing results.
- Lab and management knew about this practice.
- Created falsified training records that were requested during the inspection.
- Look for data that is too good.
- No complaints, deviations, or out of specification investigations.
- Handwritten data that looks too neat.

DI Observations – EU

Data integrity observations can be found in every audit report performed in pharmaceutical companies.

Here are some of the audit findings from EMA audit reports;

- Uncontrolled documentation was noted throughout production engineering notebooks with set-up details and passwords, crib notes on the wall of the goods in the area, and scraps of paper containing numbers of components brought onto the line.
- Printouts of particle count data from HEPA filter testing were not transferred from thermal paper to nonvolatile media to ensure the integrity of the record throughout the retention period.
- Following a software update, data was lost from an autoclave control system. The system backup was unable to recover lost data as the backup was only performed on a three-monthly basis.
- The backup CD/DVD for the autoclave control system was not stored within a controlled environment to ensure its integrity.
- Data from the integrity test was not backed up. The system was observed to overwrite previous data.
- Backups were permitted to be made on the same computer drive, which failed to ensure that a separate copy was available following drive failure or corruption.
- HPLC software in the laboratory was not configured for GMP compliance.
- Unique user passwords were not enforced.
- Users were permitted to:
- Change the default audit trail.
- Change the default "require user comments."
- Copy non-related projects.
- Use annotation tools.

3.8.7.4 How to Start a Data Integrity Audit

Let's start with the walk-through:

- What document systems are used?
- GMP computerized inventory list?
- What manufacturing systems are dependent on preset parameters?
  - SAP, MES, IPC lab
  - Checkweighers, barcodes, PLC-configured equipment
  - Sterilizers, filter integrity testers, BMS, etc.
- Paper systems – batch record, logbook, forms
- How do they manage changes to these systems?

3.8.7.5 General Guidance

- Observe conditions and practices. Do not make it a "paper audit" only.
- Look for repeating pattterns, not just isolated examples. Follow-up on corrective actions following past audits.
- Stick to current practice. Avoid citing past problems that have been corrected and have not reoccurred. Something that was not done well, months ago, may not be representative of how that activity is done today. Verify before citing it as a current problem.
- Be helpful and positive, not overly critical. You are supposed to find the problems, but for a good purpose: to get them addressed. (Like an annual physical exam.)

3.8.7.6 Key Things to Look For: GMP

- Take a general tour and be observant: Are operations adequately supervised? Who is running things on the floor? Is area access properly restricted?
- Check documents in production and lab areas to make sure they are not being filled out in advance of activities they document.
- Check for extraneous documentation – post-it notes, scratch pads, whiteboards, etc., being used as primary documents. These become your raw data if used!

Be inquisitive. Talk to people who actually do the tasks you are auditing, not just managers and subject matter experts.

- Look for the unusual, the out of place, the atypical:
  - Idle (maybe dirty) equipment in an active production area.
  - A bucket under a pipe – is it there to catch a leak?
  - Operator's manual open to the troubleshooting page – what went wrong today?
  - Note on top of a stack of documents – "John, please go back and fill in all the blank spaces."
  - Duct tape! Used to correct a multitude of sins.
  - Look into trash cans. Do you see discarded raw data? Why is it there?
- Ask for trend charts on key quality metrics. Ask what is being done about adverse trends.
- Ask for a listing of SOPs sorted by the last date of review (usually the effective date). Are there a lot of old (over two or three years) SOPs for important activities? Are these current?
- Read the key SOPs. Could you execute the procedural steps? Is the SOP clear? Watch out for:
  - Vague terms: "as necessary," "when appropriate."
  - Internal conflicts and inconsistencies.
  - Lack of definition of responsibilities.
  - Inconsistency of SOP with associated forms, other related SOPs.

- Review batch records, lab notebooks and other similar records. Do entries make sense? For example:
  - Documents show two products in the same equipment at the same time.
  - Numbers do not make common sense – always exactly in the middle of the operating range, for example.
  - Entries do not agree with source documents, or source documents are not present and cannot be located.
  - Site never seems to have a deviation of any kind.
  - Documents look "too perfect" – especially in the lab.
- Check reject and quarantine areas in the warehouse. Investigate reasons for rejects.
- Read deviation and failure investigations. Do they do a good job of identifying root-cause issues? Do they all seem to be an effort to justify what went wrong rather than understand it and get it fixed? Are remediation steps reasonable in light of the root cause?
- If labeling operations take place, check lines for proper clearance and check label storage areas. Are proper electronic or manual controls in place to prevent mix-ups and perform label count reconciliation?

3.8.7.7 Effective Interview Techniques

- Ask the same question in different ways at different times, or ask different people the same question; compare the answers you get.
- Use open-ended questions. For example, "That's interesting. Tell me some more about that." "What else?"
- Use a linked series of questions. For example:
  - Leading question: "What is your procedure for shutting down a production line due to an unacceptable condition or practice?" (assumes there is one)
  - Followed by another leading question: "When was the last time that happened?" (assumes it did)
  - Followed by open-ended: "Tell me some more about that."

3.8.7.8 Closeout and Follow-Up

- Oral review of findings prior to finalization.
- Obtain input from the auditee.
- Consider whether to modify findings based on input.
- Send draft report to auditee for comments before finalization.
- Base scope and priority of follow-up on risk assessment of findings.
- Remember to comment on the good things as well.

3.8.7.9 FDA-Style Data Integrity Audit

1. Verify the expected sequence of activities: dates, times, quantities, identifiers (such as batch, sample, or equipment numbers), and signatures.
2. Constantly double-check and cross-reference.
3. Verify signatures against a master signature list.
4. Verify against time cards and employment history.
5. Look for unofficial or private records.
6. Check inventory system and material receiving records against actual usage. Verify point 1 above.
7. Check the source of materials received label, COA, and QMS.
8. Verify times tested in the lab for in-process, finished product, and stability testing with time in the batch record. Check logbooks versus raw data.
9. Review batch record for inconsistencies.

10. Review quantity in shipping records against batch yield.
11. Check the manufacturing capacity of the line.
12. Check adequate control of lab and production records – reissues, discard of raw data.
13. Verify the use and existence of equipment in manufacturing, packaging and laboratory-check equipment logs, maintenance records, equipment purchase records, and cleaning and calibration records.
14. Use equipment use logs to reconcile the batches made.
15. Interview staff, not the managers.

The US FDA Investigators located in India and China are also (forensic) experts in data integrity!

Table 3.23 provides examples of the FDA's chromatography concerns.

Hot Questions Used by Auditors

- Do you have your source electronic data (with content and meaning in data backup and archive)?
- Do you review your source e-data (or just printouts)?
- Does your review of source e-data include a review of meaningful metadata (such as audit trails or time/date stamps)?
- SOPs on data review to include a review of audit trail.
- SOPs on data review training – users need to know the data flow.
- Do you have proper segregation of duties especially regarding system admin/engineer level access?
- Have you validated your system for "intended use," not just functional testing (especially important for commercial off-the-shelf [COTS] systems)?

Table 3.24 provides examples of practice audit checklists.

Prepare your staff. They will be asked these questions.

Auditing for data integrity goes beyond the typical internal quality auditing necessary for an effective quality system. Types of audits required in an effective data integrity program include:

- Initial gap assessment or audit where a regulated company is/is not complying with data integrity control requirements and best practices.
- Ongoing internal quality audits of established data integrity controls to ensure continuing effectiveness and compliance.

TABLE 3.23
FDA's Chromatography Concerns

Deletion of data
Folders and individual data files
Software not properly monitored
Noncompliant software used
Analysts not properly trained
- Overwriting of data
- Testing into compliance
- Altering integration parameters
- Performing sample trial/test/demo injections
- Administration and user privileges
- Lack of audit trail and data reviews

TABLE 3.24
An Audit Checklist

Concept	Question
User access	Ask Operator: What level of access do you have? Can you delete data, change programs, or change the date/time?
User access	Try to change the date/time on the system, including time zone while under an operator login.
User access	Who has admin access? Are they independent of persons responsible for the content of the records generated by the system for regulated purposes?
User access	Is a periodic review of user access performed? Are accounts deleted/disabled?
Archive	Check "Recycle Bin" on Windows desktop; no GMP data should be in there.
Archive	Are the electronic records for this system being maintained (e.g., backed-up, archived) in a secure manner?
Archive	Are the electronic records for this system maintained/archived in a manner that preserves their "content and meaning" (i.e., as verified true copies of the original electronic data set to include metadata and with provisions for software to view the data)?
Archive	Has the archival process been validated/verified?
Audit trail	Ask to see the audit trail. Can they pull it up? Ask the supervisor: What do they review? SOP on when they review? Can you trace actions to personnel?
Originality	Are supervisory or quality personnel reviewing the electronic records and relevant metadata (such as audit trails) for this system or are supervisory personnel merely reviewing the final printout?
Traceability	Do actions in logbooks/on paper match actions in the system? Did the person who ran the run record it in both places?
Accuracy	What happens if a test fails? Where does the electronic or paper record go? When can you retest? Who has to approve a retest?
Accuracy	If there are manual entries during any process, is there a witness or verification step? How do they ensure accuracy?
Accuracy	If a printout is used, is it verified against the electronic? Or is the printing process validated?

- Periodic audits of long-term data archives to verify the data deterioration and media migration controls are being followed and are effective.
- Supplier qualification audits for suppliers creating, modifying, reviewing, analyzing, transmitting, storing, and/or archiving data on behalf of a regulated company.
- Closeout gap assessment or full audit following (or close to) completion of data integrity program implementation.

Examples of data integrity auditing exercises could include:

1. Conducting a mock inspection of a specific data handling process, where the entire data life cycle would need to be explained as if it were being presented to a regulatory inspector. This can highlight any confusion about where the data resides and how it passes from one system to another, and may identify areas of weakness.
2. Picking a single result and tracing it back through to the raw data, including any laboratory notebook entries.

Verifying the data integrity and audit trail at each step, and demonstrating that all raw data, paper or electronic, is readily retrievable, fully supports the final result, and is consistent with any summary data filed with the regulatory agencies as part of a drug master file or new drug application.

3. Repeating example 2 above in the opposite direction to verify that all data has been processed and reported, and to confirm that there are no orphan data that could be indicative of trial injections or other malpractices.

Further proactive data audit activities could be based on the regulator's own guidance, e.g., the US FDA Compliance Policy Guide Manual 7346.832 on Pre-Approval Inspections (FDA, 2010) suggests that inspectors should:

Review data on finished product stability, dissolution, content uniformity, and API impurity.

Determine if data was not submitted to the application that should have been.

Look for invalidated out of specification (OOS) results and assess whether it was correct to invalidate them.

Seek out inconsistencies in manufacturing documents (e.g., identification of actual equipment used). Table 3.25 provides steps of the audit trail.

3.8.8 Logistics Service Providers Audits

Logistics service providers are assuming ever more extensive responsibilities in the distribution chain. Usually, they offer services that are characterized by a large degree of immateriality. The work performed by the logistics service provider involves an external factor, i.e., products over which it has no independent control. As a consequence, it becomes necessary to request extensive information from the contract giver on the products to be transported in order for the transport service provider to be able to deal professionally with the transport.

Furthermore, this constellation is a primary driver for the heavily customer-oriented nature of this sector. The more complex the service, for instance, that a pharmaceutical contract giver requires of his transport service provider, the more narrow and intertwined the business processes between the contract giver and contract acceptor become as a result.

3.8.8.1 The Core Tasks of Logistics

The dispatching function includes the classic tasks of a carrier such as advising, planning, and selecting the means of transportation; concluding freight agreements; issuing transport documents; and ultimately conducting the freight inspection.

The forwarding function relates to the actual transport of goods by land (truck), air, or sea, if applicable.

The transfer function refers to unloading from one truck and loading into another, possibly with intermediate storage in between.

The storage function covers collecting transport goods until the necessary load volume has been reached. This is dependent on the particular destination.

The information function involves gathering, transmitting, and processing information on shipping status, time, temperature, delays, arrival, etc. The information in this respect flows not only from the driver to the dispatcher and vice versa, but also from the dispatcher to the customer and vice versa.

Many transport service providers expand their range of services since they envisage substantial value-added potential above all in special and complementary functions. Therefore, in addition to the functions already mentioned, a wide range of services offered also includes financing; additional service tools of telematics (e.g., safety features such as "door lock" or "engine stop"); or GPS monitoring of temperatures, speeds, etc. These additional features play a role that can tip the competitive scales in the selection of transport service providers, particularly in the case of transporting high-priced pharmaceuticals.

3.8.8.2 Requirements Profile of a Logistics Service Provider

The requirements profile for transport service providers in the pharmaceuticals area is stringent and complex and in some parts surpasses requirements for carriers in the food sector. There are fundamental differences between the HACCP (hazard analysis of critical control points) concept and the GDP approaches that dominate foodstuffs. Whereas the HACCP focuses on the quality of processes, GDP – as a counterpart to GMP – concentrates on product quality and its maintenance.

TABLE 3.25
Audit Trail Audits

Guidance Document Provisions	Suggested Audit Step	Impact of Deficiency
Audit Trails		
1a. Persons must use secure, computer-generated, time-stamped audit trails to independently record the date and time of operator entries and actions that create, modify, or delete electronic records. A record is created when it is saved to durable media.	1. Review the approved functional specifications to determine if the following requirement is included: a secure, computer-generated, time-stamped audit trail is independently generated to record the date and time of operator entries and actions that create, modify, or delete electronic records. 2. Review the approved technical specifications to determine if the following requirement is designed and programmed properly: a secure, computer-generated, time-stamped audit trail is independently generated to record the date and time of operator entries and actions that create, modify, or delete electronic records. 3. Review the test cases and scripts and document test results to confirm that a secure, computer-generated, time-stamped audit trail is independently generated to record the date and time of operator entries and actions that create, modify, or delete electronic records.	The lack of an effective audit trail limits the study sponsor's ability to 1. Protect the authenticity, integrity, and, when appropriate, the confidentiality of electronic records. 2. Reconstruct and evaluate study records.
1b. Audit trails must be retained for a period at least as long as that is required for the subject electronic records and must be available for agency review and copying.	1. Determine that an approved SOP or other formal record retention policy exists that mandates the retention of audit trails for a period of at least as long as that is required for the subject electronic records. 2. Confirm compliance with the SOP or other record retention policies by reviewing documented evidence of compliance. 3. Observe the audit trails (either hard copy or electronic) and confirm the existence and effectiveness of procedures designed to copy the audit trails.	As above.
2. Personnel who create, modify, or delete electronic records should not be able to modify the audit trails.	1. Review the security/authorization profiles to confirm that personnel who create, modify, or delete electronic records do not have written access to the audit trails. 2. Review the test cases and scripts and documented test results to confirm that testing verifies that personnel who create, modify, or delete electronic records cannot modify the audit trails.	Unauthorized access to audit trail records increases the risk that the audit trail data will be corrupted and therefore not usable to (1) protect the authenticity, integrity, and, when appropriate, the confidentiality of electronic records; and (2) reconstruct and evaluate study records.

(Continued)

TABLE 3.25
Audit Trail Audits

Guidance Document Provisions	Suggested Audit Step	Impact of Deficiency
3. Clinical investigators should retain either the original or a certified copy of audit trails.	Perform a documentation review of the clinical investigators to confirm that they are in possession of either the original or a certified copy of audit trails. In the case of changes to the system, this also includes copies of source code and change control documentation, so the scope would also include developer and support personnel.	The original audit trails (or a certified copy) provide the documented evidence of study conduct. The lack of such documentation, and the ability to review the documentation, increases the risk that improper, incomplete, and/or inaccurate processing will go undetected.
4. FDA personnel should be able to read audit trials both at the study site and at any other location where associated electronic study records are maintained.	Observe the audit trails and all associated electronic study records (e.g., Case Record Forms [CRFs]), source code, and change control documentation), in all locations where they are maintained, and confirm the existence and effectiveness of procedures designed to copy the audit trails.	The original audit trails filed copy) provide the documented evidence of study conduct. The lack of such documentation, and the ability to review the documentation, increases the risk that improper, incomplete, and/or inaccurate processing will go undetected.

Other special aspects must be considered as well, depending on the shipping flow. The requirements profile is also shaped by the distribution paths and types that always entail different requirements for the transport service provider.

3.8.8.3 Quality Management System

In auditing a logistics service provider care, should be taken that the following key points are mapped in the QM system:

- The quality manual contains detailed instructions on loading/unloading, drivers (driver's manual), and training in addition to the general instructions on cleanliness and workflows.
- Regular internal audits are conducted and the results are documented in a report. The quality system is reviewed and if necessary adjusted at regular intervals.
- Quality manual, operating instructions, and audit reports are accessible to authorized third parties (inspectors, contracted auditors, customers).
- The revised EU GDP Guidelines also place great emphasis on the evaluation of contract activities and dedicate "Chapter 7: Outsourced Activities" to this topic.

3.8.8.4 Personnel and Training

- It is imperative to deploy only trained personnel. The quality of transport ultimately depends on the knowledge of and familiarity with products and workflows by the people involved.
- An initial instruction can only provide cursory insight. Continual training should include consideration of the following topics:
- Basic understanding of medicinal products and their significance: "My work is important; I move pharmaceuticals; pharmaceuticals can save lives."
- Product Information: Hazards, precautionary actions for handling.
- Conduct while transporting medicinal products: How to avoid damaging/breaking the product, what precautionary actions must be taken if damage has occurred, emergency plan for accidents, delay, and technical failure.

- Vehicle security: Unauthorized access, securing the load, parking, and conduct during breaks (especially important with valuable freight).
- Hygiene: Cleaning the loading area, procedures, controls, and personal hygiene as well.
- Freight with special transport conditions: temperature.
- The individual training components are concluded with an assessment of success; the results (score) are noted in the training documentation. Thus the personal training status is documented and is helpful in the planning and execution of the individual training blocks and sequences. It also allows conclusions to be drawn on the quality of the teaching and how the training content is received by the trainee.

3.8.8.5 Information Management

Information about all aspects of transport is indispensable. The important thing is to promptly make the data available. Track and trace modules are state-of-the-art with overland transports by truck, thanks to sophisticated telematics systems. The dispatcher knows the exact GPS position of the vehicles and notices if one departs from the planned route. In addition to temperature data, additional safety features are provided. For instance, "door lock" indicates that the loading hatch has been opened and the incident has been recorded. This data on shipment tracking can be made available to the contract acceptor in closed web areas.

- Shipment tracking is generally not as detailed in courier services. Temperature data is often not available until after delivery has taken place, when the data loggers are read out. In distribution logistics to the pharmacies this data is generally completely missing. The temperature is checked when the products are received at the pharmacy. A prerequisite of this is that the shipment of temperature-sensitive products must take place "in receptacles suitable for that purpose," in other words in validated shipment packaging and/or qualified semitrailers.
- Of course, the topics of data security and confidentiality should not be neglected when information is being discussed; after all, there is a reason why pharmaceutical products are often shipped in neutral packaging.

3.8.8.6 Different Distribution Paths and Methods

Various procedures in handling temperature-sensitive medicinal products will now be outlined and different requirements pointed out in the examples of carriage by overland truck (active cooling), distribution logistics in pharmacies (reusable packages, passive cooling), and delivery by courier (disposable packaging, passive cooling).

Overland by Truck (Refrigerated)

Carriage by overland truck is primarily for palleted products; bulk goods are the exception. Temperature control is primarily carried out with active cooling.

Critical aspects.

The temperature distribution across the loading area must be uniform. The temperature distribution is gathered by temperature mapping. This is a component of the qualification. Proper qualification of the trailer is a basic requirement for pharmaceutical transports. For this purpose, calibrated temperature loggers are spaced apart on the loading area in such a way that especially critical points such as door openings, proximity to ventilation, and ceiling area are covered by the measurements. Temperature mapping is a component of a thorough qualification process.

Temperature control is done with cooling units. A tightly knit and excellent service network in the regions covered by overland transport is crucial. Service abroad should be comparable to domestic service: service density, availability, language skills, qualified personnel, replacement trailers, etc. This is necessary for consistent and prompt handling of emergencies.

Shipment tracking is an important aspect of maintaining supply chain integrity, especially in connection with the infiltration of falsified medicinal products. The EU GMP Guide is currently being revised; the changes will probably include the traceability of shipments and products. Modern track and trace modules based on GPS/GSM contain not only precise locating functions, but also tools such as "door lock" (indicates that the loading hatch has been opened) and "engine stop" (unauthorized departure is prevented by switching off the engine).

The temperature in the loading area can also be called up. The temperature curve is then generally shown in the form of a diagram. This data can then be invoked by the contract giver through a specially protected web area. In these evaluations, care should be taken that the data cannot be manipulated, for instance, simply by being copied and pasted into other programs. For this reason, output in PDF format is preferable. Here it would certainly also be expedient to use validated software.

Distribution Logistics

Critical aspects.

Reusable packaging is cleaned on the basis of a written hygiene plan. The cleansers used do not harm the surfaces. However, constant use of the packaging leaves behind clear signs of wear; however, they are noncritical as long as functionality and safety are not affected. It is especially important to regularly change the thermal inserts. Chipped-off corners affect the cold chain duration since they promote the formation of cold bridges.

If worse comes to worst and Murphy's law moves in without being invited, a widespread emergency managing plan must be implemented. For example, this plan can include the failure of cold storage facilities (re-storing, transferring refrigerated products, time frames, etc.). But routine day-to-day problems such as delivery delays with marked lengthening of the transport time beyond the validated temperature maintaining time are also taken into account in a contingency plan (instructions to the driver, call chain).

Courier Service Providers

While palleted products play a prominent role in the overland carriage by truck, freight handled by courier services is primarily composed of small-volume packages. Passive cooling systems predominate in this segment, and this for the most part in disposable packages. They are either provided by the courier service in preconditioned form or they are held ready by the contract giver and handed over ready to ship.

Critical aspects.

A courier service is often selected when the products to be shipped are supposed to reach their destination in as short a runtime as possible. For this purpose, the sender has various services to choose from such as overnight, delivery before noon on the following day, etc. In general, these targets can be reached in Europe within 48 hours and overseas destinations within 72 hours. Staying within the runtime is the top priority when time-critical shipments are involved: the temperature maintaining time in passive packaging is laid out for a certain time, i.e., a setpoint temperature range can only be held over a defined period of time. Especially sensitive and valuable shipments are equipped with temperature loggers to monitor the temperature.

3.8.8.7 Topics for Auditing

The contract giver generally has to call the carrier's attention, first, to special features of handling the products to be transported. Second, prior to commencing a business relationship and afterward at regular intervals, the contract giver must verify the carrier's suitability by conducting an audit. Depending on the outcome of the first audit, subsequent audits may be reduced or intentionally focused on specific criteria at the auditor's discretion. The checklist is intended as a possible guide for audit questions. The requirements on qualification, alarm systems, requalification, and change control are crucial criteria to be met.

- Most GMP inspections follow the material flow of products – and start with a tour of the warehouses.
- The warehouse gives you a first impression of general procedures, cleanliness, and housekeeping.
- They might be an indicator for operations in the rest of the facility.

Get an overview of material handling, batch systems, quarantine procedures, etc.

Investigate sampling conditions and procedures for starting materials and collect information for spot checks for later evaluation (e.g., raw material batch information, finished product batches)

Quick Checks in Transport and Storage Operations

Storage/hub/depot

- Freezers/cold rooms/narcotics/quarantines
- Summer/winter qualification/temperature mapping
- Access control/pest control/cleaning
- Temperature monitoring/temperature excursions/location loggers/archiving temperature data
- Calibrated devices/calibration strategy
- Dedicated storage for pharmaceuticals

Transport vehicles

- Specification of vehicles
- Loading area: minimum and maximum load
- Thermo trucks: qualification, ATP category, cooling unit, and airflow, temperature control, and deviation management

Personnel

- Job descriptions, qualifications, responsible person, interim staff
- Training, training plan, evaluation, and documentation

Transport

- Risk analysis, validation
- Mode of transportation (air, sea, road)
- Seasonal variations
- Hubs, depots, and interim storage
- Documentation: traceability
- Subcontractors and their qualification
- Quality agreements

3.9 INTERNAL GMP AUDITS/SELF-INSPECTIONS

We are GMP auditors. We are not speaking the same language as top management. We have trouble deciding if our leaders see the same world that we are looking at. We tend to present a detailed analysis of a definite opportunity for improvement, but top management tends to ignore it. We are the voice of common sense and feel nobody of importance.

TABLE 3.26
International Guidelines

GMP-related regulations
• EU GMP Guidelines
• PIC/S Guidelines
• ICH Q7, Q8, Q9, Q10
• FDA Guidance for Industry Quality
• EU GDP Guidelines
• Quality assurance of pharmaceuticals: a compendium of guidelines and related materials Tenth Edition • Volume 2 Good manufacturing practices and inspection

Top management is "impatient" and they may (probably) have angry customers and their finance department is "on their case." What would top management's solution be? *Fix it! I don't want to know the details!*

A successful and well-managed audit has a number of benefits, in addition to meeting statutory obligations. These benefits include:

- Preparation for inspection by regulatory authorities is of great benefit to the staff so that they are exposed to GMP auditing before inspections by regulatory bodies.
- Provides objective evidence of GMP compliance, which can allow staff and managers to assess the quality of their work and to plan and carry out necessary remedial action.
- Encourages quality improvement by elimination of GMP deficiencies and by ensuring that good ideas implemented in one area can be praised and then spread throughout the organization. It is essential to note that an effective quality audit will highlight the good points as well as the bad points detected during the audit.
- Promotes self-criticism and self-improvement by detecting areas of activity that are poorly understood or have been overlooked by the staff concerned.
- This should result in cost savings if wasteful practices can be identified and eliminated.

21 CFR 210 and 211 do not directly stipulate the implementation of self-inspections; however, FDA inspectors expect them to be carried out to ensure a periodic review of the quality assurance system.

Regular self-inspections should be carried out to ensure compliance with the GMP regulations in all departments. For this reason, self-inspections can be considered an integral part of the quality assurance system. Table 3.26 provides examples of international guidelines on audits.

3.9.1 Objectives of Self-Inspections

- Compliance with existing requirements (SOPs)
- Implementation of existing, new, or modified regulations (e.g., changes in the EU GMP Guidelines)
- Identification of areas for improvement (e.g., weaknesses in documents for error prevention or descriptions in SOPs)
- Effectiveness of CAPAs and implemented changes
- Identification of potential risks (e.g., after building alterations)
- Determination of the competence level of staff and of the effectiveness of the training system
- Sharing of quality requirement know-how
- Training in inspection and audit situations

3.9.2 Contents of a Self-Inspection SOP

- Definition of inspection areas
- Inspection schedules, including specifications for the annual plan
- Definition of inspection contents, e.g. questionnaires
- Definition of responsibilities
- Selection and qualification of the inspectors
- Preparation and implementation of the self-inspection
- Documentation of the self-inspection (structure, content)
- Classification of findings
- Definition of corrective actions
- Follow-up of findings/corrective actions
- Communication of results/management review
- Evaluation of inspectors and self-inspections
- Evaluation of the effectiveness of the self-inspection system

3.9.3 Self-Inspection Areas (Examples)

- Personnel
- Premises, including changing rooms/personnel lockers
- Servicing and maintenance of buildings and facilities
- Calibration of displays and measuring systems
- Procurement and storage of starting materials, packaging materials, intermediates, and finished products
- Production processes and in-process controls
- Validation and qualification programs
- Quality control
- Documentation
- Cleaning, disinfection, and hygiene
- Recall process
- Complaints management
- Data management and data integrity
- Results of previous self-inspections and corrective actions
- Implementation of self-inspections
- Change control system
- CAPA system

3.9.4 Main Purpose of GMP Audits

- Common views and understanding necessary
- The 7 "what?" and 3 "how?" concept
- What are the critical pathways, processes and control points?
- What's the real observation = factual evidence?
- What's a comment/suggestion only?
- What's an individual failure?
- What's a system weakness?
- What's the underlying root cause?
- What's a risk in a specific environment?
- How to document and communicate identified deficiencies and risks?
- How to be part of the CAPA and continuous improvement process?
- How to take responsibility and ownership of a GMP audit?
- To assess the design (in place) and performance (in use) of a pharma QMS

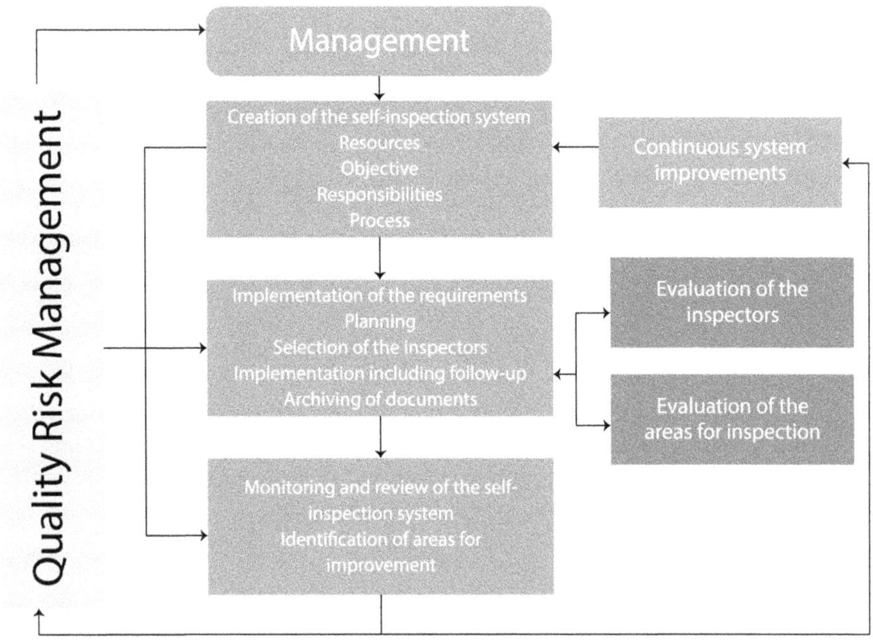

FIGURE 3.16 Risk-based self-inspections.

- Timely identification of operational deficiencies with significant potential (impact) for:
- Creating conflicting, inadequate data that could compromise the approval and maintenance of regulatory submissions (manufacturing + marketing authorization)
- Sustained disruptions of the supply chain due to imposed conditions, and restrictions by regulatory agencies as a result of governmental inspections
- To highlight and support, where appropriate, adequate corrective and preventive measures in the sense of business excellence and continuous (sustainable) improvement
- To support and further develop adequate quality systems and associated tools and practices

Figure 3.16 explains the risk-based self-inspection flow.

3.9.5 Inspection Methods

Tracing

Tracing, or following the chronological progress of something as it is processed, is a common and effective means of collecting objective evidence during an audit. In forward tracing, an auditor starts at the beginning or middle of a manufacturing process, for example, and traces forward. In backward tracing, the auditor starts at the middle or end of a process and traces backward. Backward tracing can be more revealing than forward tracing, because the auditor examines the process from the perspective of seeing the results (product or service) of the preceding activity.

A flowchart can be used as a road map for tracing activities. When tracing, the auditor should:

- Start at the beginning, middle, or end of the process.
- Choose an action, such as painting a wall, stamping, folding, or baking.
- Gather information in six areas: people, equipment, method, material, environment, and measurement for the action chosen, and record this information on a checklist.
- Follow the path of the transaction backward or forward through the process.

Discovery Method

The discovery method is sometimes called random auditing. This method investigates what is currently taking place and therefore reflects prevailing work procedures. However, the discovery method offers no discernible pattern to an audit, and an auditor can become disoriented or spend too much time in the facility.

In most situations, the discovery method should not have to be used if the audit has been adequately planned. However, this method could be effective if the auditor knows that a problem exists but has not yet located it. By going into the audit area and looking around, an auditor may be able to bring the problem to the surface.

Element Method

The element method is a commonly employed auditing strategy. The element method can refer to any element of any quality assurance standard or of the auditee's quality program. The auditor examines each element individually to see how it affects the entire system. This method is thorough and good to use on an annual basis. However, when an audit team is very large or is spending several weeks in a facility, the element method can become cumbersome.

Department Method

The department method is another way of dividing an auditing task and works especially well when accountability at the department level is important. The auditor focuses on the entire operation of one department rather than on segments of it by reviewing numerous quality elements within the department. This auditing strategy touches on almost everything in an audited facility. Here, the auditor's ability to identify applicable quality elements for that department is put to the test. Although this is an excellent strategy when the audit team wishes "to slice the apple a different way," it should not be overused. Focusing too much on one small group of people (one department) may not accurately represent what is happening throughout an organization.

TABLE 3.27
Advantages and Disadvantages of Trace Forward and Trace Back Inspections

Audit Type	Advantages	Disadvantages
Trace forward	Follows the normal flow of the system. Therefore: • Logical flow • Covers interfaces • Easy to plan • Deficiencies early in the process (which can have a major impact on later stages) are detected early in the audit Can be used "live" – it is possible to trace forward in "real-time" and inspect activities as they happen	Lacks flexibility: • Flow interrupted if people are not available in the required order • Cannot easily be used for partial audit • May quickly provide a large number of possible avenues to follow (e.g., a single batch of drug substance into multiple batches of product)
Trace back	• Can start in any department • Useful for partial audits • See the output from one department before auditing the process(es) within it, which may help provide specific points to look at • Covers interfaces • Follows the course that needs to be taken in the event of a customer complaint	• Flow interrupted if people are not available in the required order • Always historic • The early parts of the process are not seen until the end of the audit; care required to ensure that not left short of time • May provide a large number of possible avenues to follow (e.g., several excipients in a product batch)

TABLE 3.28
Announced versus Unannounced Audits

Announced	Unannounced
• Right people present	• Possible key people absent
• Allows auditee to plan	• No time for planning
• Allows "clean-up"	• "As is" situation
• Shows trust	• Demonstrates suspicion
• Length can be set in advance	• Unpredictable
• Cooperative	• Adversarial

Table 3.27 offers the advantages and disadvantages of trace forward and trace back inspections. Table 3.28 explains the differences between announced and unannounced audits.

3.9.6 Audit Sampling

- Auditing as a sampling exercise – cannot see everything.
- All sampling by nature has inherent risks – drawing conclusions from very small sample sizes.
- "Sample size of 1" when drawing conclusions is dangerous.
- Too many samples means losing the overall timeframe/objective.
- General rule of thumb: Sample two to three records to review.
- If all is OK, move on.
- If one is not OK, ask for more.
- If two or three are problems, move on.

3.9.7 Frequency of Self-Inspections

The EU GMP Guidelines do not specify the frequency of self-inspections: The decision is left up to the pharmaceutical company.

The following options are generally available:

- Predefined time intervals (e.g., annually)
- Risk-based
- Event-induced

The WHO GMP guide stipulates company-specific inspections that have to be carried out annually at least.

The qualifications required by self-inspectors must be specified in the SOP. These can normally be based on the requirements for GMP auditors. The requirements for internal self-inspectors can be less complicated and include a stronger focus on the objective of the inspection.

Table 3.29 describes the strengths of internal and external inspectors.

Table 3.30 describes the phases of self-inspections.

3.9.8 Documents in a Self-Inspection Setting (Examples)

- SOP for the self-inspection system
- Self-inspection plan (e.g., annual plan)
- Checklists, aide-mémoire
- Self-inspection reports
- Periodical evaluation of the effectiveness of the self-inspection system

TABLE 3.29
Internal versus External Inspectors Carrying Out Self-Inspections

Aspect	Internal	External
Process-specific knowledge	+	o
Company-specific knowledge	+	o
Objectivity	o	+
Impartiality	o	+
Expertise	+	+
Open communication	+	+

TABLE 3.30
Phases of the Self-Inspection Process

Planning	• Definition of the area to be inspected • Appointment of the inspection team • Definition of topics and/or key topical aspects • Creation of a checklist, if required • Notification of the pertinent persons or department
Implementation	• Inspection: on-site observations, discussion • Status records (forms, checklists) • First comments during discussions
Wrap-up	• Creation of a report containing evaluations and recommendations • Definition of measures (CAPA) • Comments from the head of the department (and senior management), if required
Follow-up	• Follow-up of actions carried out because of reported deficiencies • Evaluation of the effectiveness of the actions • Scheduling of follow-up inspections, if required

3.9.9 Checks Carried Out during Self-Inspections

- Checks for GMP compliance (SOP evaluation)
- Checks for SOP compliance (on-site observation)
- Checks of personnel qualification
- Checks of documents and electronic data, including audit trail and data integrity
- Identification of unusual or unexpected observations, e.g., defects, notes on loose sheets of paper, data sheets that have been "disposed of"
- Plausibility check of data
- Comparison of the declared (documented) status and the actual status, e.g., after system cleaning or after certain process steps have been carried out
- Identification of isolated or repeated and/or systematic errors

Table 3.31 shows how a poorly written audit finding can improve.

3.9.10 How to Report Findings

- Avoid vague observations.
  - "Bearded operator observed in the aseptic processing area wearing nothing but a head covering." (Really?)

TABLE 3.31
Example of a Poorly Written Observation and an Improved Version

Poor wording	**Not all batch records had data entered where required**
Improved wording	25 batches of product X were made in the time period January 1, 2022, to the present. I reviewed 10 of the 25 batch records and found missing data entries in 9 of the 10 records reviewed. The nine records had from one to six missing entries each. Examples of missing data entries were (1) failure to record the time of buffer preparation (four examples) (2) failure of the CIP operator to note the time the cycle was begun (four examples) (3) lack of verification signatures for process steps designated as critical (eight examples)

  - "Poor housekeeping practices noted in the buffer prep area."
- Better wording for observations.
  - "A bearded operator observed in the aseptic processing area was not wearing a facial hair restraint as required by SOP XXX."
  - "Poor housekeeping practices noted in the buffer prep area, as evidenced by [list two or three examples]."
    - Tell the reader what you saw that led you to conclude that practices were "poor" so that the reader can reach the same conclusion.

3.9.11 Benefits of the Application of Quality Risk Management to Self-Inspection Programs

- Ensuring high quality of the medicinal product by proactively identifying and controlling potential quality issues.
- Improving decision-making if a quality issue arises.
- Facilitation of better and more informed decisions that can provide the regulators with greater assurance of a company's ability to deal with potential risks and can beneficially affect the extent and level of direct regulatory oversight.

The risk associated with an area can be estimated by using different risk ranking methods or tools. For example, a simple approach could be assigned a numerical descriptive value of 1 (low) to 3 (high) to establish categories for the selected risk factors.

The estimated values for all risk factors can then be linked together in an appropriate way to yield a risk level associated with an area (e.g., low, medium, and high). The estimated values for risk factors could also be multiplied by significance-weighting factors to give a total.

Table 3.32 explains how to calculate values.

3.9.12 Example of Vertical and Horizontal Self-Inspection

A division in methodology that can be helpful is that of product and process. An auditor may choose to primarily audit a particular product and, as part of this, will look at all the processes associated with that product. Alternatively, the auditor may choose to look primarily at a process and select different products as examples to see how this process is put into practice. Figure 3.17 explains the points to be considered in horizontal and vertical audits with examples.

TABLE 3.32
Example of Categories for a Risk Factor and Their Numeric Values

Risk factor	Category (impact X likelihood) (based on data for a given time period)	Value
Results of previous self-inspections	<5 minor observations, no major or critical observations	1
	<15 minor observations, no major or critical observations	2
	>15 minor observations or any major or critical observations	3
Number and significance of deviations	<20 minor deviations not affecting the number of released batches per month; no major deviations affecting batch release	1
	>20 minor deviations not affecting the number of released batches per month	2
	<10 major deviations resulting in a delay in batch release; no major deviations resulting in batch reject or reprocessing	2
	> 10 major deviations resulting in a delay in batch release; any major deviations resulting in batch rejection or reprocessing	3

3.9.13 Organization/Responsibilities

- Organization of the Quality Unit
  - Quality Control and Quality Assurance as separate units
  - Quality Control reporting to Quality Assurance
- Independence from Production
  - Independence laid down in organograms
  - Independence demonstrated during the audit (processes)
- Responsibilities
  - Clear assignment of responsibilities, e.g., for laboratory testing, documentation and approval of test procedures and specifications, failure investigation process, release procedure, qualification and validation, complaint management, etc.
- Job descriptions
  - Should be available at least for key personnel (head of QA, head of QC, laboratory heads, etc.)
  - Should include all necessary responsibilities

3.9.14 Management Review

- Internal audit findings should be visible to the highest level of management.
- If the company has a formal management review of the quality system at prescribed intervals, internal audit findings should be one of the metrics reviewed.

3.9.15 Reporting Relationship

- Internal audit groups should not report to heads of units they are charged with auditing, if at all feasible, because this can create a conflict of interest.
- For large companies, a corporate audit function reporting as high up the corporate structure as possible is advisable.
- If the audit group reports to a site head of QA, consider bringing in a consultant once a year to perform an audit of the overall quality system, including the effectiveness of the audit group.

Table 3.33 discusses what should and shouldn't be said in an audit.

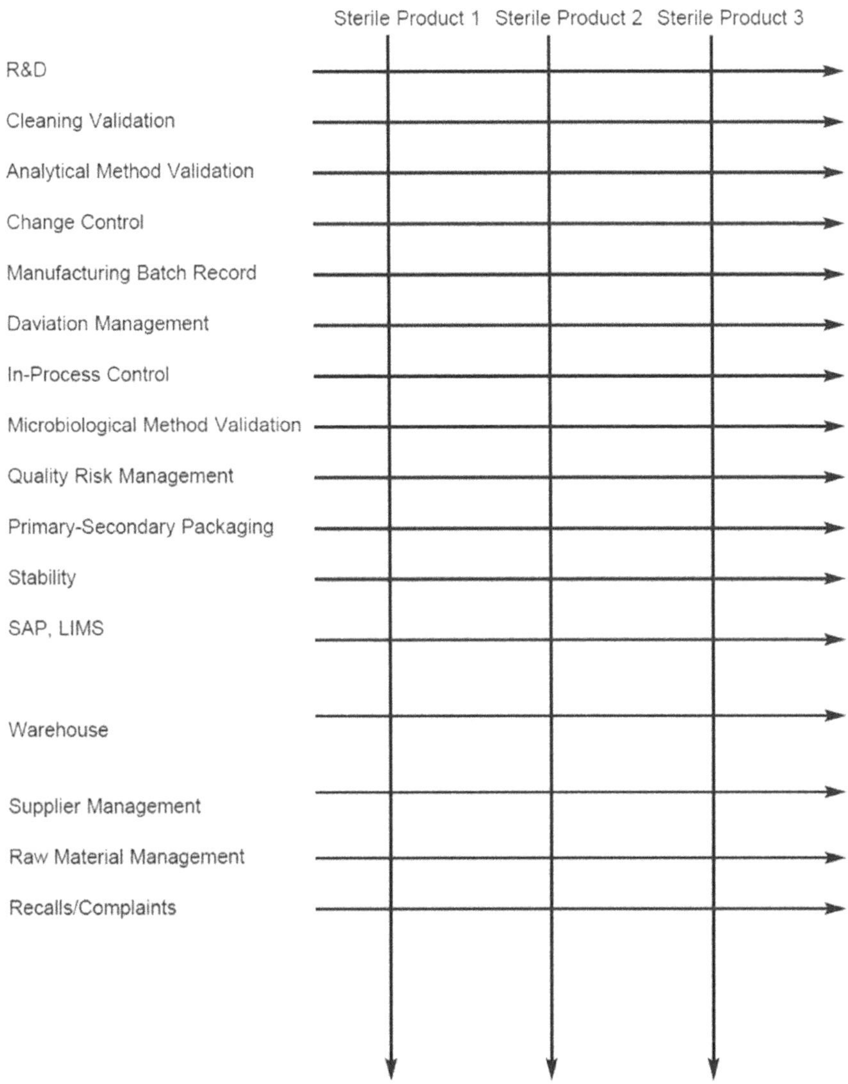

FIGURE 3.17 Horizontal versus vertical audits.

3.9.16 What About Physical Examination during the Audit?

The types of physical examination tools used by auditors can vary widely depending on the industry. GMP auditor needs to be familiar with the tools and techniques applicable to the industry or company being audited.

Physical examination tools are normally associated with product audits. For example, when auditing a packaging materials control laboratory in a pharmaceutical manufacturing plant, an auditor might use calipers, steel rules, or micrometer, to take measurements of high-density polyethylene (HDPE) closures, polypropylene (PP) or glass bottles, vials, and ampoules. GMP auditors may also observe others using these tools in order to verify that a product meets certain characteristics. In most circumstances, it is essential for tools to have been properly calibrated. If a GMP auditor uses measuring devices that require periodic calibrating, the auditing organization should follow an accepted calibration program.

TABLE 3.33
Should Say versus Shouldn't Say

The Auditor Should Say	Should Not Say
How (where, when, who, what, why) do you record the test results?	Do you record the test results in the lab log book?
How do you know this value is right?	This instrument is calibrated, isn't it?
What is the first thing you do?	First, you set up the equipment, right?
How do you know this is the current (correct) version of the drawing?	Is this drawing current (correct)?
How were you trained to perform this procedure?	Did you read the standard operating procedure during training?
What are the reporting requirements for nonconformances?	Don't you have to notify your supervisor whenever a nonconformance occurs?
What is the standard procedure for responding to customer complaints?	When a customer calls, don't you have to record the details on Form XX?
Is this equipment calibrated? How do you know?	Does this sticker indicate that the equipment is calibrated?
How do you know how to do this operation?	Do you follow the procedure for this operation?
What do you do with the finished product?	Do you place the finished product on the rack?

CASE STUDY

Using Figure 3.18, prepare at least 20 questions using the process flow in the oral solid dosage (OSD) audit.

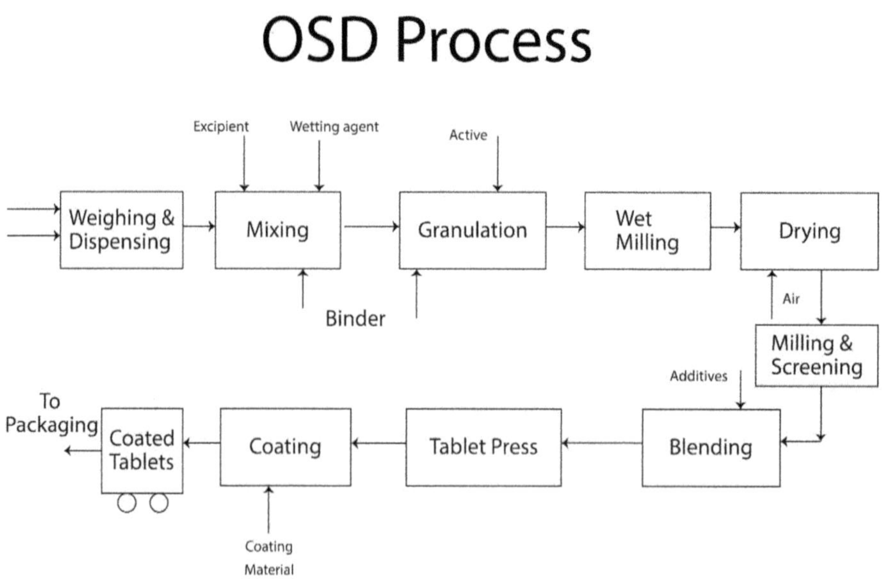

FIGURE 3.18 Develop a checklist for the OSD process.

3.10 GMP AUDIT DOCUMENTATION

- Relevant management system documents and records
- Applicable standards
- Site Master File
- Registration files
- Guidelines
- SOPs
- Protocol or project-specific requirements
- Previous audit reports
- Adverse drug reactions
- Records of the company in relation to complaints and recall

The top cause of GMP audit citations is the absence of documentation or employees not following it.

Often, the first thing GMP auditors will ask for are documents from the quality management system. The GMP auditor will seek to confirm that there is documentation covering aspects of operations that affect product quality. They will also look for evidence that the documentation is being followed (e.g., records are in place).

There's a saying that "If it isn't written down it shouldn't happen," meaning that you need to have documentation in place for all of your processes. The other part of the saying is "If it isn't written down, it didn't happen," meaning that you need to have records or other evidence in place to show that written procedures have been followed. The documents in a QMS should be used regularly, followed carefully, and referred to frequently.

In Figure 3.19, the typical document hierarchy in a quality management system is depicted.

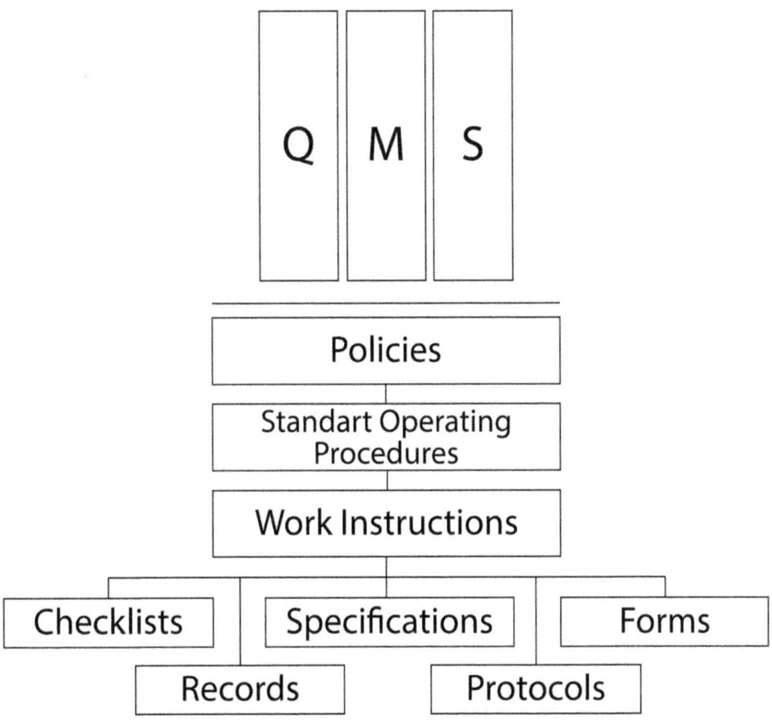

FIGURE 3.19 QMS documentation system.

3.10.1 Documentation: Specifications

- Documentation
  - Revision control (revision numbers, validity dates)
  - Change procedure (SOPs, history of changes, regulatory compliance)
  - Approval (responsibilities)
  - Reference to pharmacopoeia monographs
- Content
  - Qualitative tests
  - Quantitative tests
  - Reference to test procedures (in-house or pharmacopoeia)
  - Acceptance limits (release/shelf life, internal/external)
  - Skip lot testing (documentation/rationale)
  - Takeover of data from supplier's COA (based on supplier evaluation)
  - Takeover of data from previous manufacturing steps (bulk ⇒ filled product)
  - Official/internal tests
  - Approved suppliers
  - Storage conditions or precautions
- Test parameters
  - Description / physical characteristics (appearance)
  - Identification (two independent methods?)
  - Structural integrity (biological products)
  - Impurities (inorganic/organic, single/sum)
  - Residual solvents
  - Assay/potency (as is, in dried substance)
  - Sterility/microbial purity
  - Safety(endotoxins/pyrogens, abnormal toxicity, viral safety)
- Justification of specification
  - Pharmacopoeia monographs/relevant guidelines (ICH/VICH)
  - Analytical methods/development report
  - Initial experiences (bio batches/validation batches)
  - Statistical data review (commercial batches)
  - Stability information

3.10.2 Documentation: Test Procedures

- Content
  - Degree of details (sample preparation, equipment specified, columns, reagents, sequences, system suitability)
  - Additional documentation of details in other documents
  - Compliance with the technical dossier (CMC)
- Handling
  - Which documents are used in the laboratory (controlled copies)?
  - How is it ensured that only qualified equipment is used?
  - Is the validation covering the chosen equipment?

3.10.3 Documentation: Analytical Validation

- Procedure for analytical method validation
  - Scope (test methods, IPC methods, compendial methods)
  - Responsibilities

    - Validation requirements/parameters
    - Handling of compendial analytical methods
    - Revalidation
    - Method transfer requirements
    - Validation reports (content, format, approval)
- Validation Master Plan (VMP)
    - Analytical validation included in site VMP
    - Individual VMP for analytical laboratories
    - Responsibility/approval
    - Updating/review period
- Validation characteristics
    - Accuracy
    - Precision
    - Repeatability
    - Intermediate precision
    - Specificity
    - Detection limit
    - Quantitation limit
    - Linearity
    - Range
    - Robustness

Revalidation requirements

- After changes in the composition of the product
- After changes in the analytical procedure
- After changes in the analytical equipment

3.10.4 Documentation: Raw Data Traceability

- Traceability of test and reference/retention samples
    - Sampling
    - Receipt
    - Workflow during testing
    - Storage of reference/retention samples
- Traceability of data
    - Raw data definition
    - Documentation systems
        - Personal logbooks
        - Product-related logbooks
        - Equipment logbooks
        - Calibration logbooks/maintenance logbooks
        - Logbooks for reference standards/retention samples, etc.
- Documentation practices
    - Bound logbooks versus single pages
    - Control of issued pages, pagination
    - Completeness check
    - Documentation of alterations (pencil, correction fluid)
    - Citations/cross-references (test methods, standards, reagents, equipment)
    - Double-check calculations/data transfer

3.10.5 Documentation: Change Control

- Change control system
  - Scope (specifications, test procedures, equipment, approved suppliers)
  - Categorization of changes
  - Responsibilities (QA, QC, Production, Regulatory Affairs, etc.)
  - Documentation
  - Internal/external
  - Workflow for changes
- Compliance with marketing authorization
  - Interface with Regulatory Affairs/authorities
  - Interface with customers

3.10.6 Testing: Workflow and Documentation

- Information about new deliveries of raw material
- Information of production batches finalized
- Interface to sampling group
- Handling, documentation, and tracing of samples
- Laboratory workflow
- Performance of tests
- Documentation of tests
- Evaluation of test results
- Review of associated documentation (batch record, monitoring, deviations)
- Release decision, change of status information
- Storage of reference/retention samples

3.11 LABORATORY AUDITS

- Analytical chemistry laboratory audit
- Contracted laboratory audit
- Stability audit
- OOS and OOT audit
- Microbiological laboratory audit
- Packaging material laboratory audit

3.11.1 Analytical Chemistry Laboratory Audit

A laboratory audit is one of the most important parts of GMP audits. It is extremely important to document the data during the audit and even to record the statements of the spoken personnel. Memory becomes incredibly important, as the GMP inspector has no chance to take any notes or record audio or video for a short time, for example, when he enters a sterile test site. In GMP audits, if only the audit of laboratories is in question, advance preparation is essential for success in these audits, as in other audits.

Here are three areas in a laboratory from which information should be derived during the on-site audit:

- Personnel
- Documentation
- Observation of testing procedures

Even under the best circumstances and relationships, there is usually a limit to an auditor's contact with the person who actually performs the testing. Almost every analyst, when correctly performing a test procedure, does it in a unique way. These unique differences provide a challenge for the auditor. GMP auditors must observe the performance of the procedure. First, the GMP auditor reads and interprets the procedure. Then he/she observes the actual conditions and manipulations of the test procedure by the individual who is routinely assigned to perform that procedure.

The preparation of supplies, materials, test sites, and instruments that will be used are all significant parts of a properly performed test procedure. As the analyst performs the test, reads the results, and records the results, the auditor should note any aspects of the test that could lead to error. There are a few test procedures that are repeated on consecutive days or require more than one day to complete (e.g., microbiological testing).

As it is known, laboratory tests are carried out daily and records are kept accordingly. Let's say a reference standard calibration is included as a preparatory step in a procedure. As a GMP auditor, you should want to think practically and learn where the reference is stored and what actions are taken before and after the calibration. Not only the reference standard itself but all relevant documentation, including records, should be adequately examined and a conclusion should be reached.

The laboratory director or supervisor of the specialty area (toxicology, microbiology, chemistry, etc.) usually responds to questions from the GMP auditor. It is not only logical but imperative that audit questions be directed to the most knowledgeable person in a laboratory area. Those key persons are the only sources of complete, up-to-date information about the test procedures that are being followed. Those individuals know the background of the changes that have been made to procedures in their fields of expertise.

Written Procedures

One might ask for a list of the procedures used to control this process or alternatively, look at the SOP binder to determine if adequate written procedures.

How Laboratory Staff Are Trained and Qualified

Training can be assessed in a variety of ways, for example, by obtaining a list of all personnel working in this department and comparing this to training records for a representative number of procedures. Another way to assess the appropriateness of training is to look at the discrepancies and investigations conducted in the area. When reviewing this documentation one might look at how often retraining of the operator was conducted as part of the investigation. A high incidence of retraining could also be used to indicate the effectiveness of the training program.

Is the Total Number of Staff Adequate for a QC Lab?

Evaluation of the adequacy of staffing can be accomplished in a variety of ways. For example, one might review the organizational chart and compare the staffing represented against the number and types of tests performed. Although a preliminary evaluation may be made at this time, it can be strengthened or weakened by looking at the number of tests that are not performed on time. If a significant number of tests are not being performed when they should be, it is likely that insufficient staffing is available to perform the testing.

Another evaluation tool looks at the tests the laboratory should perform. For example, are there significant numbers of tests that should be performed but have never been implemented at the site? This can also be an indication of insufficient staffing. Review can also be undertaken by looking at whether documents are being appraised, implemented, and corrected in a timely fashion.

The list of laboratory deviations and investigations can also provide information regarding the adequacy of staffing. In those cases when there are significant numbers of errors in the laboratory — for example, documentation errors not caught, tests performed incorrectly, and so forth — one might conclude that staffing is not sufficient to ensure tests are conducted correctly, with appropriate training and supervision.

In addition to the sheer number of employees in the laboratory, another consideration is whether employees have the right type of background to perform their assigned tasks. For this type of evaluation, one might want to review the educational background training records of employees. Interviewing personnel can also be useful in determining whether they understand the science sufficiently to perform the tests correctly, and clearly understand those aspects of the testing that are critical to the proper execution and evaluation of results.

In the event that there are a large number of adverse findings in the laboratory, this may be indicative of inadequate staffing to perform the required laboratory tasks.

An evaluation of this parameter may begin during a tour of the laboratory, looking at the number and type of instruments and equipment present. Are all instruments in good working order, or are a large percentage of the items documented as being out of service? If units are out of service, the reasons for this might be investigated. Another thing to consider during the tour is the age and condition of the equipment. For example, are the doors falling off the incubators? Are there a lot of home-remedy-type fixes to the equipment? Is it rusty or damaged?

All these examples may be indicative of equipment that may not be appropriate for use. Not only should equipment be in good, working condition, but it should be appropriately sized for the number and type of tests being conducted in the laboratory. In a laboratory with 20 or more microbiologists, one microscope may be insufficient if the inspector routinely sees analysts standing in line waiting to use the microscope. Another consideration in determining the adequacy of the equipment is whether testing is performed in a timely manner. When a large number of tests are not performed on time, this may indicate inadequate equipment.

During the tour of the laboratory operation, one might evaluate whether there is sufficient space to ensure laboratory samples are not mixed up or confused during testing procedures. Another consideration is whether there is sufficient space to perform the required tests. When analysts are observed, do they have enough room to perform the desired tests? Are the hoods appropriately sized for the activities that take place within them? Are there segregated areas for tests like sterility or endotoxin testing, along with other tests where one would want to minimize extraneous contamination from the laboratory?

When touring the laboratory look at the condition of the laboratory. For example, are the walls, ceilings, and floors in good working order? Could the condition of the facility adversely affect testing? Are there designated areas for the types of tests and procedures performed in the laboratory – for example, sample storage, reserve sample storage, incubators, refrigerators, chemical storage cabinets, testing, washing, glassware storage, and pre- and post-sterilization areas? Are pipes leaking water or condensate into the area? Is there standing water or watermarks on the floor? Is there evidence of rodent or pest infestation?

Laboratory Equipment Calibration and Maintenance Programs

Evaluation of this parameter might begin with a review of the site calibration and maintenance procedure. This procedure might include laboratory equipment. If it does not include the laboratory, one might ask to see the procedure(s) applicable to the laboratory.

In some cases, companies may use outside services to perform calibrations of laboratory instruments. One might ask to see the qualifications of the service technicians performing these evaluations. For example, are all the technicians trained in the applicable GMP requirements for their activities? Another consideration is whether the site has copies of the procedures used for calibration.

If they do not have copies of the procedures, are appropriate controls established when the procedures have been changed? One should also ensure calibration adjustments made by the service technicians are appropriately documented and communicated to site personnel. Are the calibration standards used traceable to appropriate standards, for example, NIST?

One might also look at the different laboratory instruments to check whether they have calibration stickers. It is convenient at this time to determine whether the instruments are currently within calibration. Some inspectors also document instrument numbers and calibration sticker identification for subsequent comparison to the actual calibration record.

Data obtained during a tour of the laboratory may be compared to a listing of the laboratory equipment, along with associated documented calibration dates. Typically the laboratory will log when calibrations have been performed and when the next calibrations are due. One might check this listing to determine whether the calibrations are being performed on time. If calibrations are performed by on-site personnel, one might wish to tour the calibration laboratory and determine its condition, adequacy of staffing, methods, standards, procedures used, and so forth.

Computerized System Validation

A typical way to evaluate this parameter might start by asking for a copy of the computer validation policy or Site Master Plan regarding computer validation. Based upon the information in this document, one might also ask for a listing of all of the computerized or automated processes used in the laboratory, a listing of the initial and most recent qualification of these systems, and the dates when requalification will be required. Observations made in the facility should determine whether the validation policy is being followed and if it is appropriate.

If laboratory instruments are not validated, one should determine if there is a written justification or explanation for why validation is not required. (In some cases this may be addressed in the validation policy or Site Master Plan.) It may be appropriate to select representative validation protocols and review them for appropriateness and completeness.

Another consideration is whether the validations and requalifications are performed in a timely manner.

Reference Standards

The use of reference standards is a very important function in the laboratory. For these standards to work properly, they should be purchased from reputable sources, verified to meet their stated specifications, stored at the vendor's recommended storage conditions, and used within the expiration date. Failure to have appropriate procedures to ensure these activities take place may have a significant impact on the results obtained during their use.

In some cases, a company may choose to use a secondary reference standard. If so, data should be available to show that the secondary reference standard is equivalent to the primary reference standard. Many inspectors will look at the location where reference standards are stored to determine whether they are stored as required on the label, whether they are within the expiry date, and so forth.

Why System Suitability Checks on Chromatographic Systems Are Important

The data for these types of systems is typically reviewed to ensure that system suitability tests are performed at least daily. Additionally, checks should be performed as required in site SOPs. Some microbiological identification systems and other rapid microbiological methods use gas chromatography (GC). When used as part of a microbiology test, one should verify whether the checks are performed. It still is appropriate to perform daily (and perhaps more frequent) system suitability checks.

Specifications, Standards, and Representative Sampling Plans

Considerations for evaluation of this parameter identify appropriate specifications, standards, and sampling plans for the laboratory operations. One might evaluate compliance by asking for a listing of all of the representative specifications, standards, and sampling plans. This might be compared to the types of tests performed to determine appropriate procedures. The actual documents may be reviewed to determine a scientifically sound and justified methodology. Additionally, data may be reviewed to verify if procedures are actually followed.

Analysis Methods

This parameter may be evaluated by comparing procedures conducted in the laboratory against the requirements as specified in the written procedures. Additionally, one might look at various batch records to determine whether the tests performed as part of batch release were conducted as indicated in the written procedures. Another way to assess compliance is to look at the investigations and deviations log to determine whether a significant number of occurrences are due to failure to follow the written procedures for laboratory analysis.

Analytical Method Verification and Validation

Many of the methods used in a pharmaceutical laboratory are based upon procedures specified in the compendia. By definition, compendial methods are considered validated and do not require validation at the site. That being said, it still is important for the company to show that the compendial methods work with the equipment, personnel, and facilities at the site. This is frequently referred to as method verification. Compendial methods should be verified unless the compendia specify method validation procedures as part of the appropriate monograph, in which case method validation should be conducted.

All noncompendial methods should be validated. One might evaluate compliance with this parameter by asking for a listing of all the method validation and verifications conducted. This listing should be compared to the tests performed to ensure that all applicable tests have been either verified or validated. Additionally, one might wish to look at the procedures describing these requirements to ensure that revalidation or verification is performed as specified in these procedures. Another consideration is whether the validations conducted to comply with appropriate compendial and ICH requirements.

QC Laboratory Change Controls

The assessment of this parameter might start with requesting a copy of the change control procedure. This may be a site procedure or a specific procedure applicable to the laboratory. The types of changes included in this system should be equipment, software, methods, and specifications (procedures). Many inspectors will also ask for a summary of all changes made over the past 12 to 24 months. Once the summary and procedure are obtained, one might request copies of the change packages for several changes to determine whether the procedure was followed.

Another concern is whether an assessment was made as part of the change control procedure to determine the impact of the change on the validation status of the equipment, software, and/or process, as appropriate. An effective change control system should also be characterized by completing the change control documents. This is assessed by determining the percentage of changes that are not closed out. Other concerns might include determining whether changes were made to the system without following the procedure, whether appropriate controls are established and followed during the change process, and so forth.

How Do You Know the Correct Samples Are Used to Testing?

Assessment of this parameter typically starts with an evaluation of procedures used for sample receipt, tracking, and handling in the laboratory. The evaluation includes whether the samples are appropriately stored prior to testing. For example, are they stored at the recommended temperature conditions, and stored in such a way as to preserve their identity? Once the samples are received and documented in the laboratory, can their location be tracked at all times throughout the testing process? How can one ensure that the samples are subjected to the specified tests?

Are procedures in place to ensure they are tested within a specified time that has been shown to yield valid results? (For example, if a microbial sample is held too long, the results may not be representative of the results that would have been obtained if the sample was tested immediately upon receipt in the laboratory. As such, data should be available to show that the sample hold times do not adversely affect the results obtained.) Review the testing records for samples identified in the tracking log to ensure that the correct tests were performed on these samples.

Discrepancies

Determine if a written procedure exists requiring an investigation of all discrepancies.
Request a list of investigations conducted for the packaging and labeling area for the last one or two years. Establish if the investigations were completed. Review representative batch records to determine if discrepancies occurred but were not investigated.

Analytical Records

Typically this parameter is evaluated by reviewing testing records, perhaps from tests within a batch record, and comparing them to the actual raw test data. Considerations include the determination of whether the test was performed correctly. the data in the batch record were reported correctly, all data were included in the test report, data that were not included, and the data were accurate. Another similar type of evaluation may be made of data summaries prepared by the laboratory, for example, environmental monitoring data summaries. One might also verify that good documentation practices were used for all the data generated as part of the report.

Quality and Retention of Raw Data

This evaluation involves reviewing the methods used to store and document raw data. For equipment using thermal printouts, does the company make photocopies of the data to preserve the data?

In the case of chromatographs or spectra, how does the firm handle and document the reintegration of the data? Can one clearly identify the original data obtained in the evaluation? Are all raw data maintained as required in the cGMPs?

Results

Typically this parameter is evaluated by reviewing summary reports and comparing them to actual raw test data. Considerations include the determination of whether the data in the report were reported correctly, whether all data were included in the test report, whether data were not included, and whether the data are accurate. Other examples of reports requiring review include Certificates of Analysis and data transmitted to laboratory information management systems (LIMS). One might also verify that good documentation practices were used for all the data generated as part of the report.

OOS Procedure Handling

An auditor may ask to see a copy of the OOS procedure used for the laboratory. One consideration is whether all laboratory tests are covered by this procedure. In many cases, companies may not consider all microbiology-related testing as part of the OOS procedure. If this is the case, is there another procedure that addresses how those tests are handled in the event of the testing parameters not being met?
Does the OOS procedure clearly identify the tests that are, and are not, part of the procedure? Does the procedure comply with regulatory guidance on handling OOS occurrences, for example, the Barr decision?
Are the specified procedures followed? Are investigations closed in a timely manner?

Typical requests by inspectors include requesting a list of all OOS occurrences for the last 12 or 24 months, followed by a review of the types of OOS occurrences, and a request for copies of specific OOS documents.

Reserve Samples

GMPs require that reserve samples be maintained. One might request a copy of the procedure used to collect and control reserve samples. A tour of the area used to store these samples may be used to evaluate if the procedure for reserve samples is being followed. This includes ensuring that samples are stored at conditions appropriate to maintain their integrity. Another consideration is whether the procedure requires periodic review and examination of the sample in accordance with GMP requirements.

Stability Testing Program

- Is there a stability testing procedure? Does the procedure comply with the stability protocols included in the various product submissions?
- Are the current regulatory and ICH guidelines followed in the stability testing protocols?
- Are the tests performed as stated in the testing protocol? What portion of the testing is performed outside the stated time intervals?
- Are the tests performed correctly?
- Are the tests performed on the correct samples?
- Were the samples that should be tested available for testing? (Were they actually placed on stability at the correct testing conditions?)
- Are there data available to show that the methods used are stability-indicating?
- Is there a software package used to track or monitor stability testing results? If so, is the software validated? Who is responsible for the maintenance of the system?

Examples of Significant and/or Trend of Deficiencies

- There is a pattern of failure demonstrated regarding the establishment or following of a controlled system for the implementation of changes in the laboratory operations.
- There is a pattern of failure to document the investigation of discrepancies.
- Validation of computerized and/or automated processes was not conducted.
- There is a pattern of inadequate sampling practices used.
- There is a lack of validated analytical methods.
- There is a pattern of failure demonstrated showing failure to follow approved analytical procedures.
- There is a pattern of failure demonstrated showing failure to follow an adequate OOS procedure.
- There is a pattern of failure demonstrated showing failure to retain raw data.
- The methods used do not indicate stability.
- There is a pattern of failure demonstrated showing failure to follow established stability programs.

3.11.2 Contracted Laboratory Audit

Pharmaceutical companies sometimes outsource the tests they need to perform to contracted laboratories. When choosing a laboratory that complies with the GMP rules, it is necessary to fulfill the legal requirements. After a suitable laboratory has been found, the drafting contract comes to the stage of preparation. A GMP-compliant pharmaceutical and biotechnology company is expected to act in accordance with Article 7 of the EU, PIC/S GMP Guideline. Of course, before the actual signing of the contract, it must be proven that the selected laboratory can perform the required tests within the specifications in accordance with the GMP. For this purpose, the audit of the laboratory comes to the fore. Figure 3.20 offers a simple laboratory layout. Not only testability but also financial stability and reputation with customers are components that need to be examined in the laboratory. The search for a suitable laboratory can be started over the internet. The laboratory must demonstrate that it can perform all requested tests in accordance with test methods and applicable pharmacopoeias. If the tests are developed together with the customer pharmaceutical company, the

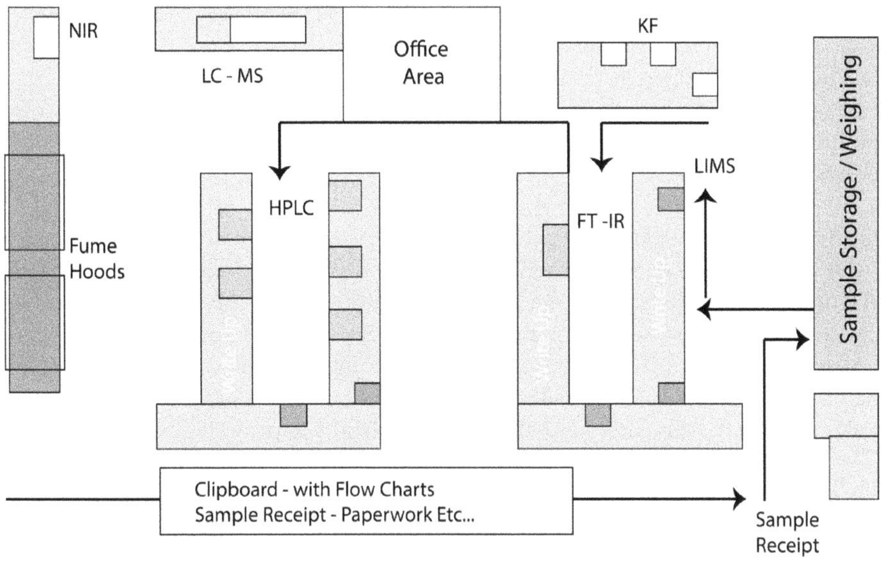

FIGURE 3.20 Laboratory layout example.

test qualification comes to the fore. If the tests are developed within the contract giver, this information flow will be transferred to the contracted laboratory. The contract giver must have robust transfer processes and provide the contracted laboratory with all necessary information flow to operate in GMP compliance. It would be appropriate to make a mutual confidentiality agreement in order to protect the knowledge of both parties. When the contract is considered commercially, issues such as price and payment terms will be included. Before working with the contract laboratory, responsible quality assurance and quality control personnel on both sides should be identified. The situation is not roughly about performing the tests by outsourcing, on the contrary, it should be essential that the interfaces that both parties will use jointly can operate without any problems. After the laboratory selection process, a number of activities will occur before the tests begin to be carried out.

Stage 1 Determining the general capabilities of the laboratory, examining the scope of service, certificates, and accreditations. This information can be found in the Site Master File of the laboratory. In fact, the option of questioning the quality system of the laboratory can be preferred.
Step 2 Drafting and mutual examination of contracts such as service and confidentiality.
Stage 3 Performing the facility audit.
Stage 4 Creation of tests in the laboratory (test, technology, method transfers).
Phase 5 Designation of the laboratory as an approved contractor.
Stage 6 Successful completion of technology or test transfer.

The sequential order of the stages may vary, but when it comes to releasing tests, it can be done first after the laboratory's GMP compliance has been verified by auditors, the relevant tests have been transferred to this laboratory, and the laboratory has completed the qualification. After the general preliminary examination of the laboratory is positive, it should be evaluated whether it can test according to the pharmacopoeias related to the GMP guidelines and directives.

The following questions are also expected to be clearly answered:

- Has the laboratory already been audited by the authorized drug control agency?
- Is there a GMP certificate?
- Does the laboratory have a license to test?

- Are there sufficient experienced and qualified (academic) personnel (e.g., pharmacists, veterinarians, chemists, engineers, biologists) in the laboratory?
- Is there a competent person or responsible person in the laboratory?
- Is the laboratory accredited according to ISO 17025?
- Is the laboratory already approved by other competent health authorities, e.g., is it FDA inspected?

Since only the ability to test can put the pharmaceutical company in a difficult situation, parameters such as who will take the test sample and according to what time schedule, how long it should be tested, and urgent actions to be taken in case of test failure should not be left to chance. One of the documents that can be requested before the audit is the Site Master File. It is already mentioned in EU GMP Part I and Section III. The laboratory must accept the audit, otherwise it must know that no test will be given to it (EU GMP 7.14).

Information about the laboratory that the pharmaceutical company should know:

- Information about the structure of the laboratory
- The full address of the laboratory
- Responsible person list in the laboratory
- Organization chart
- Up-to-date list of offered test methods with references
- List of different products and materials being tested in the laboratory (e.g., finished drug product, API, excipient, food, cosmetics, chemistry, biological, blood products, cytotoxic substances, hormones, etc.)
- Structure and task distribution of the quality assurance system
- Legal permits, accreditations, certificates (for example, GMP certificate, GLP Certificate, accreditation certificate, etc.)
- List of national and international health authority inspections
- List of external resources used by the lab

In the on-site audit, the GMP auditor will make a decision based on the evidence that the laboratory can perform all tests successfully and without errors. For this purpose, it will begin to gather information about the laboratory's quality assurance system. Sending a questionnaire by the pharmaceutical company before the on-site audit may be beneficial for the preliminary evaluation, and there is a chance for the answers given to the question list to be verified on-site. The following checklist is provided as an example for the evaluation of a contracted laboratory candidate, and detailing is the sole responsibility of pharmaceutical and biotechnology companies. The “Laboratory Controls Aide-Memoire” document published by PIC/S is a document that I can recommend you review.

What would you like to see when looking at the laboratory’s quality management?

- Creation, approval, revision, distribution and archiving of procedural instructions
- Validation master plan
- Self-inspection
- Deviation
- OOS
- CAPA
- Hygiene plan and cleaning validation procedures
- Supplier qualification (devices, materials, service providers)
- Management of raw and laboratory data
- Overview of IT infrastructure and procedures for computerized systems validation
- Stability

- Risk management
- Regular management reviews of quality
- Procedure to ensure data integrity
- Archiving of (raw) data

What would you like to see in the documentation?

- Specifications, test instructions, lab logs, and test reports (possibly not for initial qualification audits)
- SOPs and other necessary documents (e.g., device documentation)
- Procedure for generating and checking test results (double-check, IT support)

What would you like to see in personnel?

- Number of skilled staff (possibly, by education, training, and qualification)
- Induction programs for new employees
- Regular training courses on GMP and audit-related aspects
- Hygiene programs for staff
- Regular company medical checkups

What would you like to see in buildings, rooms, and equipment?

- Suitable and qualified equipment for the specified tests
- Qualified monitoring equipment for site and devices, e.g., temperature and humidity monitoring for storage of stability samples, temperature monitoring of refrigerators, freezers and incubators
- Qualification documents and, if necessary, requalification of rooms
- Documentation related to equipment qualifications and requalifications, calibrations, and maintenance (specification documents such as SOPs and work-related documentation)
- Hygiene programs for facilities and, if necessary, equipment
- Hygiene inspection when necessary
- Technical specifications and documents for cleaning and marking of equipment (cleanliness and condition of use)
- Qualification and monitoring of the laboratory water treatment plant

What would you like to see about the materials?

- Regulations regarding access control, shelf life, and use of reagents and culture media
- Monitoring of laboratory water systems, quality testing, and storage of used water if necessary
- Documentation on the manufacture and preparation of reagents or solutions made in the laboratory, with the possibility of tracing the substances used
- Regulations on qualification, tracking, storage, processing, shelf-life monitoring, and storage of standard substances
- Rules for the procurement, identification and testing, processing, and storage of reference organisms
- Specifications for labeling all materials and sample preparations used in the laboratory
- Regulations for handling toxic, hazardous, or sensitive materials
- Waste disposal requirements

What would you like to see about the samples?

- Collection, distribution, and storage of test samples
- Processing of unnecessary test samples
- Storage and monitoring of retained samples, if necessary
- Sampling if needed (only if really needed)

What would you like to see as analytical methods?

- List of analytical methods that can be performed
- Procedural instructions for creating test methods or receiving test methods from the customer
- Validation/revalidation methods of test methods, validation reports, and raw data for them
- Procedures for method transfer, transfer reports, and raw data
- Procedures for verification of pharmacopoeia procedures, documentation, and related raw data
- Requirements for collecting, documenting, testing raw data, and validating test results
- If necessary, specifications for performing stability tests (storage, test intervals)

A good and experienced auditor does not limit his audit to completing a checklist. Instead, he/she will seek to understand the processes in the laboratory and understand how individual QA procedures are linked and interact in order to be able to assess the safety of the overall system. This also includes consideration of interfaces between customer and contractor self-contained and incompatible QA systems.

3.11.3 Stability Audit

The stability of drug substances and their products is required to be ensured throughout their retest period/shelf life. Many national and international guidelines and directives have been published so that pharmaceutical and biotechnology companies can implement a successful stability program, and it is possible to find various information on concepts, procedures, and protocols for stability studies in these source documents.

There is a set of rules that must be followed in order to examine the potential effects of unstable, continuous, or occasionally unstable drugs and pharmaceutical raw materials on patient health in order to allow improvements to be made. For this reason, it should be kept in mind that stability should not only be considered as reaching a conclusion by performing a number of tests at certain times of the year on a certain number of products placed in the stability cabinets, but stability should be taken into account even during the drug discovery and development stages. For this reason, stability tests maintain their importance throughout the entire life cycle of the product. It is the pharmaceutical marketing authorization holder's responsibility to know how the API in a drug product behaves in a timely manner under specified conditions.

The purpose of stability testing is to:

- Provide evidence on how the quality of a product varies with time under the influence of environmental factors such as temperature, humidity, and light.
- Establish recommended storage conditions, retest periods, and shelf lives.

Extension of the expiration dating period should be accompanied by the full shelf-life stability data for the proposed expiry dating on a minimum of three production lots and a justification for the change in expiration date. The data should be obtained using the stability protocol approved in the application.

Stability studies are required by regulations, and, like all GMP programs, periodic evaluations are necessary to provide management with the assurance that they are functioning properly.

- Technically there are no direct regulations requiring these audits, but good management of quality systems would suggest them.
- Stability studies support shipping and warehousing procedures, assay validations, and expiration dating so they are key elements in a quality system.
- Your customers would like to know if you are monitoring the stability program.

3.11.3.1 Phase I

GMP auditors must decide whether the stability audit is an internal or external audit.
GMP auditors must keep in mind that some companies conduct their own stability testing in their QC laboratory. Many start-up companies contract with external laboratories for the stability work to avoid the expense of personnel and equipment. An internal stability testing group may be treated as though it was an external group, especially if it is located in a separate facility.

3.11.3.2 Phase II

GMP auditor may suggest dates or periods for the audit, at least one month away but two months is better. Things go more smoothly if the auditor and auditee have time to prepare and set up. Sometimes that laboratory of interest is in a completely different location from headquarters. The GMP auditor must communicate, initially, with the head of the laboratory or the head of the stability testing unit. The GMP auditor must state the standards that will be used for the audit. These will usually be the ICH guidance documents and general GMP regulations.

3.11.3.3 Auditing Considerations for Stability Studies

- What is the nature of the material that is in the stability program?
- The chemical/biochemical properties of the API are important, but so are the excipients.
- The geographic area in which the product will be sold is an important consideration.
- With regard to stability, interactions between the API and excipients may be important.
- Even with small molecules, factors such as geometric and optical isomerism can be important.
- With biopharmaceuticals factors related to half-lives, antigenicity, three-dimensional configuration, and glycosylation are important even if the potency remains the same.

In light of what is known about the components of the final product, the GMP auditor must review the stability study protocols. If the protocols do not make sense, a full audit may not be needed. A partial audit may still be useful, as any corrective action for the program should consider all of the corrections that may be needed. The review should be done by the whole audit team. The audit team needs to be a combination of subject matter experts and auditors. If the company is fortunate, there will be personnel who are both.

3.11.3.4 How to Build an Audit Team

The auditing team should include people who are technically qualified to perform the audit. The GMP lead auditor may be someone who only knows general auditing principles and procedures, but there should be at least one technically qualified person for each activity that will be audited. There are people with experience or training or both, who are qualified to audit one or more subjects. They are known as SMEs, or subject matter experts. Without technically qualified people, it is easy to be fooled into thinking that all is well, especially with complex products. GMP auditors must remember to choose SMEs who understand pharmaceutical needs.

Companies trying to save money often use only one auditor, but this can be counterproductive.

The GMP lead auditor needs to train new auditors for external and internal auditing. If specialized areas are important, GMP auditors need a specialist to work on the audit. It must be kept in mind that sterility testing needs a microbiologist. It also must be kept in mind that stability studies that require specialized instruments may need an instrument specialist.

If necessary, SMEs can be hired as third parties.

3.11.3.5 Walk-Through

GMP auditors always walk through the stability study area. Auditors must ask to follow the path of samples, starting with their receipt and ending with a result and its reporting. If possible they must observe the handling and placement of products. GMP auditors must always talk to the techs, if possible, and review the stability chambers.

- Are they well placed for ease of sample handling?
- Did they undergo a qualification study?

GMP auditors must ask for a copy of the qualification report and review it.

- Is there a calibration and maintenance schedule and SOP?

GMP auditors must look around, listen, and smell.

- Get a general feel for the area. Is it clean? Are the techs clean? Are there odd sounds? Does the area smell clean?

GMP auditors must look at the shelves and their construction.

- Are the reagents properly labeled?
- Are expiration dates shown and are the reagents in date?

GMP auditors must verify that samples are properly handled and protected.

- If you have controlled substances, security is a factor.

GMP auditors must consider laboratory safety.

- Are acids, bases, oxidants, solvents, and toxic substances properly stored and shelved (i.e., not above eye level)?
- Are there adequate safety cabinets? Are they certified?

GMP auditors must be aware of the technical staff's way of handling data. GMP auditors must try to do the audit when the group is working.

- How are "pull" dates determined?
- Is there a reminder (tickler) system?
- Are samples taken on their due dates? Are they promptly analyzed?
- How much leeway is allowed on the timing of these activities?
- Are they recording data directly in their notebooks?
- If they collect printouts and monitoring data, how are they handled?
- Remember: Data integrity starts with its collection.

Samples

GMP auditors must determine who is responsible for receiving the sample.

- Is a sample number assigned and clearly attached?
- Is there a clearly defined chain of custody?

- Sample mix-ups plague the industry.
- Are the samples being stored properly?
- Since samples are rarely tested immediately, are they stored under defined conditions with limited access?
- Can a sample be found without sorting through all of them?
- Are biological samples processed promptly?

3.11.3.6 Regulations

Section 211.165 (e) of the cGMP regulations states that the accuracy, sensitivity, specificity, and reproducibility of test methods shall be established and documented.

- Further, Section 211.166(a)(3) requires that stability test methods be reliable, meaningful, and specific, which means that the content of active ingredients, degradation products, and other components of interest in a drug product can be accurately measured without interference.
- This is often called a "stability-indicating assay."

3.11.3.7 Stability-Indicating Assay

- A stability-indicating assay is a validated quantitative analytical procedure that can detect the changes with time in the pertinent properties of the drug substance and drug product.
- A stability-indicating assay accurately measures the active ingredients, without interference from degradation products, process impurities, excipients, or other potential impurities.
- If an applicant submits a non-stability-indicating analytical procedure for release testing, then an analytical procedure capable of qualitatively and quantitatively monitoring the impurities, including degradation products, should complement it. Assay analytical procedures for stability studies should be stability-indicating unless scientifically justified.

GMP auditors must be aware of the stability-indicating assay that only measures the "intactness" of a molecule or the product potency.

- These tests are often inadequate or lack the resolving power to measure critical quality attributes (CQAs) such as glycosylation state, optical isomerism, and conformation changes.
- Naïve chemists or managers trying to save money often ignore measurements related to the biological properties of the drug products.
- With small molecules, changes in crystal form that affect dissolution rates can be a problem.

3.11.3.8 483s on Stability Issues

- Reworked lots were not placed on stability.
- Stability samples were not tested for sterility or container integrity (biological product).
- Stability testing did not include analysis of reconstituted products (several 483s).
- Stability samples were stored in glass bottles. The product is shipped to the US in 5-liter aluminum jars with aluminum caps.

483s on the stability of reference standards:

"No data to justify expiration dates and storage conditions for reference standards."

When non-compendial reference materials are used (as with most biotechs), storage and stability data must be generated to ensure the integrity of the standards over time and under conditions of use. Start collecting these data at Phase I!

483s on impurities/degradants:

"Analysts can use 'suppress unknowns' to omit to the report of unknown peaks in HPLC run."

"Unidentified HPLC peaks found during stability testing of the validation lots were not identified or evaluated."

"Systematic identification of degradation products has not been performed."

483s on stability specifications:

- Moisture content has been identified as impacting the stability of the product, yet no specification for moisture has been established.
- Lyophilized product lot xxx failed to meet the finished product moisture spec with a result of 2.54%. The specification is less than 1.5%.
- Company's response was to modify the expiration date by reducing it from 24 to 18 months and the batch was released.

483s on laboratory management:

"Consistent failure to meet stability timelines (several months overdue)."

"Sample backlog and testing discrepancies indicate stability testing program is not in control."

"Lack of personnel training in laboratory procedures and stability testing is evident from the discarding of raw analytical data, samples tested several times and supervisory personnel allowing such events to occur."

483s on stability data management:

"Tracking/documentation system for the stability program is inadequate to monitor the volume of samples."

"The batches placed on stability testing each year are representative of the batches manufactured throughout the year; failure of a single stability lot should trigger a review or testing of other batches manufactured during the year."

3.11.3.9 OOS and OOT Audit

FDA 21 CFR 211.160 and 211.165 components need to be proven by certain tests that they remain within specification, including stability properties in containers, closures, and finished drug products. Not only should the finished drug product be considered, but also APIs, excipients, packaging materials, and other components, and in-process control tests should be performed. GMP regulations require the execution of OOS procedures when a result out of specification is encountered and the root cause or causes can be determined as a result of a successful fault investigation.

For some reason, the concept of OOS has been a concept that is often referred to as a laboratory. The main purpose of laboratory failure investigations is to detect the root cause of OOS. The root cause can be a number of reasons such as sampling, sample processing, sample analysis, and manufacturing errors. All these negativities are actually an indication of a deviation. Let's say that a lot is rejected due to OOS. This should not mean that the process has been completed. On the contrary, an investigation should be carried out in order to determine whether it affects other processes, other lines, or other parties, regardless of the cause of the deviation. Rejection of a party does not mean exemption from research. The GMP expectation and related guidelines, of which the FDA guideline is the most well-known and widely used guideline published in this field, were revised in 2022 and require a written record of the investigation. For the investigation to be successful according to the OOS guidelines, it is expected to be complete, well-timed, unbiased, well-documented and scientifically based.

3.11.3.10 Common Problems Related to Laboratory OOS Investigations

The following are some common problems related to the execution of laboratory OOS investigations:

- Ongoing bias that assumes that most OOS investigations are related to laboratory errors and that the final outcome of the investigation will most likely result in the release of the product to market.
- Phase II investigations do not prompt a serious and complete review of the production batch record, manufacturing process, and process validation as appropriate.
- A pattern of not determining the root cause/probable cause of the original OOS result with subsequent retesting and release of the product to market.
- Poor analysts' interviews; answers not supported by evidence.
- Investigation is based on speculation, not scientific evidence.
- Investigation is not a stand-alone document; can't be read and understood; requires the analysts or investigator to describe verbally what happened.
- Belief that test methods have been properly validated or qualified and perform properly for their intended use; where in fact, older methods, which have not been subject to lifecycle management, just don't work consistently per their intended use.
- Expendables such as HPLC columns are assumed to be equivalent to the ones on which the method was validated; when in fact, expendables like HPLC columns for older analytical methods are no longer available and have been substituted with nonequivalent products by the vendor, despite what they may say.
- Laboratory investigations are not always performed in a timely manner.
- Confirmed OOS results do not trigger the requirement to determine if other batches are impacted (e.g., expand the investigation).
- Extraneous chromatographic peaks are observed but not investigated.
- Ambiguous or incorrectly worded analytical test methods; leave execution open to interpretation or leads to faulting execution.
- Ambiguous or incorrectly worded OOS and/or OOT standard operating procedures.
- The OOS SOP is designed around using the OOS database software as opposed to conducting the investigation itself.
- There is no capture, tracking, patterning, and trending or investigations with periodic quality reviews of OOS investigations.

3.11.3.11 OOX

Although the OOS/OOT SOP was initially used for OOS results only, it was subsequently expanded as part of a growing awareness of quality to include results that deviate from expectations (OOE), trends (OOT), calibration values (OOC), and process parameters (OOP). All of these considerations are used to identify specification noncompliance as early as possible and prevent an OOS case.

- Some are OOX because the regulation is aimed at discrepancies as well as failures to meet specifications.

Why audit the OOX program? Among the most frequent statements in 483s related to 21 CFR 192 are:

- An OOX occurred, but it was not investigate.
- It was investigated, but it was not thorough, or its explanation does not make any sense.
- The OOX procedure did not follow the requirements.
- The OOX investigation did not consider other lots that may have been affected.
- The investigation was not properly reported.
- The OOX did not result in a CAPA.
- To avoid these types of 483s, audits are needed.
- No one is perfect. Errors happen.

An SOP is needed for the OOS program.

- The SOP should be based on the guidance document.
- However, the guidance document is not detailed enough, and cannot possibly cover all aspects of a company's operations.
- The SOP is needed so the company can design a procedure that covers the guidance document in a manner that is consistent with the company's operations and products.
- Any specification or parameter that was reported to the regulatory agency as a part of a license application (marketing permit) is a potential source of an OOX.
- This includes processing parameters.
- Actually, this is not a single SOP.
- This is the master SOP, and should refer to other SOPs such as the retesting and resampling SOP, the investigation report SOP, the OOX documentation SOP, and the OOX auditing SOP.
- The auditing SOP is important in that, along with the other SOP it will set uniform procedures for employees to follow, especially when in doubt or confused.
- The auditing SOP should require auditors to be familiar with all OOX-related SOPs, the master batch record, and the batch record for the lot that produced the OOX.
- Because of the need for various types of knowledge, there will probably be more than one subject matter expert needed.
- With all of the SOPs, you must be able to have confidence that the preparation and review of the documents were done well enough that the SOPs cover all important points and that training in the use of these SOPs was adequate.
- Remember, only fools expect perfection.
- Either the master SOP or the auditing SOP should list all possible sources of OOX for the process that is used, including specifications for final products.
- Even if all the final product specifications are met, an upstream OOX may suggest a hidden problem.

FDA/OOS

This is a newly published guidance for OOS test results by the FDA for investigation of OOS/OOT test results in pharma and biotech industries.

"Investigating Out-of-Specification (OOS) Test Results for Pharmaceutical Production Guidance for Industry, May 2022, Pharmaceutical Quality/Manufacturing Standards Current Good Manufacturing Practice (CGMP), Revision 1."

This type of audit is an internal audit and generates many potential conflicts of interest. Many people become defensive and paranoid when their work is under review. Even while appearing to cooperate, they will impede an audit, sometimes unconsciously. People should not be asked to evaluate their own work. The problem is that the OOX may have arisen in an area where the company only has one real expert. Rivalries among departments and workers can come to the fore and create serious impediments to the audit. The team leader must tread carefully and know the company.

3.11.3.12 OOX Auditing SOP

Since OOX start in the laboratory, one team member should be a laboratory worker. Additionally, the GMP auditor must choose someone with actual laboratory experience. A supervisor with no real laboratory experience will not be as useful as an experienced worker with knowledge of how tests are actually performed in a real environment. One member should be an experienced production worker who has worked in several areas. This would usually be a senior worker or supervisor who has a thorough knowledge of how the final product is produced. One member should understand the structure of the documentation system. This would be an experienced worker who deals with batch records, reports, and various documents.

GMP auditors must be prepared to bring in outside experts. In a situation where there is no independent expert in the company, an external source should be considered. Do not place people in situations where they will have to review their own or their department's work. This could be a consultant or someone from a company with similar activities. Sometimes an independent reviewer with a fresh point of view can be very helpful. Keep the audit team small. You will need at least four members unless some members can perform dual functions. Teams larger than six can become unwieldy and hard to manage. The team members should be able to evaluate the risks to the patient if the OOX is real. Members should be experts on the processes in their own rights and not require training beyond what will be needed as refreshers for the audit. Team members who are inexperienced will impede the audit.

3.11.3.13 Starting the Audit

For an OOX audit, GMP auditors must obtain copies of the master batch record, the OOX audit SOP, the OOX SOP, and the regulatory guidelines on handling OOX. Depending on where the product is distributed, GMP auditors must be sure they have copies of the appropriate OOX guidance. GMP auditors must meet with the audit team, assign functions, and discuss procedures. GMP auditors must instruct team members to review the aforementioned documents so the information will be fresh.

GMP auditors must know the people who perform the annual product review for the product of interest. In addition to being a GMP auditor, you will be auditing a system, not an organization. As a result, certain auditing items such as the opening meeting and a walk-through are not necessary.

GMP auditors must schedule the audit to match the schedule of the annual product review (APR) for a product. The APR is supposed to evaluate the quality parameters for the product, so all lots with OOX should be reviewed. GMP auditors must use the list of lots with OOX and pull the batch records corresponding to those lots, and also pull a selection of lots that have no OOX associated with them. If the number of lots is large, GMP auditors must use statistical criteria to draw the appropriate number of batch records of each type.

GMP auditors must obtain a list of all lots that have OOX associated with them. The documentation specialist must take the lead here. GMP auditors must obtain copies of the batch records associated with these lots and also copies of a selected number of batch records that do not have OOX noted.

- As a team, review the OOX batch records to verify that all OOX were reported and have OOX investigation reports associated with them.
- Review the records of the non-OOX lots to verify that there were no unreported OOX.
- Obtain copies of all OOX investigation reports including copies of laboratory notebook pages that record simple OOX.

GMP auditors must verify that the OOX were noted in appropriate places and that information such as the existence of retests and their results were recorded in documents such as Certificates of Analysis. All team members should have reviewed the master batch record and have a correctly executed batch record for reference. GMP auditors must classify the OOX investigation reports. They must note missing investigation reports and record them for the final report. GMP auditor must classify the OOX by type, "location," personnel involved, frequency, etc.

- Each OOX investigation report should be tied to the batch record that contains the OOX.

GMP auditors must review the OOX investigation reports. They must classify them by whether or not a defined cause was found or if the investigation could not find a definite root cause.

- Can the data cited in the report, be verified?
- Does the report make sense? Is the conclusion supported by the data?
- Is a subject matter expert needed?

The report review does not need to be done in a meeting. Team members can review them separately.

- Members should initially work with each other to review material in their areas of expertise. For example, the laboratory worker can review the notebook pages and the laboratory investigations.
- It is important to see if a CAPA was performed and closed before the investigation report was closed.
- If there was no CAPA the reason should have been given.
- If the members come to the limits of their knowledge, outside experts should be used.
- The outside expert does not necessarily need to be physically present.
- The outside expert should be asked to answer specific questions and provide a report of their conclusions that will be attached to the OOX investigation report.
- The investigation reports should be assigned to individual team members, who will report on the results of their review during meetings where all of the reports that were reviewed will be presented to the committee.
- The team member will present the results of the member's review along with any special reasoning to the team.
- The object will be to state whether the investigation report is acceptable and that the investigation was complete.
- The CAPA might be discussed, especially if it has not been closed.
- The presenter should present a full discussion of any questionable matters and the conclusions.
- The team will vote on whether to accept the conclusions of the review.
- Normally this should be a routine matter, and many reports can be processed in a single meeting.
- The laboratory specialist should report on the laboratory notebook entries as a group.
- The main point is to discuss the reports that may be questionable.
- The conclusions of these meetings should be summarized along with recommendations for action on specific reports.

3.11.3.14 Closing the Audit

- The previously generated data and information should be summarized in an audit report prepared by the team leader.
- The report should be reviewed and signed by all team members and the head of the quality control unit.
- The quality control unit is responsible for all OOX investigations.
- The original copy of the audit report should be maintained in a specified location within the documentation system.
- A note stating that the OOX were audited should be entered into the archived copies of the batch records.
- Copies of the audit report should be sent to affected upper managers.
- The audit report should be discussed with the people responsible for the APR.
- These people need to know if the OOX in the batch records has been handled in a manner that satisfies regulatory requirements.
- Any outstanding issues or major problems should be discussed with the appropriate upper management.
- OOX investigation reports with unsupported conclusions, incorrect CAPAs, CAPAs that were not closed, and other potential problems should be discussed.
- A record of notifications should be kept by the quality control unit.
- Who was contacted, by what means, when, and by whom should be noted.
- Remember that OOX should be infrequent events.

- OOX that are attributed to statistical variation should be even less frequent.
- If specifications are set with a 99% confidence interval, only 0.3% (three parts per thousand) of OOX should occur by chance alone.
- If an OOX appears to be too frequent, it should be noted and plans for corrective action should be developed.

3.11.4 Microbiology Laboratory Audit

In the pharmaceutical and medical device industry, the microbiology laboratory is often a spin-off of the analytical chemistry laboratory. Unfortunately, the microbiology laboratory facility is often pushed into the little space that is left after the QC lab has been assigned two or three larger rooms. The challenges and risks of microbiological analyses and their impact on product quality and patient safety is underestimated or overlooked. It is not intentional but it is due to the lack of awareness and understanding of the microbial sciences as microorganisms are not visible to the naked eye. As a result, there is often inadequate space to allow the separation of "dirty activities," such as microbial culture handling and growth promotion testing, from "clean activities," such as bioburden testing or water sample testing. Process flows are often inadequate, posing a risk of cross-contamination; and refrigerators, incubators, and other storage areas double up for storage of sterile media, test samples prior to testing, and bacterial slants and tubes awaiting use in the growth promotion test. Due to this production, complaints/claims samples were contaminated in the laboratory or results were false positive. It is challenging to perform microbial data deviation investigations and identify a root cause. As per USP<1117>, analytical microbiological assay results can be difficult to interpret for several important reasons:

- Microorganisms are ubiquitous in nature, and common environmental contaminants – particularly organisms associated with humans – predominate in many types of microbiological analysis.
- The analyst has the potential to introduce contaminating organisms during sample handling or processing in the laboratory.
- Microorganisms may not be homogeneously distributed within a sample or an environment.
- Microbiological assays are subject to considerable variability of outcome. The laboratory studies should be conducted with utmost care to avoid exogenous contamination, in which the laboratory layout and workflow play an important role.

The following points should be considered in the design of the microbiology laboratory:

- How is the design and fitting of the sterility testing area?
- Where are sterility tests carried out? In an isolator?
- How is protection assessed against microbiological contamination during aseptic operations?
- Is there appropriate instructions for access to the critical areas?
- Where is growth promotion testing (GPT) performed?
- Is the area for positive tests/GPT testing separate from areas where the product is tested?
- Is the ventilation system separated from other areas? What are the pressure differentials? Are there visual alarms?
- Where are prepared materials for aseptic operations? Where are prepared culture media for testing? Is there any segregated area/room for manipulation with cultural media?
- How and where are decontaminated materials for microbiological testing? Is there some segregation of the washing area to the clean and non-clean parts?

3.11.4.1 What to Look For

- The GMP auditor must review laboratory space, design, layout, access, workflow, and organization for suitability of its purpose.
- The GMP auditor must review the cleaning and sanitization schedules, environmental monitoring programs (viable and nonviable where applicable), respective procedures, and sample records of such data.
- Verify the effectiveness of the cleaning and sanitization schedules and their linkage with the environmental monitoring program.
- The GMP auditor must review the HVAC and room qualification, validation, calibration, maintenance procedures and sample records where applicable.
- The GMP auditor must review deviations or investigations related to laboratory facility, cleaning and sanitization, or environmental issues.

Personnel

- Review:
  - Organizational structure
  - Sample job descriptions
  - Qualification for a representative number of employees (CVs, certifications)
  - Training procedures (GMP and job specific)
  - Training records (planned and for-cause)
  - Investigation follow-ups where employee training was found to be inadequate
  - Adequate personnel versus workload
  - Adequate supervision
- The GMP auditor must observe employee practices in the laboratory, specifically where aseptic techniques are employed.
- The GMP auditor must evaluate whether adequate employees and supervision are provided throughout the laboratory operations.
- The GMP auditor must evaluate how management was notified of critical laboratory issues and the level of management involvement.
- The GMP auditor must review whether laboratory employees are included in the environmental monitoring program (e.g. using surface contact plates).

Autoclave

- GMP auditor must review procedures and documentation for operation, cleaning, qualification, validation, maintenance, controls, and utilities (e.g., clean steam). associated with the autoclaves. The GMP auditor must review logs and investigations of failed sterilization cycles and the disposition of the contents of the failed cycles.
- The GMP auditor must check the temperature of the autoclave for a few historical cycles.

Isolators/Glove/Boxes/Laminar Flow Hoods

- The GMP auditor must review procedures and documentation supporting:
  - Operation
  - Cleaning and sanitization
  - Qualification/validation
- Environmental monitoring
- Calibration
- Preventive maintenance including filter changes, glove integrity and integrity of the closed areas
- The GMP auditor must review investigations related to the isolator environment, change controls, and for-cause repairs.

Incubators

- The GMP auditor must review the operation procedures, qualification, validation, calibration, cleaning, sanitization, maintenance, change controls, alarms, and periodic monitoring schedules.
- The GMP auditor must review temperature and humidity logs for a selected time frame, check for the historical log of excursions, and review investigations related to incubator performance.
- The GMP auditor must review how the temperature and humidity data is collected, e.g., location of probes, data collection, archiving, and data reviews.
- The GMP auditor must check if any alarm system is installed and whether there is proper management oversight of alarm conditions along with adequate corrective actions.
- The GMP auditor must review procedures and records regarding such activities.
- If equipment is linked to a building management system (BMS), the GMP auditor must review the BMS records for incubator records.
- The GMP auditor must review the overall condition of the incubator, its location, design, access, the way materials are staged or stored within, identification of materials, general cleanliness, state of repair, and any cleaning and sanitization schedules and corresponding records.

Refrigerators

- The GMP auditor must review procedures and documentation supporting operation, cleaning and sanitization, qualification, validation, environmental monitoring, calibration, and maintenance.
- The GMP auditor must review how the temperature data is collected, e.g., location of probes, data collection, archiving, and data reviews.
- The GMP auditor must review investigations related to refrigerator usage, change controls, and for-cause repairs.
- The GMP auditor must review the overall condition of the refrigerator, its location, design, access, the way materials are staged or stored within, identification of materials, general cleanliness, state of repair, and any cleaning and sanitization schedules for the refrigerators.

Test Instruments/Identification Systems

- The GMP auditor must check calibration records, identification, and standards used for test instruments.
- The GMP auditor must review the operation procedures, qualification, validation, calibration, cleaning, sanitization, maintenance, change control, and periodic challenge testing of such system.

Laboratory Utilities Including Water Systems

- The GMP auditor must review operation procedures, diagrams, qualification and validation documents, calibration records, cleaning and sanitization, maintenance, change controls, and periodic monitoring of these systems.
- The GMP auditor must review investigations and excursions related to such utilities and systems.

Microbiological Testing of Nonsterile Products

- The GMP auditor must review microbial specifications for nonsterile products consistent with the USP requirements or industry standards for the type of dosage forms. (quantitative – total counts, qualitative – objectionable organisms, and endotoxins).

Retest

- The GMP auditor must review instances where retests were performed.

Media

- The GMP auditor must ensure that the media used is appropriate for the test required. GMP auditors must review growth promotion characteristics for all of the media used.
- When possible, review tests performed or witness as they are performed and review the plates and tubes of the incubated media. (you will get better results if you do not contaminate the media and tubes during the test).

Stock Cultures

- The GMP auditor must review how stock cultures of reference organisms are procured and stored. GMP auditors must review the procedures for stock culture usage.

Reagents

- The GMP auditor must review the laboratory reagents' storage, handling, and identification. Identify how obsolete reagents and dilutions are disposed of.

Sampling

- The GMP auditor must review sampling procedures, including collection techniques, sampling containers, storage areas, storage conditions, time limits, transportation methods, receiving process (in the laboratory), staging, and processing.
- The GMP auditor must ensure that inhibitory substances in samples are neutralized prior to testing and sample integrity is maintained.

Methodology and Validation of Test Procedures

- The GMP auditor must determine the source of test procedures (e.g., USP and other microbiological references).
- One of the most important parts the GMP auditors should focus on is the testing methodology used in the laboratory. (Test methods should be able to detect whether there are especially *Pseudomonas cepacia* or other microorganisms that may pose a danger to the user. It can also be compared whether the test method used is the same as the one used during the application.)
- One of your responsibilities as a GMP auditor, is any equipment used in the tests performed in the laboratory. Make sure that the device, tools, apparatus, and machines are available and that the calibration, qualification, maintenance, and cleaning have been completed when the test is performed. If there are tests that prove with validation that the same or better results are obtained according to the USP, changes can be made.

Data Storage

- GMP auditors must review procedures and policies or data recording, documentation practices, storage, and archiving.
- As GMP auditors, carefully review the logbooks where all kinds of activities are recorded and pay attention to whether data integrity principles are followed. If summaries of data are not available, try to review sufficient data.
- Where an automated microbial system identifying microorganisms is used, review logs of such testing.

Contract Testing Laboratory

- As GMP auditors, see if the firm has a program to assess the quality of outsourced laboratory activities.
- The GMP auditor must determine the specific instructions given to the contractor. The GMP auditor must review if there are any quality agreements between the contract laboratory and the firm, and if not, how the quality system and compliance requirements of the firm are monitored and enforced.
- Analytical results should be reviewed, particularly for articles in which additional or retesting is conducted. As GMP auditors, we expect to prove that the results of the tests performed are available to the manufacturer.

Sterility Testing

- The GMP auditor must evaluate the facility for sterility testing.
- The GMP auditor must perform actual observation of the sterility test, if possible.(As a GMP auditor, make any observations you make in the environment in a way that does not harm the current operation).
- GMP auditors must examine the use of negative controls during sterility testing.
- GMP auditors must request lists of test failures to facilitate a review of production and control records and investigation reports.
- If there is a robotic system used in sterility testing as a GMP auditor, inquire about this system and how it is isolated from the environment.
- The GMP auditor must evaluate the time period used for sterility test sample incubation. (Most laboratories use a period of 14 days incubation.)
- Bacteriostasis and fungistasis testing must be performed. These test results must be negative. Otherwise, any sterility test results obtained on the product may not be valid.
- Carefully examine media fill, sterility test and results, and other results that prove that there are no slow-growing microorganisms in the environment.
- Examine whether the methods used for incubation are the same as those listed in approved or pending applications.

3.11.4.2 Microbiology Laboratory Staff

One of the indispensable elements of GMP audit is personnel, which is the second article of GMP. As an auditor who likes to talk to the personnel in GMP audits and ask them various questions, I can say that the first areas to be examined are to determine the size of the laboratory, the product type, the workload that the laboratory tests, and the number of personnel working. In addition to the competence of the personnel, it is necessary to question their roles and responsibilities, job descriptions, delegation, health checks and GMP training, and technical knowledge and skills. From the GMP point of view, the rule that the personnel should be fully trained in their work is a very important topic to follow in the microbiology laboratory. What could be worse than training sessions given in a one- to two-hour presentation by a person ignorant of presentation techniques in a stifling environment of the entire GMP guideline? The training should be parallel to the work being done, for example, aseptic techniques, gowning and behavior rules, cleanroom layout, hand washing and personal hygiene, and many technical issues. Therefore, as a GMP auditor, my request from you is to inquire about the training presentations, especially in this laboratory, during the audits. Who offers the training, is the trainer's qualification sufficient to provide this technical information, and does the trainer have a certificate of being a trainer and is this certificate up to date?

3.11.4.3 Laboratory Design and Sample Flow

In current cGMP guidelines, it is implicit that microbiology laboratories should be of an appropriate size to carry out all their functions. It must cover all aspects of sample handling and storage, dedicated test areas for different materials, media preparation, incubation, handling of microbial cultures, and treatment and removal of biohazardous waste. Some aspects of work may take place in demarcated areas, such as the reading of environmental monitoring samples, endotoxin testing, and the identification of microorganisms. Sterility testing must always be conducted within a purpose-built test facility. In addition to the design of the laboratory, auditors will expect to see a logical flow of samples through the testing process and adequate controls to protect the samples and activities from extraneous contamination, for example, qualified unidirectional airflow cabinets and biological safety cabinets are used to ensure contamination control.

3.11.4.4 Sample Handling

The topic of GMP sampling gains special importance when the subject is microbiology. This includes taking samples, the competence of the sampling personnel, their training, compliance with the hygiene rules, taking the sample in suitable containers with appropriate methods, storing it, finding the identifying label of the sample, quantity, batch number, etc. In addition, whether the sample will be tested with which validated or verified test method and by whom, if there is a waiting period for the test, whether this period is validated, and whether there is an SOP on what to do when nonconformity is encountered in the tests.

3.11.4.5 Culture Media

The microbiology laboratory is largely dependent on culture media. So, as a GMP auditor, it is crucial that you be able to ask targeted questions about cultural media. Culture media preparation and quality control for this experiment are separate areas of expertise. The GMP audit should be handled like any other laboratory and there should be written procedures for activities such as culture entry, reagents, chemicals, preparation, labeling, storage, and use. All kinds of containers that store such materials should be on descriptive labels, and information such as content, concentration, preparation date, preparation person, expiry date, and storage conditions should be included on the label. Checking past party records is also important in culture media. The type of water in the media, the person doing the analysis, the balance printouts of the weights taken, the autoclave cycle records, the pH settings, and the growth promotion should also be included. If the culture or reagent is sterilized, validation of the sterilization method should be questioned. If the medium is purchased from outside, the Certificate of Analysis, supplier information, and supplier qualification should also be questioned.

3.11.4.6 Reference Standards

The importance of standards is indisputable in the microbiology laboratory, as they are important in the chemistry laboratory. When reference standards are mentioned, antibiotic standards, endotoxin standards, biological indicators, and microbial cultures come to mind. Here, of course, their receipt, storage, preparation, and use must be checked. Since such standards come with Certificates of Analysis when purchased from external companies, it is essential to question these certificates. When you review a Certificate of Analysis in an endotoxin standard, you should expect the certificate to be able to indicate endotoxin potency and dilution volume.

3.11.4.7 Control of Microbial Cultures

Microorganisms are among the most common standards used in the microbiology laboratory. The control and maintenance of stock cultures are critical. Cultures should be controlled from receipt through storage and the purity and identity of the culture should be safeguarded. During an audit, the preparation, handling, and storage of cultures will come under scrutiny. This relates to concerns about misidentification and the potential of viability loss through incorrect storage. For long-term storage, there should be no more than five transfers (passages) from the original culture. This is necessary to avoid phenotypic variations.

3.11.4.8 Documentation and Electronic Systems

The actual control and information for sample handling may follow a paper-based or electronic system. If it is an electronic system, it should ensure that where appropriate, any requirements for system validation are met. The GMP auditor will focus on password access and the audit trail. Logbooks, equipment files, and laboratory notebooks will be subject to inspection. A related and important area of documentation is training files. These are invariably called upon during an audit, and here each member of the microbiology laboratory must be able to demonstrate that they are trained and assessed in the tasks that they regularly undertake.

3.11.4.9 Laboratory Equipment

The quality of microbiological test data is usually dependent upon the performance of instrumentation (for example, an incubator or an air sampler). There should be a list of all equipment available, describing instrument type, serial number, date of introduction into the laboratory, calibration and validation processes, and a preventive maintenance scheme. Within the laboratory, it's expected that the status of the instrument is verified, for example, by labeling whether it's in maintenance or within calibration. A logbook for every piece of apparatus should be in situ to explain what samples are tested on, or with the equipment, and what maintenance has been performed. This can be helpful in investigations following instrument or sample failure. Auditors will often request calibration certificates, and these must be thoroughly reviewed and approved. An example of a necessary item of apparatus is the laboratory autoclave. Autoclave load patterns should be established, assigned to cycles, and tested. Tests should include heat penetration studies and biological challenge tests. A standard audit finding is for equipment service and calibration reports to state that the equipment was adjusted or was actually out of calibration before the service and calibration activity with no follow-up being recorded. Differing kinds of apparatus would require different levels of validation and calibration, and the criticality of the instrument will determine this in conjunction with a review of the URS. It's essential that items like water baths, incubators, and refrigerators undergo temperature mapping, which, during qualification and routine operation thermometers or temperature probes, are traceable to national standards.

3.11.4.10 Cleaning and Decontamination

It is important that the results from the microbiology laboratory are representative of the samples being tested and aren't the results of cross-contamination. A way to avoid this is often through the cleaning and disinfection of premises and equipment. Additionally, if the laboratory encompasses a media preparation area, this must be maintained. Media preparation requires the employment of blending containers, glassware, balance weighing boats, spatulas, etc. Written, validated procedures should be available for the cleaning and storage of such components.

3.11.4.11 Specific Tests

GMP auditors will have specific questions to ask about certain laboratory tests.

Microbial Limits Testing

In relation to audits, an auditor may wish to review:

- Compliance with using <100 CFU of each microorganism as inoculation level and growth promotion.
- Requirement for comparison to the previous batch of media.

- Addition of "nutritive" and "selective," that is, the need to demonstrate that media not only recovers the "target" organism but also inhibits other organisms.
- Enrichment schemes, for example, for *Escherichia coli* and *Salmonella* the test uses soybean-casein media instead of lactose broth.
- Are the test organisms incubated at the correct temperature?

Preservative Efficacy Tests and Antibiotic Assays

Where preservatives are present in pharmaceutical preparations, the validation of neutralization is important for routine testing. During product development and stability programs, there could be a requirement to test the effectiveness of the preservative system. Antibiotic assays should be conducted according to written procedures. Specific requirements that would be reviewed during an audit include media preparation, temperature control (and hence incubator qualification and monitoring, etc.), control of reference standards and assay procedures (for example, plate design, such as Latin square), concentration, and volumes of material used. A thorough understanding of the calculations involved in determining potency would be required. It is insufficient for the analyst to simply put the data into a computer program and accept the result without knowing the background of the evaluation process.

Sterility Testing

In an audit, the auditor would look closely at the sterility test validation process. The facility in which the test is performed will come under scrutiny since it is expected to be of a similar quality to that in which the product was manufactured. Therefore, people and material flows would be defined for manufacturing sterile products, and the test would be performed under unidirectional conditions (within cleanrooms) or in isolators. Auditors will assess the techniques for transferring samples, media, and test equipment into the test environment. Auditors will also review environmental monitoring results, sterility test failure results, and the results of test controls, each of which partly informs about the status of the test environment.

Microbial Identification

Procedures should be in place for microbial identification methods. It is important that the laboratory can demonstrate that the staff conducting the identifications have a suitable level of knowledge and experience. An auditor will check that appropriate controls are being run with each identification test session.

3.11.5 Packaging Material Laboratory Audit

For suppliers of either printed package components or product contact components, it is imperative that processes are in place to ensure the components are meeting a well-defined acceptable standard. Product contact package components can be utilized to protect and keep the medicine from deterioration. Examples include blister strips for solid dosage forms or sealed glass vials for freeze-dried powders. In addition, certain components control the delivery of the drug such as metered dose aerosols. Printed package components such as cartons, labels, and leaflets, are designed to ensure the healthcare provider and patient are given the appropriate information for the safe and correct use of the medicine. Component supplier operations must be performed in a manner to minimize the risk of cross-contamination and mix-ups. Approved written procedures must be detailed and in place. They should include information for receiving, identifying, storing, handling, sampling, examining, and/or testing materials. These procedures must be followed.

The aim of packaging material quality control is to ensure the quality is right at the time of manufacture, hence eliminating testing on the pharmaceutical company premises. There will still be a necessity to periodically monitor components received, particularly with new suppliers, problem suppliers, or new components. Investigating packaging problems and customer complaints will also need laboratory facilities. (The same laboratory could be used for testing all raw materials received at the factory.) Therefore a suitable laboratory layout and testing procedure will be required.

Considerable thought must be given to the laboratory location and layout to ensure that efficiency and security are maintained. The main points to consider are:

- Receipt and storage of samples Inspection facilities
- Chemical and physical testing
- Return of components to stores (when full boxes have been obtained for sample removal)
- Recording and storage of results
- Reconciliation and security of components

There is only one suitable place for a testing laboratory to be located: incorporated in the warehouse receiving the packaging materials. This will have several advantages:

1. Rapid access by the laboratory staff for the sampling of component batches
2. Improved communication between stores and laboratory personnel concerning procedures and problems
3. Improved security and reconciliation when full boxes of components have to be transported to the laboratory for statistical sampling and then returned to stores
4. Rapid sentencing of batches on completion of testing

This enables a smooth flow of materials through the area from receipt to the packaging and dispatch of the product.

What to Look for in a Packaging Materials Quality Control Laboratory

- Does the layout represent the real-time activities or the layout and physical view do not match each other?
- How many staff members are working in the laboratory?
- What are their duties, roles, and responsibilities?
- What type of packaging materials are being tested in the laboratory (primary, secondary, tertiary)?
- What are the test methods to test these items?
- Are the test methods validated?
- What type of equipment is used in the laboratory?
- Are these types of equipment qualified, calibrated, maintained, and cleaned periodically?
- How are packaging material samples taken? By whom? Does the sampler have GMP, sampling, cleaning, and basic microbiological training?
- What type of software is embedded in equipment?
- Is the software validated? According to which regulations or guidelines?
- What happens after a test is completed?
- Where is data stored?
- Does laboratory staff have required GMP training?
- How is it guaranteed that the packaging material being tested matches the current artwork?
- How is artwork managed?
- How is a unique pharma code assigned to each folding box, patient information leaflet, and label?
- What happens if a glass bottle test result is out of specification?
- Does the laboratory have a valid acceptable quality level (AQL)?
- How are packaging materials received? With which kind of document of evidence?

- Does the laboratory prepare a Certificate of Analysis for each test result?
- How does the laboratory manage Certificates of Analysis?

3.12 BIOPHARMACEUTICAL FACILITY AUDITS

3.12.1 Today's Biopharmaceutical Industry

Today, biotechnological products are being used effectively in the treatment of many diseases, and the diversity and usage areas of these drugs are increasing day by day. Biopharmaceuticals (biological/biotechnological products) are products of a polypeptide or protein structure derived from living organisms (plant and animal cells, bacteria, viruses, and fungi) by biological methods (controlled gene expression, recombinant DNA technology, antibody production methods, etc.) instead of chemical compounds. These products are recombinant proteins, monoclonal antibodies, immunological medicinal products, vaccines, blood products, allergens, and gene and cell therapy products. In contrast to chemically synthesized and known drugs, most biologics are complex mixtures and cannot be easily identified or characterized. There are differences between conventional and biological drugs at many important points including structure, mode of action, production method, shape, size, body take-up routes, and even price and stocking conditions. Biotechnological products are products that cannot be copied exactly. Health authorities clearly recognize that the two drugs are different from each other, so they may have different effects, side effects, and safety features. In many countries, regulatory policies have been developed for biosimilars and are generally followed by WHO or EMA guidelines.

Biopharmaceuticals are medicines that are manufactured by biotechnology methods. They are manufactured in living cells. Biopharmaceuticals include human proteins like antibodies, hormones, cytokines, fragments, and other things. They are mostly manufactured in special cells that are grown in fermenters. Lots of these special cells are genetically modified.

Examples include but are not limited to the following:

- Enzymes
- Monoclonal antibodies
- Antibody-drug conjugates
- Fusion proteins (e.g., antibody Fc region-containing fusion proteins)
- Growth factors
- Cytokines (e.g., interleukins, interferons, tumor necrosis factors)
- Botulinum toxins

When it comes to biotechnology, it is the approach of applying biological systems and organisms to technical and industrial processes.The history of biotechnology is familiar. We know that yeast has been used for years to ferment grain to alcohol. Likewise, people engaged in farming want to obtain higher quality products by choosing the desired characteristics in plants and animals and benefit from genetic engineering. Changing the genetic material of an organism at the cellular or molecular level has been thanks to the steps taken in biotechnology in recent years. As biotechnology develops, the need for the products produced with these techniques to undergo strict controls and inspections is increasing. However, as those in the industry will remember very clearly, controlling technology with such complex technical equipment is challenging. A trained workforce also makes it difficult to maintain this technology, and their training is quite laborious and costly. However, it is striking that the production and control of substances and devices produced using biotechnology are in line with the regulations of traditional output. Therefore, the same technology is used to produce high molecular-weight substances such as antibiotics, enzymes, insulin, heparin, and albumin.

Table 3.34 explains the difference between small- and large-molecule drugs.

TABLE 3.34
Small versus Large Molecules

Small Molecules	Large Molecules
• Low molecular weight • Synthesized through chemical reactions • Example: Aspirin	• Can be greater than 150 kDa • Produced by living cells • Example: ABC mAb (monoclonal antibody)

3.12.2 Protein Drug Substance Manufacturing Processes

Protein drug substance (DS) manufacturing processes can be classified as described in the following sections.

3.12.2.1 Cell Banking

The protein drug substance is expressed, extracted, or purified using the milk or egg of transgenic animals and either the leaves or fruits of transgenic plants. Generally, however, the desired protein can be expressed from cell culture/fermentation. It can then be recovered and purified. The most commonly used cell lines are mammalian and bacterial cells. Once the expression cell line has been selected, the protein manufacturer sets up the master cell bank (MCB), which is a single cell expressing the protein or a collection of ampoules/vials containing aliquoted cells derived from the colony. The MCB should be stored under defined and verified conditions. Typically, the manufacturer creates a working cell bank (WCB) from an MCB bottle to be used for each production.

3.12.2.2 Upstream Manufacturing (Cell Growth and Crude Protein Harvest)

The upstream production step, which involves cell growth and protein harvesting, is usually extended by transferring cells from a WCB into larger vessels until the cells reach constant density under favorable conditions. The expanded cell culture is transferred to a fermenter or bioreactor to maintain both controlled cell growth and protein expression for a predetermined period of time.

Cell culture should be subject to strict controls to prevent the accidental introduction, proliferation or persistence of microorganisms during expansion studies. After the cell culture/fermentation process is completed, the purified protein is collected and both cells and cell debris are removed from the cell culture/fermentation medium with the help of steps such as centrifugation, deep filtration, and ultrafiltration/diafiltration (UF/DF). In this way, it is purified.

3.12.2.3 Downstream Manufacturing (Purification and Formulation)

The purification step is very complex. Purification may include UF/DF, affinity, column chromatography, and conventional column chromatography, e.g., anion exchange chromatography. An auditor is expected to know the rationale behind the purification step and to evaluate accordingly.

Purification separates the media components of the protein of interest from process-related impurities, such as host cell proteins and DNA, and remove aggregated, misfolded, or truncated proteins.

There may be steps such as detergent inactivation, low pH retention, viral filtration, and heat treatment to inactivate or remove adventitious agents such as viruses. Usually, after the purification step, the purified protein DS is formulated and sent in bulk to the fill/finish unit/plant.

3.12.3 Senior Management

The company's top management may undertake tasks such as establishing a system and making changes to the system.

One way to prove that the quality management system is working well is to question the extent to which top management is involved in the activities. If a product problem that can be considered important is detected during the audit, it should be considered whether the senior management is the bystander or the party directly involved in the root-cause investigation.

Top management is expected to always keep GMP-compliant plant, process, system, equipment, personnel, and area components up to date and to put product quality first.

3.12.4 Does a DS Protein Facility Have a Quality Unit?

Just as any company that manufactures pharmaceuticals in a GMP-compliant facility has a quality management system, a protein substance manufacturer company must also have a quality management system in place and functioning. The quality unit should be personally involved in the processes in the entire facility and should review and approve the procedures for functions such as production, quality control, storage, and distribution. Another important task of the quality unit is to ensure that records are kept accurate, timely, complete, and accessible at all times.

3.12.5 Internal Audits

Drug substance manufacturing companies should conduct internal audits at least once a year to check that all their activities are compliant with cGMP. The requirements that I have stated in the previous sections, such as the audit SOP, should be available and approved; the auditors should have the necessary qualifications and experience; the audits should be planned as risk-oriented; and the CAPAs should be followed up regularly after the audits.

3.12.6 Product Quality Review

In addition to its ongoing monitoring program, the firm should conduct and formally document, at least annually, a product quality review of each protein DS it manufactures. These reviews should include summary analyses of much of the data evaluated by the ongoing process performance and product quality monitoring program, and should specifically cover, among other things, the following:

- Critical in-process control and release test results
- Batches failing to meet established specifications
- Critical deviations or nonconformances, OOS test results, and related investigations
- Changes carried out to manufacturing processes or analytical methods
- Stability monitoring program results
- Quality-related complaints, returns, and recalls
- The adequacy of implemented corrective actions
- Process performance/capability
- Trending of critical quality attributes

3.12.7 Change Control

Change control, which I have mentioned throughout this book and is one of the most important components of a successful and effectively managed pharmaceutical quality management system, is also of critical importance for the company that manufactures drug substances. Change roughly consists of a series of actions that will be evaluated, approved, and implemented throughout the product's lifecycle. Drug substance products already contain quite a lot of variables, unlike products produced by traditional chemical synthesis. If changes in vital elements such as critical quality attributes and critical process parameters are on the agenda, as a GMP auditor, you should witness that changes can be implemented without errors and failures in this complex structure. While evaluating the effects and risks of changes in any of the elements, such as process, area, system, equipment, method, personnel, or material, you should also look for evidence that possible risks to the finished product are taken into account.

3.12.8 Validation and Verification

The most important of the key components of being able to carry out production and testing activities according to current GMP rules and regulations is validation.

As I have mentioned before, validations are carried out by an expert team. Verify a firm's policy and SOPs regarding validation are in place, and that the framework of the validation master plan is well structured. Change control, deviation, and quality risk management SOPs are in constant interaction with each other. Communication is required.

As you know, a management claiming to have produced its drug substance product in accordance with GMP has to prove to GMP auditors that a series of validations and qualifications have been completed beforehand and that the latest process validation has been carried out. In-process validation, for example, a series of activities must be performed according to the FDA 2011 guidelines and records must be kept in accordance with good documentation practices.

General coverage of the firm's validation programs/activities is included in the quality system section because the quality unit oversees these activities, including:

- Approving qualification/validation protocols and reports
- Ensuring that the firm practices a lifecycle approach to validation activities and ensuring that effective verification and requalification efforts occur
- Leading investigations into the failure of validated processes and test methods
- Leading new validation efforts if significant changes to existing processes and test methods are needed
- The audit team should focus on validation/verification deviations or failures and should ensure that the firm has properly evaluated its impact on product quality. This is particularly true of deviations or failures affecting cell culture, fermentation and purification processes, viral clearance/inactivation, manufacturing equipment, and analytical test methods.

3.12.9 Stability Program

As a GMP auditor, you should look for evidence that the drug substance manufacturer has a DS continuous stability program and a stability program for DS batches, and that all activities are managed according to international and local stability guidelines and directives. In case of any failure in the stability tests, you should wait for proof that necessary actions are taken, deviations are initiated, studies are carried out to determine the root cause or causes, and the health authority is informed when necessary.

3.12.10 Deviation Management

One of the important components of a GMP-compliant DS manufacturer, such as change control and risk management, in the established quality management system, is the management of deviations. As a GMP auditor, you should look for evidence that the DS manufacturer can accurately initiate, identify and manage a deviation before, during and after production. You can evaluate all records of deviations, usually initiated within at least two calendar years, according to the sampling method.

3.12.11 Reject, Abort, and Reprocess Batches

As a GMP auditor, you should want to see the actions the DS manufacturer takes regarding any rejected batch, regardless of the reason. Even if it is a rejected single batch, this may cause you to conclude that the functions of the product, such as production, testing, and quality assurance, are not functioning well or are not functioning properly. The frequency and cause of rejected and canceled batches can be investigated and documented by reconciling the date and number of WCB aliquots used for production or testing.

3.12.12 Complaint Management

Complaint management is one of the most important elements of a successful and GMP-compliant pharmaceutical quality management system. The auditor should find answers to questions concerning whether the DS manufacturer has a complaint SOP and a staff member assigned to this task in the quality unit that manages the protests, whether the complaints are categorized, whether the product batch subject to the criticism has been thoroughly examined, whether any root cause was determined, and how much is the return to the customer as it takes a while.

Based on the number of complaints received by DS manufacturers in one or two calendar years, it would be helpful to review a few complaint files by sampling again.

While I was working at a German pharmaceutical company years ago, my duties included auditing our suppliers. I was questioning how one of our suppliers handled customer complaints during an audit. In a sample complaint file I examined, I could not see a trace or a sentence regarding the root cause. After reminding them of this, I did a follow-up audit. In the complaint review document, they pointed out this time that the root cause was human error. Years passed and we were able to bring our suppliers to a certain level, and I was happy to see that the root cause, at least, had evolved from human error to systemic error.

3.12.13 Quarantine in Brief

Protein DS batches that have not yet been released by the quality unit or are pending a disposition decision for any other reason (such as a return) should be placed in quarantine – segregated from released batches by physical or other effective controls to prevent their use until a disposition decision has been reached.

3.12.14 Recalls in Brief

The firm should have a written procedure documenting the circumstances under which a batch recall should be considered. The procedure should define the parties involved in the recall evaluation and describe how the recall should be initiated, who should be notified, and how the recalled material should be quarantined and disposed of.

3.12.15 Hot Topic: Data Integrity

I tried to gather the issue of data integrity under a separate heading in this book. It is among the most important duties for GMP auditors to examine how pharma and biotechnology companies comply with the ALCOA+ principles. Unfortunately, I will not have a very different statement about data integrity when it comes to the DS manufacturer, because the aforementioned ALCOA+ principles apply to these manufacturers.

3.12.16 Facilities and Equipment

All manufacturers should have suitable equipment and facilities to guarantee that their drugs are high quality.

Manufacturers of protein DS qualify and maintain several pieces of critical equipment not typically seen at other types of establishments, including fermenters/bioreactors, centrifuges, depth filtration equipment, UF/DF equipment, and column chromatography equipment. Although qualification activities associated with using this equipment are typically assessed before application approval, the inspection team needs to assess the firm's verification activities related to equipment maintenance and requalification/continued performance as well as the qualification of new pieces of equipment.

Although not necessarily unique to protein DS manufacturers, several utilities generally found in these manufacturing facilities should be maintained and requalified as appropriate. Utilities include water for injection (WFI) and purified water systems, process gas systems, and clean steam systems. Maintaining WFI and purified water systems is particularly important because water is a critical material for biotechnological manufacturing and large volumes of it must be generated.

The inspection team must confirm that manufacturers carefully consider the routes/mechanisms of microbial contamination and product cross-contamination and effectively control both to provide suitable facilities and equipment for protein DS manufacture. This means, among other things, that heating, ventilation, and air conditioning (HVAC) systems should be appropriately designed and maintained; products and processes should be appropriately segregated; and equipment should be appropriately cleaned and sanitized. Although the manufacturer's microbial and product cross-contamination control strategies (including HVAC design and cleaning/sanitization validation) are typically evaluated before application approval, the inspection team must assess the firm's maintenance of microbial and cross-contamination control, including verification activities.

Many protein DS manufacturing processes support microbial proliferation because the media and the buffers used in production are conducive to microbial growth. Therefore, manufacturing processes should be designed with stringent controls to prevent adventitious microorganism introduction/ingress, limit microbial proliferation and persistence, and remove microorganisms from the process at certain points. The inspection team should confirm that the manufacturer is maintaining adequate microbial controls over its processes. The inspection team should verify that in-process (i.e., bioburden and endotoxin) action limits established within the quality system are not exceeded and, if excursions occur, they are appropriately investigated and corrected. Additionally, the inspection team should confirm appropriate microbial control of raw materials, including water used for manufacturing (e.g., verify proper maintenance and control of water production systems).

The inspection team should note:

- From a microbial control perspective, facility and equipment requirements for upstream processes generally do not need to be as stringent for microbial fermentation as they are for mammalian cell culture. Similarly, establishments using completely closed systems may have less stringent room air quality requirements than establishments that do not.
- Because cross-contamination prevention is complex, multifaceted, and dependent upon proper control of both equipment and facilities, it is addressed in its own section within this system.
- Most protein DS establishments do not manufacture highly potent or toxic products and do not use spore-forming microorganisms for production. However, if the establishment being inspected does manufacture these products or use these microorganisms, specific and thorough coverage relevant to the facilities and equipment system is warranted.

3.12.17 Equipment in Brief

As a GMP auditor, you must verify that the physical integrity of the equipment is maintained; that its installations are carried out according to the specified parameters; and that its qualification, calibration, maintenance and cleaning are carried out regularly. You should also want to see evidence that appropriate action, including CAPAs, has been taken when equipment deviations occur.

3.12.18 Qualification, Calibration, Maintenance, Cleaning, Sanitization, and Sterilization of Equipment

As a GMP auditor, you should look for evidence of how the DS manufacturer manages old and new equipment. For example, the equipment may have been purchased many years ago and perhaps not qualified, so you should expect them to demonstrate how the equipment has been managed to date and always performs correct production or testing without qualification or validation. Even if the equipment is qualified, you should question how its qualified status is maintained. Another area to question is whether the equipment has been requalified or validated in the event of a spare part change or relocation.

As a GMP auditor, you should want to see how equipment is calibrated. For example, a DS manufacturer may have the calibrations performed by its own personnel or may outsource this service entirely. In either case, you should know which documents to review and what questions to ask. Your other important task is to ask for evidence of what action the firm took for equipment that is directly or indirectly associated with production deviations or variances.

Protein DS manufacturers should be using validated cleaning procedures for nondisposable equipment that directly contacts protein DS.

Generally, original cleaning validation should be completed before an application or manufacturing supplement is approved. Therefore, on surveillance inspections, the inspection team should focus on the execution of those validated cleaning procedures as well as the firm's continued verification activities.

In general, manufacturers should use sterile equipment for cell culture/fermentation processes, although in some situations, it may be acceptable to sanitize microbial fermenters. Equipment for purification processes may be sterilized or sanitized. The inspection team must conduct inspections on ongoing sanitization and sterilization, including verification and requalification at the frequency described in the quality system or as indicated by changes in the facility microorganisms or frequency of contamination events. Requalification should be considered following major equipment repairs or changes or the failure of sanitization/sterilization processes.

3.12.19 Disposable Equipment

Disposable equipment should be integral and single-use only. The firm should diligently seek to resolve leakage issues by determining actual or potential root causes and implementing appropriate CAPAs. The firm should adequately evaluate the leakage's impact on the affected product.

During the examination of disposable equipment, it would be appropriate to question what precautions the supplier company takes regarding extractables/leachables and what risk assessments it makes, along with issues such as how the component is sterilized, how it is packaged, and how it is transported before it comes to the facility.

3.12.20 Computerized Systems in Brief

cGMP-related computerized systems should be validated. The depth and scope of automated system validation depend on the aplication's diversity, complexity, and criticality. The firm should have suitable written procedures available for operating and maintaining of computerized systems.

3.12.21 Critical Utilities

3.12.21.1 Water Systems

Water is one of the most important inputs in the production of a pharmaceutical product. However, it should be kept in mind that water can be a problem in itself when it comes to DS product production. The risks of water systems should be evaluated before installation and it should be checked whether any contamination occurs at each stage, because inadequate installation, qualification, cleaning, and maintenance of water systems can cause microbial film formation and contamination. This is not a desirable situation in any production facility. You should look for evidence that all activities such as maintenance, cleaning, calibration, and qualification in water systems are carried out according to guidelines, pharmacopoeia, and directives. In fact, you may have to stay in a facility for days even for the audit of water, steam, gas, and HVAC systems, which are considered utilities. It is useful to keep in mind that water is both included in the production as input and acts as a cleaning material.

3.12.21.2 Process Gases

Compressed air, oxygen and nitrogen are the main gases used in the production of a DS. These gases can be used in both upstream and downstream processes. Installation, calibration, qualification, and maintenance of gas systems must be handled carefully and must be managed in a way that does not cause any contamination.

3.12.21.3 Clean Steam

Because clean steam can encounter product contact surfaces, the clean steam condensate quality specifications should not be inferior to compendial (USP) standards for WFI.

3.12.21.4 HVAC Systems

HVAC systems are essential for providing suitable environments for protein DS production, especially for open operations performed when a component or protein DS is exposed to the immediate environment and therefore susceptible to contamination from the surrounding air. Of particular concern for mammalian cell culture is the susceptibility to adventitious agents (e.g., viruses, mycoplasma) and the potential to spread these agents throughout the manufacturing environment. Manufacturing processes involving mammalian cell lines typically include viral clearance/inactivation steps, and HVAC systems should function to maintain the segregation of pre- and post-viral clearance/inactivation steps. HVAC systems are generally a point of focus during product-related audits. Surveillance inspections should focus on the maintenance of HVAC systems and the demonstration that these systems continue to function as intended.

3.12.22 How is the Facility Cleaned and Disinfected?

Facilities should be appropriately cleaned and disinfected to ensure sufficient microbial control. The firm should have established and validated cleaning and sanitizing procedures and be able to provide justification for established residual limits using validated analytical testing methods.

3.12.23 Monitoring the Facility in Terms of Environmental Controls

The firm's environmental monitoring (EM) data should demonstrate the HVAC system's effectiveness as well as the adequacy of the facility's cleaning and sanitization programs. EM frequency and acceptance criteria should be suitable for each stage of production and the types of operations conducted in specific areas. The EM data should be adequate to support root-cause investigations for contamination events.

3.12.24 Pest Control in Brief

Pest control is an important consideration in preventing contamination of raw materials, equipment, and protein DS.

3.12.25 Cross-Contamination

See EU, PIC/S GMP, Annex 1 concerning a contamination control strategy.

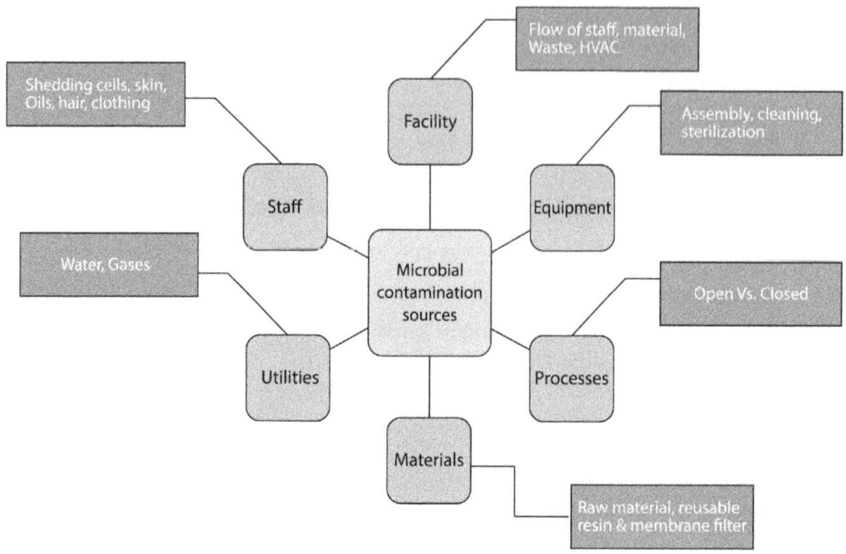

FIGURE 3.21 Microbial contamination sources.

For pharmaceutical manufacturers and their suppliers, contamination of any kind that leads to product or production losses represents a significant risk. As recent events in the past, such as foreign particulate contamination, have shown, this can lead to supply bottlenecks for individual medicinal products or groups of medicinal products. Figure 3.21 summarizes sources of microbial contamination.

Manufacturers should design their production facilities, equipment, and processes and implement quality risk management (QRM) to ensure appropriate contamination control to minimize or detect contamination. Since measures affect different stages of a manufacturing process and often fall under the responsibility of different departments (e.g., quality control, quality assurance, or manufacturing), it may not always ensure that the data obtained in the process, e.g., from the original qualifications and validations, process controls, and ongoing environmental monitoring, are linked with each other. This also applies to corrective and preventive actions that are often taken as a result of deviations and trend analyses but are neither integrated into a strategy for a holistic view, nor is there a linkage of all critical control points and the evaluation of the effectiveness of all controls (design, procedures, technology, and organization). However, a holistic view is proposed in Annex 1 (2022) for particulates, microbial, and pyrogen contamination.

The objective of a contamination control strategy, for aseptically filled products, is sterility assurance and the production of a sterile product (a product absolutely devoid of viable microorganisms). This outcome is intrinsically dependent upon a process specifically purposed to impart that desired state, consistently. Unlike terminal sterilization, where the sterility assurance level can provide a statistical understanding of the probability of non-sterility, with aseptic processing, where the imperative is to prevent microbial ingress, the same statistical assurance cannot be provided. Thus the manufacturer is therefore reliant upon a good contamination control strategy (CCS).

The preparation of a CCS is a major challenge to all stakeholders in sterile product manufacturing, as Annex 1 only sets out the requirements at a principal level and informs less on scope, content, and documentation structure. Further, the CCS has a new narrative to document the approach taken as a "complete story" and is not expected just to be a list of existing documents that may lack connection or continuity of the approach to contamination control. One of the challenges is how much detail is required to provide this CCS story.

TABLE 3.35
CSS Roadmap

Step 1 — Choose to familiarize yourself with Annex 1
Step 2 — Check how well your current pharmaceutical quality management system supports Annex 1
Step 3 — Identify a leader and team members in the facility for CSS applications, taking into account their expertise and educational background
Step 4 — Identify gaps between existing operations and Annex 1
Step 5 — Put patient health, product safety and risks at the center of improvements according to the difference analysis result
Step 6 — Start creating CSS by sharing gap analysis results with all your stakeholders
Step 7 — Make sure to activate CAPAs for improvement actions and determine risk-based changes for long-term CAPAs.
Step 8 — Implement CAPAs including validation and qualifications; complete training according to SOPs
Step 9 — Evaluate the results and effectiveness of CAPAs and look for unexpected results
Step 10 — Take the life cycle approach and collect all your monitored data and review and report trend analyses to maintain or change CCS success and efficiency

Considering all critical control points and evaluating the effectiveness of design, procedural, technical, and organizational controls, CCS should be implemented facility-wide to determine the monitoring methods used to manage product quality and safety risks. The development of CCS requires detailed technical and process knowledge.

Table 3.35 presents the CCS roadmap. Contamination – From a broad perspective, it is the presence of all kinds of microbial, physical, biological, and chemical foreign substances in the process, starting from all inputs into production and going to storage and even distribution, even though they are undesirable.

- How can we get value from all this data we are collecting?
- What do regulators want us to do with the resulting information?
- How can this data help us improve our processes and systems?

Protein DS manufacturing processes should be designed and executed to prevent cross-contamination between different product processes; areas that do and do not use animal-derived components; pre- and post-viral activities; and upstream and downstream operations. Manufacturers should integrate facility and equipment cGMP concepts to effectively prevent product cross-contamination. These concepts include:

- Appropriate segregation or containment of different manufacturing processes or process steps
- Appropriate flow of personnel, products, raw materials, waste materials, and equipment
- Adequate procedural controls
- Appropriate changeover activities
- Adequate validation and verification for the cleaning of shared product contact equipment (with inactivation procedures as necessary)
- Adequate personnel training

When evaluating a firm's cross-contamination prevention measures, the GMP auditor must evaluate not only the details of the individual measures but also how these measures fit together to form a single, robust cross-contamination control system. Because of the inherent complexity of cross-contamination prevention, there are many opportunities for failure, which could pose significant patient safety risks.

When evaluating cross-contamination prevention, GMP auditors should keep the following principles in mind:

- For campaign-based manufacturing activities (time-based segregation of different manufacturing processes in the same manufacturing area), appropriate changeover procedures between product manufacturing campaigns are critical for the prevention of product cross-contamination.
- For firms conducting different manufacturing activities simultaneously in physically separated areas, the awareness and control of potential crossover points and other opportunities for cross-contamination (e.g., in shared washing areas, personnel movement) are of key importance.
- Firms engaged in concurrent manufacturing (conducting multiple manufacturing processes in the same area) should be able to demonstrate there are sufficient controls to prevent cross-contamination.
- Firms manufacturing protein DS using equipment that is not product-dedicated (i.e., shared), regardless of how manufacturing activities are segregated or contained, should also ensure adequate cleaning of such equipment.
- Firms should ensure that personnel are adequately trained to prevent cross-contamination and that appropriate procedures are in place to direct the flow of personnel, products, raw materials, waste materials, and equipment.

3.12.26 Storage and Handling of MCBs and WCBs

Protein quality characteristics are strongly tied to the cells from which they are produced; therefore, the cell banks should be appropriately controlled and defined.

3.12.27 Handling of Raw Materials

Critical raw materials are those that may impact the quality of protein DS and thus impact product safety and efficacy.

- The GMP auditor should confirm that the manufacturers have a robust pharmaceutical quality system (PQS) ensuring appropriate microbiological control of raw materials, including water.
- The GMP auditor should also verify that the firm adequately monitors and documents all aspects related to critical raw material, including, for example, sourcing and vendor qualification, management of inventory, and testing and release of these materials.

3.12.28 Cell Banks in Brief

Manufactureres should appropriately store, maintain, and handle MCBs and WCBs to ensure the uninterrupted production of protein DS of consistent quality, Firms should restrict cell bank access, properly segregate different cell lines, store cells at appropriate temperatures, monitor the storage areas for temperature excursions, and account for aliquot removal. Liquid nitrogen freezers using liquid/vapor phase storage are often used. Because the catastrophic failure of a cell bank could halt the production of a sole-source or medically necessary product, manufacturers should store cell banks in multiple distinct locations, preferably at separate facilities. Cell banks should be tested for identity, purity, and stability as described in ICH guidance for industry Q5D.

3.12.29 Production System

A firm's production system includes production processes, in-process sampling, and in-process controls. From a risk perspective, the GMP auditor should attend to the following when covering a protein DS manufacturer's production system:

- Process consistency.
- Clearance of impurities, including high-risk process- and product-related impurities.
- Meeting in-process limits, including in-process bioburden and endotoxin action limits, and release specifications.
- Stringent controls to prevent adventitious microorganism contamination throughout the production process.
- Viral clearance/inactivation (if applicable).
- Process changes.

The GMP auditor should verify that protein DS are manufactured in strict conformance with approved, validated processes to ensure the desired clinical outcome. Process- and product-specific manufacturing details can be found in relevant biologics license applications (BLAs). GMP auditors should reference these applications throughout the inspection. However, the GMP auditor should not assume that aspects of the control strategy defined as critical by the firm, including those defined as critical in the application, are the only aspects required to ensure the safety and efficacy of the protein DS. For example, all process parameters and in-process controls included as established conditions in the BLA ensure product safety and efficacy. Similarly, all attributes included in the control strategy may be important, regardless of whether the firm or applicant has listed them as critical quality attributes. Use of the term "critical" throughout this compliance program may not align with the firm's use of the term. When auditing production processes, GMP auditors should note that high-risk impurities differ from process to process and depend on things such as the products themselves as well as the expression systems and raw materials employed. Potentially relevant high-risk impurities include, but are not limited to, host cell proteins, host cell DNA, residual antifoam and antishear materials, metals and other ligands from affinity chromatography columns, cyanide from PEGylation processes, and product-related impurities (e.g., protein aggregates, high and low molecular weight moieties).

3.12.30 Personnel

Production personnel should be qualified, trained, and attentive to hygiene.

3.12.31 Master and Batch Production Records in Brief

Firms should have appropriately controlled master and batch production records that accurately reflect their manufacturing processes. When auditing a firm's records (production, laboratory, or otherwise), the GMP auditor should focus on record/data integrity. Although the GMP auditor has occasionally observed intentional record falsification, data integrity issues stemming from poor documentation practices are significantly more prevalent (e.g., operations are not recorded contemporaneously, records are not complete and accurate). Therefore, GMP auditors should review documentation practices to determine whether they are appropriate.

3.12.32 In-Process Sampling and Controls in Brief

As discussed in ICH Q7, manufacturers should establish procedures that "monitor the progress and control the performance of processing steps that cause variability in the quality characteristics of intermediates and APIs," with related sampling plans and procedures "based on scientifically sound sampling practices."

3.12.33 Excursions, Deviations, and Failures in Brief

The GMP auditor should review an appropriate number of records documenting investigation excursions, deviations, and failures across multiple systems.

3.12.34 Process Validation Management

Prospective validation, which is typically covered during pre-approval inspections, may include, but is not limited to, studies for individual unit operations, in-process hold times, buffer and media hold times, viral clearance, impurity clearance, column resin lifetime (at reduced scale), and shipping.

Concurrent validation studies include studies for UF/DF membrane lifetimes, column resin lifetimes (at the manufacturing scale), and reprocessing steps. The firm's approach to process validation should be consistent with a lifecycle approach to validation.

3.12.35 Reworking and Reprocessing Management

Reworking: It is the exposure of an active substance or semi-product that does not comply with the standard or specification to one or more process steps different from the production processes created to have acceptable quality (For example, recrystallizing with a different solvent).

Reprocessing: reprocessing an API or intermediate, including one those that do not meet standards or specifications, and repeating a crystallization step or other suitable chemical or physical manipulation step (for example, distillation, filtration, chromatography, grinding) that is part of the established manufacturing process. If the step is not completed after an in-process control test, continuing the processing step is considered part of the normal process and does not count as rework. Because of the more significant potential for unintended impacts on product quality, GMP auditors should encounter protein DS reprocessing much less frequently than small molecule API reprocessing. The GMP auditor should **rarely** meet protein DS reprocessing steps that have not been specifically approved in the relevant BLA or appropriately reported to the FDA. Because viral filters occasionally fail, manufacturers may be approved for refiltration of the protein DS (following successful validation).

The GMP auditor should critically evaluate any reprocessing that may indicate that the process is not operating within a state of control, regardless of whether the applicant has submitted such a supplement.

3.12.36 Material Weighing and dispensing

Material weighing and dispensing should be controlled. Operations should be performed with sufficient accuracy and in a manner that does not affect the material's suitability for use or cause cross-contamination.

3.12.37 Media and Buffer Preparation and Holding in Brief

Media and buffers should have adequate control procedures.

3.12.38 Cell Culture Management

Protein DS manufacturing processes generally begin with the thawing of a WCB vial, the contents of which are used to start one or more seed trains, which are, in turn, used to start one or more inoculum trains. The inoculum trains are then used to begin growth/expression in production-scale fermenters/bioreactors. Throughout the expansion process, as the cell culture volume increases, several types of cell culture vessels may be used, each of which has its own process controls. Vessels include shaker flasks, spinner flasks, roller bottles, cubes, and wave bags as well as small-scale fermenters/bioreactors. The most commonly used reactor is a stirred-tank reactor, but others, such as airlift bioreactors, hollow fiber bioreactors, and ceramic cartridge bioreactors, may be used. Additionally, roller bottles and disposable bioreactors may be employed. Cells may grow in suspension (most common), on microcarriers, or, on rare occasions, in a fixed solid phase. Production-scale growth/expression may occur, for example, when using:

- A single-batch closed system.
- A fed-batch system (nutrients added periodically).
- Perfusion – Cells are held at high density while waste and the desired protein are continuously removed and nutrients are continually added.

Systems that are not closed should maintain microbial control, with the level of control commensurate with the level of risk. Very long production operations, such as those involving perfusion, are considered high risk from a microbial contamination perspective.

3.12.39 Steps Taken after Post-Expression, Harvest, and Recovery?

Following production-scale expression, the contents of the fermenter/bioreactor (known as unprocessed, unclarified, or unpurified bulk) will be subjected to one or more harvest or clarification steps to separate cellular matter from the cell culture medium/supernatant and produce the clarified bulk. Steps may include centrifugation (generally continuous), depth filtration, standard membrane filtration, or tangential flow filtration. Harvest and clarification controls may include centrifuge type, centrifuge flow rate, centrifuge bowl speed, centrifuge temperature, filter type, depth filtration differential pressure, filtration temperature, filtration flux, filter reuse cycles, harvest vessel temperature and hold time, harvest bioburden, and harvest endotoxin. Harvest and clarification parameters should always be consistent with application commitments.

Generally, the supernatant of the fermentation of nonmicrobial (e.g., mammalian) and yeast expression systems will contain the excreted protein DS, whereas, for bacterial expression, the cellular/pelleted matter will contain the desired protein (although there are exceptions). If the cellular matter contains the protein DS, recovery of the desired protein requires cell disruption (e.g., treatment with lytic enzymes or physical disruption) followed by clarification. For proteins that have been expressed in the form of insoluble inclusion bodies (generally larger proteins expressed in microbial systems), the protein may require denaturation, followed by controlled refolding, chemical reduction, and concentration, to produce an active form of the protein.

3.12.40 Ultrafiltration/Diafiltration?

UF/DF may be used at any point in the production process where volume reduction or buffer exchange is desired. As such, it may be used before, in between, or after column chromatography steps. UF/DF membranes may be reused but should be product-dedicated, and membrane lifetimes should be validated concurrently during commercial production. The effectiveness of cleaning, sanitization, and storage should be verified at scale.

3.12.41 Column Chromatography Management

Chromatography resins (e.g., affinity, cation/anion exchange, hydrophobic, size exclusion) are generally reused but should be product-dedicated. Chromatography column housings (hardware) need not be product-dedicated; however, the firm should have adequate cleaning validation and changeover procedures for non-dedicated column housings. Resin lifetimes can be initially validated using small-scale models, but concurrent validation at the commercial scale is expected to be ongoing at the time of approval. The effectiveness of cleaning, sanitization, and storage should be verified at scale.

If the firm uses specialty capture/affinity resins, such as protein A, these should be a point of focus when covering purification. Specialty capture resins are often susceptible to degradation/

ligand leakage. Therefore, the GMP auditor should verify that resins currently in use are within their validated cycle lifetimes and that recent operations have not demonstrated signs of resin deterioration. The GMP auditor should also cover the firm's activities to control or monitor the leakage of ligands into the product stream, such as material/vendor qualification and testing and, if applicable, routine monitoring or testing. Additionally, because the sanitization and storage conditions for resins that are susceptible to degradation may not be as harsh as those used for other resins, the inspection team should confirm adequate ongoing microbial control of these resins.

3.12.42 Viral Clearance/Inactivation Steps

All products manufactured in mammalian expression systems should have processes dedicated to viral clearance or inactivation. Recently approved products will almost always contain viral filtration steps in their manufacturing processes because of the proven effectiveness of this method. Common viral inactivation treatment methods include low pH, heat, detergent, and solvent. To ensure product safety, viral clearance/inactivation processes should always be conducted in strict accordance with procedures approved by the quality unit and any changes managed through a robust PQS.

3.12.43 Bulk Drug Filtration and Filling

Most protein DS are not sterile. However, they are filtered to reduce bioburden. The bulk fill process should be conducted under conditions designed to ensure microbial control during storage and handling. Factors affecting microbial control include the use of open or closed systems, the growth-promoting potential of the bulk, the microbial control risks during the drug product manufacturing process, and the storage temperature of the protein DS. Protein DS not stored frozen have the highest risk; therefore, open filling operations for these DS should be conducted under conditions designed to maintain microbial control (e.g., safety cabinet or ISO 5 type of environment). Protein DS intended to be stored frozen should be frozen according to the approved application.

3.12.44 Protein DS Bulk Storage

Typically, protein DS are frozen (<–15°C) during storage. However, there may be cases where some bulk protein DS may not be stable when frozen and are held at 2°C–8°C, requiring additional consideration when looking at microbial control.

For any approved storage temperature, the protein DS should be:

- Stored in the appropriate containers as indicated in the application.
- Protected from light if indicated in the application.
- Stored in a controlled and monitored freezer or other cooling device at a temperature as indicated in the application.
- Adequately labeled to prevent mix-ups.
- Stored in an area with controlled access and under appropriate conditions so that quality, purity, and strength are not affected.

3.12.45 Laboratory Control System

Laboratories overseeing the quality assurance and control of protein DS and their in-process intermediates comprise several analytical disciplines, including chemistry, biology, and microbiology. The firm should have appropriate procedures in place to routinely track method performance parameters and effectively manage signals/patterns of potential concern to prevent drift in assay performance away from validated operating conditions throughout the method's life cycle. The

GMP auditor will encounter familiar testing platforms, such as high-performance liquid chromatography, as well as protein-specific testing platforms, such as polyacrylamide gel electrophoresis, isoelectric focusing, and cell-based assays, commonly called bioassays, designed to determine the potency of the protein DS.

Because of the need for specialized experience and equipment, it is common for some testing, such as potency testing, mycoplasma testing, and viral safety testing, to be conducted at establishments other than the protein DS manufacturer. Animal testing is rarely required, but when it is, it would likely not be conducted at the protein DS establishment.

In general, GMP auditors should approach protein DS testing operations as they would other cGMP testing operations. Many of the same expectations and principles apply, regardless of the specific test methods.

Specifications are not intended to fully characterize a drug substance (e.g., a complete characterization of a protein's posttranslational modifications), and meeting them does not guarantee quality.

3.12.46 Sampling in Brief

GMP auditors should evaluate how the firm conducts sampling based on approved procedures.

3.12.47 Test Methods in Brief

GMP auditors should evaluate the firm's procedures to establish test methods.

3.12.48 Record Keeping in Brief

GMP auditors should cover recordkeeping using questions linked to data integrity when covering laboratory recordkeeping because the questions in that section are applicable to all cGMP recordkeeping activities.

3.12.49 Laboratory Equipment, Reagents, and Standards in Brief

GMP auditors should evaluate a sample of laboratory equipment, reagents, and chemicals to assess the firm's qualification, maintenance, and storage procedures.

3.12.50 Test Method Validation and Postapproval Changes in Brief

GMP auditors should ensure that test methods have been validated and procedures approved by the quality unit make postapproval changes and are reported appropriately.

3.12.51 Out-of-Specification Results and Invalid Tests in Brief

GMP auditors should review recent OOS test records. Invalid tests should not be part of the OOS records. A test should be considered invalid if, for example, the system suitability assessment has failed multiple times, highlighting potential issues with test method robustness.

For OOS investigations, retesting may be allowed under strictly defined conditions; however, procedures should be in place to determine whether retesting is appropriate, how retesting should be performed (e.g., sampling, duplicates), and how results should be interpreted.

3.12.52 Stability Testing and Reserve Samples in Brief

The GMP auditor evaluates the adequacy of the containers used to evaluate the stability of the protein DS as well as the firm's recordkeeping practice related to stability testing results and triage and management of pertinent deviations and failures.

3.12.53 Tests Performed in the QC Lab

3.12.53.1 Potency Assays/Bioassays

Potency assays should be conducted for product release and stability testing as well as for comparability testing (comparing material from different manufacturing processes) following a manufacturing change. Potency assessments can take many different analytical forms; however, they all examine a product's biological function. Potency assays may involve measurements of enzymatic activity or protein binding or they may be more complicated. For example, some potency assays are cell-based (commonly called bioassays) and are designed to measure activities such as cellular activation, cellular uptake, or cellular viral protection. Regardless of assay type, potency assays should have controls typical of analytical methods, such as system suitability.

3.12.53.2 Protein Reference Standards

Protein reference standards are generally prepared in-house by the firm or supplied by the applicant of the product to be tested. To ensure the suitability of these standards, manufacturers should create them according to defined procedures and sufficiently characterize them. Reference standard changes should be conducted per an approved protocol and are typically reported in an annual report. In the absence of an approved protocol, they should be handled through the submission of a supplement. In ensuring the suitability of reference standards for protein DS, the inspection team should confirm that reference materials have been or are being qualified or requalified by procedures used for routine DS release as well as additional characterization assays. The firm should also strictly follow storage conditions, usage conditions, and handling instructions for reference standards to avoid adding impurities and to ensure accurate analysis.

3.12.53.3 How Are Microbiological Tests Performed?

Protein DS manufacturing processes are particularly conducive to the growth of microorganisms, and because the resulting products are generally administered parenterally, it is essential for manufacturers to have adequate microbial testing and controls. Microbiological testing and controls generally include:

- Screening raw materials for adventitious agents
- Testing cell cultures before harvest for bioburden, mycoplasma, and other adventitious agents (e.g., viruses)
- Testing buffers and in-process materials for bioburden and endotoxin and observing established limits
- Testing bulk protein DS for bioburden and endotoxin and observing established specifications

3.12.53.4 Bacterial Endotoxin Testing

GMP auditors should verify that the firm's validated bacterial endotoxin test is conducted as approved in the BLA and under the manufacturer's quality management system. The failure of an in-process material, raw material, or protein DS to comply with endotoxin action limits or specifications requires a thorough investigation into the root cause of the problem.

3.12.53.5 Bioburden and Mycoplasma Testing

Using validated and approved test methods, manufacturers should perform bioburden (microbial limits) testing on raw materials (as appropriate), in-process materials, and the protein DS. They should have and follow written procedures describing bioburden sample collection and the bioburden test method itself.

Cell banks and unprocessed bulk should be specifically tested for mycoplasma if appropriate. Mycoplasma testing is expected for mammalian cell cultures. The expected mycoplasma control strategy for each protein DS should be detailed and may include the use of established compendial test methods. Because some raw materials may be a source of mycoplasma, manufacturers should test incoming raw materials as appropriate or have other controls in place (e.g., high-temperature short time [HTST] processing or the use of 0.1 micron cell culture media filtration).

3.12.53.6 Identification of Microorganisms

Identification of microorganisms recovered from the product and the manufacturing environment can provide important information for investigations into the origin and source of non-host cell contaminants in protein DS manufacturing operations, thereby aiding the prompt resolution of microbial issues. Accurate identification of contaminates and their source (e.g., equipment cleaning, sanitization process) may be important in reaching conclusions concerning potential product impact.

3.12.53.7 Viral Safety Testing

The GMP auditor expects protein DS manufacturers to perform viral safety testing if appropriate. Because of the severe health outcomes of viral contamination of an injectable product, testing for viruses at various processing points is critical for DS protein derived from human or animal cell lines.

The primary guidance for viral safety testing of biotechnological products is ICH Q5A. Per ICH Q5A, viral safety should be ensured through raw material controls, appropriate viral clearance after cell culture and during purification, the testing of cell banks (including end-of-production cell banks), and the testing of unprocessed bulk. ICH Q5A is currently under revision.

Animal-derived raw materials are often tested based on their source (e.g., calf serum is tested for bovine viruses). Tests are often performed by raw material vendors; however, protein DS manufacturers should not rely on COA testing results unless they have appropriately qualified the vendors, with appropriate activities to ensure continued vendor qualification, based on a risk assessment.

3.12.54 Packaging and Labeling System

To provide assurance of drug quality, manufacturers should appropriately package, label, and ship the protein DS they manufacture.

Bulk containers of protein DS should be free of leaks. If a manufacturer experiences leakage, the event should be investigated, the potential product impact should be evaluated, and the manufacturer should be able to provide sound, scientific reasoning for the related disposition decision. Additionally, because of the potential for theft and counterfeiting, manufacturers should ensure that containers have been appropriately secured (e.g., using anti-tampering devices).

Manufacturers should confirm that printed labels contain the correct information and conform to specifications in the master production record. They should destroy excess, obsolete, and outdated labels, and should have procedures to reconcile discrepancies between the numbers of labels issued and used, investigating discrepancies if appropriate. Because a single batch of bulk protein DS is often filled into multiple containers, GMP auditors should confirm that manufacturers consistently label these containers with the appropriate batch number. To avoid potential mix-ups, manufacturers should also segregate batches of bulk protein DS from one another by physical or other suitable means.

Finally, manufacturers should ship their protein DS in accordance with validated shipping procedures, using qualified shippers.

3.12.55 Sampling in Brief

Samples may be collected to document suspected contamination or adulteration of raw materials, in-process materials, and protein DS encountered during an inspection; however, physical sample analysis is not expected to document cGMP deficiencies.

Following are some examples of findings encountered in audits of DS producer companies.

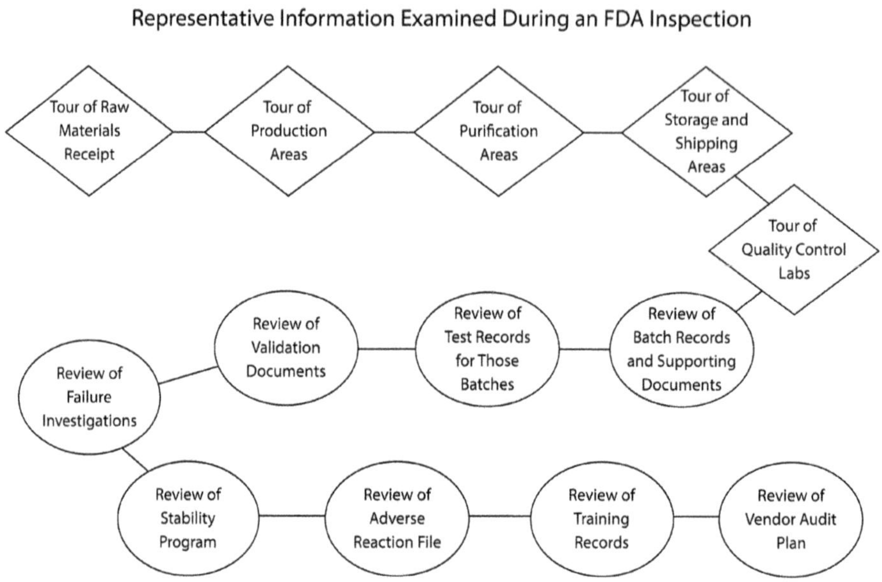

FIGURE 3.22 FDA facility tour.

3.12.56 Critical Findings

- Release of DS that does not conform to final specifications or is clearly of unacceptable quality
- Failure to conduct an investigation (including an OOS testing result investigation) when evidence suggests that the safety, purity, or potency of a released DS has been adversely impacted
- Multiple instances of inadequate OOS investigations, followed by material acceptance
- Unjustified use of a critical raw material that does not meet specifications
- Failure to ensure that materials are free from adventitious agents as required
- Release of DS that was produced using a WCB that was neither approved in the application nor created and qualified in accordance with an FDA-approved protocol
- Serious data integrity concerns (e.g., apparent changes to records to make unacceptable released material appear acceptable)

Figure 3.22 shows the steps of the FDA's facility tour.

3.13 AUDITING VACCINE MANUFACTURING FACILITIES

Although we have been vaccinated against many diseases since childhood, we have understood very clearly since December 2019 that vaccines are an integral part of our daily lives. After COVID-19 turned into a global pandemic, all eyes turned to big pharmaceutical companies and drug options that could prevent COVID. Although an oral drug was considered an option much later, it did not take long to realize that the fastest and most effective solution was to produce a vaccine. Although big pharmaceutical companies announced that they started to develop vaccines one after another, each country was alone with its own destiny, and each country began to question whether it would develop vaccines with its own capabilities, leaving aside its economic power and seeking practical ways to ensure this rapid transition in existing established vaccine facilities. As a specialist and GMP auditor, who followed Turkey's vaccine initiative with great curiosity and attention, like in other countries, many questions inevitably began to circulate in my mind and search for the most appropriate answer.

Yes, COVID-19 is a pandemic with very sad results, but as the pandemic is about to complete its fourth year, we observe that at least cases have decreased on a country basis, but different variants continue to threaten the health of people.

We have witnessed that the most effective way to stop the pandemic, at least to slow it, is vaccine production. However, we also realized that no country was fully prepared for the pandemic, regardless of its economic power, technical infrastructure, and trained manpower. I just want to go back a few months before the pandemic started. I was giving a series of sterile manufacturing techniques training to a pharmaceutical company based in Turkey, and I was literally shuttled between the company's three facilities. When one of the training topics came to EU, PIC/S GMP Production 5.71, I asked the participants if the company had any pandemic emergency action plans. One of the most critical questions asked by both local and international health authorities in inspections is when there is an emergency and the company is affected by this situation, is it possible to access the medicine at any cost, by keeping the patient's health in the foreground?

The responses I received did not surprise me because it is more frightening to see that countries do not have such a plan, let alone a local pharmaceutical company's pandemic prevention plans.

That's why vaccines can turn such a critical and bad chance into a ray of hope all of a sudden.

But after the COVID-19 vaccines were produced, there was news of some contamination, spoilage, late shipments, and problems in accessing raw materials. Some companies chose to have the vaccine production done by a contracted manufacturer with the quality agreements they made with their own know-how, and as a result, some problems of access to vaccines emerged.

I don't need to tell you how difficult it is to develop, produce, and ship any vaccine, not just the COVID-19 vaccine.

So what is the primary goal for developing and producing a vaccine?

Before moving forward, we need to focus on the API (antigen) of a vaccine candidate.

3.13.1 What Is an Antigen?

Antigens are proteins on the surface of a pathogen that causes antibody production by the immune system. Bacteria and viruses are covered with antigens and the body recognizes these antigens in the natural infection process. Depending on the pathogen, the antigen coat may consist of several or several thousand proteins. On the other hand, vaccines are usually made using only a few antigens from a bacteria or virus. This is because some antigens are better at igniting the immune system than others. The ability to cause disease is called virulence, and the ability to generate a protective immune response is called immunogenicity. The number of antigens in vaccines is actually quite small compared to the number of antigens that encounter the immune system during a real infection (or just from the environment on a daily basis). For example, the pertussis-causing pathogen contains more than 3000 antigens; The vaccine used to prevent pertussis contains only 3–5 different antigens.

3.13.2 Vaccine Production Process

Although antigen types differ among different vaccines, the overall vaccine production process is similar for all types:

Upstream

An antigen is produced in the upstream process. It may be the pathogen itself, or it may be an antigen produced from a recombinant protein. The antigen induces an immune response. There are three types of antigens:

- The virus is produced by primary cells (e.g., chicken fibroblasts) or a continuous cell line (e.g., Vero).

- Bacteria are grown in fermenters.
- Recombinant protein is produced by bacteria, yeast, or cell culture.

Downstream

The antigen is separated from the impurities in the downstream process. This increases product safety and stability.

- The antigen, if necessary, is released from the substrate (e.g., cell lysis) and isolated from the bulk medium.
- Depending on the characteristics of the antigen and the impurity, purification technologies may be used: chromatography, ultrafiltration, precipitation, enzyme digestion, or other processes.

Formulation

All the ingredients that make up the final vaccine are combined in the formulation process.

- The formulation is designed to maximize stability while creating a format that provides effective delivery and the preferred method of clinical administration.
- The formulated vaccine may contain an adjuvant to potentiate the immune response, stabilizers to extend shelf life, and/or preservatives for the delivery of multidose vials.

3.13.3 Vaccine Contents

- Adjuvant (aluminum salts, aluminum hydroxide, aluminum phosphate or potassium aluminum sulfate, squalene, MF59, monophosphoryl lipid, liposome, virosome, flagellin, TLR agonists, cytokine)
- Preservative (thiomersal, 2-phenoxyethanol)
- Antibiotics (neomycin, polymyxin B, gentamicin, kanamycin)
- Stabilizers (gelatin, sugars, amino acids, proteins)
- Residues from the production process (cell culture residues, antibiotics, inactivation residues, undesirable agents/vectors)

3.13.4 Vaccine Production Facility Audit

Before vaccines can be produced, marketed, and used, the vaccine manufacturer must obtain authorization, registration, licensing, evaluation, and approval. For vaccine production, the quality assurance team is responsible for controlling the quality, safety, potency, and efficacy of the vaccine; and supervising and monitoring production processes, distribution, and marketing.

GMP is designed so that the risks involved in any pharmaceutical manufacture are minimized. This international practice requires the manufacturer to ensure that products are safe, pure, and effective.

Like any other pharmaceutical or biotech production environment, vaccine production requires dedicated cleanroom areas. The production of biological drugs poses challenges to cleanroom design. Vaccine production facility design and construction are no exception to this rule.

3.13.5 General Cleanroom Assessments for GMP Facility Design

- Different types of vaccines
- Vaccine manufacturing plant detailed layout
- Types of equipment required for vaccine production
- Different steps of producing a viral vector vaccine
- Vaccine-related products manufactured in cleanrooms

All stages of vaccine development and production require cleanroom environments to provide the critical environment required. Appropriate temperature, humidity, airflow, air pressure (positive or negative depending on biosafety level [BSL]), and other environmental requirements are imperative to ensure a safe vaccine without any risk of contamination.

Here are a few points to consider for a vaccine manufacturing facility design or biomanufacturing facility layout:

- The type of vaccines or ingredients produced
- BSLs depending on the kind of vaccines or viral vector used
- Workflow patterns
- ISO classification/GMP requirements for vaccine cleanroom design
- Process stage (R&D, preclinical and clinical stages, large-scale production)
- Dimensions, equipment inside the room, number of people working inside
- Vaccine production capacity/number of expected doses
- Facility expansion
- Standard facilities (plumbing, electricity, process gases, etc.)

3.13.6 Vaccine Types

We often talk about the vaccine itself and the disease it protects us from. Still, vaccine production requires proteins, nucleic acid, viral vectors, cells, adjuvants, plasmid DNA, and many other components. Many of these vaccine components need to be produced in cleanrooms. It is also important to note that the design of the vaccine manufacturing facility will vary depending on the vaccine type. For example, a viral vaccine manufacturing facility has a different format than a plant-based or mRNA vaccine facility, as the workflow, equipment, and cleanroom classes will differ.

A GMP vaccine facility can produce:

- Inactivated vaccines
- Live attenuated vaccines
- Messenger RNA (mRNA) vaccines (COVID-19 vaccines)
- Subunit, recombinant, polysaccharide, virus-like particles (VLP), and conjugated vaccines
- Toxoid vaccines
- Viral vector vaccines
- DNA vaccines

3.13.7 Vaccine Production Facility Layout

Vaccine products range from living organisms (attenuated salmonella typhi) to inert proteins (for example, virus-like particles used to protect against hepatitis B infection). Such an assortment significantly impacts GMP plant design, process validation, and control.

1. *Cleanroom gowning room/gowning area.*

As with all cleanroom expectations, the facility where the vaccine will be produced also needs an apron area to eliminate the risk of contamination. Multiple airlocks, material airlocks (MALs), and personnel airlocks (PALs) are included in the design.

Table 3.36 describes the cleanroom grades.

2. *Cell culture production.*

Extracting cells, tissues, or organs from an origin like an animal or plant and placing them in an artificial medium to promote growth. The cell culture cleanroom is an ideally reserved space.

TABLE 3.36
Cleanroom Grades

Activity	Open	Dedicated Area
Harvest	Class C	Class D
Inactivation	Class C	Class D
Purification	Class C	Class D
Final sterile filtration	Class A in B background	Class D
Formulation	Class A in B background	Class D
Filling	Filling the bulk tank with open connections from B to A Filling Process in A to B	Closed filling bulk tank: D Filling the isolator or Class III biosafety cabinet: A to D
Labeling	Clean unclassified (CNC)	Clean unclassified (CNC)
Packaging	Clean unclassified (CNC)	Clean unclassified (CNC)
QC Test	Sterility test: A in B	Sterility test: isolator in D

Cleanroom size will be affected by the quantity and size of bioreactors or cell culture fermenters. Bioprocessing capacities range from 50 to 2000 liters and even more.

3. *Vaccine production.*

This cleanroom area of the viral vaccine production facility is used to collect and neutralize the pathogens in the virus. Suppose vaccines containing viral vectors are produced in the facility subject to inspection, depending on the virus type. Vaccine facility design and layout must consider BSL and GMP requirements. The BSL will affect the creation of the HVAC system and vaccine manufacturing plant as the cleanroom may need negative pressure for BSL-3 and BSL-4. The cleanroom rating for these vaccine activities is usually Grade C (ISO7).

4. *Vaccine purification.*

Vaccine purification is one of the most challenging steps of the process as it usually requires more than one purification step. New cutting-edge technologies now make it possible to accelerate the purification step in DSP. So, the facility that manufactures vaccine products requires chromatography equipment for this step. The vaccine formulationwill take place in this cleanroom, where adjuvants, stabilizers, and preservatives are added. Environmental monitoring is also something to consider in GMP facility design. Considering the workflow and manufacturing processes is critical when designing a vaccine manufacturing facility.

5. *Aseptic filling/finishing.*

The vaccine manufacturing facility design should include a dedicated area for aseptic filling/finishing for sterilization and packaging of newly produced vaccines. The aseptic fill/end room is a crucial area of the GMP facility to ensure vaccine safety in vials and syringes. In this area, the HVAC will be designed and operated to hold the core, aseptically using positive pressure extending away from the centre. Biosafety cabinets or laminar airflow (LAF) will be required to achieve an ISO 5 (Class A) area for critical operations of the filling process with an ISO 8 (Class D) cleanroom background.

6. *Labeling and packaging.*

Although this area does not require a specific cleanroom class or ISO class, it is an important part of vaccine manufacturing facility design. It is very important to ensure that the appropriate vaccine

TABLE 3.37
Cleanrooms and Equipment

Vaccine Cleanroom Considerations	Vaccine Production Plant Equipment
• Air locks (MAL and PAL)	• Bioreactors and bioprocess equipment
• Personnel workflow patterns	• WFI (Water for Injection)
• HVAC system	• Ultra-low-temperature freezers
• Positive/negative pressure	• Centrifuges
• Environmental monitoring	• Filling and capping machine
• Low wall turns	• Filtration units
• Environmental monitoring	• Separators
• ISO class/GMP grades	• Reactors (media preparation)
• Heavy-duty panels for the wall system	• CIP systems
• Recessed finish	• Fermenters
• Gowning protocols and areas	• Storage equipment
• N+1 redundancy utilities	

goes into the correct bottle with the correct label. The clean unclassified (CNC) or unclassified (UNC) cleanroom class conforms to WHO guidelines depending on whether the vaccine cover is made. Usually, capping is done at the previous stage of vaccine production. The vaccine manufacturing facility will need cold storage equipment such as packaging, bottle labelers, and freezers (newer vaccines need to be stored at –70°C).

7. *Quality control.*

Quality control is necessary to test the potency of manufactured vaccines. Living organisms and biological drugs are much more complex than chemical drugs. The reaction of living organisms in vials may differ in each batch of vaccine production and therefore must be carefully tested. The quality control laboratory needs a Class A cleanroom with a Class B background in an open space. An isolator with a Class D background will work in a segregated and dedicated QC lab.

Table 3.37 explains cleanroom and equipment requirements.

3.13.8 Personnel

To prevent cross-contamination:

- Is there a procedure to prevent the simultaneous processing of other live or infectious material by the same persons?
- Do staff move to other areas within one working day?
- Logs and records.
- Is there a procedure to prevent simultaneous processing of inactivated and non-inactivated products by the same persons?
- Do personnel move from areas with non-inactivated products to areas with inactivated products?

3.13.9 Questions to be Asked by a GMP Auditor

Qualifications

- Is the staff dedicated/qualified?
- Is the background/education appropriate for the activity?

- Do you have the education (qualification/continuous)?
- Are medical checkups/X-rays done regularly and according to the risk of infection (BCG)?
- Is the immunological status controlled?

Hygiene

- Is there a hygiene concept including changing clothes, masks, gloves, and disinfection?
- Is showering available under certain conditions?

Areas

- Is the room classification appropriate for the activities?
- Is the design of the rooms and equipment suitable for the activities?
- How are the pressure levels (positive, negative, holding) defined?
- Are negative pressure areas or safety cabinets used for aseptic processing of pathogens surrounded by a positive pressure sterile zone? Are the rooms dedicated to the product?
- Is the HVAC system adequate?
- Is there a space and room concept for the entire company?
- Is there a hygiene concept for areas and rooms?
- Is there a concept of environmental monitoring?
- Are pressures monitored?
- Is fumigation possible?
- Are procedures and management in place in case of loss of integrity and damage?
- Are open or closed systems used?

Environmental Control

- Are equipment and environmental particle and microbial contamination controlled?
- Cross Contamination
- Is access to rooms/facilities limited to authorized persons only?
- How do you prevent cross-contamination by air?

Equipment

- Is the equipment dedicated or multiproduct?
- Is the equipment being removed from the room for cleaning? If so, is it predisinfected and disinfection validated?
- Is production on a campaign basis or continuous?
- Is the same equipment used for both decontamination and sterilization?
- Are streams of contaminated material and equipment separate from those that have been sterilized?
- Are cross-campaign and wastewater decontaminations validated and periodically revalidated?

Processes

- Is the lot definition available and is it in accordance with the marketing authorization?
- Are storage conditions defined for all intermediates, active substances, and drug products?
- Are specifications set for yields?
- Are all process parameters covered (e.g., pH, temperature, time, flow rate)?
- Are protocols available for buffer preparations?
- Are expiration dates and storage conditions specified?

- Is status and identity tagging sufficient?
- Are tampons quality control tested and released before use? Is the biological load measured?
- Are endotoxins measured?
- Where are tampons manufactured?
- Is the tampon produced on-site?
- Are the tampons sterilized in place?

Water

- Is the biological load measured?
- Are endotoxins measured?
- Is the water used sterile?
- Is the water quality monitored regularly?

Gases

- Are gas specifications and tests available?
- **Disposal of Waste Material**
- Are procedures and documentation available?
- Is the waste material disinfected with a validated method?

Cell Banks and Cell Banking

- Have cleaning and decontamination procedures been validated?
- Are they being monitored?

Field and Line Cleaning

- Are procedures and documentation available?

Containers

- Is there a cleaning, sterilization, and testing procedure?

Culture Medium

- Preparation, labeling, sterilization, sampling, and testing procedure?
- Is there a certificate if it is material of animal origin?

Pre-Master Cell Bank

- Specifications, analysis, certificates, tests, origin?

3.13.10 Monitoring

- Including HVAC, LAF
- Incubation (T°, RPM)
- In process control (IPC)
- Vaccination
- Viability
- Parameter showing the suspension step
- Growth control
- Microbiological control

3.13.11 Uniform Composition of Each Container: Aliquoting Conditions

- Pooling of cells for banking if multiple containers are used
- Smooth suspension
- Closure validation, verification
- Labeling (validated to prevent loss of information on the container)
- Sampling
- Consensus
- Lot number control if combining

3.13.12 Freezing and Storage

- Time limit between aliquoting and freezing, documentation
- Conditions (T°, time limits)

3.13.13 Qualification before and after Freezing (Characterization, Testing)

- Minimum identity before freezing
- Minimum purity before freezing
- Minimum viability after freezing

3.13.14 Maintenance of Main and Working Cell Banks

- Freezer or nitrogen tank (liquid or gas phase)
- Records (limits, corrective action procedure)
- Alarm system (registers, 24-hour connection)
- Contamination risk (control of nitrogen level, validation)
- Risk of a mix-up (specific tanks for commercial production, map and description of stored containers)
- Same handling of all containers during storage (procedure)

3.13.15 Protection against Disasters

- Redundancy, remote facilities (procedure, description)
- Backup power
- Automatic liquid nitrogen filling systems (alarm system, contracted with supplier)

3.13.16 Vial Usage Records

- Containers are nonreturnable once removed (procedure)
- Inventory

3.13.17 Facilities and Equipment

- Are dedicated facilities used for BCG vaccines and live organisms for *Bacillus anthracis*, *Clostridium botulinum*, and *Clostridium tetani* and tuberculin production until the inactivation process is completed?
- Single harvest or continuous harvesting (simultaneous fermentation and harvesting)?
- Are the structure, material, and material coating (surface, roughness, finish, weld seam treatment, etc.) of the following components and fittings adequate and confirm the GMP rules:

    - Fermenter (open, closed, or enclosed system?)
    - Piping (dead legs)
    - Valves, ventilation filters
    - Manometers
    - pH-/oximeters
    - Thermocouples, temperature sensors
    - Pipes and valves for filling and emptying
- How is the addition of the following objects recorded and documented?
    - Water
    - Medium
    - Buffers, acids
    - Cell substrates
    - Induction agent gases

3.13.18 Antifoam

- Campaign fermentation or continuous fermentation?
- Does the process follow an automated procedure?
- Does the addition of all necessary components continue automatically?
- Is the following objects guaranteed to be added aseptically:
    - Cell substrates
    - Water
    - Medium
    - Buffers
    - Gases

3.13.19 Sterilization in Place (SIP)

- Do you have validation documents?
- Is tracking data available (e.g., about the temperature record)?

3.13.20 Non-Sterilization in Place (SIP)

- Are there validation documents of the sterilization equipment/method used?
- Is there evidence of sterility (e.g., media hold test)?

3.13.21 Inoculation

- Is there information about seed culture/cell bank management?
- Where is the cell substrate produced?
- Is there an expansion in the culture (in pre-culture/intermediate fermenter or main fermenter)?
- Is there an inoculation procedure?
- Is there an assessment of the risk of contamination during inoculation and has the company addressed this risk?
- Is the biological load of the seed material measured?
- Are the endotoxins of the seed material measured?
- Is the virus content of the seed material checked?

3.13.22 Medium

- Is there a growth promotion test set up from batch to batch?
- Are there protocols proving that all ingredients are provided in the correct quantity and quality (animal origin ingredients: evaluated for TSE risks)?
- Is the medium produced directly in the fermenter or is the medium produced in a formulation tank?
- Is the medium being replenished from an external source (e.g., media bag, supplier container)?
- Are there data proving that medium transfer does not affect media sterility?
- Is the medium sterilized in situ?
- Is there data proving the sterility of the medium (e.g., medium retention test [if performed], filter integrity test in case of filtration, temperature curves in case of heat sterilization)?

3.13.23 Induction Agents

- Is the biological load measured?
- Are endotoxins measured?
- Where are the manufactured agents?
- Is it produced on-site?
- Are they sterilized in situ?

3.13.24 Antifoam

- What type of antifoam is used?
- Is the biological load measured?
- Are endotoxins measured?
- What are the specifications?

3.13.25 Fermentation

- Is there a match between the process specifications (e.g., cell doubling number, yield, etc.) and the data of the inspected lot?
- Is there evidence that sampling poses no risk of contamination?
- Is there an inactivation process?
- Are intermediate products stored?

3.13.26 Harvest

- Is there evidence that the harvest is not at risk of contamination?

3.13.27 Tracing

- Are all critical operating parameters monitored during the process as follows:
  - Process time
  - Heat
  - pH
  - pO2

- pCO_2
- Pressure
- Agitation rates
- Addition of gases
- Addition of buffers, acids
- Bioburden
- Viral content
- Endotoxins
- Viscosity
- Are other parameters of the fermentation process monitored:
  - Contamination
  - Cell identification
  - Cell growth
  - Cell efficiency
  - Cell viability
  - Cell ratio (co-culturing of two different cells)
  - Cell aggregate formation

3.13.28 Extraction and Isolation

- Centrifugation
- Filtering
- Collapse
- What is the filter life and how is it evaluated?
- Adsorption to the filter

3.13.29 Storage and Expiry Date of Intermediate Products

- Is the storage temperature defined?
- Is the expiration date documented?

3.13.30 Qualification

- How is equipment cleaned and validated?
- Are there any product-specific tests? Are these tests validated?
- Is the holding time of dirty and clean equipment defined and covered by cleaning validation studies?

3.13.31 Viral Removal Steps

- Are critical process steps performed within their validated parameters?
- Are pre- and post-virus cleaning steps carried out in separate air handling units and in separate areas?
- Is the equipment divided into pre and post-virus removal steps?
- Do staff move from pre-viral to post-viral?

3.13.32 Purification Column Resins

- Were the resins tested for the following?
  - Chemical/biological aspects

    - Physical aspects
    - Functional aspects
- Is the life of the resins/maximum number of runs defined and what is the basis?
- Have HETP (height equivalent to a theoretical plate) and asymmetric measurements been made?
- Have leaks been tested?
- Is the consistency of purification profiles a performance criterion?
- Are resins specific to a production step of a product?

3.13.33 Chromatography Systems

- Is the size of the column resin volume defined or calculated?
- Are the flow and pressure defined during filling?
- Are there control and preventive parts replacements?
- Visual inspection of the resin or other pre-use checks of the column?
- Are the cleaning procedure and cleaning agents used explained?
- What are the storage conditions (e.g., temperature, time, storage solutions)?
- How is the preparation, use, and disassembly of the system provided?
- What are the specifications for critical parameters (e.g., linear fluid flow, column bed height, gradient slope, temperature)?
- Are the product collection criteria clearly defined?

3.13.34 API

- Is there traceability to cell banks?
- Is the API characterized by chemical and biological methods?
- Are the specifications defined (identity, purity, potency, yield, etc.)?
- Is an ongoing stability program established?
- Is the consistency of the first-produced bulk final API batches documented? (characterization tests, in-process controls, specifications)?
- Is a procedure in place on how to select the reference material?
- Is a fully characterized API batch retained as reference material?

3.13.35 Finished Product

- Quality (e.g., appearance, particles, pH, humidity)
- Identity
- Protein concentration/content purity/contamination (viral, pyrogens, microbial, chemical)
- Efficiency (potency)
- Sterility
- Is an ongoing stability program established?
- Interbatch consistency of first-produced batches
- Is the temperature monitored or transport validated?
- Is there a system for distribution traceability?

CASE STUDY

Using Table 3.38, classify nonconformities encountered at the vaccine facility.

TABLE 3.38
Nonconformities

Nonconformity	Major/Minor (Justify)	Documentation to Request
1. Fermenter with improper CIP design (no/inadequate/unvalidated spray ball, pump).		
2. Open system. A lot of flexible piping is involved. The critical operation is performed manually (e.g., critical valves in CIP-SIP). Use of culture bottle with no sterile sampling device).		
3. Critical service failures during production.		
4. Lack of integrity testing before/after the process (including filters). Mechanical seal failure, loss of pressure, and condensate. Media hold test failure.		
5. No data acquisition (SCADA) / trend analysis.		
6. Vessel inside and outside corrosion; quality of welding, the material of construction.		
7. Dead legs with improper CIP-SIP.		
8. Lack of vendor documents and certifications including DQ and FAT.		
9. Lack of exhaust filter and incinerator.		
10. Untrained/poorly trained operators.		

3.14 CELL AND GENE THERAPY

3.14.1 How Are Starting Materials for ATMPs Defined?

Advanced therapy medicinal products (ATMPs) are the leading group representing and leading innovative treatments in biotechnological drugs. The USA and EU countries and health authorities are working day and night to expand this type of treatment. While advanced treatments are examined in four sections in the EU, the USA adopts the classification model as two main groups. For a drug product produced from cells or tissue to be classified as advanced therapy in both the USA and EU, it should be known that cell processing is a manipulation that changes biological properties. As the name suggests, these products are considered inflammable and complex biological products that require extensive and complex preclinical and clinical developments.

An ATMP is a medicinal product that is any of the following:

- Gene therapy medicinal product
- Somatic cell therapy medicinal product
- Tissue engineering product

The reason for the emergence of ATMPs is to give hope to patients who cannot be treated with many traditional therapeutic approaches. Such therapeutic gene- or cell-based drugs act as a way forward for treating diseases for which there is no cure yet. In addition to the development and production of innovative medicines, the licensing processes can be very laborious because they are individualized drugs.

Producing products that require advanced knowledge and technological infrastructure also raises awareness of infrastructure requirements.

The answer to the question of what is ATMP starting material can be defined as all the materials from which the active substance is produced and extracted. It should be noted here that the donation, procurement, and testing of tissues or cells used as starting material is governed by Directive 2004/23 or 2002/98/EC, although it is unrelated to GMP.

What are the necessary regulations for ATMPs and how they should be applied? Chapter IV of the EU GMP guide is about ATMPs and details are explained in this guide.

PIC/S was also recently released to ensure global compliance with GMP requirements for ATMPs. It is possible to obtain descriptive information about plants, equipment, and production in ATMPs. As the GMP guidelines of all health authorities have been harmonized with PIC/S, it has been made easier for pharmaceutical manufacturers to obtain permission in international markets. ATMP can be classified into three major groups.

(i) Gene therapy medicines in which genetic sequences are designed to modify, control, inhibit or overexpress a specific target gene sequence.
(ii) Somatic cell therapy medicines, can be highly variable, in which the cells can be from the autologous or allogeneic origin and can be in different stages of differentiation such as adult cells, fully differentiated, self-renewing stem cells, or progenitor cells.
(iii) Tissue-engineered medicines, which are products that have cells or tissues that have been modified with a combination of different techniques of cellular and molecular biology, biomaterials and engineering, with their goal being to repair, regenerate, or replace human tissue to restore tissue function or replace unhealthy tissues. Also, some ATMPs are called combined ATMPs when they contain one or more medical devices as an integral part of the medicine.

When transferring the manufacture of ATMPs to the pharmaceutical industry, there will be concerns about implementating GMP standards. Given the unique characteristics and the high patient-specific nature of the ATMPs, the common GMP standards cannot simply be applied. Therefore, even though the development of ATMPs plays a crucial role in treating rare or degenerative diseases, bringing these medical products to market takes work.

Figure 3.23 explains cell and gene therapy.

3.14.2 Crucial Aspects of Manufacturing ATMPs in a Modern GMP Environment

Despite a bumpy history, ATMP development is an active, ongoing process. However, translating the manufacture of these therapeutic products into a GMP environment still poses significant challenges.

The intrinsic variability of the starting material as well as of the finished product requires manufacturing procedures that necessarily deviate from those of traditional biological products. The extreme diversity of the starting material obtained from a patient is due to – among other things – the person's age, health, and illness. This requires impeccable batch identity and tracking. Moreover, ATMP therapies require a complex trial design, as a unique patient population is targeted. Therefore, unlike common biopharmaceuticals, we need flexible, well-monitored and controlled processes to cope with the complexity of ATMPs. Despite the introduction of centralized procedures for the manufacture of ATMPs, ATMP developers are asking for more specific guidance regarding the GMP requirements that govern the manufacturing process.

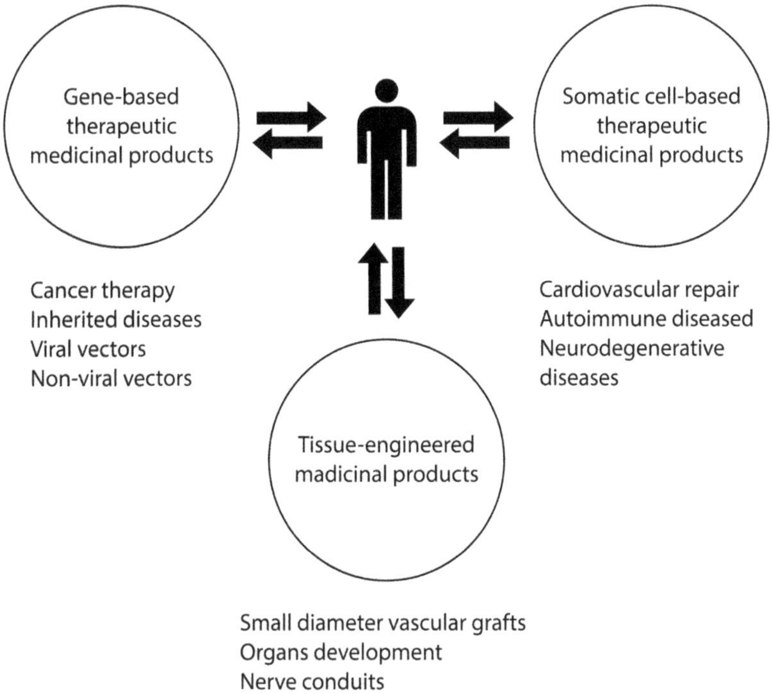

FIGURE 3.23 Cell and gene therapies.

3.14.3 Transport from Patient to the Site of Manipulation

One of the first obstacles to the production of ATMPs under GMP conditions mentioned in the literature is the transportation of the collected sample from the patient to the site of manipulation. The context of the intraoperative environment somehow escapes GMP regulation, which contradicts the need to avoid contamination. A possible solution to this problem could be operating in a closed system, but the very definition of a closed environment is controversial.

3.14.4 Sterility of the Final Product

Another reported case where producers of ATMPs have to deal with the lack of unified GMP requirements is that of Sterility. As with ATMPs, the presence of whole live cells prevents filtration through sterilization-grade filters, so the final product cannot be sterilized. Since final sterilization cannot be performed, the Sterility of the final product must be ensured through routine sterility testing during manufacture and aseptic process simulations.

3.14.5 Logistics Related to ATMP Manufacturing

The logistics associated with the production of ATMPs are also very critical, as the stability of the human sample and of the final ATMPs decreases significantly with the duration of transportation and is strictly temperature dependent. These and many more aspects of the production of ATMPs must be strictly regulated to meet GMP standards.

As challenging as it may be, developers of ATMPs have powerful tools to support the manufacturing process in a GMP-compliant manner. The installation of cleanrooms and the application of a risk-based approach through the method of failure modes and effects analysis (FMEA) can ensure the safe disposal of processed tissues and cells at every step of the production of ATMPs. In-process controls (IPCs) are also essential tools to control the safety and quality of the entire production process.

It is often impossible to inspect every quality expectation of a cellular therapy product. While 100% inspection of parts or widgets is possible in some industries, it is not possible in cellular therapy product manufacturing. It is not possible to conduct a laboratory test that detects adequate assessment of donor eligibility and suitability. When dealing with lot sizes of one, individual product per patient, random sampling is not an acceptable method to determine compliance. Each product must be developed with quality built in as it is manufactured. Testing alone should not be relied on to ensure quality. Compliance with procedures and policies designed to ensure quality is built in is documented through recordkeeping practices. This is often referred to as the "batch record." An audit of records in real time prior to distribution to determine if all steps and requirements are complete, accurate, and fulfilled can serve as a surrogate to demonstrate that the required action steps were taken and therefore quality expectations are met. If the audit finds a lack of compliance, then corrective action can be taken to prevent an unfortunate outcome. If an investigation of a problem-prone area detects an area of weakness where a problem might occur, preventive action may be taken to clarify expectations before an error occurs. As with all other procedures, there should be a written document to explain the audit process and elements. While the quality plan addresses the need to audit and management's desire for audits, an audit plan should be developed so that everyone in the organization can understand the process. The plan should cover the elements of auditing: types of audits, what aspects will be audited, who will audit, the audit schedule, how audit findings will be reported, to whom findings will be reported, and what will be done with the findings.

3.14.6 Focused Audits

Focused audits usually look at a specific step, procedure, or record to determine compliance with the written directions to staff. Examples of focused audits include such things as reviewing temperature monitoring records for completeness, donor eligibility, and suitability records for completeness, or review of critical calculations for accuracy. Just as the name suggests, these audits focus on a single, measurable element.

3.14.7 Process Audits

Process audits evaluate the consistency of obtaining the expected result. Often these audits cross into many independent areas of the organization and review how well the process is controlled when more than one manager is responsible for the final result. An example of a process audit may be to examine the effectiveness of finding suitable, eligible, and compatible donors; harvesting enough cells during collection for a therapeutic dose for transplant; processing the cells in a manner that allows viable cells to engraft promptly.

3.14.8 What Tests Should Be Performed in QC Labs?

Sterility. It is recommended that microbiological testing should be performed on cell banks, in-process intermediates, and the final product(s). The FDA indicates that the 21 Code of Federal Regulations (CFR) 610.12 method should be used. This method is described in the United States Pharmacopoeia (USP) <71>. If an alternative method is employed, its suitability should be determined and must be equal to or greater than the assurance provided by the recommended way. Sterility tests should also be performed at critical points during manufacturing. For cryopreserved products that are not manipulated before administration, the testing should be performed before cryopreservation. If the products are thawed and manipulated, e.g., washed, the sterility testing should be repeated. These should include a gram stain to provide immediate results before administration, a 14-day culture, a provision to report the results to the recipient's physician as soon as they are obtained, and a plan to deal with positive results from the 14-day culture.

Mycoplasma. The FDA identified the sources of mycoplasma infections as any animal serum used during manufacturing and the culture environment, mainly if open systems are used. The FDA recommends testing at a manufacturing stage when contamination is most likely detected, e.g., after pooling cultures but before cell washing.

Adventitious agent testing. The FDA advises consultation of its "Points to Consider in the Characterization of Cell Lines used to Produce Biologicals" and the International Conference on Harmonisation (ICH) Guidance Q5A "Guidance of Viral Safety Evaluation of Biotechnology Products derived from Cell Lines of Human and Animal Origin" in determining what testing may be required for other adventitious agents.

Identity testing. Identity testing is performed to verify that the product is correct and distinguish it from other products manufactured in the facility. This may be accomplished using a variety of tests.

Purity. Purity is defined as relative freedom from extraneous material in the final products, whether or not that material is harmful to the intended recipient or deleterious to the development.

Residual contaminants. The final product should be tested for residual proteins and peptides used during manufacturing and reagents used during culturing and purification, e.g., cytokines, growth factors, antibodies, beads, and serum. The final product should also be tested for cell debris and other immunophenotypes.

Endotoxin. The traditional method for testing for pyrogenicity was the rabbit pyrogen test. This, however, has been largely replaced by the Limulus amebocyte lysate test.

Potency. Potency assays should be performed on the product at all stages of clinical trials. However, by the start of phase 3, the assay should consist of in vivo or in vitro tests that measure appropriate biological activity. This assay must be validated before product licensure. Suppose it is not impossible to develop a quantitative biological assay. In that case,a quantitative physical assay can be used if performed in conjunction with a qualitative biological assay and correlates with it.

Viability. There should be a specified release criterion for cell viability. The FDA usually requires a benchmark of 70%. If this cannot be achieved, then it should be demonstrated that the dead cells do not adversely affect the safe administration of the product or its therapeutic effect.

Cell dose. The minimum number of viable and functional cells to be administered should be specified. It is recommended that the FDA be informed of the maximum dose that has been established based on preclinical experiments.

Product stability. The stability of the product must be evaluated during the early phases of the clinical trial to determine that it will be stable over the entire trial period. The ICH has published guidelines to help with these studies: "ICH Q5C Stability Testing of Biotechnological/Biological Products" and "ICH Q1A(R2) Stability Testing of New Drugs and Products."

In-process stability testing. The stability of frozen products should be regularly assessed.

Final product stability testing. The stability of the final product between the time of product formulation and administration should also be established. This should be done at the appropriate temperatures and at time points consistent with the anticipated holding times. It should include stability during the shipment of the product to other sites, preferably under stressed conditions.

3.14.9 Suggested Design of a High-Level GMP Facility Floor Plan (Partial)

Figure 3.24 presents a simple facility layout. The floor plan shows an arrangement of three manufacturing cleanrooms that are used by staff requiring secondary gowning. The staff directions of travel are shown by the light gray arrows and the relative pressure relationships by the dark gray arrows. The shaded boxes are pass-throughs.

3.14.10 Seed Lot and Cell Bank System

"When seed lots and cell banks, including master and working generations, are used, they should be established under appropriate conditions, including compliance with GMP." Table 3.39 recommends a variety of questions regarding seed lots and cell banks.

3.14.11 Automated Production of ATMPs

"If the output of an automated production system meets the definition of ATMP, the requirements of Regulation No 1394/2007 apply. Accordingly, in the case of authorized ATMPs or ATMPs used in a clinical trial setting, GMP requirements apply."

"The use of automated equipment may ease compliance with certain GMP requirements and may also bring certain advantages in respect to product's quality."

"The manufacturer is responsible for the quality of the ATMP, and therefore, has to ensure the suitability of the automated equipment for the specific intended purpose." Table 3.40 lists recommended questions about automated production.

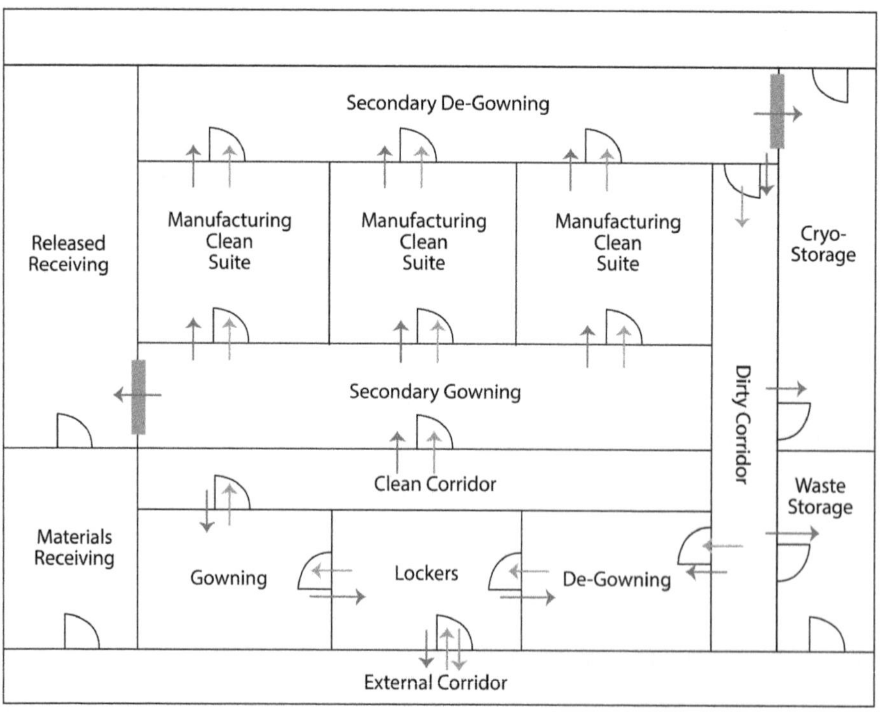

FIGURE 3.24 Facility layout.

TABLE 3.39
Question Examples Regarding Seed Lots and Cell Banks

	Question	Comments	Notes
8.11	Is all handling of cell seeds and cell banks carried out in a GMP-compliant manner?		☐ Fulfilled ☐ Control needed ☐ Action needed [room for notes]
8.12	Has the limited number of cell doublings and passages been defined and has this been specified in the MA or clinical trial?	Suggested documents: • Product specification	☐ Fulfilled ☐ Control needed ☐ Action needed [room for notes]
8.13	Is there documentation available from stages prior to the master seed or cell bank generation (initial sourcing and early development) that could support traceability including used components?	Note: See 8.14 for exemptions.	☐ Fulfilled ☐ Control needed ☐ Action needed [room for notes]
8.14	If the seed lot and/or cell bank in use was established prior to Regulation 1394/2007 (meaning that all information is not available), has an extensive characterization been performed and has this strategy been approved?		☐ Fulfilled ☐ Control needed ☐ Action needed ☐ Not applicable [room for notes]
8.15	Are the seed lots and cell banks stored and used under conditions that prevent mix-ups and cross-contamination?	Note: For example, store in sealed containers, in the vapor phase of LN_2. Guidance documents: ICH Q5D10	☐ Fulfilled ☐ Control needed ☐ Action needed [room for notes]
8.16	Do you have a stock inventory?	Note: For cell stock, it is desirable to split the stock and store splits at different locations in a controlled environment.	☐ Fulfilled ☐ Control needed ☐ Action needed [room for notes]
	Do you monitor storage temperatures?		☐ Fulfilled ☐ Control needed ☐ Action needed [room for notes]
8.16	Do you retain temperature records?		☐ Fulfilled ☐ Control needed ☐ Action needed [room for notes]

3.15 PLASMA PROTEIN THERAPY MANUFACTURING FACILITY AUDIT

Plasma-derived therapies and their recombinant analogs (collectively called plasma protein therapies) are unique among pharmaceuticals and biologics. Their production begins with a biological starting material, human plasma, rather than a synthetic or chemical, which is the starting material for most pharmaceuticals. Each therapy has a unique biochemical profile as due to differences in production and processing methods that lead to differing clinical responses and efficacy among patients. No two plasma protein therapies are exactly alike and no two patients will respond in precisely the same way to a single treatment. Plasma protein therapies are not interchangeable and have been defined by regulators as sole-source biologic products. Therefore no generics or substitutions exist. Further,

TABLE 3.40
Question Examples Regarding Automated Production of ATMPs

	Question	Comments	Notes
17.12 17.13	Can you demonstrate the automated equipment's suitability for its intended use?	Note: If the automated equipment has a CE mark relevant for ATMP production, and the equipment is used as intended the level of effort to demonstrate suitability is reduced.	☐ Fulfilled ☐ Control needed ☐ Action needed [room for notes]
17.15	Did you qualify the equipment according to section 10.1?	Suggested documents: • URS for each piece of equipment • Qualification plan for equipment	☐ Fulfilled ☐ Control needed ☐ Action needed [room for notes]
17.16	Has the manufacturer provided enough information for the ATMP producer to fully understand the functioning of the automated equipment?		☐ Fulfilled ☐ Control needed ☐ Action needed [room for notes]
17.18	Do you have an SOP for each piece of equipment?		☐ Fulfilled ☐ Control needed ☐ Action needed [room for notes]
17.19	Do you maintain the equipment as instructed to ensure function?	Suggested documents • Validation Master Plan • Program for services/calibration	☐ Fulfilled ☐ Control needed ☐ Action needed [room for notes]
17.20	Is there a program for services/ calibration of the automated equipment?		☐ Fulfilled ☐ Control needed ☐ Action needed [room for notes]
17.21	Have you validated the equipment for aseptic processing?		☐ Fulfilled ☐ Control needed ☐ Action needed [room for notes]

because of their biological nature and special handling requirements, manufacturers must work with specialty distributors familiar with and able to meet the storage and handling requirements to ensure safety and efficacy. The complexities of producing plasma protein therapies place it in a unique class of biologics from the starting material through manufacturing and final distribution to patients.

3.15.1 Plasma-Derived Therapies

Human plasma is rich in several proteins. Proteins extracted from plasma, including albumin, clotting factors, immunoglobulins, and alpha-1 proteinase inhibitors, are used to make therapies that treat rare, chronic, often genetic diseases such as hemophilia, primary immunodeficiencies, and alpha-1 antitrypsin deficiency, and acute conditions such as burns and shock. The production of safe, high-quality, plasma-derived therapies depends on the willingness of people to donate plasma. Source plasma can be given at 380 FDA-licensed and International Quality Plasma Program (IQPP)-certified plasma collection centers throughout the US through d plasmapheresis. In Europe, source plasma can be given at 27 IQPP-certified centers regulated by their appropriate national authorities. During this process, whole blood is separated into cellular and other components using specialized equipment called a plasmapheresis device. This sterile, self-contained, automated process separates red blood cells and other cellular components in the blood, which are then returned to the donor. The first plasma donation can take up to three hours and involves completing health screenings, donor education, and donation. Subsequent plasma donations take approximately one-and-a-half to two hours. For the first donation to produce therapies, the donor must make a second donation and both must test negatively for transmissible diseases. Only then can those donations be used to manufacture medicines, and does that donor become qualified. This Qualified Donor Standard is part of the industry's IQPP voluntary standards program for plasma collectors that adds another layer of safety to existing regulatory requirements. US regulations allow a person to donate plasma two times in seven days, with at least two days in between donations. The safety and well-being of the donor and the plasma that is donated are the highest priorities for manufacturers of plasma protein therapies. Therefore donor health and plasma safety are assessed at every donation.

Further, before plasma can enter the manufacturing process, it is held in inventory for 60 days as part of the industry's voluntary international standards program for manufacturers, Quality Standards of Excellence, Assurance and Leadership (QSEAL). This robust standard allows for identifying, retrieving, and destroying of any plasma donation resulting from post-donation information. Given the comprehensive screening criteria for plasma donors, there could be many reasons for retrieving and destroying a plasma donation. For example, a donor may have received a tattoo or piercing that would have disqualified them at the time of the original donation, or perhaps the donor failed to report foreign travel to certain parts of the world that would have disqualified them. Regardless of the reason, the QSEAL Inventory Hold Standard offers a robust and vital quality control measure for producing plasma protein therapies.

3.15.2 Fractionation

The process of producing plasma-derived therapies is called "fractionation." During this process, plasma is pooled, purified, and processed to extract specific plasma proteins that have proven health benefits. The therapeutic proteins are extracted or "fractioned off" from a plasma production pool of multiple donations in a particular order. As part of the fractionation process, the combination of time, temperature, pH, and alcohol concentration allows the extraction of the specific therapeutic proteins. Different purification methods and viral inactivation and removal processes are then used based on the individual properties of the protein classes.

Manufacturers expend substantial research and technology investments to increase the number of proteins extracted from plasma, known as the "yield," and to identify enhancements that can create even more effective therapies. The complex steps and regulations required to collect donated plasma and complete the highly technical, innovative manufacturing processes that ultimately result in a final therapy result in long time intervals between donation and the last product release – between seven and nine months. This sets the production of plasma protein therapies apart from chemical pharmaceuticals and other biologics whose manufacturing processes are much more truncated and whose direct manufacturing costs are a significantly smaller portion of the overall cost. Figure 3.25 presents the plasma fractionation phases.

Critical to the collection phase:

- Physical examination

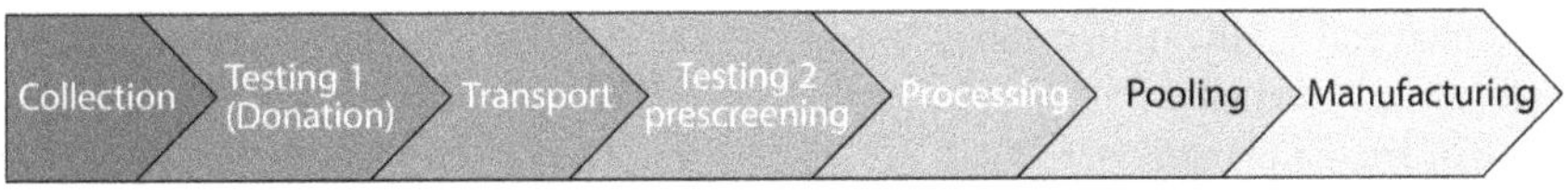

FIGURE 3.25 Plasma fractionation phases.

- Medical history/evaluation
- Donor safety criteria
- Donation volume/frequency
- Anticoagulant/storage requirements

Critical to the testing 1 phase:

- Samples
- Required tests
- Test systems

Critical to the transport phase:

- Transport conditions

Critical to the testing 2 prescreening phase:

- Samples
- Pooling algorithm
- Required tests
- Test systems

Critical to the processing phase:

- Storage requirements
- Look back
- Post-donation reports
- Selecting/sorting

Critical to the pooling phase:

- Required tests
- Samples
- Test systems
- Authority (pool test data)

Critical to the manufacturing phase:

- Virus inactivation
- Assessments
- Environmental control
- Enhanced batch review
- Authority batch release
- Pharmacovigilance

Table 3.41 Presents the differences between GMP and GPG (good practice guidelines).

3.15.3 Steps of the Process under Control

1. Acceptance and verification criteria
2. Freezing of plasma units
3. Storage of frozen plasma

TABLE 3.41
GMP versus GPG

GMP	GPG
1. Pharmaceutical quality systems: pharmaceutical quality system, GMP for medicinal products, quality risk management	1. General principles 1.2. Quality systems 1.3. Good practice 1.4. Quality risk management
2. Personnel	2. Personnel and organization
3. Premise and equipment	3. Premises 4. Equipment and materials
4. Documentation	5. Documentation
5. Production	6. Blood collection, testing, and processing
Good distribution practice (GDP)	7. Storage and distribution
6. Quality control	11. Quality monitoring and control
1. Product quality review	
7. Outsourced activities	8. Contract management
8. Complaints and product recall	9. Nonconformance 9.2. Complaints 9.3. Product recall
9. Self-inspection	10. Self-inspection, audit, and improvements

4. Biological parameters in plasma (QC)
   - Protein content (labile/nonlabile)
   - Microbiological contamination
5. Temperature during the chain of transport

3.15.4 Production

- Prepared in a way that removes cells and cellular debris
- Separated from cells to prevent microbial contamination
- No antibacterial or antifungal agents were added
- Bags registered for blood and blood components storage; sealed to prevent contamination

Table 3.42 explains the clinical applications of plasmas.

3.15.5 Inspection before Freezing

- Volume
- Essentially free of red blood cells (no cloudy reddening visible) and without visible hemolysis
- No clots or other evident abnormalities
- No leakage (gentle squeezing)
- Color (from yellow to light green)

Table 3.43 presents an example of a plasma center audit checklist.

TABLE 3.42
Plasma Proteins for Clinical Use

Plasma Protein	Indications for Clinical Use
Albumin	Restoring of blood volume in trauma, burns, and surgery patients, hypoalbuminemia
Immunoglobulins: Intravenous Subcutaneous Specific	Passive prophylaxis, primary immunodeficiency disease, secondary immune deficiency diseases, Kawasaki's disease, chronic lymphocytic leukemia or HIV infection during childhood to prevent bacterial infections, bone marrow transplantation to prevent graft versus host disease and bacterial infections Immune thrombocytopenic purpura, Guillain–Barré syndrome, polyneuropathies and other neurological disorders
Coagulation factor VIII	Hemophilia A, genetic deficiency
Von Willebrand factor	Genetic deficiency
Coagulation factor IX	Hemophilia B, genetic deficiency
Coagulation factor X	Genetic deficiency
Coagulation factor XIII	Genetic deficiency
Prothrombin complex concentrate	Hemophilia B, anticoagulant overdose
C1 inhibitor	Hereditary angioedema
Antithrombin III	Congenital deficiency, disseminated intravascular coagulation
Fibrinogen	Congenital deficiency, massive hemorrhage
Alpha-I antitrypsin	Hereditary deficiencies, emphysema, and COPD cirrhosis

3.16 RADIOPHARMACEUTICAL MANUFACTURER AUDITS

While writing this chapter, I proudly remembered once again that I served as the lead auditor and consultant in obtaining licenses from the Ministry of Health for five radiopharmaceutical products belonging to TAEA, one of Turkey's proud projects.

Before briefly explaining the audit of radiopharmaceuticals, I would like to clarify that these products do not have long shelf life or storage conditions like other medicinal products due to their half-life. I can divide radiopharmaceuticals into those with marketing authorization and those that will be used in clinical trials. Another critical difference is that they contain radiation, unlike chemical synthesis or biotechnologically produced products. These products are manufactured, analyzed, and shipped in facilities under high protection due to the high radiation risk.

Short-lived radiopharmaceuticals are usually hospital based. An instrument called a cyclotron is generally used to produce products grouped as PET. If a radiopharmaceutical is to be made for humans, it should be expected that this production facility will have the features to allow aseptic fillings. Another important detail is that the protecting personnel from ionizing radiation is a priority.

As you know, radiopharmaceutical products are evaluated as medical drug products. Guidelines to keep in mind for these products are:

- US Pharmacopoeia (USP), European Pharmacopoeia (Ph. Eur.), Japan Pharmacopoeia, International Pharmacopoeia, British Pharmacopoeia (BP)
- EU, PIC/S GMP and Annexes (especially Annex 1, Annex 3, Annex 11, Annex 15)
- FDA guidelines
- ICH guidelines

PET is a medical imaging modality involving a unique radiopharmaceutical drug product containing a positron-emitting isotope. The majority of PET drugs are injected intravenously into patients for diagnostic purposes. Many PET drugs are produced using cyclotrons and other production equipment

TABLE 3.43
Plasma Center Audit Checklist

A – Qualified Donors, Donor Record File (DRF) Review & Donor Privacy

#	Audit Question	Yes	No	Ranking
1	For Applicant Donors, does the center conduct screening including a physical examination completed by a physician or physician substitute in accordance with all applicable regulatory and IQPP screening and testing criteria?			Major
2	Does the center reclassify an Applicant Donor to a Qualified Donor based on the successful passing of the following within the minimum time interval between donations and no later than six months after the previous screening? a) Physical examination b) NDDR check c) Two donor screenings d) Tested non-reactive for two sets of testing for HIV, HBV and HCV (based on all applicable regulatory and IQPP requirements) from donations and/or sample-only collection			Critical
3	Does the center have a system in place to control Applicant Donor units and ensure they are not shipped for use in manufacturing of therapeutic products?			Major
4	If a Qualified Donor does not return for donation within six months of their previous donation, is the donor reclassified as an Applicant Donor?			Critical

Auditor Comments on Section A:

E – Qualified Donors, Donor Record File (DRF) Review & Donor Privacy and Training

#	Audit Question	Yes	No	Ranking
1	Do the center records reflect that the corporate training guide is being implemented and that the records are up to date?			Major
2	Do all center employees, where applicable, have documented annual cGMP and Exposure Control Plan (Biosafety Practices and Procedures) training?			Minor
3	Is there documentation on file (such as a diploma, GED equivalent, college/university transcript or professional license) demonstrating that center employees (with a functional job related to donor screening, plasma collection, product handling or other similar functions) have attained the minimum level of education required in the standard?			Minor
4	Does the center maintain job descriptions for every position in the donor center, and have existing personnel signed the job description associated with their position?			Minor

Auditor Comments on Section E:

at locations close to the patients to whom the drugs are administered (e.g., in hospitals or academic institutions). A cyclotron is a particle accelerator that accelerates charged particles using a high-frequency, alternating voltage (potential difference) to prepare a specific radionuclide through a particular nuclear reaction. PET drug production is a related series of processes and operations that result in the preparation of a PET drug product. Primary functions or steps in PET drug production typically include chemical syntheses, aseptic filling, labeling, and testing. Aseptic processing and sterility controls within the systems are critical processes and controls for coverage in the various designs. PET drugs are intended for diagnostic use and not to provide a therapeutic effect; however, many PET drugs provide their diagnostic development by binding to receptors, a type of pharmacological activity.

The GMP auditor should assess the risk of intense ionizing radiation during the inspection of the PET production facility. Using a dosimeter to estimate how much radiation they will be exposed to would be wise.

Compared to other pharmaceutical product facilities, PET production facilities employ several operators, each with a different task, to deliver the drug to the patient quickly and safely.

If you are aiming for an effective audit as a GMP auditor, you can contact the facility in advance and let them know that you want to monitor real-time activities. Since PET production can be carried out at any time, the GMP auditor may have to schedule the inspection during the earliest hours of the day.

3.16.1 Production Environment

When we scrutinize radiopharmaceuticals, we realize that most human applications are made with aseptic filling. Maintaining product quality is a top priority, as terminal sterilization such as autoclaving is rarely required. I have included guidelines for these activities that should be referenced at the beginning of the chapter. Clean areas are needed to protect products from contamination. Information on what conditions should be provided in cleanrooms is presented in various chapters of this book.

Class C air quality is sufficient in a facility that uses an automatic radiosynthesis module to produce radiopharmaceuticals. Still, Class A environment is required for critical processes such as preparing final vials. A laminar airflow workbench or a pharmaceutical isolator may work to achieve this.

In the radiopharmaceutical production process, the GMP auditor needs to know that the liquid or gaseous products produced in the cyclotron are transferred to special compartments called hot cells. It would be best to look for evidence of a high air evacuation rate, as polluted air must be prevented from entering the production area. You should also see that the air pressure level is appropriate to protect the production area from airborne contamination. The final airlock's negative pressure (–15 Pa) will also cover the surrounding areas against a gaseous radioactivity leak.

Figure 3.26 displays a simple layout.

3.16.2 Radiation Protection

As I have stated, the exposure of personnel to ionizing radiation must be subject to strict controls by the manufacturer. Significant amounts of lead or tungsten shielding are required due to high-energy gamma emissions. For this purpose, many different lead-protected cabinets (hot cells) or isolators are manufactured. The shielding used may vary depending on the activity and structure of the radionuclide. For example, 75 mm shielding for a hot cell can reduce the radiation level by almost 300,000 times. The first measure of personnel protection from ionizing radiation is possible by commissioning fully automatic systems.

3.16.3 Staff and Workflow

In the aseptic filling environment, it should be regarded with the awareness that the primary source of contamination is humans. One way to limit this situation is to wear unique clothing for personnel

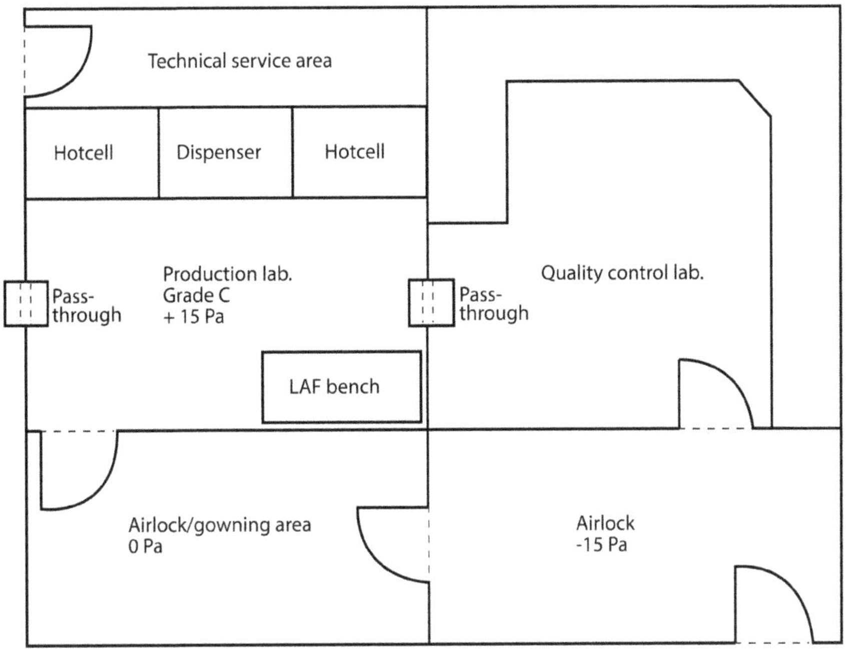

FIGURE 3.26 A simple layout.

entering the environment. Additional clothes should be worn to enter the production cleanroom. Protectors such as cleanroom clothes, shoes, and hair caps should generally be chosen. It is desirable to have a sink in the locker area and an airlock for hand washing and disinfection. As with the principles in other cleanrooms, systems should be in place to ensure that only one door is opened at a time. Monitoring equipment should be available in such areas (locker rooms and airlocks) against the risk of radioactive contamination, and the qualification, validation, and calibration maintenance of this equipment should be done periodically.

3.16.4 Quality Control Laboratories

Establishing a laboratory adjacent to the production site makes sense so that quality control tests of radiopharmaceuticals with short half-lives can be carried out immediately after production. Various equipment and test methods are available for radiopharmaceutical quality control testing. A minimal amount of product is required for each test, thus avoiding radiation exposure to personnel to some extent.

For a test such as pH, for example, a small lead-protected area will suffice. Shielding of HPLC devices may be necessary.

3.16.5 Methods and Equipment for Quality Control of Radiopharmaceuticals

There are two essential reasons why radiopharmaceuticals are tested immediately after production. The first is to prove that the radiopharmaceutical is given to the patient in a safe dose, and the second is to allow the intended PET scan to provide accurate information to the physician. Fluorodeoxyglucose (FDG) is the most commonly used product in radiopharmaceuticals that you may encounter in such a facility as a GMP auditor. PET is the most widely used radiopharmaceutical in cardiology, neurology, oncology, and inflammation. A cyclotron is needed for FDG production. FDG is sterilized immediately after manufacture,diluted to clinically practical doses, and released for intravenous administration by a qualified person (QP) if final control tests meet product criteria and pharmacopoeia requirements.

Requirements for various radiopharmaceuticals are defined in Ph. Eur. and USP. For others, the conditions may be based on production process knowledge. As a GMP auditor, you need to know both the production method and possible deviations in this process to understand the quality control methods for FDG release fully.

3.16.5.1 QC Equipment

Different tests are carried out in radiopharmaceuticals due to their different structures. The quality control laboratory has various equipment, such as an electrochemical detector, ultraviolet light detector, HPLC, GC, TLC (thin-layer chromatography), flame ionization detector, pH meter dose calibrator, osmometer, and endotoxin test module. Given an example of FDG, the test methods to be performed before intravenous administration for human use should be validated and the equipment's DQ, IQ, OQ, and PQ stages should be completed.

3.16.5.2 Aseptic Filling

The sealed sterile vial is assembled under aseptic conditions just before production for PET drugs produced in vials. As a GMP auditor, you must examine the stages of actual output to product filling, including environmental monitoring, and ask the necessary questions. Due to the continuous environmental monitoring of the aseptic filling area, these data should be easily accessible and able to evaluate trends. The GMP auditor may also wish to see evidence of the results of microbiological monitoring during the aseptic process. The GMP auditor should ask whether the requirements of Annex 1 to EU, PIC/S GMP, published in August 2022, are met and closely monitor the site and staff. Although there is no ISO classification for hot cells, it must be verified by a GMP auditor that it is a clean and controlled area. GMP auditor may wish to retrospectively or prospectively examine the production batch record of a selected batch of radiopharmaceutical products. To investigate microbiological data, a representative may request records for several months. In case of any deviation or OOS/OOT situation during the controls, the company should also question what its reaction is. During the batch inspection, evidence of laboratory procedures, test methods, sampling, analytical method development and validation, and stability-related activities may be requested to see a record.

3.16.5.3 Endotoxin Test Module

As is known, when a drug product is not produced under sterile conditions, it may contain sources of contamination such as endotoxin. A GMP auditor should understand that the endotoxin level limit is <2.50 endotoxin units/ml for the release of radiopharmaceuticals. The GMP auditor may also want to see that the validation of the test performed with the qualified Endosafe test module has been completed and reported. The GMP auditor should question whether the horseshoe crab is used in the test called the LAL test. If another alternative test is used instead of the horseshoe crab, it should first be proven that this test gives equivalent results, that the test method is included in the pharmacopoeia, and whether a health authority accepts it. The GMP auditor should also question whether a bacterial endotoxin test is performed on each batch of PET product. Once the endotoxin test has been initiated, the product may be released under control. Still, values must be proven within specifications before the product is presented to the patient.

3.16.5.4 pH Analysis

Most radiopharmaceuticals are given to patients by the intravenous route. While the product is being given to the patient, necessary tests must be completed so as not to damage the blood vessels. A value between 4 and 8 is an acceptable limit, although a neutral solution is generally preferred.

3.16.5.5 Sterility Test

The GMP auditor should inquire whether sterility testing is initiated within 30 hours of production completion. Sterility testing can be completed after the product is placed on the market. The GMP auditor should inquire about how the release is carried out during weekends or holiday periods. In addition, if 30 hours are exceeded, the auditor will want to see the actions taken by the manufacturer and their results.

If the sterility test is outsourced, the GMP auditor will want to see the documents related to this laboratory. The records include information such as the laboratory's license, the quality agreement between the company and the laboratory, and the test method transfer. If the sterility test is performed by the laboratory personnel in the PET manufacturing site, the GMP auditor will want to question their competencies, education, training, and experience.

The PET manufacturer must submit to the GMP auditor documents proving that it has valid and validated methods, qualified equipment, trained and experienced personnel, testing procedures, and required areas to perform sterility testing.

When examining the sterility test, the GMP auditor should start from the sampling stage and evaluate the steps up to the end of the test step by step and ask to see evidence documents when needed.

In case of failure in the first sterility test, the GMP auditor may request records of the actions taken by the company. An OOS/OOT and deviation procedures must be in place for all tests of the PET manufacturer, including sterility testing.

3.16.6 Sterilizing Filters

The GMP auditor must evaluate how the firm determines the integrity of the membrane filter used for aseptic fill. Filter units used to sterilize PET drugs should be subjected to the manufacturers' recommended integrity tests such as a bubble point test described in the application or SOP. The filter should be documented to be intact and not damaged. The GMP auditor must determine that the filter integrity test was performed after the completion of filtration of the PET drug product and before the release of the PET drug for human administration. When sterilizing filters are used in manufacturing PET drugs, it is recommended to perform the following during an inspection:

1. The GMP auditor must evaluate the firm's procedure in response to a filter that fails post-filtration testing.
2. The GMP auditor must determine if investigations were performed following filter failures.
3. If a bubble point test is used to verify the integrity of the sterilizing filter, the GMP auditor must evaluate the operator's competence in performing a bubble point test.

3.16.7 Media Fill

Media fill is critical to the validation to ensure the facility can prooduce sterile PET drugs by aseptic processing. When encountered during an audit, the GMP auditor must perform the following:

1. Determine if a media fill procedure qualifies for the aseptic filling procedure.
2. Evaluate if the media fill simulates the aseptic processing for a routine production run. (The use of vials assembled before actual packing should be evaluated under the worst-case storage period in the media fill study.) Report if the media fill procedures were followed and the results, including:
   - Number of runs performed
   - Number of vials/syringes filled per run
   - Sizes of vials, syringes, and fill volume
   - Media used
   - Positive controls
   - Incubation periods
   - Incubation temperatures
   - Results

3. Evaluate whether the firm investigated the result of failed media fills and instituted corrective actions.
4. Review results of media fill performed; verify that results are acceptable.
5. Evaluate if all appropriate personnel were included in the media fill program.
6. Evaluate the growth promotion testing of media used in the media fill simulations. Conducting a growth promotion test on media-fill media is necessary to verify that the media supports bacterial growth. Commercially prepared media should be used within the label's shelf life and stored according to the label's recommendations. Only qualified vendors should be used.
7. Upon receipt, incoming environmental monitoring plates should be checked to confirm identity and physical integrity. Only qualified vendors should be used.
8. Review the temperatures and incubation time records used to incubate media fill samples.
9. Evaluate whether the microorganisms from positive vials were identified according to genus.
10. Evaluate if the firm determines that such microorganisms correlate to those found during environmental monitoring.

3.16.8 Qualification and Validation

All equipment must be validated before the QC equipment can be used to perform QC analysis to release radiopharmaceuticals. This means that the whole process is controlled, and the qualification of the process is continuously verified. To do this, validation protocols on different levels are written and performed. The validation protocol consists of three topics: installation qualification, operation qualification, and performance qualification (also named IQ, OQ, and PQ, respectively). If the process is validated and approved, the equipment can be used for QC analysis of radiopharmaceuticals.

For PET facilities, the same area or room may typically be organized for multiple operations, such as manufacturing, laboratory, and aseptic controls. The same individuals may be performing various functions.

Personnel operating in the nuclear pharmacy may be conducting batch release or quality operations, which would be regulated. This system includes the measures and activities that provide an appropriate physical environment and resources for producing the PET drug.

It includes:

a) Facilities adequate to ensure the orderly handling of materials and equipment, prevention of mix-ups, and prevention of contamination of equipment or product.
b) Equipment procedures for maintenance, cleanliness, and suitability for intended purposes. Equipment should be properly installed, maintained, and capable of repeatedly producing valid results.
c) Equipment construction and maintenance contact surfaces are not reactive, additive, or absorptive.
d) The firm's adherence to written procedures should be verified through observation whenever possible. These areas may indicate deficiencies in this system and other systems that would warrant more in-depth coverage.
e) Although the cyclotron is included in the definition of a PET drug, the cyclotron used to manufacture the radioisotopes for PET drugs is not to be physically audited due to the complexity of the instrumentation and health hazards. The maintenance records of the target window (including target window rebuild) may be reviewed to identify problems encountered, and corrective actions. Usually, attention is devoted to this issue if a quality problem has been placed.
f) All testing, measuring, and monitoring equipment (e.g., thermometers, thermocouple systems, pressure gauges, pH meters) used in production should be maintained and calibrated.

This would include the automated radiochemical synthesis unit used to produce the PET drug product. Depending on the application and the process, specific production equipment may be single use, such as resin cartridges used in purification. In some applications, the connective tubing may be single use and must be replaced for each production run.

g) It is checked to ensure cleanliness and suitability immediately before use. Records of these checks must be kept.

3.16.9 Equipment

Equipment should be appropriately qualified for its intended use. This includes user requirement specifications, design qualification (if applicable), installation qualification, operational qualification and performance qualification. Equipment and devices, as appropriate, should be calibrated and maintained. Consideration should be given to reducing the risk of product contamination, minimizing the risk of staff radiation exposure, and optimizing ergonomics, facilitate equipment operation, maintenance, and cleaning. Records should be retained.

Equipment maintenance, qualification, and calibration operations should be recorded and the records maintained.

Computerized systems, such as those controlling equipment, should be included in validation.

The dose calibrator (the activity meter) should be qualified using suitable reference standards. If such a reference standard recognized by a national authority is unavailable, dose calibrator manufacturer recommendations or published literature may be used when deciding the appropriate dial setting.

An extensive range of commercial systems is available for the automated preparation of radiopharmaceuticals. Most modern systems are cassette based and thus utilize single-use disposable plastic synthesis cassettes consisting of manifold stopcocks connected to tubing, syringes, and reagent vials. The system manufacturer often provides synthesis cassettes, and can be either sterile or certified as low bioburden. Using these systems avoids any cleaning considerations and facilitates fully automated preparations with minimal radiation exposure to personnel. Finished products should, whenever possible, be transferred automatically from the production hot cell to avoid opening the hot compartment to retrieve the product, with the associated radiation exposure to personnel. This can be achieved by automated removal of the product vial to a lead container or by transfer of the product in a capillary tube to a separate hot cell, which will be the case if automated dispensing of the product is required.

3.17 MEDICINAL GAS MANUFACTURER AUDITS

While writing this book, listing the topics in light of some current events was one of the factors that made it easier for me to do my job. Especially in the chaos created by the COVID-19 pandemic, it is an admirable attitude for the World Health Organization to work on new guidelines by prioritizing patient health. Although one of these guides remains as a draft, the GMP guide on medical gases gains more importance than ever, because, with COVID-19, the oxygen needs of patients and the inspections of these facilities have also been brought to the agenda.

Therefore, urgent actions are required to increase the oxygen production needed. When businesses use industrial oxygen for oxygen production, they must consider all risks because medical gas standards have been set and exceeding this will raise doubts about the purity of the oxygen. Contamination of living and nonliving particles, among other impurities, can put all patients in need of oxygen, including COVID-19, at risk.

Although the EU and health authorities such as the PIC/S and FDA have guidelines for medical gases and even aide-mémoire, WHO has acted knowingly about the sensitivity in this regard and has started preparing to publish a comprehensive guide.

Even though both the EU, PIC/S and WHO guidelines have published guidelines for medical gas manufacturers, it is strongly recommended that these organizations benefit from many other GMP guidelines to ensure efficient and error-free production. For example, GMP for APIs, pharmaceutical water, data integrity, good quality control laboratory practices, and good storage and distribution practices should also be followed.

As GMP auditors, a robust quality management system, one of the first criteria we look for in auditing other fields, is a requirement we expect from manufacturers of medical gas products. We require documentation to prove that all quality management system components have been designed, defined, and implemented, and that top management is involved in all quality activities.

As GMP auditors, we expect senior management to be responsible for the quality of medical gases produced, pass quality control tests, release, store, and distributed, and ensure sustainable quality in the returning of medical gases after use to containers.

As GMP auditors we want to see that there are approved procedures and the necessary documentation regarding complaints and recalls, that the job descriptions of the quality management system, allocation of duties including delegation, and that operations are carried out according to written and approved procedures. We also want to see that the correct containers and labels are used, all risks arising from the production, testing, release, storage, and distribution of medical gases are determined, and necessary precautions are taken.

Additionally, we request that calibration, qualifications, and validations are completed and continuously monitored, that the finished product is released within specifications, that any deviations and OOS/OOT are investigated for the root cause, and that the relevant CAPAs are completed. Finally, we request that the impact and risks of changes be evaluated and that the effectiveness of the changes be measured.

As GMP auditors, we expect the personnel involved in the production, control, certification, release, storage, and distribution of medical gases to have sufficient training, education, and technical skills per national and international GMP guidelines. We also request proof that they have undergone health checkups at regular intervals before they start work and while they are actively on duty. In the GMP training received by the personnel, it should be ensured that they learn about the risks of the product on the patient and the importance of their work. Suppose an outside service is outsourced to maintain or clean cylinders or valves. In that case,the employees providing this service are expected to be at least as qualified as the medical gas manufacturer personnel.

GMP auditors must know that inadequately trained personnel could inadvertently fill the wrong gas into a storage tank or connect the harmful gas to a gas supply system, which can result in severe injury or death.

GMP auditors must review that personnel, including drivers, who connect portable cryogenic containers to gas supply systems are trained appropriately in the specifics of those supply systems.

GMP auditors must know that drivers delivering multiple gases (including numerous medical and industrial gases) are trained to identify and distinguish each gas accurately.

Note: The FDA recommends that cGMP training be provided annually and that manufacturers keep training records that include time and attendance entries.

The specifications created under a GMP-compliant documentation framework must be made for medical gas production, control, storage, and distribution. Both SOPs and work instructions should be designed in an easily understandable language that fully reflects the activity to be performed, sent for review,approved, and put into effect. We may request to see evidence of completion and registration of all required training within a minimum of one week and a maximum of two weeks before to the effective date. The responsible persons should approve documents and invalid documents should be meticulously collected. Only authorized persons should delete, rewrite, and modify documents. Any employee should be able to access the documents with read permission and should not make any changes. There should be parts such as aim, scope, responsibilities, application, document history, and approvers in the documents. Documentation can be managed as paper, electronic, or hybrid. Documents should have a review date, and a reminder should be sent to the document author automatically for documents whose review date is approaching. If this system cannot be provided, it should be ensured that it is operated manually. If there will be any data entry into documents, this entry should be made in a readable, clear, and indelible manner. It should be carried out per the principles of data integrity and good documentation practices. Records should be kept regularly and access should be made only by authorized persons. A labeling procedure covering all labels should be established and used throughout the company. The information on the label must be descriptive, complete, accurate, and legible. The labels on the medical gas cylinders should include information such as the full name of the medical gas, information recommended in the pharmacopoeia, unique lot number, test date, name and address of the manufacturer, warnings, precautions, product code, expiry date, storage, and transportation precautions.

As GMP auditors we want to see that records are available and accessible for every batch of gas produced.

We would like to see SOPs and records for at least the following:

a) Equipment
b) Analytical apparatus and instruments
c) Maintenance and calibration
d) Cleaning and sanitization
e) Personnel matters such as training, clothing, and hygiene
f) Qualification and validation
g) Self-inspection
h) Complaints
i) Recalls
j) Returns

We expect the medical gas manufacturer to have a sampling SOP and this document should contain information such as sampling instructions, qualification of the sampling staff, and sampling methods. We want a procedure for how and by whom batch numbers are issued for medical gases. Suppose the medical gas manufacturer does not use an electronic system and assigns numbers to each gas lot manually. In that case, we request that the numbers be given only once and that a risk assessment be made in case of any confusion in the lot numbers. We ask that all test records be retained, and we particularly want to see the approved release and rejection procedure for finished products. We expect the medical gas manufacturer to classify their equipment as critical, major, and minor; and especially that the calibration, qualification, maintenance, repair, and essential cleaning of equipment is performed by qualified personnel, and records are kept.

We also require the medical gas manufacturer to have a complaint management SOP. In this procedure, we request to see that the manufacturer has a classification for the complaints, how and by whom the complaints will be received, assigns a number to the form opened for the complaint only by the responsible person, and initiates a detailed investigation about the product subject to the complaint.

We also want to determine the possible root cause or causes and whether appropriate CAPAs have been implemented.

We may request documentation to prove that the possibility of recall of the batch, or any other batches that may have been affected, considered. After the complaints are closed, we expect the company to provide an analysis of whether similar complaints occur again.

We require the medical gas producer to have a recall SOP. In this procedure, we would like a detailed explanation of how medical gases can be recalled under what conditions and by whom. If medical gas is recalled, we are investigating the assurance that these records are kept. We expect the firm to simulate recall.

We demand an SOP where the medical gas producer company defines the medical gas returns. If the medical gases are to be returned after they are distributed, the medical gas manufacturer should determine how and by whom this process will be done. The company should have sufficient space for the returned medical gases; and the returned, rejected, quarantined, or released medical gases should be kept in a separate area and have legible identification labels on them. Records of any returned medical gas should be maintained.

We require the medical gas manufacturer to have an approved SOP that details self-inspections, client audits, supplier audits, and health authority inspections.

The procedure should explain how and by whom the audits will be carried out and contain information on how preparations for health authority inspections should be made. The procedure should define the technical and personal characteristics of self-inspections, and supplier audits should be planned as risk-based, and auditors should have them. The procedure should clearly state which departments, processes, areas, and systems the internal audits cover.

We want to make sure that self-inspections cover:

- Personnel
- Premises
- Maintenance
- Equipment
- Production
- Quality control
- Documentation including labeling
- Sanitation and hygiene
- Calibration, qualification, validation
- Release, rejections
- Recalls, complaints, deviations, OOS/OOT
- Previous self-inspection, health authority inspection, client audits records

Whether it is a self-inspection or supplier audit, we want to see evidence that audit findings are written clearly and comprehensibly in the report prepared at the end of the audit.

If qualified external auditors outsource self-inspections, we request the evaluation results of whether these auditors are of the required qualification.

Regarding audits of suppliers or contracted products, services, and analysis purchasers, we want to see evidence that the qualification process has started due to the audit and that the suppliers are evaluated against the list of approved suppliers.

Medical gas production facilities must be designed, constructed, commissioned, and maintained to meet current GMP criteria. As a GMP auditor, we would like to see that the necessary precautions are taken to prevent contamination and cross-contamination, which is the most crucial factor in the design of the facilities, and that cleaning and maintenance are planned in a way that does not affect the quality of the product. The medical gas production facility must have sufficient space for production, quality control, and storage.

Knowing that more than one different gas is produced at the facility, GMP auditors will want to see that there are marked areas for different gases during the audit's tour. It will also state that there should be easily readable status information on each gas cylinder or cryogenic container. For example, "full cylinder," "waiting for inspection," "pending filling," "quarantined," "certified," and "rejected." Indicative expressions such as "ready for delivery," and "empty" must be included in the labels. The GMP auditor also wants to see that appropriate means are used, such as marked floor areas, partitions, barriers, signs, and labels for complete separation. If manual systems are used for product separation, proof documents regarding the risk assessments required to prevent mix-ups may be requested. If an electronic system is used, the medical gas manufacturer must prove that this system has been validated. Cylindrical or cryogenic containers, whether empty or filled, should be stored weatherproof. When medical gases have special storage conditions, the manufacturer must prove that they provide these particular areas and carry out the necessary controls and measurements.

As the GMP auditor, we would like to see that the equipment and tools to be used in the medical gas production facility are selected, installed, qualified, and commissioned per all the operations to be performed; their maintenance and cleaning are performed periodically; and the records of the processes are kept. We expect it to be proven that the placement, installation, and use of both equipment and utilities will not cause any contamination or cross-contamination and that dust and dirt accumulation is kept to a minimum. The most crucial factor to be considered in the design of the equipment is to prove that the medical gas manufacturer transfers the correct gas to the suitable container. We would also like to see if there are cross-connections between pipelines carrying different gases. Let's say that cross-linking is inevitable, in which case we can request proof that there will be no cross-contamination between different gases through qualification and necessary controls.

GMP auditors must review that automatic, mechanical, and electronic equipment, or other types of equipment, are checked according to a written program designed to ensure proper performance.

GMP auditors must review that equipment is qualified at the temperature and pressures used during filling.

GMP auditors must review that manifold valves are qualified for use (e.g., appropriately designed to prevent mix-ups during medical gas filling operations and shown in order to avoid contamination of medical gas).

GMP auditors must review that other valves critical to preventing drug contamination, such as check valves used in filling systems, are qualified for use.

GMP auditors must review that the manufacturer checks the performance of vacuum gauges daily to ensure that the needle on the gauge returns to zero when there is no vacuum or pressure (above atmospheric pressure).

GMP auditors must review that the manufacturer calibrates vacuum and pressure gauges annually against an established standard.

GMP auditors must review that the manufacturer calibrates thermometers according to the equipment manufacturer's instructions at least annually.

GMP auditors must review that computerized systems, including hardware and software are validated for their intended use in manufacturing, processing, and holding medical gases.

GMP auditors must review that automated filling systems are validated to provide assurance that the filling is done properly.

GMP auditors must review that a risk assessment is performed to determine the potential of the computerized system to affect product quality, safety, and record integrity. To address risk, GMP auditors must check that manufacturers use an audit trail as part of these systems.

We can examine whether the manifolds have unique connections during the facility tour. If connections meet different standards in the same filling area, we would like to determine with documentation how and how often the controls are made. During the facility tour, we can examine whether the tanks and tankers are allocated to a single, defined gas quality. If the tanks and tankers are not unique, we should see the precautions the medical gas producer takes against the risk of contamination. Suppose a standard system supplies gas to the medical and industrial gas manifolds. In that case, we may question whether there is a validated method to prevent backflow from the industrial gas line to the medical gas line. We also want to examine the characteristics of the fill manifolds and see the documentation as to whether they are assigned to a single medical gas or a specific medical gas.

To review the exceptional situations in the medical gas facility, and what kind of procedure is applied in these cases? We can request information. We may ask questions about the conditions under which the quality of industrial gas is expected to meet GMP standards. Maintenance, calibration, and cleaning of the equipment should be carried out at specific intervals, and if necessary, preventive maintenance programs should be applied. We can request documents proving that maintenance, repair, calibration, and cleaning are done in a way that does not affect the medical gas quality. Before putting the equipment into use, we may request documents proving that the necessary precautions have been taken against contamination that may adversely affect the quality of the finished product. All records should be appropriately stored and easily accessible. The change procedure should follow any changes in analytical testing,and the methods should be validated.

As GMP auditors, we expect the medical gas producer to have a qualification SOP as well as a validation master plan. We may want to see evidence documents that there are approved plans for qualification and validations and that URS, risk management, and DQ, IQ, OQ, and PQ protocols and reports are available in the qualification processes. We expect qualifications and validations to be based on risk assessment and at least include equipment, facilities, utilities, processing, filling, storage, and distribution. We may request to see evidence documents that records of all qualifications and validations are kept in a secure environment.

GMP auditors must know that medical gases are often produced in enclosed equipment.

During the medical gas production, controls and measures that will eliminate the contamination risks that may arise from personnel or the environment should be defined and implemented. It may be requested to see the proof document that records are kept for all kinds of production data, for each produced cylinder/mobile cryogenic container lot.

It may be desirable to see evidence that records contain, as appropriate, relevant information such as:

- Name of the product
- Batch number
- Identity information of people who perform each production step
- Equipment used (filling manifold)
- Quantity of cylinders/mobile cryogenic containers before filling, including individual identification references and water capacity(s)
- Operations before filling
- Basic parameters required to ensure correct filling under standard conditions
- Results of appropriate checks to ensure containers are filled
- Characteristics of the finished product and results of quality control tests (including reference to the calibration status of test equipment)
- Quantity of rejected cylinders/mobile cryogenic containers with individual identification references and reasons for rejection
- Details of any problem or unusual event and signed authorization for any deviation from instructions
- Batch label
- Characteristics of the finished product and results of quality control tests (including reference to the calibration status of the test equipment) by the person responsible, date and signature

The GMP auditor wants to see that each filled cylinder is traceable to production and filling. The auditor may request documentation to prove that both cylinders and mobile cryogenic containers are controlled, prepared, filled, and stored in a way that prevents mixing. The medical gas producer should know the distinction between industrial and medical gases and prove that it has labeled, color-coded signs or allocated separate areas to recognize this distinction easily. The GMP auditor can request a document showing that the cylinder or mobile cryogenic containers used in medical and industrial gases have not been changed in these areas unless they comply with the medical gas specifications and are made per GMP standards.

Suppose there is continuous production, such as the air separation method in the medical gas production facility. In that case, the manufacturer should provide trend reports and documentation that this production is continuously monitored in quality. Evidence that bulk transfers of active agent gases meet the exact requirements of medical gases should be provided with documentation. Likewise, assurance must be provided that the requirements for filling into cylinders or cryogenic containers are the same as for medical gases.

The medical gas manufacturer must prove that it records each batch of gas transferred to the tankers.

The GMP auditor will want to see that the records contain at least the following information:

a) Name of the product
b) Batch number
c) Identification reference for the tank (tanker) for which the lot is certified
d) Date and time of the filling operation
e) Identity of the person(s) filling the tank(s)
f) Identity of the person(s) performing each key step (e.g., line clearing, receipt, prefill preparation, filling)
g) Reference to the supply tanker (tank), as appropriate, to the source gas as applicable
h) Details of the filling process
i) Equipment used (e.g., filling manifold)
j) Prefilling operations carried out
k) Essential parameters necessary to ensure correct filling under standard conditions
l) A sample of the batch label
m) Specification of the finished product and results of quality control tests (including reference to calibration status of test equipment)
n) Details of any problem or unusual event and signed authorization for any deviation from the filling instructions
o) Certification statement, date, and signature of the authorized responsible person

As a GMP auditor, we can ask questions about transfers and deliveries of cryogenic and liquefied gas.

The medical gas manufacturer must demonstrate that it has designed the cryogenic or liquefied gases according to validated procedures so as not to cause any contamination before and during the transfer from the primary storage. We would like to see documentation of proof that the transfer lines are equipped with a check valve or other suitable alternative. The medical gas manufacturer must document that the flexible connections and connecting hoses and connectors are flushed with the relevant gas before connection. It may be desirable to see documents proving that the transfer hoses used to fill tanks or tankers have product-specific connections. The medical gas producer must prove that the sample has been tested to ensure the quality of the gas delivered so that gas can be delivered to tanks containing gas of the same quality.

The medical gas manufacturer must fill and label cylinders and mobile cryogenic containers according to approved procedures.

Before filling cylinders or mobile cryogenic containers, it may be desirable to see proof that gases have been identified, checked to specifications, and approved for filling. When the medical gas producer produces with a continuous process, it should be able to demonstrate that it has made in-process control to prove that the gas is within specifications.

The GMP auditor may request to see proof that cylinders, mobile cryogenic containers, and valves comply with the relevant technical specifications. GMP auditors may also want to see the color code on the cylinders according to the relevant standards.

Medical gas producers must have documentation proving that cylinders, mobile cryogenic containers, and valves have been checked and properly maintained before first use in production. It must be proven that the controls and maintenance processes carried out do not affect the quality and safety of the medical gas. In the hydrostatic pressure test performed on the medical gas producer cylinders, at least drinking-water-quality water should remain. Valves must undergo visual inspection prior to installation to show that they are not contaminated with any contaminants.

3.17.1 How to Do an Internal Visual Inspection

Conduct an internal visual inspection:

a) When the medical gas manufacturer is new and entering the medical gas service for the first time.

b) After any hydrostatic statutory pressure test or equivalent test in which the valve has been removed.
c) Every time the valve is replaced.

The medical gas manufacturer must demonstrate that it is responsible for the maintenance and repair of cylinders, mobile cryogenic containers, and valves. If the medical gas producer outsources these activities to approved subcontractors, it must be able to prove that it has made the relevant technical and quality agreements. It should be able to demonstrate that the medical gas producer audits organizations that perform such activities. If applicable, a system should be in place for the traceability of cylinders and mobile cryogenic containers.

The medical gas manufacturer must perform a number of tests according to approved procedures before filling:

a) In the case of cylinders fitted with a minimum pressure retention valve, for positive residual pressure in each cylinder.
b) In the case of cylinders that are not fitted with a minimum pressure retention valve, to make sure it is not contaminated with water or other contaminants.
c) Ensuring that all previous batch labels have been removed.
d) The removal and replacement of damaged product labels.
e) A visual external inspection of each cylinder, mobile cryogenic vessel and valve for dents, arc burns, debris, other damage, and contamination with oil or grease; cleaning should be done if necessary.
f) On each cylinder or mobile cryogenic vessel outlet connection to determine that it is the proper type for the particular gas involved.
g) For the date of the next test to be performed on the valve (in the case of valves that need to be periodically tested).
h) On cylinders or mobile cryogenic vessels to ensure that any tests required by national or international regulations (e.g. hydrostatic pressure test or equivalent for cylinders) have been conducted and are still valid.
i) That each cylinder is labeled as required.

The medical gas manufacturer must define a batch for filling operations.

The medical gas manufacturer must demonstrate that cylinders and cryogenic containers returned for refilling are protected from contamination. Validations of these procedures, which include purging and evacuation operations, are expected to be completed. It may be desirable to see documentation of proof that each cylinder or cryogenic container has been properly filled. The medical gas manufacturer must provide documentation confirming that the filled cylinder has been tested for leaks before the gasket or protective device is installed. The medical gas manufacturer must prove that a protective cap is attached to the cylinder valves to protect the outlets from contamination after filling. The GMP auditor may want to see that appropriate labels are attached to each cylinder and mobile cryogenic container. If the medical gas producer is going to produce a medical gas by compressing two or more gases, it must provide the necessary mixture validation to ensure that the gases are mixed in accordance with each cylinder and that the mixture is homogeneous.

The medical gas manufacturer must provide documentation proving that it has tested each batch of medical gas (cylinder, mobile cryogenic container, tank) according to marketing authorization, specifications, and pharmacopoeias.

3.17.2 Sampling

The medical gas manufacturer must demonstrate that it has an approved sampling blade for the appropriate testing of medical gases.

If a single medical gas is mentioned:

a) Filled via a multicylinder manifold, the gas from at least one cylinder from each manifold filling cycle should be tested for identity and assay each time the cylinders are changed on the manifold.
b) Filled into cylinders one at a time, the gas from at least one cylinder of each uninterrupted filling cycle should be tested for identity and assay.

In the case of uninterrupted filling, there should be documentation that the same personnel, equipment, and batch of gas should be used in the same shift.

When a medical gas is produced by mixing two or more gases from the same manifold into a cylinder, the medical gas manufacturer must verify that the gas in each cylinder passes the assay and identity test.

The GMP auditor may want to see proof that the identity test for excipients is performed in one cylinder per manifold filling cycle. Fewer cylinders may be allowed to be tested if the medical gas manufacturer has provided validation of the filling system.

If continuous in-line testing is performed on the mixture to be filled, the medical gas manufacturer must provide documentation proving that premixed gases follow the same principles as single gases. The medical gas manufacturer must prove that the water content has been tested when necessary.

The GMP auditor may want to see documented evidence that the final test on mobile cryogenic containers includes assay and identity tests on each container. If the medical gas producer is going to test on a batch basis, it is expected to prove that it is done only on the condition that the critical properties of the gas remain in each container before refilling.

The GMP auditor will look for evidence that the medical gas manufacturer has followed the precautions taken when mobile cryogenic containers are hot or returned with residual product from the market.

The GMP auditor will question in which cases the sample is taken for testing and in which cases the proof of analysis is sufficient.

The GMP auditor will want to see the records performed manually.

Records must contain at least the following information:

a) Name of the medicinal gas
b) Batch number
c) Reference to specifications and test methods approved during marketing authorization
d) Reference (limits) to test results obtained and any specifications
e) Dates and reference numbers of the tests performed
f) Information about the personnel performing the test - signature, initials, date, etc.
g) Names, departments, signatures and date information of people verifying the activities of people performing the test
h) A clear statement of release or rejection (or other status decision) and the date and signature of the designated responsible person

The GMP auditor will want to see the documents regarding what information should be included if the records are made automatically:

a) Name of the medicinal gas, time and date, and the identity of the person initiating the test. Where access to the sampling and analysis system is controlled, the initials of the person initiating the test may be automatically recorded. The person initiating the test is not required to be part of the Quality Control department.
b) Batch number.

c) Test results, reference to the specification limits, and a statement of passed or rejected.
d) A clear statement of the change of status of the product being tested.

GMP auditors may want to see proof documents that the medical gas manufacturer has adopted the life cycle approach, established continuous improvement programs, and implemented it in areas such as equipment, instruments, facilities, processes, and products.

GMP auditors may want to see evidence documents or records of how the medical gas producer's storage areas are prevented from entering unauthorized persons. Auditors will also want to see that the storage areas are designed in a way that allows the storage of different medical gases. GMP auditors may want to see documents proving that the warehouse areas are properly designed, commissioned, maintained and cleaned, and that the necessary ventilation is provided.

If special storage conditions are required, the medical gas manufacturer must demonstrate that it provides, controls, monitors, and maintains these areas. It should provide documentation to prove how often and by whom the warehouse areas will be cleaned and that the records of the cleaning are kept. It should be able to present supporting documents that necessary precautions are taken for pest control and that necessary precautions are taken against any activity.

It has to submit documents proving that the environment is created to separate the unused or broken and damaged cylinders from the usable ones and that such cylinders are stored separately. The medical gas producer must have documentation proving that it periodically performs stock controls and will take necessary actions in case of conflict.

GMP auditors will want to see proof that the medical gas manufacturer's filled gas cylinders are delivered to customers in a clean and safe manner and are carried in accordance with the conditions specified on the labels; that the identity of the product, batch, and container is protected; and that the labels are legible.

The medical gas manufacturer is expected to demonstrate that it has detailed distribution records to allow for recall when necessary.

GMP auditors may ask questions about how the medical gases are transported in which types of vehicles, and how the vehicles are cleaned and maintained. In addition, it can be questioned which procedures are used for the cleaning of tankers and vehicles and which records are kept.

GMP auditors may inquire what procedures the medical gas manufacturer has for shipping medical gases and how shipping records are kept.

The medical gas manufacturer is expected to provide at least the following information in its transport-related procedure:

a) The identity of the medicinal gas is not lost.
b) There is no risk of contamination of the medicinal gas.
c) Precautions are taken against damage and theft.
d) Environmental conditions are maintained if required.

Table 3.44 offers types of medicinal gas firms.

3.18 AUDITING LAUNDRY SERVICE PROVIDERS

In pharmaceutical and biotechnology companies, employees' clothes can be washed, and sterilized; they are chosen as reusable or disposable. The clothing selection process starts with a URS-like document prepared just before equipment or device is purchased. After deciding whether the clothing should be reusable or disposable, cleaning the clothing begins to occupy the agenda. My purpose here is not to list numerous parameters such as the comfort of disposable clothing, ease of use, and low contamination risks.

As a GMP auditor, there are two points that we need to focus on. The first is that the company wants to clean or sterilize the fabric clothes with an in-house laundry process, or the cleaning and

TABLE 3.44
Medical Gas Firm Categories and Risk Factors Initial Manufacturers

Facility Type	Risk Factors
Chemical synthesizer or processor Maybe original manufacturers of carbon dioxide, nitrous oxide, or helium	• Carbon dioxide manufacturers are at higher risk due to the use of varied sources of starting material, including waste streams from other industries. • Nitrous oxide manufacturers are uncommon, with only a few in the U.S. cGMP issues have included failure to establish specifications for, and control over, raw materials, and incomplete process validation. • Helium manufacturers are uncommon. cGMP issues have included failure to test products for impurities that can be reasonably anticipated based on the origin of the source gas entering the process, in accordance with the USP General Notices.
Air separation units Original manufacturers oxygen and nitrogen	Historically, the cGMP issues have been a failure to validate automated control systems and failure to establish effective Quality Unit control over the release of a product. There is an additional risk factor if the firm installs healthcare facility medical oxygen systems, which requires employee training to ensure that connections are done correctly.

sterilization of the fabric clothes is outsourced. Both cases involve relatively high risks in themselves. I want to focus on leaving fabric clothes at the mercy of a cleaning company by using outsourcing. Pharmaceutical manufacturers do not show awareness to audit suppliers, especially sterile API suppliers, that claim to clean an element with a high risk of contamination, such as clothing. If a service is outsourced, then the audit of the service supplier should naturally come to the fore. So, on what criteria are these companies selected and can they even be promoted to approved suppliers? Auditing such companies with the GMP concept and quality point of view is a meticulous process. Operate the "know the product, the process" logic that I strongly recommend you do before each GMP audit.

How are the clothes prepared for shipping, when we consider, the customer? Is there any coding, identification, or tracking system available? Is a lot number assigned to the clothes? Has the validation been completed if a computerized system is used in follow-up activities? Does size seperate clothes? In which environment and with what kind of packaging are the clothes stored? What exactly is the service expected from the supplier? Is it only to remove dirt, and stains, or at the same time repair ripped, punctured, or torn parts as part of the service?

Does the supplier's dirt, stain, oil, etc., have a system to classify foreign materials? Are newly purchased clothes sent for cleaning before use? Has a quality agreement been signed with the clothing cleaning company? Are the roles and responsibilities of the parties clearly defined in the contract? Does the supplier company personnel undergo special training in sorting, grouping, preventing mix-ups, and prechecking dirty clothes?

What is the first thing to do when dirty clothes come to the cleaning company? What is the repair limit for torn, punctured, and ripped clothes? After how many washes or repairs are clothes marked as unusable? What types of equipment, cleaning agents, and methods are used to wash clothes? Have these methods been validated? How are qualification, calibration, cleaning, and maintenance of washing machines provided? Is there any monitoring, such as of temperature, humidity, air velocity, and number of air changes in the area, where the washing process is carried out? How and by whom are the sequential activities such as washing, testing, drying, and packaging carried out? What type of water is used in the washing process? For example, according to which criteria was the use of pure water determined? Which type of sterilization method was chosen for sterilizing clothes and why?

What are the sterilization conditions by autoclaving or radiation? Is the sterilization process validated?

Under what conditions and how are the clothes that have been repaired, cleaned and sterilized packed? In which type of packaging is the finished garment put? What quality control tests do the clothes go through after they come from the customer? Have the relevant test methods been validated? Is there GMP training for the personnel performing the tests? How is the calibration, maintenance, qualification, and cleaning of test equipment provided? Are final checks done on the packaged clothes? If yes, which controls are carried out and by whom?

How is the shipment carried out? What features are required in the delivery vehicle?

3.19 AUDITING PALLET MANUFACTURERS

Another audit that pharmaceutical manufacturers often overlook or neglect is pallet supplier audits. When evaluated from the GMP point of view, pallets do not have strategic importance such as primary or secondary packaging materials. Still, this approach may be inconvenient for the pharmaceutical company in many cases. Due to 2,4,6-tribromoanisole (TBA) contamination in 2010 and 2011, big pharmaceutical companies such as Abbott, J&J, and Pfizer suffered losses of millions of dollars and had to withdraw their products. The culprit of this event was the pallets used to ship and store pharmaceutical products. You must recognize these suppliers in a world where pallets are so important. As GMP auditors, we know that pharmaceutical companies should use pallets made of light metal materials such as plastic and aluminum and that wooden pallets should not be taken into classified areas. Why? Wood can host many pests, such as insects, microorganisms, rodents, and reptiles. The entry of such pests into a facility where pharmaceutical products are produced, my God, is unimaginable.

Although they are used for change in warehouse areas, wooden pallets must be kept away from these pests. There are several methods chosen for this. The most widely used and common among these are heat-treated pallets. So, as GMP auditors, we need to know what the heat treatment is, how it is done, whether there are any risks, who can do this job, and how to understand whether a pallet has undergone heat treatment.

Let's take a tour of the world of pallets and review the types of pallets, the materials they are made from, and their journey along with the heat treatment.

3.19.1 Shipping Pallets

Shipping pallets are packaging materials that provide support on the pallet during transportation (i.e., marine shipment, land transportation). They are rigid, flat panels that provide stability to the goods during lifting (using a forklift or pallet jack), stacking, and transportation. They come in standardized sizes, with varying designs and materials that suit the needs of the shipping and warehousing service providers. They are only a part of the whole packaging. After the goods have been placed on the pallet, they are further secured by wrapping and strapping.

3.19.2 Materials and Sizes of Shipping Pallets

3.19.2.1 Wooden Pallets

Wooden pallets are the most commonly used shipping pallets. They are best used for transporting furniture, and sealed and non-biodegradable products. They are durable, can handle heavier loads, are cheaper, and quickly repair.

Wooden pallets are the most preferred pallet type in transportation. They are generally preferred because they are durable, can handle heavy loads, have price advantages, and are easy to repair. They have enough friction force to allow the forklift to lift them but they also have disadvantages. In addition, they are expensive and more difficult to transport because they are heavy. Due to the porous structure of wood, it will be a suitable host for all kinds of insect infestations,. Due to its organic structure, it is vulnerable to moisture, corrosion, and chemical and microbial contamination. If wooden pallets are used in warehouse areas of pharmaceutical and biotechnology companies, they must be frequently disinfected and varnished to increase their resistance to contamination.

3.19.2.2 Plastic Pallets

Plastic pallets, preferred by pharmaceutical and biotechnology companies, are produced from polymer-based plastics using a wide variety of production technologies. They are more durable than wooden pallets and are even lighter than wooden and metal pallets. The latter feature provides ease of transportation. The demand for plastic pallets has increased because they are more resistant to odor, chemical, and microbiological contamination. A plastic pallet can be easily cleaned and sterilized due to its non-absorbent properties. Plastic pallets are also produced using recycled plastics.

- Examples of the most commonly used polymeric plastic materials include the following:Polyethylene terephthalate, or simply PET,can be preferred because it is lightweight, affordable, and recyclable. It is resistant to gas, moisture, or liquids.High density polyethylene, or HDPE for short, has a high strength-density ratio and maintains this durability at high temperatures. Another prominent advantage is that it is resistant to impacts and can maintain its integrity in case of accidental fall during transportation or storage. Its chemical resistance is so high that it can be easily used even in the transportation of hazardous industrial chemicals.Polypropylene, or PP. has a more complex structure than HDPE. PP maintains its durability even if it is constantly twisted or bent.

3.19.2.3 Metal Pallets

Metal pallets are often made of aluminum, carbon steel, and stainless steel. Metal pallets are commonly used in shipping military equipment, automotive, food, beverage, and pharmaceutical products. These pallets can easily carry heavier loads than wooden and plastic pallets and are not easily affected by weather conditions. Additionally, due to their structure, they are resistant to chemicals and microbial contamination. For this reason, transportation can be carried out on long journeys without any damage to its structure.

However, they have the most initial expense, cannot be recycled, and are difficult to repair due to the nature of metallic objects. Due to their heavier weight, utilizing these pallets makes transportation costs more expensive.

3.19.2.4 Paper Pallets

Paper pallets are usually made of cardboard or corrugated sheets. They are lightweight, customizable, environment-friendly, and easy to manufacture. However, they have relatively lower strength, which limits their ability to carry heavier loads and provide more protection to the products. They become sensitive when exposed to moisture and temperature changes, and can cause fires when ignited.

3.19.2.5 Pallet Sizes

The dimensions of the pallets are determined and designed to ensure that the materials to be transported do not encounter any difficulties during transportation. Such a standard size determination effort allows for the use of a common language and fair transportation in international trade.Table 3.45 explains pallet dimensions, while Table 3.46 shares pallet types and size information.

There are a number of organizations or associations that determine the dimensions of shipping pallets in various parts of the world.

International Standards Organization (ISO)

Table 3.45 presents the six ISO-approved pallet dimensions, which are used as standard pallet sizes by many regions. They are detailed in the ISO Standard 6780:2003 Flat Pallets for Intercontinental Materials Handling – Principal Dimensions and Tolerances.

TABLE 3.45
Pallet Dimensions

Dimensions in mm (W × L)	Dimensions in inches (W × L)	Region Most Used In
1016 × 1219	40 × 48	North America
1000 × 2000	39.37 × 47.24	Europe, Asia
1165 × 1165	45.9 × 45.9	Australia
1067 × 1067	42 × 42	North America, Europe, Asia
1100 × 1100	43.3 × 43.3	Asia
800 × 1200	31.5 × 47.24	Europe

TABLE 3.46
Different Pallet Types and Dimensions

EURO Pallet Type	Dimensions (W x L × D)	ISO Alternative
EUR 1 or EUR	800 mm × 1200 mm	ISO 1
EUR 2	1200 mm × 1000 mm	ISO 2
EUR 3	1000 mm × 1200 mm	
EUR 6	800 mm × 600 mm	ISO 0, half the size of EUR
	600 mm × 400 mm	Quarter the size of EUR
	400 mm × 300 mm	One-eighth the size of EUR

European Pallet Association (EPAL)
EPAL or EUR pallets are available in sizes that meet the specifications required for export among European Union member countries. EPAL pallet dimensions also meet ISO standards in some cases.

3.19.3 Wood Pallet Treatments

As a result of various risk assessments, it has been revealed that untreated wooden packaging materials can harbor pests that will cause economic and environmental damage. The ISPM 15 standard serves precisely this purpose and recommends measures to minimize the risk of pests.

The risk increases with the transportation of wooden packaging materials from one country to another, from one region to another through various trade channels. Risks become inevitable after a point, as the volume of wooden packaging materials transported by trade is relatively high. The International Plant Protection Convention (IPPC) draws attention to the need to take necessary measures.

The processes defined in ISPM 15 do not provide absolute protection against all wood pests.Still, implementing these measures provides a safer trading environment worldwide where many risks are mitigated.

Today, three main methods exist for eliminating the harmful risks of wooden packaging materials, including heat treatment, fumigation, and methyl bromide under the specified conditions.

- Heat treatment using conventional steam or dry kiln heat chamber. A minimum core temperature of 56°C for half an hour should be finished for this treatment type. This will be performed by exposing the timber to steam or using a kiln-drying heating chamber. Wood that has passed through traditional heat treatment bears the “HT” treatment code inside the IPPC brand.

- Heat treatment using dielectric heating. A minimum core temperature of 60°C for 1 minute continuously should be achieved using microwaves or radio waves. The moisture content, material, length, and density of the timber, and frequency of microwaves and radio waves ought to be considered to fulfill the treatment requirement. Wood that has undergone dielectric heating treatment bears the of "DH" treatment code inside the IPPC brand.
  - Chemical fumigation. Raw wood that will undergo chemical fumigation must meet the minimum concentration time (CT) and final residual concentration throughout its entire profile, which may be located in Annex 1 of ISPM 15. Wood packaging material exceeding 20 cm in cross-section in its smallest measurment should not undergo chemical fumigation.
- Methyl bromide treatment. Treatment time is less than 24 hours with a minimum temperature of 10°C. However, all European member states have banned methyl bromide fumigation because its capability to harm the ozone layer. Wood that has undergone methyl bromide treatment bears "MB" in the IPPC marking.
- Sulfuryl fluoride treatment. The treatment time must be directly proportional to the targeted temperature. If the moisture concentration of a material is above 75% on a dry basis, this process should not be performed. There is an IPPC mark and SF letters on the edge of the treated wood.

3.19.4 GMP Auditor Must Verify That a Robust Temperature Mapping Study Is Applied in Heat Chambers

Heat treatment is achieved by controlling the temperature within the heat chamber. The GMP auditor must know that chamber temperatures required for effective treatment depend on the:

a. Kind and condition of the treatment chamber
b. Volume and direction of airflow through the wood stack
c. Moisture content of the ambient air surrounding the wood stack during treatment
d. Initial temperature of the wood
e. Moisture content of the wood
f. Density of the wood
g. Dimensions of the wood
h. Species of wood being treated
i. Amount of heat being applied to the chamber, which is determined by the heating system used.

The airflow within the closed chamber depends on the:

- Capacity of the equipment in the chamber to move the air
- Dimensions of the wood being treated
- Size of the air plenum
- Degree of separation between pieces of wood within the stack.

3.19.5 Heating Chamber

A heat chamber may be constructed in a variety of materials. The materials used in construction should not affect its operation. An array of heat sources may be used including natural gas, oil, electricity, solar power, and biofuels.

Most heat chambers are used for kiln-drying work by the "fresh air/discharged air" principle. Air is heated and vented into the chamber with the use of fans. The heated air equilibrates in the chamber resulting in uniform chamber temperatures. In many chambers, fans moving the air are located in the ceiling while in others fans operate at one side of the chamber. In either case, the heated air is forced to move through the stockpiled wood.

For heat chambers that have an artificial heat source such as oil or gas, the heat chamber should be well insulated to minimize heat loss, to keep the heat distribution in the chamber as homogeneous as possible, and to ensure consistency between treatments. Use of insulation barriers both in the walls and below the flooring may be required. Insulation may be detrimental for heat chambers that utilize solar heat.

The construction of the chamber may influence the efficacy of the treatment.

Some criteria that should be fulfilled are:

- Generally, heat chamber doors should not be damaged and should seal to prevent leakage of heat from the chamber.
- The chamber itself should be constructed in a manner that minimizes heat loss.
- Airflow should move consistently through the wood stack, and equipment to direct the airflow, such as baffles, should be available and used.
- Fans should be used to circulate the air in the chamber.
- Fans should correspond to the requirements of the chamber and should be working according to the manufacturer's specifications. If more than one fan is used, all should operate in a manner that maximizes airflow in the same direction.
- Venting used in the chamber should ensure a uniform temperature distribution.
- Temperature sensors, including cables, should be in proper working order.
- Valves and motors used to reverse or change air flows should operate suitably.
- Pooling of moisture on floors may be an indicator that the facility is inadequately measuring moisture content, has insufficient air circulation, or other issues that require resolution.

3.19.6 Loading a Heat Chamber

The way a heat chamber is loaded influences the airflow through the wood stack and therefore the location of cold spots in the chamber and the wood located in these cold spots. To ensure proper airflow through the wood stack, the following should be considered:

- The wood stack should be raised off the ground to allow for effective airflow under the wood and to avoid cooling influences from the ground.
- The stack should not be overloaded so as to prevent airflow over the top of the stack.
- The air plenum should contain adequate free space to permit sufficient, uniform airflow through the wood stack.
- The material to be treated should be uniform (e.g., only pallets or only boards) to ensure a homogenous heat distribution. Mixed loads such as pallets and boxes may make it difficult to achieve the recommended temperature and may require multiple temperature sensors to confirm that appropriate treatment has occurred.
- Wood stacks of sawn wood should be stockpiled using spacers or stickers between boards. Spacers must be placed parallel to the airflow direction. Some heat chambers may require special perforated stickers to guarantee the needed airflow.
- In cases where the chamber is not loaded throughout the whole cross-section, baffles need to be installed to guide the airflow through the wood stack. Where baffles are not used, the air will move along the path of the lowest air resistance. In this situation, the chamber operator is likely to underestimate the time required to reach core temperatures, since the chamber is likely to heat much faster than the wood.

3.19.7 Air Circulation

Air circulation fans help to ensure controlled movement of heated air within the chamber. Airflow can be measured with the use of anemometers. These may be fixed units monitored by chamber systems or hand-operated units that record air flows infrequently to determine if fans are operating within desired parameters. A minimum airflow of 0.5 m/second (100 ft/minute) is recognized as essential for normal chamber operation.

Fans should be installed to ensure airflow in a common direction. Airflow reversal during treatment helps to ensure uniform heating on all sides of the wood, but should not be a requirement. Airflow reversal ensures that wood on both sides of the chamber receives heated air at its maximum temperature.

3.19.8 Venting

Venting of the heating chamber may be used to exhaust excess moisture released during treatment. However, early in the heating process, it may be desirable to retain moist air to aid in the temperature rise of the wood stack resulting in a reduction in the total heating time.

3.19.9 Humidification

The moisture content of the wood affects the wood's ability to heat. The moisture within the wood rises to the surface, as the wood is heated, cooling the surface and requiring longer heating. Effective heat treatment is therefore dependent on a number of properties that influence the moisture content of the wood such as the:

- Thickness of the wood
- Density of the wood
- Direction of the grain of the wood (wood is more permeable in the longitudinal direction)
- Structural irregularities of the wood

Humidification systems using steam injection or units that spray water into the heating chamber may be helpful in ensuring effective heating as air passes through the wood stack. Treatment schedules should account for variations in the thickness, density, and initial moisture content of the wood being treated. For example, treatment times for wood of higher densities or greater thickness should be longer than treatment times for less dense or thinner pieces of wood.

3.19.10 Proper Treatment of Wood/Wood Packaging Material

Treatment schedules may be regulated by automatic or semiautomatic systems that monitor temperatures and humidity within the chamber. More basic kilns require monitoring sensor data typically collected on a data recorder. Sensors should be routinely calibrated by independent testing authorities or others per the manufacturer's specifications. This is necessary to verify that the system's operation is consistent from one treatment to another and within defined parameters of sensor accuracy. Simple calibration of sensors may be accomplished using water baths of different temperatures (including those likely to be present during the treatment) and a second precalibrated temperature measuring device. Variations in sensor accuracy should be accounted for in treatment procedures such that any deviation is negated by appropriate changes to the duration or final heating temperatures required during treatment. For example, sensors may vary 1°C–2°C once calibrated. This variation could be included when developing temperature and time combinations used for treating the wood. For example, a slight time or temperature adjustment to the schedule can be used to ensure that regardless of the known extremes of variation in a given sensor, the wood reaches and maintains 56°C for at least 30 continuous minutes throughout the profile of the wood. Although this may result in overtreatment of some pieces of wood, producers can be certain that all wood has reached the phytosanitary requirements.

3.19.11 Heat Chamber Controllers

Heat chamber controllers are computer systems **Should you ask questions about the validation of these systems?)** that respond to temperature sensors, wood probes, and other kiln equipment to ensure that wood treatment follows the chamber operator's specifications. Controllers will automatically close baffles, reverse fans, etc. in response to timed events or to maximized treatment conditions. Most heat chamber controllers are located in a building adjacent to the chamber and are capable of identifying problems in chamber operations, and either notifying the operator or resolving the issue independently. For example, some sophisticated controllers will restart treatments if a malfunction occurs (e.g., power failure, faulty sensor, etc.). Chamber controllers also record treatment data (**Should you ask questions about data integrity?**) to verify that treatments have been completed according to the operator specifications. Where facilities use published schedules to achieve treatment, the schedules should provide guidance in dealing with equipment failures. If these schedules do not provide guidance, the treatment should be restarted once the equipment is repaired.

3.19.12 Temperature Measurement

Facilities vary in the approaches used to measure temperature during wood treatment. Some heat chambers use sensors (**Should you ask questions about the calibration of these sensors?**) inserted into the wood to measure core temperatures during each treatment. Others measure chamber air temperature, relative humidity, cold spots within the heat chamber and other factors to estimate the wood core temperature. This latter system bases the treatment temperature of the wood on calibration testing carried out during initial replicated verification tests of core temperatures as compared with chamber temperatures, humidity, and other factors. The initial series of test treatments uses sufficient temperature sensors placed into the wood at various points throughout the chamber (including and particularly to determine the cold spots). These sensors are inserted into the core of a particular wood species of a specific dimension. The temperature curve of the sensors is then compared with the rates of change in chamber temperatures, relative humidity, etc. to establish a "heating curve" based on these factors. (**How are these temperature curves managed?**) Future treatments can then be performed by measuring more easily obtained factors such as chamber temperatures, relative humidity, etc. provided no changes to the operating conditions are made, including changes in wood species, dimension, initial moisture content, initial wood core temperate (to adjust treatment times for frozen wood for example), etc. **What are the critical process parameters? What are the critical material attributes? How about process validation?** Other facilities may utilize established time/temperature schedules published in research documents that recommend specified ambient air temperatures, relative humidity curves, etc. for a particular species and dimension of wood. These schedules often overtreat the wood to account for variations in chamber type, operating conditions, etc. but achieve the minimum core temperature and time requirements.

Temperature recording devices may vary from simple physical chart recordings of temperature to elaborate systems that utilize computerized programs and data loggers. Records may be paper charts or, increasingly, computer databases that record treatment information electronically. Records of sensor readings during treatments should be maintained for review by the national plant protection organization (NPPO) or designated authority for some time consistent with the period in which the treated wood is to be utilized in international commerce (e.g., one year). Measuring and recording systems should be calibrated regularly (e.g., annually) by recognized individuals (including manufacturers) or organizations per the manufacturer's specifications.

3.19.13 Temperature Sensors

At least two sensors should be used where the heat treatment is determined based on temperature sensors inserted into the wood. **How was this number (at least two sensors) determined?** They should be placed in wood located in the chamber's cold spot. **How does a GMP auditor know that point x is the coldest area in the chamber? Did the supplier perform any risk-based temperature mapping?** The wood piece where the sensor is inserted should be the biggest one located furthest from the heat source, as this needs more time to heat throughout the whole profile.

Using multiple sensors ensures that any mechanical failure in a sensor during treatment is detected.

This should be used for heat treatments without moisture reduction and during kiln-drying processes that include ISPM 15 treatments. As the target criterion for ISPM 15 is the temperature, other measurements such as wood moisture content, do not confirm of treatment. **Why?** If the airflow in the chamber is routinely reversed during treatment, an increased number of sensors needs to be used to account for a change to the location of the cold spot or the presence of multiple cold areas.

3.19.14 GMP Auditor Must Verify Temperature Sensors Are Regularly Calibrated

Both the chamber and wood temperature sensors need to be calibrated regularly. From a technical point of view, it seems to be reasonable that the calibration should be carried out at least once a year. Generally, the complete measuring chain (sensor, cable, data logger, etc.) has to be calibrated, not only the isolated sensor. Calibration has to be carried out in accordance with the manufacturer's specifications, guidance from approved calibration and testing companies, or using procedures approved by the NPPO. The calibration should include at least three tests of temperatures to establish a calibration curve. The temperatures used during testing should represent temperatures used during the treatment process (e.g., 20°C, 56°C, and 80°C). Ice or boiling water may not be appropriate to develop a calibration curve that represents the operational temperatures of a sensor in use.

3.19.15 Wood Temperature Sensors

A sufficient number of core temperature sensors should be used to measure and record the wood temperature. The use of 5 to 13 sensors is recommended by some NPPOs during the initial approval (testing) process of a facility. **How are the numbers 5–13 chosen?** The size of the heating chamber; the species, density, and sizes of the wood being treated; the source of heating; the size and number of the cold spots within the chamber; fan speeds; or other factors will influence the number of temperature sensors necessary to be certain that wood has been treated effectively. The use of sensors enables the performance of the chamber to be evaluated at a number of locations and thereby determine the cold spot. To effectively heat treat all the wood in the stack, wood located at the cold spot must reach and hold 56°C for a minimum of 30 *continuous* minutes. Wood located in other areas of the chamber should achieve 56°C for a minimum of 30 minutes earlier during the treatment process. **The use of multiple core temperature sensors is not necessary for each treatment once the chamber has been calibrated**. Nevertheless, a minimum of two sensors should be used, so that the failures of one sensor would be discovered immediately. The use of a sensor inserted into the core of the wood of the largest thickness, positioned in the cold spot, or several sensors placed at the cold spots will provide ongoing assurances of achieving phytosanitary requirements.

3.19.16 How to Measure Temperatures in the Cold Spot

Often the airflow within the heating chamber is irregular as a result of the location of the wood stack, variations in individual fan speeds, cracks or leaks in the chamber walls or doors, or other factors. These can result in the ambient air temperature being less than uniform within the heating chamber but uniform from treatment to treatment. Schedules should account for areas in the chamber where the wood is slow to heat to the prescribed temperature. This may be achieved by placing temperature sensors specifically in the cold spot. The cold spot may also be influenced in size or number by the species, dimension, and density of the wood being treated.

3.19.17 IPPC Marking on Wooden Pallets

Approval and handling by valid suppliers are regulated and inspected by the NPPO of the importing or exporting countries. An IPPC mark must be added to the wood treated by the provider to certify that it is subject to the procedures specified by ISPM 15.
So what kind of elements does this sign consist of?

- Corporate logo of IPPC
- Country code consisting of two letters determined by ISO (XX)
- Manufacturer code provided by NPPO to treatment provider (000)
- Code for treatment (YY)

Figure 3.27 displays the IPPC logo and heat-treated marking on a pallet example.

The image provided here is an image belonging to the IPPC and its use in different formats is allowed, provided that it contains all components.Pallets that have been treated in accordance with ISPM 15 standard requirements and have not been repaired or modified can be used without recycling. However, if one third of the wooden pallet is replaced with new wood during the repair, then the pallet is considered newly produced and subjected to treatment.

3.19.18 What Does the IPPC Logo Tell Us?

Additional information includes the number of the pallet manufacturer's certification number and the name of the regulator responsible for heat-treating the wood. The pallets also have two letters that indicate the method of treatment: "HT" indicates wood that was heat treated; "MB" refers to fumigation using methyl bromide.

Although other manufacturers may add additional details to heat-treated pallets, the ISPM 15 authentication marks are sufficient to prove that they are safe for use in cross-border shipping. You

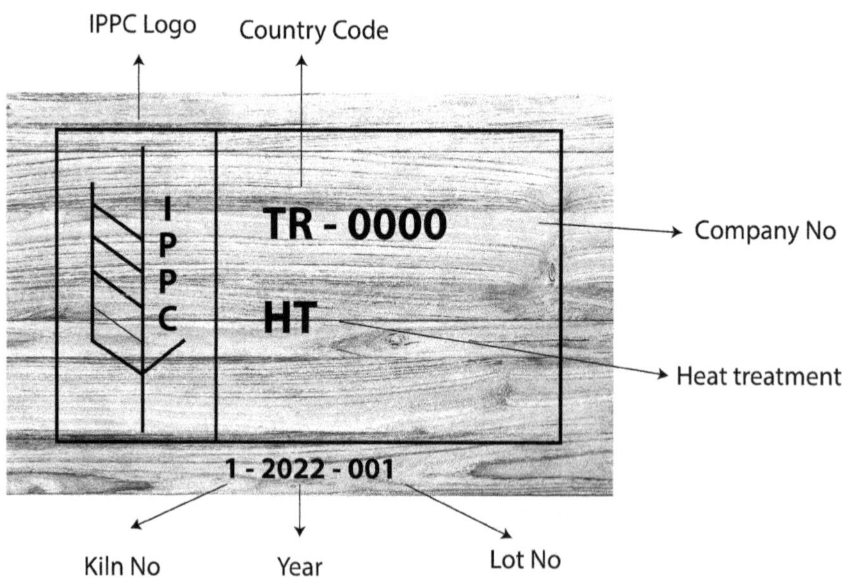

FIGURE 3.27 IPPC logo and heat-treated pallet example.

will be denied entry to most countries if you don't comply with the requirements. It is important to only purchase pallets from reputable companies. These logos are important to look for when using wood pallets to ship products domestically or internationally.

3.19.19 How to Clean Non-wooden Pallets

If the pallets used by pharmaceutical and biotechnology companies are plastic, metal, or paper-based, they are exempt from the regulation and treatment is not required. The pallets on which food, beverages, medicine, and agricultural products are transported and stored must be checked regularly and the necessary treatments must be completed.Cleaning methods such as high-pressure washing (What type of water is used, at which temperature? What does high pressure mean? Is this method automatic, semi-automatic?), power washing, and manual washing are employed in plastic pallets.

3.19.20 Verify That Plastic Pallets Are Used in Cleanrooms

The use of plastic pallets is quite common in cleanrooms in pharmaceutical and biotechnology companies. When it comes to a controlled environment, they have an advantage over all other types of pallets, including wood. Even though wooden pallets have been treated, they are sensitive to chemical and biological dirt and moisture, and these disadvantages pave the way for the formation of bacteria, yeast, mold, and fungi. Another disadvantage of wooden pallets is that they are a source of contamination such as sawdust and splinters. When the inability to clean and disinfect properly is added to all these disadvantages, the answer to why plastic pallets are preferred is found. The use of metal pallets is permissible only if certain specific conditions are taken into account in the risk assessment. Plastic pallets continue to be a priority not only for cleanrooms but also for transporting products in cleanrooms. Demands are increasing due to their easy-to-clean, non-absorbent, and easy-to-sterilize structure. When evaluated holistically, plastic pallets become extremely important in preventing contamination not only in cleanrooms but also in the entire supply chain.

3.19.21 Methyl Bromide Poses High Risks

Methyl bromide is the chemical used for wood packaging fumigation. Bromomethane is also known. This chemical was used as a pesticide fumigant and refrigerant. It is used in many different industries and products around the globe. Although it was once more widespread, the usage of methyl bromide is now much less common.

Exposure to this chemical can burn skin, damage the lungs, irritate eyes, damage kidneys, and cause negative neurological effects There are also short- and long-term health issues for humans. Ozone depletion can also be caused by methyl bromide. This allows for more UV radiation from the sun to reach our planet's surfaces.

In 2005, methyl bromide was banned in all applications of the European Union. It was allowed to be used in shipping sterilization and quarantine, but it was banned completely in 2010. The United States Environmental Protection Agency also eliminated methyl bromide in 2005. But, it has not been banned from use in shipping cases.

3.19.22 Verify Refurbished Pallets Are Also QC Checked

For refurbished pallets to be suitable for pharmaceutical transportation, they must meet the proper controls. These standards involve avoiding chemical treatment, handling raw materials, and pallet

TABLE 3.47
Comparison of Different Types of Pallet Materials

	Pros	Cons
Wood	• Lowest up-front cost • General purpose pallet • Varying strengths • Absorptive	• Splinters and nails easily break loose • Jagged edges can damage the flooring • Absorptive nature allows for bacteria growth
Plastic	• Highly durable • Will not splinter • Easy to clean • Nonporous • Resists contamination build-up • Long-term cost savings	• Low friction between load and pallet • Burns hotter than wood in the event of a fire
Metal (aluminum)	• Durable pallet type • Unlikely to splinter • Easy to clean • Resists contamination build-up • Long-term cost savings	• Light • Expensive • Hardness of metal can damage flooring
Stainless steel	• Strongest pallet type • No additional coatings are needed overtop stainless steel • More durable than aluminum, plastic, and nearly any other pallet material	• Like aluminum, stainless steel pallets are very expensive • Extremely heavy • Strictly for internal use; transfer loading is required • Conductive to electricity • May scratch or damage floors

components properly to prevent cross-contamination, and thorough inspections. They undergo inspection when they come in and throughout the disassembly, reassembly, and repair process. Any contaminated pallets are set aside for cleaning. They must be clear of contaminants such as grease, chemicals, dirt, and insects. They will also go through the ISPM 15 process, which involves heating the wood to a temperature that kills potential pest infestation.

3.19.23 Verify Mold Growth Is Always Prevented

One of the primary reasons that mold is unacceptable in pallets containing pharmaceuticals is that buyers will likely reject the products. Therefore, mold on your pallets is simply a bad practice. It is best to dry the wood pallets to a moisture content under 19%. Hardwood pallets can be air- or kiln-dried after their assembly to reach the desired moisture content – one that will be too low to allow any mold to grow. A company can also treat the wood with a mold inhibitor to further prevent mold growth. Table 3.47 compares different pallet materials.

3.19.24 GMP Auditor Must See Where Wooden Pallets Are Stored

Another helpful way to ensure wooden pallets will be acceptable for pharmaceutical transportation is to store them properly. The environment needs to have a dry, well-ventilated atmosphere to prevent moisture exposure. Store the pallets indoors as much as possible, and make sure the space is free of pests. Make sure the pallet providers use clean trailers and flatbeds when delivering the pallets.

CASE STUDY

Which of the following marks are in accordance with ISPM 15?

Yes ☐ **No** ☐

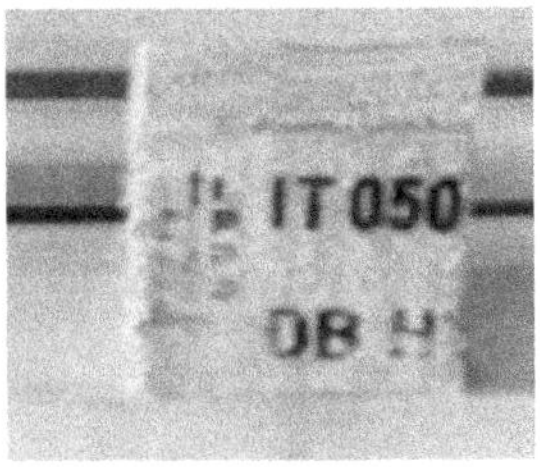

Yes ☐ **No** ☐

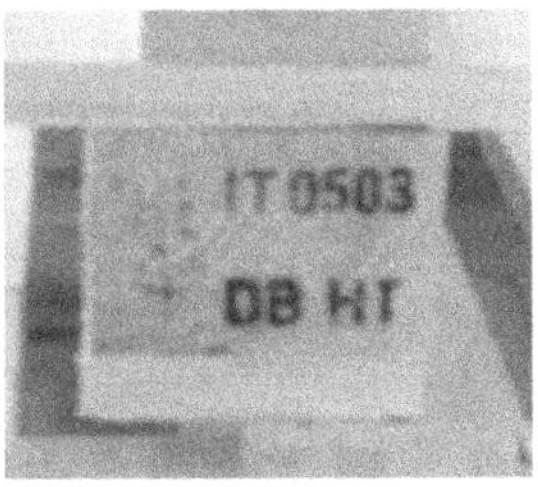

Yes ☐ **No** ☐

Yes ☐ **No** ☐

Yes ☐ **No** ☐

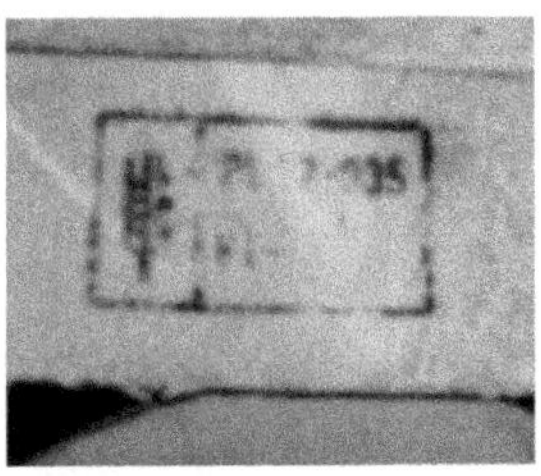

Yes ☐ **No** ☐

3.20 AUDITING SOFTWARE DEVELOPERS AND SUPPLIERS

A partnership based on trust and cooperation between pharmaceutical and biotechnology companies and software suppliers, which are subject to the health authority, is one of the biggest problems encountered. From the perspective of pharmaceutical and biotechnology companies, there are severe doubts that the causes of the problems are with the supplier, not themselves. In contrast the opinion that the issues in the software supplier are overstated by the pharmaceutical companies, which are their customers, is common. The effective supplier involvement advocated by the GAMP guideline is unfortunately not viable when it is a one-sided approach. Supervising the software supplier is one of the effective ways for the pharmaceutical and biotechnology companies to evaluate the improvement opportunities in their favor and to decide which activities will be carried out only on the supplier's side without repetition. Of course, this approach will also remove the obstacles to successful results intensively studied projects. Both pharmaceutical and biotechnology companies and software suppliers should read the newly released ICH Q9 and GAMP. With the transition from manual to electronic systems in the pharmaceutical industry, the importance of software suppliers has begun to increase. However, while the significance of computerized systems and software is starting to be understood, it has been determined that the link between suppliers and user operators could be more robust. The principle of honesty, which is the right thing to do, should be adopted very clearly, especially on the supplier side. As it is known, health authority guidelines and directives have been advocating and even mandating the auditing of software suppliers for many years. Effective communication can often improve collaborative approaches between a software supplier and a client pharmaceutical company.Supplier audits are the basis of concrete and constructive cooperation for both parties. However, pharmaceutical and biotechnology companies acting as customers and software companies acting as suppliers may have some prejudices for justified reasons.

It is in the interests of both parties to answer some questions before planning the audits.

- How well are the software and authority requirements and expectations understood by pharmaceutical and biotechnology companies?
- Can it describe precisely the service it needs?
- How close are software suppliers to their customer's expectations and demands?
- How aware are software suppliers of health authority expectations?
- Is the risk-based approach on the agenda of both parties?
- Do pharmaceutical and biotechnology companies see software suppliers as fraudsters who constantly manipulate and ignore risks, or as solution partners or even subject matter experts who will contribute to their processes?
- Can both parties focus on the content rather than the form of the subject service?

Both parties must ensure that the software system meets the expectations of the health authority; provides operational benefits; and passes the necessary tests, works, and all validations.

For one thing, pharmaceutical and biotechnology companies should be aware that no supplier can meet the needs of their customers alone. Responsible to the authority, the pharmaceutical company can remedy deficiencies and weaknesses in the software supplier with the full support of a GMP-compliant quality management system. However, when it comes to the design and implementation of the system, sometimes this balance fail to be achieved. The first and primary expectation of the company that buys the software is that the relevant software meets all legal requirements, and it is a big mistake to think that the recognition of the software supplier in the industry and the fact that it has many customers is the key to its success. The pharmaceutical company should define user requirements in a clear, precise, understandable, and validated way for the URS. Unfortunately, the software's impact on the system is not up for debate. From the perspective of pharmaceutical and biotech companies, attempts to purchase GXP-compliant software often fail because they have to deal with immature and non-expert suppliers. Such a situation mainly occurs within the framework of the quality management system of the customer company in software supplier audits, and the supplier audit comes to the point of teaching the supplier its job. Such cases bring additional costs to pharmaceutical and biotechnology companies.

From the health authority perspective, health authority requirements and guidelines for the pharmaceutical and biotechnology industries are constantly being updated, and expectations increase with each update. Patient safety comes first and both parties must understand both product safety and data integrity principles. However, we see that the software suppliers interpret the authority expectations according to themselves and transfer these approaches to the software they sell. Software suppliers misunderstand the innovations in the guide and legislation due to a lack of education and f quality awareness, and many problems have arisen, primarily due to the misunderstanding of 21 CFR Part 11, EU GMP Annex 11, and GAMP 5. On the other hand, companies that are accountable to the authority often make unreasonable demands from their software suppliers due to their lack of resentment. For this reason, software suppliers need to understand the changes in the guidelines and directives and reflect them on their products. Pharmaceutical and biotech companies should also maintain expertise in procedures.

When we consider it in terms of risks, risk assessments must be applied both in the pharmaceutical company, which is the customer, and in the software company, which is a vendor or supplier.

When asked typical risk management questions, the answers should intersect with patient safety, product quality, and data integrity for both parties. So, software companies should also support their customers with software architectures, in their areas of expertise. The pharmaceutical company is responsible for system security, product quality, and data integrity.

As a GMP auditor, I recommend you thoroughly read the ASTM E2500 standard.

What do we evaluate in terms of tests?

Reviewing the design of the software is the first verification step. It should be verified by the software supplier that the developed software meets all the features of the design and is robust. Then, the client company will make its evaluations. The time and effort spent on testing activities can lead to the perception that nothing will go wrong. Pharmaceutical companies, too, can only expect the software supplier to perform 100% non-reflective testing when it inadequately implements its risk management or fully understand its processes. If the test strategy is not based on reliable tests, detailed tests may result in success during the software commissioning phase.

As GMP auditors who will audit software suppliers, I have a reminder for you. No user, no matter how smart they are, can test a software system and its manufacturer. This rule has never changed, and it doesn't change quickly. That's why you should look at the testing strategy in a GMP audit, and question the quality of the test records. You must thoroughly absorb the supplier's overall testing strategy to conduct a meaningful audit.

So, it is essential to have an approach for selecting vendors and performing quality systems, data governance, and data integrity audits of vendors who supply hardware, associated software, and supporting services for GMP systems.

GMP system selection requires carefully reviewing specific requirements (such as user, functional, specification, and data integrity). The user requirements for an off-the-shelf or configurable system can further define the GMP applicability. Custom applications, either internally or externally developed, may require an iterative process to establish the user requirements before the finalization of the software design. It is advised to perform a gap assessment to evaluate the proposed GMP system to ensure alignment with 21 CFR Part 11, EU Annex 11, and the applicable GMP regulations and guidance. If electronic signatures are used, they must comply with the regulations; please see the full audit manual for specific details. The gap assessment should consist of core questions with both a yes/no structure and a description of the reasoning that will determine the criticality assessment.

3.20.1 Defining System Criticality

Questions that may be useful in defining system criticality may include, but are not limited to, the following:

High criticality proposed system considerations

- Systems to maintain records used for release decisions
- Systems to maintain records of process steps
- Systems that control specifications
- Systems that hold copies of manufacturing records
- Systems electronically linked with another critical GMP system, including but not limited to:
- Systems that will transmit or transfer data to other GMP systems
- Lab execution system (LES), lab information management system (LIMS), manufacturing execution system (MES)
- Systems to manage enterprise resource planning (ERP), including but not limited to SAP, JD Edwards, and Oracle
- Systems to manage and document GMP processes including but not limited to:
  - Data/document control
  - Training
  - CAPA
  - Deviations/investigations
  - Change management

Medium criticality proposed system considerations

- Systems used for in-process testing
- Process capability modeling and simulation
- Publishing software

Low criticality proposed system considerations

- Systems electronically linked with another noncritical system, such as software used to compile information for management review
- Systems used for information only (FIO) data management and storage
- FDA electronic submissions gateway
- Web interface software

The selection of auditors to audit a GMP-applicable software supplier should include personnel with a clear understanding of the software validation process, testing, and documentation. This does not necessarily require an IT expert, but an ability to ensure the software life cycle used by the Supplier is robust and the processes to evaluate changes will also minimize issues in other modules of the applications. Based on the specific application(s) or services provided, the lead auditor should consider the requirement for an SME to assist with audit execution.

These documents serve as a preliminary vendor evaluation.

3.20.2 Audit Preparation

1. Document the rationale for the Supplier's risk level and type of audit (i.e., on-site or virtual).
2. Auditors should be selected based on their knowledge and expertise of the system being evaluated.

3. Ensure the audit team has the appropriate documented education, Training, and experience for the scope of the audit.
4. Establish the appropriate supplier or service provider (S/SP) quality contact through the project manager or outsourcing.
5. Confirm the requirements for a confidentiality disclosure agreement (CDA), nonDisclosure agreement (NDA), or quality agreement.
6. Determine materials/services/systems that will be within the scope of the audit.
   ** If She supplier has multiple sites, determine the systems/services that will be purchased from each site, and that will be within the scope of the audit. Note if these are different sites, determine whether a corporate or local quality system governs them.
7. Select audit date(s) with the S/SP and agreement on the number of auditors. Communicate the materials/services/systems that will be within the scope of the audit.
8. Open the audit record through the appropriate quality system (i.e., Trackwise or paper-based). The audit preparation and execution should follow the auditing firm's procedures and policies.
9. Perform a search of all deviations/changes/complaints/incidents related to the Supplier's material/service/systems.
   **For a time-based surveillance audit, review the last audit report. Review any observations from that report and verify that the S/SP has completed the corrective action plans.
10. Develop an audit agenda,provide it to the S/SP within an appropriate time frame before the audit, and request an update of any previously completed corrective actions.
11. Request vendor documentation such as certifications, SOP index, organizational chart, and any other documentation the S/SP will provide.

A risk-based approach should be followed when determining whether to use an on-site or remote (e.g., web conference) audit. There are different factors to consider when deciding between conducting an on-site or remote audit. The first factor is based on the criticality of the system(s)/service(s) to be purchased. The second factor is the availability of the documentation to support the audit (i.e., on-site only or electronically provided).

If the criticality of the system is low or medium and the vendor can provide all required audit documents electronically, a virtual audit approach is appropriate. If the criticality of the system is high, an on-site audit should be considered, even if the vendor can provide the audit documents electronically. The rationale should be documented and justified for both on-site and virtual audits.

Audits are documented consistently with the firm's procedures whether they are on-site audits or remote audits. Remote audits should include the completion of a questionnaire by the auditee and a meeting or follow-up communication or performance of the audit via Skype. The decision to accept, conditionally accept, or reject the vendor system(s) in question is based on the nature of the observations and the responses provided by the auditee. It is often necessary to "conditionally accept" a vendor's system when they are the sole Supplier of an essential product. Vendors may be conditionally accepted based on their proposed remediation of gaps identified during the audit to mitigate or minimize the risk. The oversight activities may include additional testing performed by the purchaser, more frequent communication/evaluation-site audits, or specific commitments from the vendor that the purchaser will verify.

3.21 AUDITING CLOUD SERVICE PROVIDERS

With the developing technology, the increase in device usage and the number of new users has led to the accumulation of more data. While saving this data for later use is essential, the low storage capacity caused some problems. Since the expansion of storage and storage space in smart devices such as computers, tablets, and phones brings high costs, "cloud computing" technology has been developed, which can solve these problems.

3.21.1 What Is Cloud Computing?

Many people benefit from the services provided by this technology without knowing what cloud computing is. Cloud computing is the name given to computing services with an internet base, where users can access at any time and share computer resources with other users on computers, tablets or other devices. Cloud computing systems can be used wherever there is internet access. For example, cloud computing is used when sending or receiving email to another user via the internet. In addition to the email service, many services such as videos or movies watched utilizing the internet, listening to music, playing games on online platforms, and storing and sharing photos are supported by cloud computing technology.

Amazon, one of the world's largest cloud infrastructure services providers, revised its data centers and launched its first cloud technology service, Amazon S3, in 2006. Then, in 2008, the consulting and research company Gartner revealed that cloud computing can take the relationship between users and suppliers to a different dimension in the IT services industry. Cloud technology has been used all over the world since then.

Cloud computing has taken the IT world by storm. Often viewed as the utopia of utility computing, cloud computing offers flexibility and financial benefits second to none. It also lowers the entry point to high-performance computing, allowing organizations to leverage computing power that they have need the capital budget and operational expertise to acquire.

Cloud computing is regarded as an evolutionary rather than a revolutionary step. In other words, cloud computing hasn't drastically altered existing technologies, somewhat, it has succeeded due to the collaboration of several existing technologies.

3.21.2 Where Is Cloud Computing Used?

Cloud computing is used in data backup, email services, software development, data loss prevention, extensive data analysis, and various internet applications. Today, many institutions or organizations benefit from the advantages of cloud computing technology. Many actions such as watching movies, playing games, listening to music, and storing images or other files over the internet have cloud computing technology. Cloud computing technology is widely used in many fields from small businesses to government institutions, nongovernmental organizations, and commercial organizations. Some of the actions that these organizations with cloud computing technology can do are as follows:

- Generating new application ideas and services
- Data storage and backup
- Hosting the websites of the institution or company
- Creating data analytics
- Developing new models by examining data with artificial intelligence technology and machine learning
- To strengthen the image quality of the images on the devices

3.21.3 Cloud Models Figure 3.28 introduces different cloud service providers.

- *Private cloud.*

The cloud infrastructure is operated solely for an organization. The organization or a third party may manage it, which may exist on-premise or off-premise.

- *Community cloud.*

Several organisations share the cloud infrastructure and support a specific community with shared concerns. The organizations or a third party manage it and may exist on premise or off-premise.

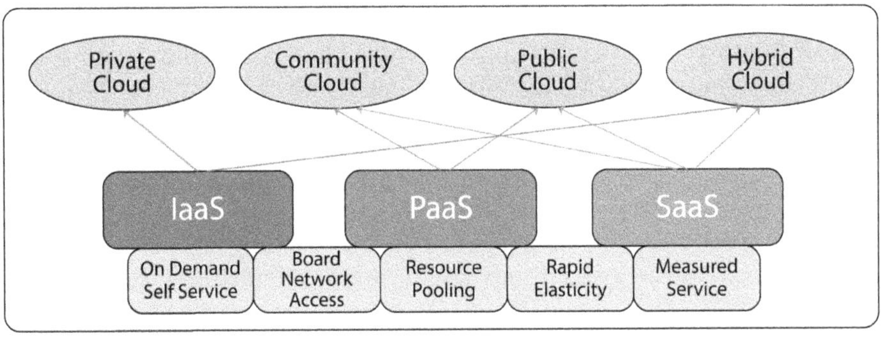

FIGURE 3.28 Cloud service types.

- *Public cloud.*

The cloud infrastructure is made available to the general public or a large industry group and is owned by an organization selling cloud services.

- *Hybrid cloud.*

The cloud infrastructure is a composition of two or more clouds (private, community, or public) that remain unique entities but are bound together by standardized or proprietary technology that enables data and application portability.

3.21.3.1 Infrastructure as a Service

Providers of Infrastructure as a Service (IaaS) give their clients access to many types of infrastructure. Usually, the supplier divides a very large physical infrastructure resource into smaller virtual resources so that the customer can access them. Occasionally, a full virtual computer with an operating system is offered as a service. In some cases, the only service offered is storage, or it could be a virtual computer that is empty of an operating system. The cost of the necessary license is either added as a premium or combined with the service fee in situations where the operating system or other software is provided.

IaaS companies frequently supply services to other cloud providers. As service providers, many existing platforms use IaaS providers to obtain additional capacity as needed. Offering EC2 IaaS, one of the more well-known IaaS suppliers is Amazon.

3.21.3.2 Platform as a Service

Middleware is added to the IaaS-provided software stack by Platform as a Service (PaaS) providers. Software like a DB2 database and runtime environments like the Java Runtime Environment (JRE) or Websphere application server are examples of middleware. More complex programs cannot operate without this middleware, which also gives the application access to a rich operating environment. There are two ways that PaaS providers can help with the additional capacity required for a large multitenant system. They occasionally give the customer virtual machines in the vein of IaaS. In other situations, they offer an interface that allows data to be uploaded from a database or apps to be running environments.

The Windows Azure platform from Microsoft is a well-known illustration of a PaaS. Every approach has benefits and drawbacks. The provider usually has more control and a stronger boundary between tenants when using an IaaS-style solution. However, because common overheads like the operating system and the virtual machine itself are shared by several tenants, this method is less effective.

3.21.3.3 Software as a Service

Providers of Software as a Service (SaaS), often referred to as Application as a Service (AaaS) providers, usually provide their clients an advanced online interface. The consumer, in most circumstances, is fully abstracted from the subtleties of the application running behind the scenes. A shared application, platform, and infrastructure layer are left behind when tenant separation occurs, typically at the application layer. Popular examples of SaaS include Google Apps and Salesforce.co m. SaaS providers often extend the capacity of their systems through scale-up or scale-out methods, depending on the characteristics of the application. As their capacity requirements increase, SaaS applications that expand up are typically migrated to larger platforms. Scaling SaaS services usually operate on massive server clusters.The supplier expands the cluster's machine count as more capacity is needed.

3.21.3.4 Auditing the SaaS Vendor

A common practice among regulated companies is to perform an audit of the potential supplier's quality systems prior to the selection of a software system for traditional implementation.

- Preselection and ongoing audits are of greater significance to SaaS suppliers than conventional software vendors.
  - Greater reliance is placed on a SaaS supplier's quality systems.
  - Focus on SaaS supplier ongoing IT governance.
  - How do we ensure data integrity is maintained throughout the data life cycle?
  - Have we the skills to perform the audits?
- Audit should be directed to areas in which the SaaS supplier maintains control of specific processes, such as:
  - Location and control of data centers used to store the software and client data.
  - Change control of hardware and software for ongoing updates or upgrades.
  - Communication methods and timelines for customer notification of software updates or upgrades.
  - Physical and logical security, such as user identification, passwords, and roles and privileges.
  - Backup and recovery.
  - Data security and privacy.
- Many cloud service providers regularly undergo independent audits performed by qualified third-party accredited assessors for:
  - ISO (27001, 27017, 27018)
  - IT security
  - ISO 9001
  - SOC 1, 2 (Type I and II) System and Organisation Controls
  - Health Information Trust Alliance (HITRUST)
  - US Federal Risk and Authorization Management Program (FedRAMP)
  - Payment Card Industry (PCI)
  - HIPAA
  - GDPR
- Cloud service providers willingly grant access to independent third-party audit reports.

Creating a Vendor Audit Diagnosis Map

- Analyze vendor answers to critical questions "à la minute"
- Obtaining objective evidence: Effectiveness

Check of Vendor Deliverables

- Examining vendor deliverables against answers

What Is a Diagnosis Map?
A diagnosis map is a tool that can be created before a virtual or in-person vendor audit, which has a decision tree with a line of questions or evidence requirements based on the vendor's answers.

Domino Questions
Domino questions are questions that stem from the vendor's answers. These questions might take the audit down a different path.

- No mock intrusion testing.
- How is the effectiveness of intrusion detection determined?
- Can we see the last report from intrusion detection?
- How is the vendor certain the items in the report capture all incidents?

Within the diagnosis map, define the following:

Main Audit Questions

- Following Yes–No branches
- Domino questions off Yes–No branch
- Required evidence to prove vendor's answers are accurate
- Process if evidence (documentation) is not available or insufficient
- Process if SMEs are not available or cannot provide evidence
- End of map will disclose if an audit observation is in order

Answers to Critical Questions "À La Minute"
"À la minute" is a restaurant term meaning "prepared very quickly and served immediately".

Using your diagnosis map as *mise en place*, which translates to "put in place," in a SaaS vendor audit, you can:

- Quickly ask domino questions based on the vendor's answers.
- Assess which documents are needed to provide supportive evidence.
- Determine which SMEs should be requested into audit to answer questions to your satisfaction.

Let's consider some of these critical questions; use a diagnosis map approach; and discuss the domino questions, documentation requirements/analysis, and SMEs vital in conducting an effective SaaS vendor audit.

Vendor Assessments/Audit
Question: What kind of intrusion detection system do you have?
Answer: We use XXXX System.

Domino Questions:

- How did you choose this system? How did you audit the system vendor and validate the tool?
- What kind of reports does the system produce and who reviews them?
- What kind of alerts are sent and to whom?
- Has there been mock intrusion testing to determine effectiveness?
- How often are any intrusion variant updates sent to your system?

Question: What kind of intrusion detection system do you have?
Answer: We use the XXXX system.
Documentation/SMEs

Documentation

- Last month's/year's report from the system
- IT Security, GDPR, Disaster Recovery SOP
- Training records for the IT security team
- Any CAPAs or postmortems on attempted or successful intrusions/GDPR notification
- Mock intrusion testing final report

SMEs

- IT security officer/manager
- Infrastructure/data center manager
- GDPR officer/team

Question: Are your primary and secondary data centers in a co-location?
Answer: Secondary data center is in a co-location.

Domino Questions

- What is the address of the primary/secondary data center?
- How is their data center secured and segregated from the others?
- What are the physical, technical, and administrative measures for data center security?
- What are the fire suppression, HVAC, air exchange, ISP, electrical, etc. measures, and are these providers on an approved vendor list?
- How are visitors and work personnel processed for access to the co-located data center?

Analyzing the Answers: Evidence
Documentation

- Approved vendor list
- IT Security and data center security SOP
- Training records for the data center team
- Access log to secondary data center
- Log of last pest control, HVAC, etc. activities

SMEs

- IT security officer/manager
- Infrastructure/data center manager

Question: Do you have a master training plan?
Answer: Yes, we have a training program.

Domino Questions

- How often is the training plan reviewed, revised, and approved?
- How is training distributed and documented for the entire workforce (employees, contractors)?

- What are the repercussions for an employee who is overdue on training items?
- What is the effectiveness check on training such as exams or quizzes? What are the repercussions of failures?
- What are the credentials of the employees in the training department?

Analyzing the Answers: Evidence

Documentation

- Current training plan
- Training records/reports of key personnel
- Training program SOP
- CVs of training personnel
- Validation documentation of LMS

SMEs

- Training department personnel
- Approver of training plan
- HR personnel

Auditing a cloud service provider could be difficult, with the rare exception of those that are exclusively focused on the pharmaceutical sector or associated services. A typical cloud service business will behave with agility, not precisely tied to locations, typically with fast-changing staff and focused on a best-effort strategy, while the regulated user is focused on a middle- to long-term approach, relying on some predictability.However, ensuring data integrity, accessibility, readability, and confidentiality requires some formal processes and controls, which need to be verified during an audit.

3.21.4 Audit Objectives

The following should always be kept in mind when auditing a cloud service provider:

- Encourage self-assurance in accuracy and working capability.
- The assessment of the service provider's quality management system (QMS) and the required candor and openness throughout this evaluation help to foster trust in the customer-provider relationship going forward.

However, since no organization is flawless, the audit results are typically as follows:

- Provide rational evidence for confidence.
- Address areas of concern, with related corrective measures (action plan).

An audit makes sense for assuring future collaboration only if the audited organization accepts the audit findings and agrees to change its approach through the implementation of appropriate corrective and improvement measures. Via this key topic, the clouds can be separated from the clouds.

3.21.5 How to Audit

A GMP-compliant firm can manage IT services itself or outsource these services. If you are considering using an external service provider, it should be your most natural right to know whether some criteria are met before you start using it. First, you should confirm that whatever the service provider is doing, it does the job well. In addition, you should make sure that it has a sound quality management system, that all activities implemented in the system are carried out according to written procedures, that the results of the activities are kept regularly, that the work done is carried out by trained, skillful, and qualified personnel. It is not possible to know all this at once, so you should at least do due diligence of this service provider.

There are not many options you have to perform an audit.

- You can send a questionnaire to get a first impression of your supplier.
- You can send a questionnaire and also make a teleconference remotely.
- Perhaps another method you will choose is that after reading the answers to the questionnaire, you can choose the path of an on-site audit and verification of the questions.

3.21.6 Timing of the Audit

When you outsource cloud computing audits, you must first ensure that this resource is qualified. In other words, if you are outsourcing your API, excipient, packaging material manufacturers, or contract manufacturers, the expectations will not change when it comes to cloud computing.

For example, let's say you want to have an IaaS audit. If you are performing this audit on site, one or two days may be a suitable time for you. However, when there is a SaaS audit, the audit should not be less than five days when it is considered that this supplier supports a complex process such as validation. In some cases, this period may be extended to seven days when the audit is carried out with a single auditor. It is inevitable for the GMP auditor to receive support from the IT department or information security specialists, since it may be risky for the auditor to audit alone, depending on the complexity of the service.

Let's say that a SaaS audit will be conducted, in this case, when the company is notified in advance of the list of documents to be reviewed during the audit, the company can easily determine who will answer the questions in this context.

This is a sample list of documents that will be examined during the SaaS audit:

- QMS
- CSV
- Training
- Incident management
- Change management
- Configuration
- Software release
- Release testing
- Patch testing
- SDLC process
- Traceability of requirements
- Project methodology
- Functional testing
- Third-party management
- Security of third parties
- Backup/restore, disaster recovery and business continuity test evidence
- Service Level Agreements (SLAs)
- Security/access control
- Data encryption

3.21.7 Questionnaire

The questionnaire method is a good and quick way to get a first impression of a supplier, but what if the supplier answered yes to all questions? You have to trust the person's way of doing business and honesty with whoever is filling out these questions. Choosing this type of way can often yield useful results when well-managed, for example, you can choose the most critical ones among the given answers and request evidence documents, records, and images related to these answers. When evaluated from your point of view, that is, the pharmaceutical company may forget these answers and dive into other intensive work. Regardless, you need to examine the answers one by one and to reach the evidence in order to determine the risks and obtain detailed information.

3.21.8 Not Only Questionnaires But Teleconference

Another method to choose is to go over the responses via a teleconference after the questionnaire responses have arrived. Thanks to the teleconference, you can ask questions that you did not ask, request details of your dissatisfaction with the answers, review the documents to be examined through a platform offered by the supplier, and try to reach a final decision about the supplier. However, some suppliers may not want to share such information on the pretext that it violates their corporate procedures or intellectual property. In my opinion, don't waste your time and start looking for companies that are ready to share all kinds of information with you and have the same documents as they do.

3.21.9 Not Only Questionnaire But the On-Site Audit

In both options, I tried to explain the answers given to the questionnaires and the next step you should take, but for some reason, I am not comfortable. I have many questions in my mind that will convince me that I need to go into more detail. Companies that own a hosting facility and data center may ignore the fact that the issue is a serious business such as pharmaceutical production. Because such companies are not even aware of the FDA, EMA, or WHO guidelines or the publications of institutions such as ISPE and PDA. All they know is that they have ISO 9001, 27001, 90003, and similar standards, under the misconception that it somehow also includes GXP. Another handicap is that the hosting facility and data center are in separate locations or even in separate countries. Since we cannot separate the two important components, it should be auditing both locations according to GXP rules.

As I said, it is quite good for these companies to have these standards on a minimum scale, but when it comes to data integrity, data security, and the fact that auditors hold pharmaceutical companies accountable in health authority audits, when it comes to guidelines such as 21 CFR Part 11, 21 CFR Part 58, and EU, PIC/S GMP Annex 11, unfortunately, we don't have many options. One of the most common ways such companies use subcontractors in some of their activities for this. The parts that the GMP auditor will look at, whether they are using it or outsourcing it, are the quality management system, personnel records, and training. As a GMP auditor, the most critical point you need to focus on is training records according to GXP. They probably won't be able to provide you with some documents, or records, but it is very important that you measure the company's awareness of GXP. Let's say you have some staff training records, you have to wait for them to prove who gave the training based on what qualifications.

3.21.10 Audit Findings

- Installation qualification documents are not preapproved prior to execution.
- IQ/OQ documents are not executed against equipment specifications to demonstrate fitness for the intended use.

- There is no specification and associated qualification testing, with associated quality assurance oversight, of the hypervisor layer.
- GXP awareness training was given only to staff working greater than 80 hours per year on the qualified infrastructure.
- Qualification documents are electronically signed in an EDMS. The electronic signatures are not compliant with the requirements of FDA 21 CFR Part 11, 11.50.
- The EDMS was incompletely validated by the vendor of the software because only two test cases, both designed to pass, were found in the validation documents. There were no test cases for security, audit trail, or other Part 11 functions.

3.22 AUDITING ARTWORK DESIGN AGENCIES

A problem that pharmaceutical and biotechnology companies have difficulty solving and cannot completely eliminate is the 483 reports and recalls related to packaging materials. Pharmaceutical packages are grouped as primary, secondary, and tertiary as stated in the previous sections. Primary packages usually contain label information or some information embedded in the package (for example, the batch number, expiration date, code number, etc. in the blisters). When it comes to secondary packaging, it is possible to count materials such as folding boxes, leaflets, and labels. Whether primary or secondary, the information printed on all packaging materials is lacking, inaccurate, or lacking the required information, occupying the agenda of many health authorities. So, through which stages is the information such as text, numbers, braille, and 2D barcode on the packaging materials printed on the packaging? Who are the parties involved in this process? What are the duties and responsibilities of a license holder? What are the issues that companies that perform the printing process should pay attention to? All this happens thanks to a successful and robust artwork program. Pharmaceutical companies and their partners should be extremely careful and prevent invalid and incomplete information from being transferred to the packaging. Artworks are managed by the company having an in-house graphic design department or outsourcing this work directly to a design agency or printing houses from which it receives printed packaging material services. In either case, it's not hard to notice that there are some critical checkpoints. As GMP auditors, we expect an SOP to perform the artwork no matter which method is chosen. In this SOP, the process of starting the artwork and archiving it by invalidating it should be written in a detailed and understandable language. Roles and responsibilities should be defined in the SOP. There should be restrictions on who cannot directly initiate artwork. The computer systems where the artwork will be carried out should be questioned in terms of both data integrity and validations. We should pay attention to the completeness of the necessary GMP and technical training of the person who makes the artwork, that is, the person who carries out the design in programs such as Adobe Illustrator, Adobe InDesign, and Photoshop. We must again verify whether the version numbers of the programs used, and the software that has been changed or patched, are handled by a change control program. We must have access to information about why the need for artwork design change arose. After the artwork is prepared, to which departments is it sent first, who reviews it, and reports or approves comments? How is the flow managed? What warning does the designer receive when the reviewer does not approve the new design? In fact, if there is software that manages these activities in the background, validation of this software should be questioned. This critical activity should not be outsourced without researching the knowledge and experience of the company selling the software on GXP regulations and its recognition in the industry. Figure 3.29 explains who checked which areas of a folding box and approved the information on the folding box elements, and Figure 3.30 explains an example model process.

In terms of Figure 3.29, several departments such as Marketing, Regulatory Affairs, Manufacturing, Engineering, Quality Control, and Quality Assurance are responsible to check every single part of a folding box. For example, when the manufacturing department must check the dimension of the folding box, the Quality Control department must check its 2D barcode, pharmacode, dimension, color, text content, etc.

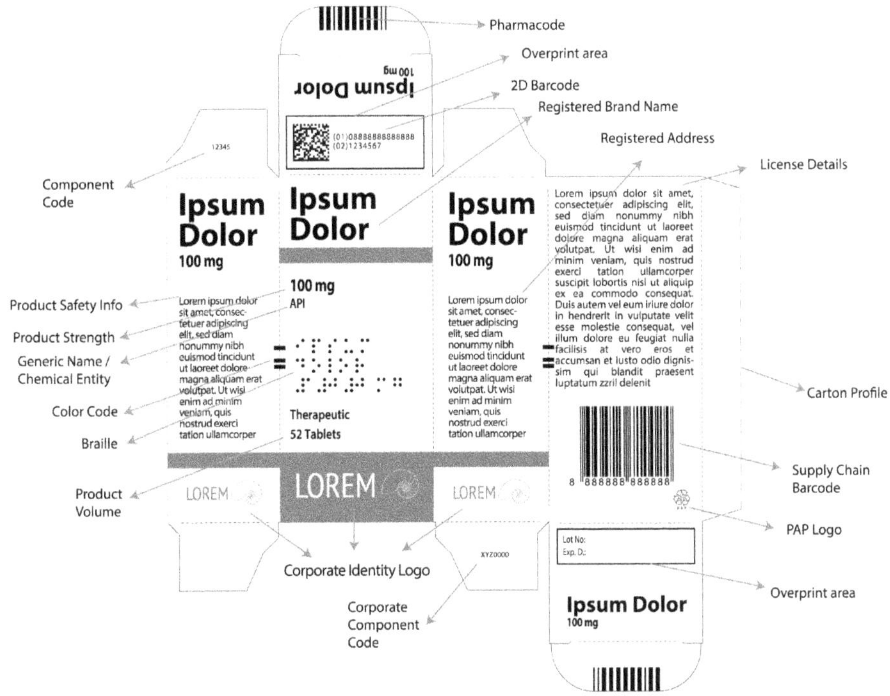

FIGURE 3.29 Folding box elements.

3.22.1 Why Is Artwork Important?

- We can only sell our products if they are properly packaged.
- We can only ship our products when the text on the packaging is correct.
- Our patients' lives rely on this text being absolutely correct.
- Our corporate reputation and profit are directly impacted when the text is incorrect.

In order to be successful, a number of elements need to be in place to have an effective artwork change capability. Table 3.48 provides examples of questions to be asked in an artwork service provider or graphics department audit.

3.22.2 Artwork Team Responsibilities

- Single point of contact for pack change
- Pack change planning and coordination
- Managing artwork service provider
- Approval of printer proof and packaging samples

3.22.3 Leadership–Global Logistics Responsibilities

- Strategy and standards
- Capability development
- Process ownership
- Performance management

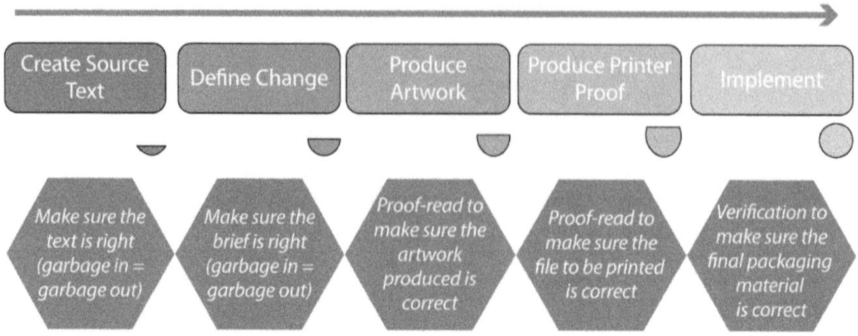

FIGURE 3.30 Example model process with defined control points.

TABLE 3.48
Questions to be Asked during the Audit of an Artwork Services Provider/Department

Element	Question
Quality management system	Is the end-to-end artwork capability managed under a formal quality management system?
End-to-end coverage	Does the scope of the end-to-end capability cover all labeling and artwork creation?
Effective impact assessment	Does a trigger for change result in all the relevant artworks being created or changed?
Critical control points	Are critical control points defined?
Complete artwork	Is a complete version of the artwork created and approved?
Adequate process and task definition	Is the overall process and are individuals' tasks defined to ensure complete and consistent working?
Clear roles and responsibilities	Are roles and responsibilities clearly defined, including at key approval points?
Adequate artwork quality checks	Are there adequate quality checks of all text and graphics at every document revision?
Audit trail	Is an audit trail created and held for an acceptable period?
Document version management	Is there effective document version management in place?
Document form consistency	Are there methods in place to ensure different forms of each document are the same?
Correct document use	Is everyone using the correct version of a document?
Education, training, and competency	Is ongoing education, training, and competency assessment in place?
Service provider management	Are service providers formally managed under a quality management system and subject to routine audits?
IT tools management	Are critical Information technology tools managed under a quality management system?
Ensuring what you see is what you get	Do all IT tools used during review and approval ensure the correct image is displayed?
Effective planning	Is work planning and adequate resource-level management in place?
Working environment	Do individuals have an appropriate working environment?
Continuous improvement	Is there an issue management and continuous improvement process in place?
Governance	Is there adequate cross functional/organizational governance in place?

3.22.4 Regulatory Responsibilities

- Provision of accurate and approved product text
- Production of a complete right-first-time brief of the regulatory driven change
- Final approval of the completed artwork

3.22.5 Packaging Vendor Responsibilities

- Production of a complete right-first-time brief for vendor-driven changes
- Input to change assessment
- What is impacted?
- Can it be made?
- Can it be made then?
- What is the write-off impact?
- Input to design, e.g., cutter guides
- Maintenance of vendor bill of materials (BOM)
- Implementation of approved change in an agreed timescale

3.22.6 Pack Tech Responsibilities

- Creation of component specifications
- Management of a component range
- Leadership and introduction of packaging legislation

3.22.7 Supply Chain, Commercial, Licensees, Co-Marketeers Responsibilities

- Provision of required Supply Chain, Regulatory or Commercial information
- Final approval of the completed artwork

3.22.8 Planning Responsibilities

- Integration of changes to the supply schedule
- Agreement/management of material write-offs
- Production of a complete right-first-time brief for rework changes

3.22.9 Artwork Service Provider Responsibilities

- Execution of individual pack changes
- Creation of artwork and drawings
- Proofreading

3.23 AUDITING SINGLE-USE SYSTEM MANUFACTURERS

3.23.1 Single-Use Bioreactor versus Stainless System Bioreactor

Continuous argument over advantages and disadvantages of using these two different systems is not new. It dates back to almost the 1990s. One of the biggest challenges at the beginning of deciding on which method is the most suitable one for a biopharmaceutical facility is to really understanding the secrets behind them.

Several indicative factors are available to make the right decision to proceed. If the facility has been manufacturing small volumes of biopharmaceuticals, then a single-use system (SUS) could be a real benefit to its business activities. Or if the facility has been manufacturing large volumes of biopharmaceuticals, then its objective could be different than that of small-volume facility. New applications or replacement of such complicated systems to existing ones also depends on different actors like manufacturer, supplier, regulatory authorities, etc.

It must be well understood at the very beginning that every single one of these stakeholders' needs are different and could be tough.

3.23.2 Why SUS?

As the biotechnology industry grows, more flexibility and responsiveness is to be expected in production facilities. There are also trends to shorten production times and to offer products more quickly, and rigorous authority expectations and sanctions and capacity competencies are the most important handicaps against companies.

Single-use systems are discussed in terms of their ability to meet many needs:

- Flexibility
- Decrease in capital investment
- Increasing speed
- Security

The emergence of new opportunities in the biopharmaceutical industry such as the emergence of innovative companies, the establishment of new firms, the maturing of contracted production approaches, monoclonal antibody production, and vaccine production, enables the development of single-use technology (SUT).

Today, disposable systems are primarily used for product development and clinical phases. However, such productions are usually in small batch sizes and require more flexible platforms due to their shorter campaign production.

It must be taken into account that SUT is an important partner for these small-scale purposes. The emergence of SUS dates to the 1970s. This industry, which initially started to produce small components, such as filters and bags, has later developed and nowadays extends to buffer, media preparation and storage, cell culture, downstream processing, formulation, and filling. Table 3.49 introduces the components of the SUT.

TABLE 3.49
Different components of Single-Use Technology

Components		Systems
Bioprocess containers/bags	Assemblies	Bag carriers
Bottles	Mixer	System frames
Filters	Bioreactor	Bag lifts
Tubing	Storage	Instruments and sensors
Fittings/connectors	Filtration	Pumps/motors
Gaskets/O-rings	Sampling	Scale/load cells
Valves	Transfer	Tubing hangers/racks
Clamps	Fill/finish	Tubing welders
Fasteners		Control valves
Pump liners		Electrical control enclosures
Centrifuge liners		Software
Chromatography columns		
Filling needles		
Impellers/mixers		
Sensors		

Despite such positive developments, scale up is regarded as a handicap. Advantages and disadvantages of SUT are conveyed in the following sections.

The roles of the supplier and end user are strongly linked in the SUT space. While this section is primarily targeted for use by the single-use product end user, intermediate suppliers may also find parts of this section useful in defining the quality of components and the auditing of their respective suppliers. The process of establishing consistent quality along the entire supply chain allows for more robust final quality to the end user and, ultimately, to patients.

3.23.3 Defining Supplier Audit Requirements

Single-use suppliers, manufacturers, and assemblers should be audited on design, manufacturing process, and supply chain management until delivery to the end user. Figure 3.31 highlights the importance of audits in single-use technology (SUT) systems. The frequency and duration of such audits should be included in the quality agreement. The supplier audit should cover both manufacturing and technical aspects, with special focus on the following aspects:

- Compliance to ISO 9001 certification (and other applicable standards)
- Management and control of raw materials and components
- Manufacturing process overview
- Sourcing and supply chain strategy
- Manufacturing environment monitoring (microbiological and particulates trending)
- Sterilization process (frequency, procedure, and records of dose audits for irradiation)
- Quality management systems (including quality policy and change management processes)
- SOPs
- Document controls
- Complaint management (with tracking, recording, and corrective and preventive action processes)
- Change management process (with specific focus on major versus minor status)
- Operator training plan and records
- Batch record tracking

The frequency of audits should be established in the quality agreement and determined based on the risk of the supplier and type or criticality of the item/service. Important criteria used to define the frequency and depth of audits should include the criticality of the product being supplied and the relationship with the supplier. A frequent audit schedule is warranted for suppliers of products that are key components in the production of therapeutics. Suppliers of SUT products are often classified

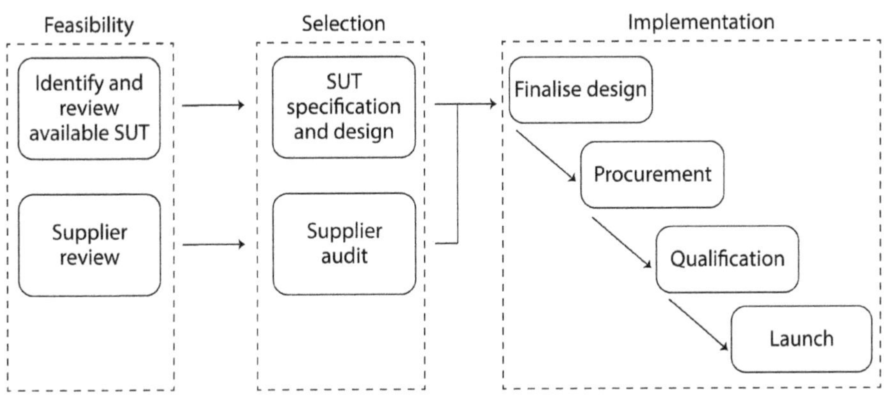

FIGURE 3.31 SUT implementation and key areas.

in this category. The relationship with the supplier is also important in defining the audit depth and schedule. A less frequent audit schedule could be used for an established supplier that has reliably provided several products over multiple years. Alternatively, a new supplier should be audited frequently until an appropriate level of confidence is established. For example, the frequency may be once a year for new suppliers (where there is a lack of historical performance data) or for highly critical items/service. Meanwhile, the frequency may be less frequent (e.g., every two to three years) for well-known suppliers whose long-term reliability and performance have been documented.

Additional considerations for supplier audits include:

- Obtaining feedback from the end user's manufacturing staff
- Obtaining feedback from the end user's purchasing staff
- Environmental monitoring SOPs and reports
- Technical and manufacturing staff training SOPs
- Inventory control SOPs
- Change control SOPs
- Inspection of incoming parts SOPs
- Inspection during manufacturing SOPs
- Packaging SOPs
- Quality control SOPs
- Supplier audit SOPs
- Health and safety SOPs and records
- Job tracking file review
- Quality agreement
- Checklists for audits
- Frequency of audits
- Impact of noncompliance with quality criteria

The following guidelines can help an auditor conduct a thorough audit of SUS areas. This section's ISO standards are:

- Requirements for Quality Management Systems under ISO 9001: 2015
- Cleanrooms and Related Controlled Environments, ISO 14644-2 (2015)Since SUSs are not regarded as medical devices, ISO 13485 Medical Devices are not included in the selected recommendations. The goal of pharmaceutical and biotechnology firms is to make sure auditees comply with relevant cleanroom standards and quality management systems, even though many of their suppliers hold ISO 13485 certification. Although it is not necessary, supplier certification to ISO 13485 will be noted in the audit reports. The evaluation of an auditee's adherence to relevant standards is known as an audit. The specifications that a manufacturer must adhere to are provided in this handbook (where appropriate).The production components unique to each SUS product type are contained in this handbook. It is not appropriate to directly reference observations to these as they are not regulations. Nonetheless, these components might be a part of the SUS manufacturing processes, and these manufacturing processes would be in charge of managing and controlling the components for the producer. The risks related to SUSs in the contamination control strategy (CCS) are delineated in Table 3.50.

The GMP auditor should verify that SUS are designed to reduce the need for manipulations and complexity of manual interventions.

- The GMP auditor should verify that sterilization processes for SUS is validated and shown to have no adverse impact on system performance.

TABLE 3.50
Specific Risks to Consider Concerning SUS as Part of the CCS

- The interaction between the product and product contact surface (such as adsorption, or leachables and extractables); the fragile nature of the system compared with fixed reusable systems.
- The increase in the number and complexity of manual operations (including inspection and handling of the system) and connections made
- The complexity of the assembly
- The performance of the pre- and post-use integrity testing for sterilizing grade filters paragraph.
- The risk of holes and leakage
- The potential for compromising the system at the point of opening the outer packaging
- The risk of particle contamination

The selection and application of these systems depend heavily on the evaluation of disposable system providers, including sterilization.

When it comes to sterile SUS, the GMP auditor should confirm that the supplier's qualification process includes sterility assurance verification and that each unit's proof of sterilization is examined upon receipt.

- The GMP auditor should verify that the adsorption and reactivity of the product with product contact surfaces are evaluated under process conditions.
- The GMP auditor should verify that the extractable and leachable profiles of the SUS and any impact on the quality of the product especially where the system is made from polymer-based materials are evaluated.
- The GMP auditor should verify that an assessment is carried out for each component to evaluate the applicability of the extractable profile data.

For components considered to be at high risk from leachables, including those that may absorb processed materials or those with extended material contact times, the GMP auditor should verify that an assessment of leachable profile studies, including safety concerns, is taken into consideration.

If applying simulated processing conditions, the GMP auditor should verify that these accurately reflect the actual processing conditions and be based on a scientific rationale.

When single-use components are subjected to more harsh circumstances (such as freezing and thawing procedures) during normal manufacturing or transit, special attention must be paid to their structural integrity.

- The GMP auditor should verify that acceptance criteria is established and implemented for SUS corresponding to the risks or criticality of the products and its processes. Table 3.51 displays some example questions during an audit regarding SUS components.
- The GMP auditor should verify that on receipt, each piece of SUS is checked to ensure that they have been manufactured, supplied, and delivered in accordance with the approved specification.
- The GMP auditor should verify that a visual inspection of the outer packaging (e.g., appearance of exterior carton, product pouches), label printing, and review of attached documents (e.g., certificate of conformance and proof of sterilization) is carried out and documented prior to use.
- The GMP auditor should verify that critical manual handling operations of SUS such as assembly and connections are subject to appropriate controls and verified during aseptic process simulation (APS).

TABLE 3.51
Example Questions during an Audit Regarding Single-Use System Components

Design/Drawing Management

- Are custom items manufactured?
- If so, how is the design process managed?
- Is there a procedure? Is the auditee able to show control of custom processes?
- How do quotation drawings, drawings approved by customers, shop floor drawings, and quality specification drawings relate to each other?
- How are they controlled and synchronized?
- Are signatures required from the customer to designate approval? Customers should have ability to approve final design. If not, why?
- Are signatures required from the customer to designate approval?
- How is confidential design information destroyed? This information is intellectual property, and the auditee needs to be aware that they have the responsibility to manage it securely.
- Are specifications in place for each item? All products should have specifications that set product requirements and critical to quality attributes. If so, are they managed through document control and change control processes?
- Is it possible to continue production with N-1 revision of a drawing, waiting for the approval of the revision N?

Tubing

Note: Silicone liquid raw materials feed and mixing systems, and silicone "gum" handling need to be evaluated. For example, some companies use a mixture of liquid resins, similar to epoxy. How are the amounts calibrated? Gum resin (blocks) need to be stored to prevent contamination.

- Is there segregation of peroxide-cured and platinum-cured silicone?
- Are there changeover procedures for any tubing types that might be produced on the same equipment?
- If water-cooling bath is used, is there management of the process? Does it come in contact with inner surface of tubing? Water used should be adequately controlled and tested.
- Are heat treatment ovens used, for example, in curing?
- Are online size monitoring inspections in place, e.g., for wall thickness?
- How often is the screener in the extruder replaced? Is this preventive maintenance (PM) documented? How are the screens controlled and documented within the batch record?

Sterilization

- Where sterilization is required, is there a validated sterilization master plan and protocol of all control parameters for the sterilization process?
- If sterilization is required, are there procedures in place?
- If a contracted service is used, has the provider been qualified?
- Are validation procedures in place for the method of sterilization being used? If sterilization services are contracted out to a third party, have they validated their equipment/process?
- Are sterilization contractors audited regularly?
- Are load configurations and dosage defined?
- If the sterilized SUS pieces are returned to the supplier, how are they handled/stored? Are nonsterile/sterile flows/ products clearly identified and physically segregated?

3.24 AUDITING STERILIZATION SERVICES

3.24.1 What Is Sterilization?

Sterilization processes are used to destroy or eliminate the microorganisms that are present on the surfaces of materials in order to aid in providing a safe pharmaceutical product for the patient. All sterilization processes are intended to result in materials or surfaces that have become sterile or free from viable microorganisms. While this dictionary definition of sterility sounds good, in the real world we do not have a way to routinely test or evaluate for the absence of all viable microorganisms in all samples.

When products are labeled sterile, the common assumption is that the vendor is claiming to meet the dictionary definition of sterility. This is further complicated by the fact that most sterile products are released using a national standard — the compendial sterility test methodology. It is common to assume that passing this test indicates that there are no viable organisms present.

3.24.2 Application of Gamma Radiation

The primary application of gamma radiation is for medical devices, ranging from sterile dressings, tubes, catheters, syringes, infusions assemblies, and implants; and single-use disposable technologies, such as bags for holding products or devices for making aseptic connections.

The main reason why gamma radiation is selected as a sterilization method is due to its relatively high penetrability and as there is only a small temperature rise (typically <5°C) associated with its use. This means that the technology is suitable for sterilizing heat-liable and heat-sensitive products that could not be processed by steam sterilization. Another reason for the wide use of the method is because the technology allows for high-volume processing.

3.24.3 Process Requirements

The product is put into totes, which are specialized containers made usually of metal, and subjected to gamma sterilization. The validation process determines how much product can fit inside a tote. Simply, the tote is then exposed to the sterilization source for a period of time. Arguably, gamma radiation is the simplest of all sterilization technologies, because there are very few variables to be controlled.

3.24.4 Dry Heat Sterilization

Dry heat sterilization is sterilization by thermal (heat) conduction. The dry heat sterilization process is accomplished by conduction, which is where heat is absorbed by the exterior surface of an item and then passed inward to the next layer. Eventually, the entire item reaches the peak temperature needed to achieve sterilization. The ideal temperature and duration for dry heat sterilization are 160°C for 2 hours or 170°C for 1 hour, which is also the theoretical temperature at which microbial kill becomes effective. Dry heat causes proteins to coagulate, which kills bacteria. Within the biopharmaceutical sector, dry heat sterilization is a process that is more commonly deployed as a depyrogenation step, which results in the inactivation of bacterial endotoxin as well as the destruction of microorganisms. The temperatures and times required for dry heat sterilization are far higher than autoclave temperatures. This is because, in contrast to saturated steam, the mechanisms of microbial inactivation resulting from exposure to dry heat are very different and much slower. As a result, sterilizing components using dry heat is less effective.

Figure 3.35 presents dry heat sterilization example.

3.24.5 Steam Sterilization

The most widely used sterilization method in the pharmaceutical industry is steam sterilization in autoclaves, moist heat in the form of saturated steam under pressure. This is primarily applicable to the terminal sterilization of products, stainless steel items, and equipment not intended for single use. Sterilization occurs as the latent heat of condensation is transferred to the load causing it to heat rapidly. Steam sterilization is non-toxic, inexpensive, rapidly microbiocidal, sporicidal, and efficient at heating and penetrating fabrics. It is because of the ability of heat to penetrate that steam sterilization is widely used as a terminal process for drug products in glass ampoules, vials, syringes and plastic containers. It is also used for sterilizing closures, filters, manufacturing equipment, cleaning equipment, and product holding vessels.

3.24.6 Gaseous Sterilization

There are different types of gaseous sterilization. Sterilizing gases include formaldehyde, ethylene oxide (EO), propylene oxide, ozone, peracetic acid, vapor hydrogen peroxide, and chlorine dioxide. Most common to sterile manufacturing is EO, which is used to sterilize many plastics, and vapor hydrogen peroxide, which is used to decontaminate barrier systems (i.e., isolators).

Gaseous sterilants are effective surface sterilizing agents, in that they will sterilize the outside of a device, or the primary packaging in which the device is held. Unlike other forms of sterilization such as radiation or heat, the agent does not normally penetrate as well into the item being sterilized. The key parameters affecting the effectivity of gas sterilization are active concentration, temperature, duration of exposure and relative humidity. Aside from steam (moist heat) sterilization, EO sterilization is the most widely used method of sterilization in the medical device and biopharmaceutical sectors. EO is commonly selected for objects sensitive to temperatures greater than 60°C and/or to radiation.

3.24.7 Ethylene Oxide

Ethylene oxide (C_2H_4O), sometimes called oxirane, is an organic alkylating agent, which functions as a very potent and highly penetrating gas. It functions as a so-called cold sterilant, in contrast to methods of sterilization by heat. As an industrial sterilant, EO is effective when sterilizing paper, cloth, and most types of plastics. It is compatible with most materials, even when repeatedly applied. EO is most commonly applied to medical devices and its use has increased with the growth of single-use sterile disposable materials, although not to the extent as with gamma radiation.

EO is capable of destroying most viruses, bacteria and fungi, including bacterial spores. The alkylating properties of EO provide it with its microcidal properties.

3.24.8 Chlorine Dioxide Gas

Chlorine compounds are a significant type of microbiocidal halogen, and chlorine is widely used as a sporicidal disinfectant and as a sterilant in liquid form. When in gaseous form, chlorine is normally in the form of chlorine dioxide, where it is a powerful surface sterilant. However, it has a poor ability to penetrate.

Chlorine dioxide (ClO_2) gas is a single-electron, transfer-oxidizing agent, similar to liquid bleach. This makes it a highly active oxidizing agent and can thereby destroy the cellular activity of proteins. Its deleterious effect on bacterial endospores is believed to be directed primarily toward the cell membrane rather than DNA. As a means of gaseous sterilization, it is not as widely used as EO or ozone. Chlorine dioxide has been used to successfully sterilize medical devices and laboratory equipment, and to decontaminate cleanrooms and clean zones. There are some applications with barrier devices, including isolators, as an alternative to the more commonly used hydrogen peroxide vapor.

3.24.9 Hydrogen Peroxide Vapor Sterilization

Hydrogen peroxide vapor is a gaseous form of hydrogen peroxide (H_2O_2). It is used as a low- temperature antimicrobial gas for the decontamination of enclosed and sealed areas such as laboratory workstations, isolators, cleanrooms, and aircraft interiors. Within the biopharmaceutical and healthcare sectors, the primary use is with isolators. Decontamination in isolators with vaporized H_2O_2 has a number of advantages, such as the fact that the degradation products (water and oxygen) are non-toxic and can be removed easily after decontamination. Furthermore, there is a high degree of compatibility of H_2O_2 with materials commonly used in process areas, and the agent is of relatively low cost. H_2O_2, in the gaseous state, is the prominent gaseous sterilant where the requirement is to sterilize or sanitize surfaces rather than to penetrate. This is an important point, as the gaseous agent will not penetrate most surfaces and it cannot sanitize the material held within. For effective sanitization, the physical properties of the item being sterilized must be relatively smooth, impervious to moisture, and be of a shape that permits all surfaces to be exposed to the sterilant. Alternatives to H_2O_2 are peracetic acid, although now less commonly used, and chlorine dioxide gas, which is little developed.

3.24.10 Sterilization by Filtration

Filtration is a means of sterilizing fluids (liquids or gases) through the removal, rather than the destruction or inactivation, of microorganisms. Liquids that would be damaged by heat (i.e., those containing proteins such as large-molecule drug products) irradiation or chemical sterilization, can only be sterilized by filtration. Thus the sterilization of liquids is a key step for aseptic manufacturing, as the means of passing the bulk product to the point where it will be dispensed into bottles. Due to the small pore size of the filters, the term "microfiltration" is sometimes referenced in literature. Sterilization, in relation to filters, generally refers to the removal of bacteria and fungi. In processing of certain biologics, viruses must be removed or inactivated using nanofiltration. This requires a nanofilter with a pore size of 20–50 nm. The ideal characteristics of a sterilizing grade filter is in selecting a filter that is compatible with the process.

Sterile filtration of liquids and gases in pharmaceutical manufacture is almost always performed using membrane fllters. These are thin uniform porous sheets, which act as sieves that trap particles larger in size than the pores in the membranes. Other factors which prevent the passing through of particles include inertial impaction to the walls or surfaces of the pores and lodgment in crevices within the depth of the membrane.

3.24.11 Terminal Sterilization, Decontamination, Disinfection, and Cleaning

Terminal sterilization is the application of a lethal sterilizing agent to finished product within a sealed container to achieve a predetermined sterility assurance level (SAL) of 10^{-6} or better. The SAL is the probability of a single item in a batch being nonsterile after being subjected to a sterilization process.

Decontamination is a process that eliminates viable bioburden via use of chemical agents.

Disinfection is the process by which surface bioburden is reduced to a safe level.

Cleaning is the removal of contamination from an item to the extent necessary for further processing or for intended use.

Figure 3.32 shows the steps from cleaning to sterilization.

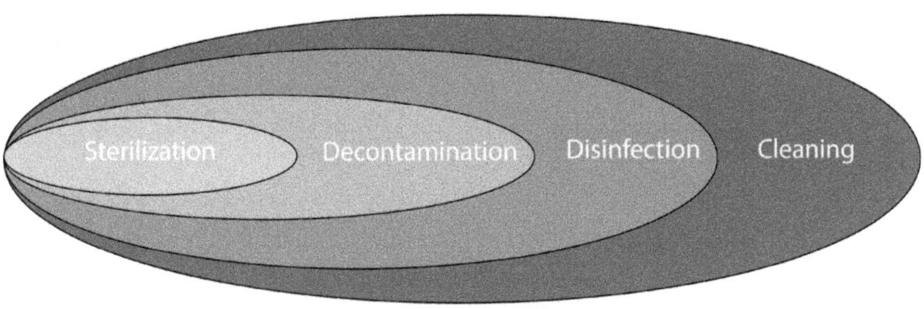

FIGURE 3.32 Steps from cleaning to sterilization.

3.24.12 How to Audit Terminal Sterilization

There are two broad methods to produce a sterile drug product, terminal sterilization and aseptic processing. There are various methods of terminal sterilization including moist heat sterilization, dry heat/depyrogenation, irradiation and ethylene oxide. Terminal sterilization is always the preferred method over aseptic processing when possible. However there are situations when terminal sterilization cannot be performed and one must rely on aseptic processing. Aseptic processing does present a higher risk of microbial contamination of product than terminal sterilization. Figure 3.33 reviews aseptic filling versus terminal sterilization.

When conducting an audit of the terminally sterilized product process, consider the following factors:

Facility Considerations

- Has the firm performed a risk assessment of its facility and equipment?
- Does the firm understand the areas of risk as they relate to contamination of drug product?
- Are the facility, the equipment layout, and the air handling system designed and suitable for preventing viable and nonviable contamination?
- Is the material and personnel flow unidirectional (dirty to clean)?
- Is there trend data to demonstrate the cleanroom quality?
- Does the facility have smooth cleanable surfaces? Are the materials nonporous?

Figure 3.34 reviews the different sterilization methods in a sterile medicinal product manufacturing site. Figure 3.35 reviews the dry heat sterilization method in a sterile medicinal product manufacturing site.

Utilities

- Are there detailed P&IDs/as-built diagrams?
- Is there a risk assessment that includes slopes, dead legs, nonsanitary fittings, and leaks?
- Is there appropriate sampling at the points of use?
- Is there inline monitoring for total organic carbon (TOC) and conductivity?

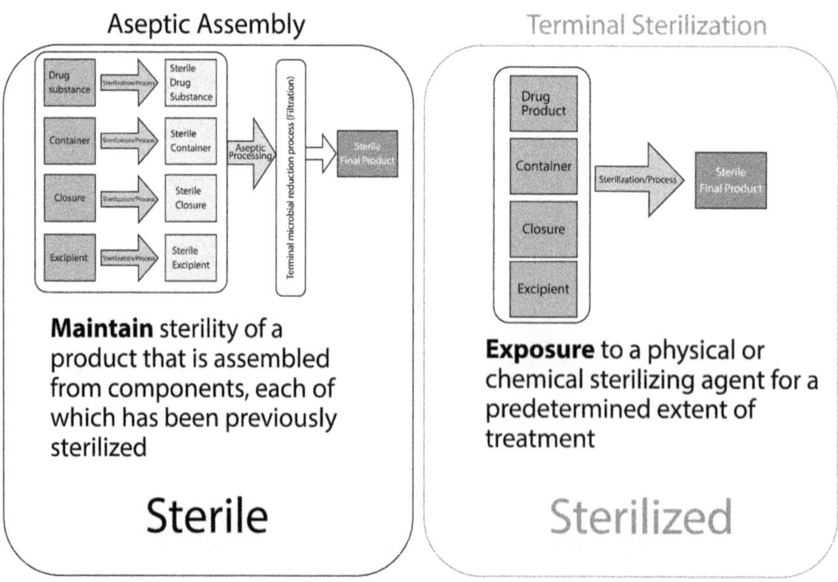

FIGURE 3.33 Aseptic filling vs Terminasl Sterilization

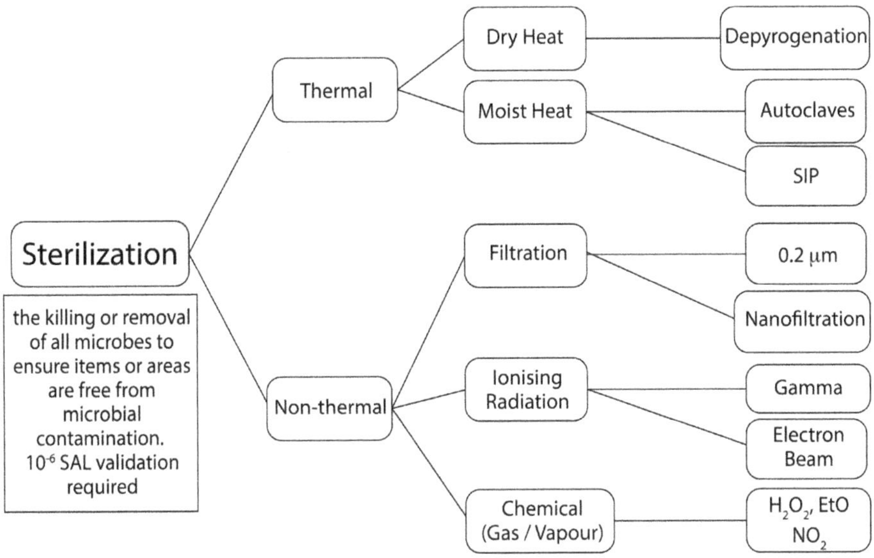

FIGURE 3.34 Different types of sterilization methods

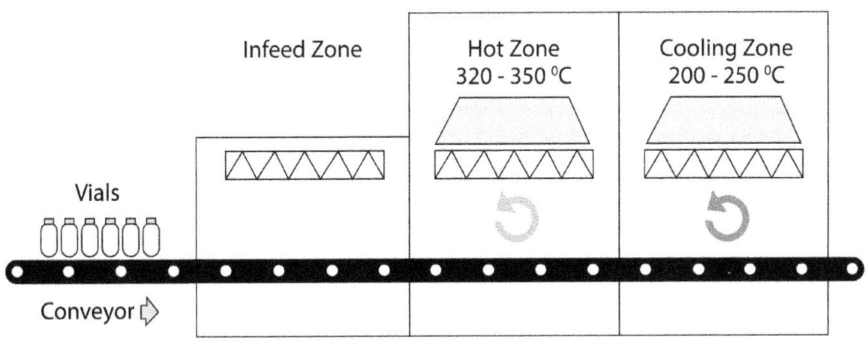

FIGURE 3.35 Dry heat sterilization.

- Is there trend data for chemistry, microbiological, and endotoxin tests?
- Are appropriate alert and action levels established?
- Is there a scheduled sanitization?
- Is passivation performed when needed and what material is used to passivate?
- Is the facility controlled and classified?
- Are the HEPA filters integrity tested?
- Does the testing include air velocity measurement?
- Are there appropriate pressure differentials, and temperature and humidity set points?
- Have airflow pattern (smoke) studies been conducted under dynamic conditions to verify the unidirectional airflow and air turbulence within the critical area where sterilized drug product, containers, and closures are exposed to environmental conditions?

Qualification and Validation

- First and foremost, have all the sterilization processes been validated and has all the equipment such as autoclaves and ovens been qualified?

- What type of sterilization cycles were used, for example, was it bioburden-based or an overkill?
- Does the validation documentation describe the equipment and include the IQ/OQ/PQ data?
- Are there procedures for revalidation? What is the time period and is it based on risk?
- Does the validation documentation include empty chamber and loaded chamber heat distribution studies?
- Was there an identified worst-case load?
- Were biological indicators (BIs) used to validate the cycles?
- What type of indicator was used?
- What organism was used (genus and species) and is it appropriate for the type of sterilization?
- Is there a verifiable spore count and what is the approximate D-value of the BI?
- How many BIs were used per sterilization load?
- Are there any worst-case locations and were the BIs placed in these locations?
- Is there a diagram of the distribution of the BIs in the loading pattern used?

3.25 LYOPHILIZATION

Lyophilization is derived from the Greek, meaning "solvent-loving, liquid-attracting," and refers to the ability of a dry product to be rehydrated at a high level. Operationally, freeze-drying can be explained as a controllable method of rehydrating perishable products through vacuum desiccation (Figure 3.36). Lyophilization is expressed as a multistage operation where each step is critical. To be successful in this process, the main components must be well understood and kept under tight control. Stable preservation of biological materials is a fundamental requirement for biomedical science, agriculture, and biotechnology. Lyophilization is widely used in the pharmaceutical and biotechnology industries to improve the stability of formulations. It is also an important and well-known process that improves the long-term stability of labile drugs, especially therapeutic proteins. After the lyophilization process, the dried and porous product can be converted back to its original structural form before the process by quickly absorbing the solvent with its characteristic lyophilic feature. The porous product cake formed as a result of the process has a low moisture content and can be stored for a very long time under specified storage conditions until its desired use. Water, which is indispensable for life, supports the continuation of metabolism and all vital activities as a common solvent that supports biochemical activities in the cell. Freeze-drying products are primarily composed of water, that is, solvent and other materials dissolved or suspended in water. In the absence of water, life as we know it cannot continue due to the conclusion of the dormant or death phase in living cells and the inhibition of biochemical activities in cellular extracts. In this respect, the event that should be followed in the sublimation and desorption stages for successful results is the removal of water from the components of the product exposed to the lyophilization process. Therefore, it is necessary to reduce or immobilize the water content in the stored samples to stabilize the perishable products. The key part in the lyophilization process is that the lyophilization process is considered a stabilization process. The stabilization process, on the other hand, involves greatly changing the natural kinetic clock of a substance. For example, let's assume that the effectiveness of a vaccine decreases to 90% as a result of 2 days of storage at +4°C. If the kinetic clock can be slowed and the effect of 1 second time can be extended to 1 hour by reducing the amount of moisture in the vaccine, then the stability of the vaccine can be up to 20 years instead of 2 days. The striking example here indeed highlights the undeniable potential of benefits the lyophilization process can offer researchers and manufacturers when properly executed.

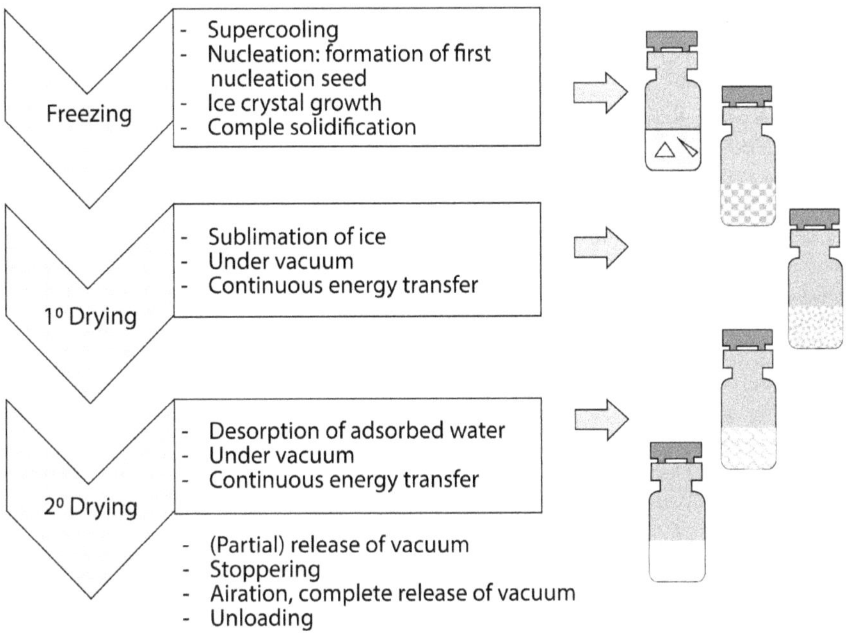

FIGURE 3.36 The Freeze drying process.

3.25.1 Why Drying?

The following steps are typically involved in the lyophilization process:

- Dissolving the drug and its excipients in an appropriate solvent, usually water for injection (WFI).
- Filtering the bulk solution with a 0.22 micron bacteria-retentive filter to sterilize it.
- Filling each sterile container individually and aseptically sealing the containers halfway.
- Carrying the containers with a half-stoppered lid to the lyophilizer and inserting them into the chamber while maintaining aseptic conditions.
- Prefreezing the solution in a separate room or freeze-drying the solution by setting the partially-stoppered containers on cooled shelves.
- Heating the shelves and creating a vacuum inside the chamber to release the frozen water.
- Complete stoppering of the vials commonly by hydraulic or screw rod stoppering mechanisms inserted in the lyophilizers.

Lyophilization is an essential process step, and any actions that may compromise the material's or product's sterility must be seen as continuations of the sterilized product's aseptic processing.

- Between filling products for lyophilization and the process's completion, the GMP auditor should confirm that the lyophilization equipment and its procedures are made to guarantee that product or material sterility is maintained during lyophilization by preventing microbial and particle contamination.
- The GMP auditor should verify that all control measures in place are determined by the site's CCS.
- The GMP auditor should verify that the sterilization of the lyophilizer and associated equipment (e.g., trays, vial support rings) are validated, and the holding time between the sterilization cycle and use appropriately challenged during APS.

TABLE 3.52
Points to be Considered for Loading and Unloading of Lyophilizer

- The loading pattern within the lyophilizer should be specified and documented.
- The transfer of partially closed containers to a lyophilizer should be undertaken under grade A conditions at all times and handled in a manner designed to minimize direct operator intervention. Technologies such as conveyor systems or portable transfer systems (e.g., clean air transfer carts, portable unidirectional airflow workstations) should be used to ensure that the cleanliness of the system used to transfer the partially closed containers is maintained. Alternatively, where supported by validation, trays closed in grade A and not reopened while in the grade B area may be used to protect partially stoppered vials (e.g., appropriately closed boxes).
- Airflow patterns should not be adversely affected by transport devices and venting of the loading zone.
- Unsealed containers (such as partially stoppered vials) should be maintained under grade A conditions and should normally be separated from operators by physical barrier technology or any other appropriate measures.
- Where seating of the stoppers is not completed prior to opening the lyophilizer chamber, product removed from the lyophilizer should remain under grade A conditions during subsequent handling.
- Utensils used during loading and unloading of the lyophilizer (e.g., trays, bags, placing devices, tweezers) should be sterile.

- The GMP auditor should verify that the lyophilizer is sterilized regularly, based on system design.
- The GMP auditor should verify that resterilization is performed following maintenance or cleaning.
- The GMP auditor should verify that sterilized lyophilizers and associated equipment are protected from contamination after sterilization.
- The GMP auditor should verify that lyophilizers and associated product transfer and loading/unloading areas are designed to minimize operator intervention as far as possible.

Table 3.52 presents important things in loading and unloading of lyophilizers.

- The GMP auditor should verify that the frequency of lyophilizer sterilization is determined based on the design and risks related to system contamination during use.
- The GMP auditor should confirm that lyophilizers are sterilized prior to each load when they are manually loaded or unloaded without the use of barrier technology separation.
- A GMP auditor should confirm that the frequency of sterilization is appropriate and recorded as part of the CCS for lyophilizers that are protected by closed barrier systems or loaded and unloaded by automated systems.
- The GMP auditor should verify that the integrity of the lyophilizer is maintained following sterilization and during lyophilization.
- The findings of the integrity testing for the lyophilizer filter should be included in the batch certification or release, and the GMP auditor should confirm that the filter is sterilized before every system usage.
- The GMP auditor should verify that the frequency of vacuum/leak integrity testing of the chamber is documented and the maximum permitted leakage of air into the lyophilizer is specified and checked at the start of every cycle.
- The GMP auditor should verify that lyophilization trays are checked regularly to ensure that they are not misshapen or damaged.

3.26 AUDITING STERILE MEDICINAL PRODUCTS

The topics I will address in this section will focus on the audits of drug products that are both aseptically processed and terminally sterilized. Product types such as ophthalmic drugs, small-volume and large-volume parenteral products, and biological products will be examined.

I will also share my thoughts and important points about EU GMP, PIC/S GMP Annex 1, which was published in August 2022.

One of the difficulties encountered in the guidelines published by all health authorities or guiding institutions is to move slower than the expected speed in harmonization. The FDA's guideline for sterile medical drug products (Guidance for Industry Sterile Drug Products Produced by Aseptic Processing – Current Good Manufacturing Practice, September 2004) needs to be harmonized with Annex 1 without delay.

Before starting the audit, that is, before designing a detailed plan, we should discuss the issues that the GMP auditors who will carry out the audit should have absolute knowledge about sterile pharmaceutical products; have completed the necessary technical training; and microbiologists, chemists, and biologists should be included in the audit team, since the subject is sterile medicinal drug products.

Specific guidelines must be followed throughout the production of sterile goods in order to reduce the possibility of microbiological, particle, and endotoxin/pyrogen contamination. As per the relevant portions of the GMP rules, facilities, equipment, and processes must be suitably constructed, qualified, and/or validated. If necessary, they should also be submitted to continuous verification. To improve the protection of the product against potential extraneous sources of endotoxin/pyrogen, particulate, and microbial contamination of personnel, materials, and the surrounding environment, as well as to aid in the prompt detection of potential contaminants in the environment and the product, the use of appropriate technologies (e.g., restricted access barriers systems [RABS], isolators, robotic systems, rapid/alternative methods, and continuous monitoring systems) should be taken into consideration.

3.26.1 Aseptic Processing

The aseptic process carries a considerable risk of microbial contamination compared to terminal sterilization. If you ask the drug manufacturers, you will see that they would prefer terminal sterilization if they had the opportunity.

The name of aseptic filling already indicates that the filling process is done under aseptic conditions. Before a sterile drug product is produced, it is expected that all components entering the drug product must be subjected to the sterilization process and all sterilized components are then brought together in an area with extremely strict environmental conditions to finish the drug product. Therefore, unlike terminal sterilization, there is a variable that we have to deal with in aseptic operations. It is possible to say that all components that are brought together after sterilization are directly exposed to the risk of microbial contamination in case of manual or mechanical intervention. For this reason, aseptic fillings are carried out in an environment characterized as cleanrooms from past to present, where access is limited and pressure differences, air exchange rates and numbers, and living and nonliving particle counts from the environment, surface, and air are controlled. Although cleanrooms are controlled environments, the inclusion of even a limited number of personnel in the environment after a point becomes a source of contamination in itself. Cleanrooms are divided into certain classes according to the criticality of the product produced, and the maximum number of living and nonliving particles to be found in each room has been determined. In case these values are exceeded (OOS) or if they are on the threshold of exceedance (OOT), necessary precautions must be taken. RABS and isolator technology have emerged instead of approaches such as cleanrooms as technology develops, the complexity of pharmaceutical products increases, and the expectations of health authorities and end user patients rise. One of the biggest advantages of RABS and Isolators is that operations are carried out with limited or no need for personnel intervention. Aseptic interventions such as blow–fill–seal also reduce the risk of contamination.

- The GMP auditor should verify that the aseptic process is clearly defined.
- The GMP auditor should ensure that the risks associated with the aseptic process, and any associated requirements, are identified, assessed, and appropriately controlled.

The site's CCS should clearly define the acceptance criteria for these controls, requirements for monitoring, and the review of their effectiveness.

- The GMP auditor should ensure that methods and procedures to control these risks are described and implemented, and accepted residual risks are formally documented.
- The GMP auditor should verify that precautions to minimize microbial, endotoxin/pyrogenic, and particle contamination are taken, as per the site's CCS, during the preparation of the aseptic environment, during all processing stages (including the stages before and after bulk product sterilization), and until the product is sealed in its final container.
- The GMP auditor should ensure that the presence of materials liable to generate particles and fibers are minimized in cleanrooms.
- When should a manufacturer consider the use of equipment such as RABS, isolators, or other systems? And why?
- Where the final formulation cannot be filtered, what should be considered?
- The GMP auditor should verify that the unwrapping, assembly, and preparation of sterilized equipment, components, and ancillary items with direct or indirect product contact are treated as an aseptic process and performed in grade A with a grade B background.
- The GMP auditor should ensure that the filling line set-up and filling of the sterile product are treated as an aseptic process and performed in grade A with a grade B background.

Table 3.53 presents cleanroom grades.

- The GMP auditor should verify that preparation and filling of sterile products such as ointments, creams, suspensions, and emulsions are performed in grade A with a grade B background when the product and components are exposed to the environment and the product is not subsequently filtered (via a sterilizing grade filter) or terminally sterilized.

TABLE 3.53
Examples of Operations and Grades for Aseptic Preparation and Processing Operations

Grade A	• Aseptic assembly of filling equipment. • Connections made under aseptic conditions (where sterilized product contact surfaces are exposed) that are post the final sterilizing grade filter. These connections should be sterilized by steam-in-place whenever possible. • Aseptic compounding and mixing. • Replenishment of sterile bulk product, containers and closures. • Removal and cooling of unprotected (e.g., with no packaging) items from sterilizers. • Staging and conveying of sterile primary packaging components in the aseptic filling line while not wrapped. • Aseptic filling, sealing of containers such as ampoules, vial closure, transfer of open or partially stoppered vials. • Loading of a lyophilizer.
Grade B	• Background support for grade A (when not in an isolator). • Conveying or staging, while protected from the surrounding environment, of equipment, components and ancillary items for introduction into grade A.
Grade C	• Preparation of solutions to be filtered including sampling and dispensing.
Grade D	• Cleaning of equipment. • Handling of components, equipment and accessories after cleaning. • Assembly under HEPA-filtered airflow of cleaned components, equipment and accessories prior to sterilization. • Assembly of closed and sterilized SUS using intrinsic sterile connection devices.

- The GMP auditor should verify that aseptic connections are performed in grade A with a grade B background unless subsequently sterilized in place or conducted with intrinsic sterile connection devices that minimize any potential contamination from the immediate environment.
- The GMP auditor should verify that aseptic manipulations (including nonintrinsic sterile connection devices) are minimized through the use of engineering design solutions such as preassembled and sterilized equipment.
- The GMP auditor should verify that product contact piping and equipment are preassembled and sterilized in place.
- The GMP auditor should verify that there are an authorized list of allowed and qualified interventions, both inherent and corrective, that may occur during production.
- The GMP auditor should verify that interventions are carefully designed to ensure that the risk of contamination of the environment, process, and product is effectively minimized.
- The GMP auditor should ensure that the process of designing interventions includes the consideration of any impact on air flows and critical surfaces and products.
- The GMP auditor should verify that aseptic technique are observed at all times, including the appropriate use of sterile tools for manipulations.
- The GMP auditor should verify that nonqualified interventions are only used in exceptional circumstances, with due consideration of the risks associated with the intervention and with the authorization of the Quality Unit.
- The GMP auditor should verify that any nonqualified interventions are thoroughly assessed by the quality department and considered during batch disposition.
- The GMP auditor should verify that interventions and stoppages are recorded in the batch record and that each line stoppage or intervention is sufficiently documented in batch records with the associated time, duration of the event, and operators involved.
- The GMP auditor should verify that the duration of each aspect of aseptic preparation and processing are minimized and limited to a defined and validated maximum time.

Table 3.54 presents holding time essentials.

- The GMP auditor should verify that aseptic operations (including APS) are observed on a regular basis by personnel with specific expertise in aseptic processing to verify the correct performance of operations including operator behavior in the cleanroom and address inappropriate practices if detected.
- The GMP auditor should verify that open primary packaging containers are maintained under grade A conditions with the appropriate background for the technology and that final containers are closed by appropriately validated methods.

TABLE 3.54
Things to be Considered in Holding Time

- The holding time between equipment, component, and container cleaning, drying and sterilization.
- The holding time for sterilized equipment, components, and containers before use and during filling/assembly.
- The holding time for a decontaminated environment, such as the RABS or isolator before use.
- The time between the start of the preparation of a product and its sterilization or filtration through a microorganism-retaining filter (if applicable), through to the end of the aseptic filling process. There should be a maximum permissible time for each product that takes into account its composition and the prescribed method of storage.
- The holding time for sterilized product prior to filling.
- The aseptic processing time.
- The filling time.

- GMP auditor should verify that blow–fill–seal (BFS), form–fill–seal (FFS), small and large-volume parenteral (SVP and LVP) bags, glass or plastic ampoules, the critical parameters and variables that affect seal integrity are evaluated, determined, effectively controlled and monitored during operations.

Glass ampoules, BFS units, and small-volume containers (≤100 ml) closed by fusion should be subject to 100% integrity testing using validated methods.

- When is a reduced sampling acceptable?
- The GMP auditor should verify that visual inspection is not considered as an acceptable integrity test method.
- The GMP auditor should verify that the sample sizes are based on information such as supplier management, packaging component specifications, and process knowledge.
- The GMP auditor should verify that containers sealed under vacuum are tested for maintenance of vacuum after an appropriate predetermined period prior to certification/release and during shelf life.
- The GMP auditor should verify that the container closure integrity validation takes into consideration any transportation or shipping requirements that may negatively impact the integrity of the container (e.g., by decompression or extreme temperatures).
- The GMP auditor should verify that where the equipment used to crimp vial caps can generate large quantities of nonviable particles, measures to prevent particle contamination such as locating the equipment at a physically separate station equipped with adequate air extraction are taken.

Critical defects should not be identified during any subsequent sampling and inspection of acceptable containers. Any critical defect identified subsequently should trigger an investigation as it indicates a possible failure of the original inspection process.

What type of precautions are taken when inspection is performed manually?

- The GMP auditor should verify that inspection rates are appropriately controlled and qualified.
- The GMP auditor should verify that operators performing the inspection undergoes visual inspection qualification (while wearing corrective lenses, if these are normally worn) at least annually.
- The GMP auditor should verify that the qualification is undertaken using appropriate samples from the manufacturer's defect library sets and taking into consideration worst-case scenarios and including consideration of eyesight checks. (e.g., inspection time, line speed where the product is transferred to the operator by a conveyor system, container size, or fatigue).
- The GMP auditor should verify that operator distractions are minimized and frequent breaks, of an appropriate duration, are taken from inspection.
- The GMP auditor should verify that the process is validated to detect known defects (which may impact product quality or safety) and be equal to, or better than, manual inspection methods.
- The GMP auditor should verify that the performance of the equipment is challenged using representative defects prior to start-up and at regular intervals throughout the batch.

3.26.2 Quality Management System

The manufacture of sterile products is a complex activity that requires specific controls and measures to ensure the quality of the products manufactured.

The pharmaceutical quality system (PQS) for sterile product manufacture should also ensure that: An effective risk management system is integrated into all areas of the product life cycle with the aim to minimize microbial contamination and ensure the quality of sterile products manufactured.

The manufacturer has sufficient knowledge and expertise in relation to the products manufactured and the equipment, engineering and manufacturing methods employed that have an impact on product quality.

Root cause analysis of procedural, process, or equipment failure is performed in such a way that the risk to the product is correctly identified and understood so that suitable CAPAs are implemented.

Risk management is used in the creation and upkeep of the CCS to detect, evaluate, and, if necessary, decrease or eliminate contamination risks.Risk management should be documented and should include the rationale for decisions taken in relation to risk reduction and acceptance of residual risk.

Processes associated with the finishing, storage, and transport of sterile products should not compromise the sterile product. Aspects that should be considered include container integrity, risks of contamination, and avoidance of degradation by ensuring that products are stored and maintained in accordance with the registered storage conditions.

Those in charge of certifying or releasing sterile products have the right access to production and quality data, as well as sufficient training and expertise in producing sterile products and the related essential qualities.This is in order to allow such persons to determine if the sterile products have been manufactured in accordance with the registered specifications and approved process and are of the required quality.

Senior management should effectively oversee the state of control throughout the facility and product lifecycle. Risk management outcomes should be reviewed regularly as part of the ongoing quality management, during change, in the event of a significant emerging problem, and during the periodic product quality review.

3.26.3 Deviations, Root-Cause Analysis

All nonconformities, such as sterility test failures, environmental monitoring excursions, or deviations from established procedures should be adequately investigated before certification/release of the batch. The investigation should determine the potential impact on process and product quality and whether any other processes or batches are potentially impacted. The reason for including or excluding a product or batch from the scope of the investigation should be clearly justified and recorded.

Auditing of the quality system requires two phases. The first phase is to evaluate whether the Quality Unit has fulfilled its responsibility regarding the review and approval of procedures and assured their suitability for use. The second phase is to assess the data collected by the firm to identify potential quality problems. For sterile manufacturing operations, this latter objective involves large amounts of data that link to the other inspectional systems. The comprehensive review of such data by the firm is an essential element in ensuring that products are produced with a high degree of sterility assurance. It is therefore important to review the firm's system for using the data to assess the state of control of their manufacturing operations and facility. The data summaries and trend reports maintained by the Quality Unit should be reviewed during every inspection.

3.26.4 Materials Handling

The areas are applicable to sterile drug products and should be covered if this system is selected. In sterile operations, the quality attributes of each of the materials (ingredients, WFI, containers, closures) have a bearing on the critical attributes of the finished product.

The GMP auditor should review the firm's procedures for receipt, handling, sampling, testing, approval, and storage of manufacturing materials to verify their fitness for use. Emphasis should be placed on incoming materials that are represented to be sterile and/or pyrogen-free. It is important to properly monitor and analyze raw materials and packaging materials to make sure that the levels of endotoxin/pyrogen and bioburden are appropriate for use.

3.26.5 Quality Risk Management

QRM concepts should be applied to the management of processes, equipment, facilities, and industrial operations in order to preventatively identify, assess, and regulate possible hazards to quality. When using alternative strategies, they should adhere to Annex 1's goals and be justified with the right kind of analysis, risk assessment, and mitigation. The design of the building, its machinery, and its processes should be done in a way that makes sense. Next, carefully thought-out procedures should be put into place, and lastly, monitoring systems should be used to show that the procedures and design have been carried out correctly and are still performing as intended. Sterile conditions cannot be guaranteed by monitoring or testing alone.

3.26.6 Records Management

1. Periodic product evaluations, complaints, adverse events, investigations, product retain evaluations, complaints, rejected lots, stability, and returned goods that indicate possible product contamination or risks to patients (for example, hazy or cloudy product, foreign matter/particulates in injectable products, cracked, and leaky containers).
2. Discrepancy and failure investigations, such as:
   - All initial positive sterility tests and endotoxin and media fill failures regardless of final disposition.
   - Unexpected results or trends.
   - All failures that occurred during validation or revalidation of sterilization/depyrogenation processes.
   - All investigations involving media fills/process simulations.
   - Environmental (microbial/viable and particle/nonviable counts) and personnel monitoring results that exceed alert or action levels.
   - Process deviations or equipment malfunctions that involve critical equipment, such as sterilizers and lyophilizers.
   - Out of specification (OOS) results for assay, impurities, particulate matter, or reconstitution time, if applicable.
   - Product rejects (rejects determined during manufacturing and quality control test).
3. Trends reports/summaries of quality indicators:
   - For aseptic processing, a summary of all media fills performed since the last audit.
   - Environmental monitoring trend data (microbial and particle counts).
   - Personnel monitoring trend data.
   - Summary of water system test results.
4. Summary of change controls for critical utilities and equipment implemented since the last audit, for example:
   - Sterilizers, lyophilizers, depyrogenation equipment
   - Aseptic processing line
   - Clean steam generator, process gas system
   - WFI and/or Purified Water system
   - Air handling systems
   - Automated building management system

Every audit of a sterile drug manufacturer should include a review of the type of information listed above and an observation of the manufacturing operations occurring in the critical areas. The information can be used to select other system(s) to be covered during the audit.

In addition, the review of summary data and observation of operations can focus the audit on potential problem areas and it provides an overview of the effectiveness of the quality system. The audit of the quality system may necessitate follow-up within another system. However, this coverage does not constitute or require complete coverage of these systems.

3.26.7 Facilities and Equipment

The principle objective of an effective sterile drug manufacturing operation from a facility and equipment standpoint is to provide suitable protection for the product.

- Review and evaluate the firm's rationale for, and adequacy of, the facility and equipment design (Reference: Section IV of FDA's 2004 Aseptic Processing Guidance, EU, PIC/S GMP Annex 1).
- Evaluate the data that provide information relevant to the state of control of the facility and equipment.

In addition to the review of design elements and data, the GMP auditor should look for visible deficiencies in the facility and equipment, such as cleanliness, equipment deterioration (e.g., warping, corrosion), inaccessible and/or difficult-to-clean surfaces, and changes to critical equipment or systems that have not been qualified which may impact product quality.

- The GMP auditor should look for aberrant events due to facility deterioration, a pattern of recurring and uncorrected maintenance issues, and increase or changes in production output that exceed the capacity of the facility and equipment.

3.26.8 Contamination Control Strategy

In order to manage risks to the quality and safety of pharmaceutical products, a CCS should be applied throughout the facility to identify all essential control points and evaluate the efficacy of all controls (design, procedural, technical, and organizational) and monitoring mechanisms. Strong assurance of contamination prevention should be established by the CCS's combined approach. The CCS should encourage ongoing development of the manufacturing and control processes and be actively reviewed and revised as needed. Its efficacy ought to be considered in the recurring management evaluation. It might not be necessary to replace control systems that are already in place and being properly maintained, but they should be mentioned in the CCS and the relationships between the various systems should be recognized.

Contamination control and steps taken to minimize the risk of contamination from microbial, endotoxin/pyrogen, and particle sources include a series of interrelated events and measures. These are typically assessed, controlled, and monitored individually, but their collective effectiveness should be considered together.

It takes in-depth technical and process knowledge to develop the CCS. Particulates (e.g., glass and other visible and subvisible particles) and microbiological and cellular waste (e.g., pyrogen, endotoxin) are potential sources of contamination.

The CCS should consider all aspects of contamination control with ongoing and periodic reviews resulting in updates within the pharmaceutical quality system as appropriate. Changes to the systems in place should be assessed for any impact on the CCS before and after implementation. Table 3.55 presents a list of elements during a CCS study.

The manufacturer should take all steps and precautions necessary to ensure the sterility of the products manufactured within its facilities. Sole reliance on sterility or other quality aspects should not be placed on any terminal process or finished product test.

TABLE 3.55
Elements to be Considered within a CCS

- Design of both the plant and processes including the associated documentation.
- Premises and equipment.
- Personnel.
- Utilities.
- Raw material controls, including in-process controls.
- Product containers and closures.
- Vendor approval, such as key component suppliers, sterilization of components and single-use systems (SUS), and critical service providers.
- Management of outsourced activities and availability/transfer of critical information between parties, e.g., contract sterilization services.
- Process risk management.
- Process validation.
- Validation of sterilization processes.
- Preventive maintenance – maintaining equipment, utilities and premises (planned and unplanned maintenance) to a standard that will ensure there is no additional risk of contamination.
- Cleaning and disinfection.
- Monitoring systems, including an assessment of the feasibility of the introduction of scientifically sound, alternative methods that optimize the detection of environmental contamination.
- Prevention mechanisms: trend analysis, detailed investigation, root cause determination, corrective and preventive actions (CAPA) and the need for comprehensive investigational tools.
- Continuous improvement based on information derived from the above.

3.26.9 Facilities

GMP auditor must evaluate the design and layout of the facility (e.g., personnel/material flow, cleanroom design).

Specifications for cleanroom areas (layout, air filtration, appropriate air classification, pressure differentials between rooms and areas, temperature, and humidity) should be appropriate, and based on the risk of product contamination with particulate matter and microorganisms.

The GMP auditor should review the certification and qualification of the cleanroom areas to verify the areas meet design criteria and specifications. Certification and qualification typically include data in support of the following: airflow pattern studies; HEPA filter integrity testing; air velocity measurement; nonviable particle; and verification of appropriate pressure differentials, temperature, and humidity setpoints.

The GMP auditor should evaluate the airflow pattern (smoke studies) conducted under dynamic conditions to verify the unidirectional airflow and air turbulence within the critical area where sterilized drug products, containers, and closures are exposed to environmental conditions. Cleanrooms and change rooms should be maintained to an appropriate cleanliness standard and supplied with air that has passed through filters of an appropriate efficiency.

The GMP auditor should pay special attention to facilities that are performing construction in the clean areas or in the vicinity of cleanrooms. Because microbes (e.g., fungal spores) can be liberated from the movement of walls and other construction activities, determine if the facility is returned to acceptable environmental control through proper measures (environmental monitoring, media fills) before production can resume.

The GMP auditor should verify that environmental monitoring of nonviable particles is occurring during operations including sites where there is the most risk to exposed products, containers, and closures.

The GMP auditor should check if pressure differentials, temperature, and humidity are continuously monitored during routine production.

The GMP auditor should determine if continuous monitoring systems are alarmed to alert operators of excursions.

The GMP auditor should check if excursions from acceptable ranges are investigated to determine the impact on the product and that needed corrective actions are taken.

The GMP auditor should evaluate the program for periodic testing/recertification of the HEPA filters in critical areas to maintain appropriate airflow. The tests typically include integrity testing of the HEPA filters and air velocity checks.

- The sanitization/disinfection of cleanroom areas, processing lines, and nonautoclavable equipment, materials, and components should be reviewed.
- The GMP auditor should focus on the areas where the sterile product is exposed up to and including sealing operations. These critical areas represent the highest risk to the product. The suitability, efficacy, and limitations of disinfecting agents and the adequacy of procedures should be reviewed, including the data that establishes the expiry of the disinfection solution.
- For multiuse facilities and non-dedicated equipment, the GMP auditor should evaluate the adequacy of the changeover procedures and cleaning to prevent cross-contamination between products.

For the manufacture of sterile products, there are four grades of cleanroom/zone:

Grade A: The critical zone for high-risk operations (e.g., aseptic processing line, filling zone, stopper bowl, open primary packaging or for making aseptic connections under the protection of first air). Normally, such conditions are provided by localized airflow protection, such as unidirectional airflow workstations within RABS or isolators. The maintenance of unidirectional airflow should be demonstrated and qualified across the whole of the grade A area. Direct intervention (e.g., without the protection of barrier and glove port technology) into the grade A area by operators should be minimized by premises, equipment, process and procedural design.

Grade B: For aseptic preparation and filling, this is the background cleanroom for grade A (where it is not an isolator). Air pressure differences should be continuously monitored. Cleanrooms of lower grade than grade B can be considered where isolator technology is used.

Grades C and D: These are cleanrooms used for carrying out less critical stages in the manufacture of aseptically filled sterile products or as a background for isolators. They can also be used for the preparation/filling of terminally sterilized products.

- The GMP auditor should verify that in cleanrooms and critical zones, all exposed surfaces are smooth, impervious, and unbroken in order to minimize the shedding or accumulation of particles or microorganisms.
- The GMP auditor should ensure that reducing the accumulation of dust and, to facilitate cleaning, that there are no recesses that are difficult to clean effectively, therefore projecting ledges, shelves, cupboards, and equipment should be kept to a minimum.
- The GMP auditor should verify that doors are designed to avoid recesses that cannot be cleaned. Sliding doors may be undesirable for this reason.
- The GMP auditor should ensure that materials used in cleanrooms, both in the construction of the room and for items used within the room, are selected to minimize the generation of particles and to permit the repeated application of cleaning, disinfectant, and sporicidal agents where used.
- The GMP auditor should verify that ceilings are designed and sealed to prevent contamination from the space above them.

- The GMP auditor should observe that sinks and drains are prohibited in the grade A and grade B areas. In other cleanrooms, air breaks are fitted between the machine or sink and the drains. Floor drains in lower-grade cleanrooms should be fitted with traps or water seals designed to prevent backflow, and should be regularly cleaned, disinfected, and maintained.
- The GMP auditor should be aware that the transfer of equipment and materials into and out of the cleanrooms and critical zones is one of the greatest potential sources of contamination.
- The GMP auditor should ask questions about whether any activities with the potential to compromise the cleanliness of cleanrooms or the critical zone are assessed and if they cannot be eliminated, appropriate controls are implemented.
- The GMP auditor should observe that the transfer of materials, equipment, and components into the grade A or B areas is carried out via a unidirectional process. The auditor also should be aware that where possible, items are sterilized and passed into these areas through double-ended sterilizers (e.g., through a double-door autoclave or depyrogenation oven/tunnel) sealed into the wall. The GMP auditor should ask questions about procedures and related activities if sterilization upon transfer of the items is not possible.
- The GMP auditor should ensure that the removal of items from grade A and B areas (e.g., materials, waste, environmental samples) is carried out via a separate unidirectional process. If this is not possible, time-based separation of movement (incoming/exiting material) by procedure should be considered and controls applied to avoid potential contamination of incoming items.
- The GMP auditor should observe that airlocks are designed and used to provide physical separation and to minimize microbial and particle contamination of the different areas and are present for material and personnel moving between different grades.
- The GMP auditor asks questions about whether airlocks used for personnel movement are separated from those used for material movement.
- The GMP auditor should verify that time-based separation of movement (personnel/material) by procedure is considered.
- The GMP auditor should observe that airlocks are flushed effectively with filtered air to ensure that the grade of the cleanroom is maintained. The auditor also should observe that the final stage of the airlock is, in the "at rest" state, of the same cleanliness grade (viable and total particle) as the cleanroom into which it leads.
- The GMP auditor should ask questions about whether management provides separate change rooms for entering and leaving the grade B area. How does the management of the sterile facility know that the risk of contamination is high and separate change rooms for entering and leaving production areas are used?
- The GMP auditor should verify that handwashing facilities are provided only in the first stage of the locker room and that they are not in the locker rooms that reach class B directly.
- The GMP auditor should only confirm that the materials and equipment, which are included in an approved list and evaluated during the transfer of the transfer process, are transferred to A or B classes through an air lock or transition covers.
- The GMP auditor should also witness that the equipment and materials (designed to be used for use in class A) when appropriate are protected.
- The GMP auditor must verify that the manufacturer has implemented and recorded appropriate risk assessment and reduction measures according to CCs and contains a specific disinfection and monitoring program approved by the quality assurance.
- The GMP auditor must confirm that the transport of materials or equipment from low-graded or higher degrees to clean areas are subject to cleaning and disinfection compatible with risk and CCS.

Where applicable, the GMP auditor should test that the input and output doors are not opened at the same time for the transition caps and airlock (for material and personnel).

- The GMP auditor must verify that a passing system is used for airlock to A and B classes. For the airlock to class C and D areas, a minimum visual and/or audio warning system should be operated.
- The GMP auditor should question whether a time delay has been created between the closure and opening of locked doors.
- The GMP auditor should verify that clean rooms provide a positive pressure and/or an airflow that maintains a flowed air flow according to the background environment of a lower degree under all operational conditions.
- The GMP auditor should verify at least 10 pascal (guidance value) air pressure differences in adjacent rooms of the degrees.
- The GMP auditor will also want to learn which additional measures are taken to protect the critical region. The GMP auditor should question how specific materials (e.g., pathogenic, highly toxic, or radioactive products or live viral or bacterial materials) are changed in places that should be changed. They will also want to know what modification covers.
- The GMP auditor should observe that airflow patterns within cleanrooms and zones are visualized to demonstrate that there is no ingress from lower-grade to higher-grade areas and that air does not travel from less clean areas (such as the floor) or over operators or equipment that may transfer contamination to the higher grade areas.
- The GMP auditor should verify that if unidirectional airflow is required, visualization studies are performed to determine compliance.
- The GMP auditor should ask questions about the transfer of filled, closed products to an adjacent cleanroom of a lower grade via a small egress point.

Where applicable, the GMP auditor should verify that airflow visualization studies demonstrate that air does not ingress from the lower-grade cleanrooms to the grade B area.

- GMP auditor should require to know whether corrective actions such as design improvement are implemented in case of any contamination risk.

The auditor will also require to see the evidence that airflow pattern studies are performed both at rest and in operation (e.g., simulating operator interventions).

- The GMP auditor may want to see that video recordings of the airflow patterns are retained. They also may want to see that the outcome of the air visualization studies is documented and considered when establishing the facility's environmental monitoring program.
- The GMP auditor should ensure that indicators of air pressure differences are fitted between cleanrooms and/or between isolators and their background. They also may want to see that set points and the criticality of air pressure differences are considered within the CCS. Are air pressure differences identified as critical continuously monitored and recorded?

Is there any warning system that in place to instantly indicate and warn operators of any failure in the air supply or reduction of air pressure differences (below set limits for those identified as critical)?

- The GMP auditor should verify that the warning signal is not overridden without assessment and that a procedure is available to outline the steps to be taken when a warning signal is given.
- The GMP auditor should verify that if alarm delays are set, these are assessed and justified within the CCS.

Other air pressure differences should be monitored and recorded at regular intervals.

- The GMP auditor should verify that facilities are designed to permit observation of production activities from outside the grade A and B areas (e.g., through the provision of windows or remote cameras with a full view of the area and processes to allow observation and supervision without entry).
- The GMP auditor should verify that these requirements are considered when designing new facilities or during the refurbishment of existing facilities.

3.26.10 Equipment

Equipment used in the manufacture of sterile drug products may include the following:

- Production equipment
- Container/closure processing equipment (e.g., stopper washer, glassware depyrogenation equipment)
- Support system/material system-related equipment (e.g., WFI system and related equipment, process gas-related equipment)

3.26.11 Aseptic Processing Equipment

- The GMP auditor should determine that all equipment that comes in direct contact with the product (e.g., filters, transfer lines, holding tanks, stopper bowls, filling line equipment) and sterile components (e.g., stoppers) are sterilized and protected from contamination prior to and during use. Equipment logs or other related information may provide insight into significant maintenance or other problems that may increase the exposure of batches to contamination risk.
- The GMP auditor should verify that a written, detailed description of the equipment design is available (including process and instrumentation diagrams as appropriate).
- The GMP auditor should verify that this forms part of the initial qualification package and be kept up to date.
- The GMP auditor should verify that equipment monitoring requirements are defined in "user requirements specifications" during early stages of development and confirmed during qualification. The auditor should observe that process and equipment alarm events are acknowledged and evaluated for trends. The frequency at which alarms are assessed should be based on their criticality (with critical alarms reviewed immediately).
- The GMP auditor should verify that equipment, fittings and services are designed and installed so that operations, maintenance, and repairs can be performed outside the cleanroom. What happens if maintenance has to be performed in the cleanroom, and the required standards of cleanliness and/or asepsis cannot be maintained?
- The GMP auditor should verify that the cleaning process is validated to be able to:
  - Eliminate any residue or debris that could reduce the disinfecting agent's effectiveness.
  - Reduce the amount of particles, microbes, and chemicals that contaminate the product both during the process and prior to disinfection.
- Does the facility sterilize direct and indirect product contact parts? Direct product contact parts are those that the product passes through, such as filling needles or pumps. Indirect product contact parts are equipment parts that do not contact the product, but may come into contact with other sterilized surfaces, the sterility of which is critical to the overall product sterility (e.g., sterilized items such as stopper bowls and guides, and sterilized components).

- The GMP auditor should verify that all equipment such as sterilizers, air handling systems (including air filtration), and water systems are subjected to qualification, monitoring and planned maintenance. What happens upon completion of maintenance?
- What type of precautions are taken where unplanned maintenance of equipment critical to the sterility of the product is to be carried out?

A conveyor belt should not go over a divider separating a processing area with lower air quality from a grade A or B area unless the belt is continuously sterilized, for example, in a sterilizing tunnel.

- The GMP auditor should verify that particle counters, including sampling tubing, are qualified. The GMP auditor should verify that portable particle counters with a short length of sample tubing are used for classification purposes.
- What type of sampling heads are used in unidirectional airflow systems?
- Why should they be oriented appropriately and positioned as close as possible to the critical location?

Stopper washer

- The GMP auditor should verify the qualification of the equipment, cycle validation and supporting data, equipment preventive maintenance (maintenance requirements and frequency), quality of water used for washing, and associated water sampling/qualification data. The appropriateness of the air supply used in any drying operations should also be verified.

Capping Equipment (vials)

The vial cap provides the final closure element of a sealed vial. The capping machine folds and crimps the cap (aluminum) over the neck of the stoppered vial. The cap on the vial protects the stopper from external damage, while firmly holding the stopper in the fully seated, sealed position.

- The GMP auditor should evaluate the established processing settings (crimp angles, pressures), and preventive maintenance schedules of the capping machine. Air supply quality to the capping units should also be evaluated.

3.26.12 Post-Fill Visual Inspection/Automated Inspection Equipment

A 100% inspection of the final filled and sealed product may occur via a manual, automated, or semiautomated inspection process. Manual and semiautomated inspection processes involve specified viewing fields and calibrated light sources. Semiautomated processes may use conveyor belts and rotational units that present the filled product to an operator for visual inspection. All conveyor and rotational speed set points should be verified against established parameters. Automated inspection systems may inspect for one or all types of defects in a given filled product. Defect categories with relevant action levels should be defined. The qualification of the equipment and the challenges performed to verify equipment functionality prior to routine use should be evaluated as well as the training program for operators performing manual visual inspections.

3.26.13 Sterilizers

The GMP auditor should cover the installation and operation qualification of equipment and the performance qualification of the process (IQ, OQ, and PQ); and operation, calibration, and preventive maintenance of representative types of equipment used to sterilize finished dosage forms, filling equipment, containers, closures, etc. Such equipment includes autoclaves, dry heat ovens, dry heat tunnels, steam-in-place equipment, and chemical sterilization systems (i.e., hydrogen peroxide, peracetic acid). The GMP auditor should include a physical examination of the equipment.

- The GMP auditor should review the engineering specifications, which may be described in the equipment's DQ (design qualification). DQ is performed prior to the IQ and OQ, and verify that the sterilizer is maintained, calibrated, and drained properly and that it has appropriate measuring devices (temperature sensors, pressure gauges, etc.).

Records of unplanned maintenance, as well as preventive maintenance, should be reviewed to ensure all significant changes have been evaluated and qualified as appropriate. Equipment logs should also be reviewed. For example, repeat sterilization of loads because of cycle failures can indicate a serious problem with a sterilizer. The impact of resterilization on product quality should be evaluated. (Performance qualification is covered under the production system.)

The equipment can be computer-controlled or operated in a manual mode. For a computer-controlled system, a programmable logic controller (PLC) or a more complex Supervisory Controlled and Data Acquisition Management System (SCADA) may require an assessment to determine if the computer control and/or monitoring system are Part 11 compliant.

3.26.14 Barrier Technologies

Isolators or RABS, which are different technologies, and associated processes should be designed to provide protection through the separation of the grade A environment from the environment of the surrounding room. The hazards introduced by the entry or removal of items during processing should be minimized and supported by high-capability transfer technologies or validated systems that robustly prevent contamination and are appropriate for the respective technology.

Isolators

- GMP auditors should evaluate the design and control elements that maintain the separation or isolation of the product. Pressure differential, glove integrity, and protection of the transfer (i.e., entry, exit) ports are key elements for the isolators. The transfer of containers, closures, and supplies (including environmental monitoring supplies) into an isolator should be carefully controlled. Another critical element for these systems is the effectiveness of the chamber decontamination program. Current methods (e.g., vaporized hydrogen peroxide, steam hydrogen peroxide, peracetic acid) used to decontaminate isolator barriers are capable of *surface* sterilization but lack the penetrating capabilities of steam sterilization.
- GMP auditors should be mindful of the limitations of these surface sterilants, including their inefficiency in penetrating obstructed or protected surfaces. Validation of the decontamination of the interior (surfaces) of an isolator should demonstrate a 6-log reduction of the biological indicator (BI).

Quantitative measuring devices (e.g., near-infrared) or chemical indicators (qualitative test) can be used to determine the worst-case location for decontamination validation using BIs. Factors to be considered in decontamination validation include the location of the BI and the type of surfaces where the BIs are inoculated. Utensils and equipment surfaces inside the isolator that have direct contact with sterile products and components should be sterilized to render them free of microorganisms. The sterilization validation should achieve a minimum of a 6-log reduction of the BI.

- The GMP auditor should verify that the design of open isolators ensures grade A conditions with first air protection in the critical zone and unidirectional airflow that sweeps over and away from exposed products during processing.

- The GMP auditor should be aware that the design of closed isolators ensures grade A conditions with adequate protection for exposed products during processing.
- The GMP auditor should know that airflow may not be fully unidirectional in closed isolators where simple operations are conducted.
- The GMP auditor should be observing that any turbulent airflow does not increase the risk of contamination of the exposed product.
- The GMP auditor should know when negative pressure isolators are only used when containment of the product is considered essential (e.g., radiopharmaceutical products).
- The GMP auditor should verify that specialized risk control measures are applied to ensure the critical zone is not compromised.
- The GMP auditor must be trained to learn that the background environment for open isolators generally corresponds to a minimum grade C. They should verify that the background for closed isolators corresponds to a minimum of grade D.
- The GMP auditor should observe that the decision on the background classification is based on risk assessment and justified in the CCS.
- When conducting a risk assessment for the CCS of an isolator, the GMP auditor should be aware that important factors to take into account include the bio-decontamination program, the level of automation, the impact of glove manipulations that could jeopardize the "first air" protection of critical process points, the potential loss of barrier/glove integrity, the transfer mechanisms in place, and maintenance or setup activities that might necessitate opening the doors before the isolator's final bio-decontamination.
- The GMP auditor also should verify that when additional process risks are identified; a higher grade of background is considered unless appropriately justified in the CCS.
- The GMP auditor should be aware that airflow pattern studies are performed at the interfaces of open isolators to demonstrate the absence of air ingress.
- The GMP auditor should verify that leak testing of the glove system is performed using a methodology demonstrated to be suitable for the task and criticality.
- The GMP auditor should ask about the intervals of the testing.

When is generally, glove integrity testing performed? Let's say at the beginning and end of each batch or campaign.

- When is additional glove integrity testing is necessary?
- How frequently is an integrity/leak test performed?
- Does glove integrity monitoring include a visual inspection associated with each use and following any manipulation that may affect the integrity of the system?

Decontamination methods (cleaning and bio-decontamination, and where applicable inactivation for biological materials) should be appropriately defined and controlled. The cleaning process prior to the bio-decontamination step is essential; any residues that remain may inhibit the effectiveness of the decontamination process. Evidence should also be available to demonstrate that the cleaning and bio-decontamination agents used do not have an adverse impact on the product produced within the RABS or isolator.

- The GMP auditor should verify that the bio-decontamination process of the interior is automated, validated, and controlled within defined cycle parameters and includes a sporicidal agent in a suitable form (e.g., gaseous or vaporized form).
- The GMP auditor should verify that gloves are appropriately extended with fingers separated to ensure contact with the agent.
- Do the methods used (cleaning and sporicidal bio-decontamination) render the interior surfaces and critical zone of the isolator free from viable microorganisms?

Restricted Access Barrier System (RABS)

In general, a RABS is a fill-finish line in a rigid wall enclosure that provides full physical separation of the filling line from operators. It is important to note that the inside surfaces of the RABS are disinfected with a sporicidal agent, but this is not accomplished using the automated decontamination cycles employed for isolators. This requires firms to carefully supervise disinfection procedures and ensure the ongoing effectiveness of the disinfection program. Operators use glove ports, half suits, or automation to access areas within the enclosure during filling. There are two types of RABS: "open" and "closed." The doors to a closed RABS are never opened during an operation. Whereas an open RABS is designed to operate with doors closed at all times, on rare predefined circumstances the doors of the enclosure can be opened to perform certain interventions. If doors are routinely opened during a filling operation, the system is not considered a RABS because it no longer restricts access to the critical areas. Typically, the cleanroom surrounding the RABS is controlled as a Class 10,000 (ISO 7) area and operators are fully gowned. The CCS should take into account their use. Any alternative methods for using isolators or RABS should be supported by evidence.

The design of RABS should ensure grade A conditions with unidirectional airflow and first air protection in the critical zone. A positive airflow from the critical zone to the supporting background environment should be maintained.

The background environment for RABS used for aseptic processing should correspond to a minimum of grade B, and airflow pattern studies should be performed to demonstrate the absence of air ingress during interventions, including door openings if applicable.

When auditing a RABS:

- The GMP auditor should determine that the gloves and gauntlets attached to the glove ports are sterile when installed. After installation, the gloves should be disinfected or changed at appropriate intervals to minimize the risk of contamination.
- The GMP auditor should verify there is a well-defined written procedure that describes what is done when an open-door intervention is performed. All open-door interventions should be documented and described in batch records, and followed by disinfection.
- RABS entry is often accompanied by an appropriate line clearance, which should be clearly documented in batch records.
- The GMP auditor should determine that all fluid pathways and product contact parts such as stopper bowl, feed, and placement systems are sterilized prior to the filling of each batch.
- The GMP auditor should observe how sterile components and supplies are transferred to the RABS.
- The GMP auditor should verify that the transfer system prevents exposure of sterile surfaces to less clean environments.
- The GMP auditor should verify that non-product contact surfaces within the RABS undergo thorough disinfection with a sporicidal agent before each batch. The effectiveness of the overall disinfection program should be validated and routinely evaluated by the environmental monitoring program.
- The GMP auditor should verify that gloves used in the grade A area are sterilized before installation and sterilized or effectively bio-decontaminated by a validated method prior to each manufacturing campaign.
- If exposed to the background environment during operation, is any disinfection implemented?
- Is any checklist available at the facility for the visual examination of gloves before each use?
- Is any integrity testing being performed at periodic intervals?

- The GMP auditor should verify that the sporicidal disinfection includes the routine application of a sporicidal agent using a method that has been validated and demonstrated to robustly include all areas of the interior surfaces and ensure a suitable environment for aseptic processing.

3.26.15 The Reactor, Centrifuge, Dryer, Mill

This type of equipment can be used to aseptically manufacture sterile bulk APIs. The equipment and all transfer piping must be sterilized prior to processing. This is typically done with sterilize-in-place (SIP) systems that use clean steam or a chemical sterilant.

- The GMP auditor should review the validation, cycle controls, and routine monitoring of the SIP system. The equipment and all transfer piping must remain integral (no fluid or air leaks) and sterile throughout the entire manufacturing process.
- The GMP auditor should determine how the firm verifies the integrity of the equipment train throughout the process. If a piece of equipment is opened in the process (e.g. adding seed crystals), verify the area surrounding the open operation is robustly protected from contamination risks with a Class 100 (ISO 5) air system as well as implement a carefully designed aseptic operation.

3.26.16 Processing Equipment

Depyrogenation equipment may include a dry heat oven and/or depyrogenation tunnel. Depyrogenation of stoppers can also be accomplished by dilution via a washing process. The final rinse of the washing process uses WFI.

3.26.17 Support Utilities

- The GMP auditor should verify that the nature and extent of controls applied to utility systems are commensurate with the risk to product quality associated with the utility.
- The GMP auditor should verify that the impact is determined via a risk assessment and documented as part of the CCS.

Table 3.56 presents higher-risk utilities.

- The GMP auditor should verify that utilities are designed, installed, qualified, operated, maintained, and monitored in a manner to ensure that the utility system functions as expected.
- The GMP auditor should verify that results for critical parameters and critical quality attributes of high-risk utilities are subject to regular trend analysis to ensure that system capabilities remain appropriate.

TABLE 3.56
Higher-Risk Utilities

- Directly contact product, e.g., water for washing and rinsing, gases, and steam for sterilization.
- Contact materials that will ultimately become part of the product.
- Contact surfaces that come into contact with the product.
- Otherwise directly impact the product.

TABLE 3.57
Pipes, Drains, and Slopes

• Pipeline flow direction, slopes, diameter, and length • Tank and vessel details • Valves, filters, drains, sampling, and user points

- The GMP auditor should verify that records of utility system installation are maintained throughout the system's life cycle.
- The GMP auditor should verify that such records include current drawings and schematic diagrams, construction material lists, and system specifications.
- The GMP auditor should verify that pipes, ducts, and other utilities are not present in cleanrooms. If unavoidable, then they should be installed so that they do not create recesses, unsealed openings, and surfaces that are difficult to clean.

Table 3.57 presents pipes, drains, and slopes.

- The GMP auditor should verify that installation allows cleaning and disinfection of outer surface of the pipes.

Water Systems

Water for injection (WFI) is an ingredient in many sterile drugs, including injectable products and sterile ophthalmic products. It is also used in depyrogenation (or endotoxin removal) of equipment and stoppers, and in cleaning operations. The quality of the water, and its endotoxin levels and controls, used in the upstream process should also be evaluated in order to ensure the removal of bacterial endotoxin to the appropriate level downstream. Purified water can be used for some sterile noninjectable solutions.

- The GMP auditor should observe and understand elements of the generation and distribution systems.
- The GMP auditor should evaluate the water system's as-built diagrams and inspect for leaks, pipe slopes (via the isometric diagrams and verification of the degree of the slopes), so-called dead legs, and nonsanitary fittings in the distribution system.
- The GMP auditor should evaluate how microbial alert and action levels are established.
- The GMP auditor should evaluate sampling sites, procedures, frequencies, and tests performed.
- The GMP auditor should review procedures for preventive maintenance and calibration of critical instruments, including scheduling and equipment update procedures.
- The GMP auditor should review raw data to verify that all of the above is completed per established procedure.
- The GMP auditor should review and observe routine monitoring (in-line TOC and conductivity) of the water system.
- The GMP auditor should review trend data for chemistry, microbiological, and endotoxin tests.
- The GMP auditor should review the investigation of results that are at or over alert and action levels.
- The GMP auditor should verify that water treatment plant and distribution systems are designed, constructed, installed, commissioned, qualified, monitored and maintained to prevent microbiological contamination and to ensure a reliable source of water of an appropriate quality.

- The GMP auditor should verify that measures are taken to minimize the risk of the presence of particulates, microbial contamination/proliferation, and endotoxin/pyrogen (e.g., sloping of piping to provide complete drainage and the avoidance of dead legs). What happens when filters are included in the system?
- The GMP auditor should verify that the water produced complies with the current monograph of the relevant pharmacopoeia.
- The GMP auditor should verify that water systems are qualified and validated to maintain the appropriate levels of physical, chemical, and microbial control, taking the effect of seasonal variation into account.
- The GMP auditor should verify that water flow remains turbulent through the pipes in water distribution systems to minimize the risk of microbial adhesion, and subsequent biofilm formation.
- The GMP auditor should verify that the flow rate is established during qualification and be routinely monitored.
- The GMP auditor should verify that WFI is produced from water meeting specifications that have been defined during the qualification process, stored, and distributed in a manner that minimizes the risk of microbial growth (e.g., by constant circulation at a temperature above 70°C). GMP auditor should verify that WFI is produced by distillation or by a purification process that is equivalent to distillation. This may include reverse osmosis coupled with other appropriate techniques such as electrodeionization (EDI), ultrafiltration, or nanofiltration. When hydrophobic bacteria-retentive vent filters are installed in WFI storage tanks, the integrity of the filters should be checked both before and after usage to ensure that they do not become contaminated.
- The GMP auditor should verify that controls are in place to prevent condensation forming on the filter (e.g., by heating).
- What type of measures are taken to minimize the risk of biofilm formation?
- How are sterilization, disinfection, or regeneration of water systems carried out?
- The GMP auditor should verify that disinfection of a water system with chemicals is followed by a validated rinsing/flushing procedure.
- The GMP auditor should verify that water is tested after disinfection/regeneration.
- The GMP auditor should verify that chemical testing results are approved before the water system is returned to use and microbiological/endotoxin results are verified to be within specification and approved before batches manufactured using water from the system are considered for certification/release.
- The GMP auditor should verify that regular ongoing chemical and microbial monitoring of water systems are performed to ensure that the water continues to meet compendial expectations.
- The GMP auditor should verify that alert levels are based on the initial qualification data and thereafter periodically reassessed on data obtained during subsequent requalifications, routine monitoring, and investigations.
- The GMP auditor should verify that a review of ongoing monitoring data is carried out to identify any adverse trend in system performance.
- The GMP auditor should verify that sampling programs reflect the requirements of the CCS, including all outlets and points of use, at a specified interval, and ensure that representative water samples are obtained for analysis on a regular basis.
- The GMP auditor should verify that sample plans are based on the qualification data, and potential worst-case sampling locations and that at least one representative sample is included every day of the water that is used for manufacturing processes.
- The GMP auditor should verify that alert level excursions are documented and reviewed, and include an investigation to determine whether the excursion is a single (isolated) event or if results are indicative of an adverse trend or system deterioration.

- The GMP auditor should verify that each action limit excursion is investigated to determine the probable root causes and any potential impact on the quality of products and manufacturing processes as a result of the use of the water.
- The GMP auditor should confirm that WFI systems have continuous monitoring systems, such as conductivity and total organic carbon (TOC) monitors, since these could provide a more accurate picture of the overall operation of the system than discrete sampling. Sensor placements ought to be determined by risk.

Steam as a Direct Sterilizing Agent

- The GMP auditor should verify that feed water to a pure steam (clean steam) generator is appropriately purified. The GMP auditor should verify that pure steam generators are designed, qualified, and operated in a manner to ensure that the quality of steam produced meets defined chemical and endotoxin levels.
- The GMP auditor should verify that steam used as a direct sterilizing agent is of suitable quality and should not contain additives at a level that could cause contamination of product or equipment.
- The GMP auditor should verify that the generator supplying pure steam used for the direct sterilization of materials or product contact surfaces (e.g., porous hard-good autoclave loads), and steam condensate meet the current monograph for WFI of the relevant pharmacopoeia (microbial testing is not mandatory for steam condensate).
- The GMP auditor should verify that a suitable sampling schedule is in place to ensure that representative pure steam is obtained for analysis on a regular basis.
- The GMP auditor should verify these parameters include the following (unless otherwise justified): noncondensable gases, dryness value (dryness fraction), and superheat.
- The GMP auditor should verify that other aspects of the quality of pure steam used for sterilization are assessed periodically against validated parameters.

Process Gas

Process gas and related equipment controls may be covered in conjunction with the facilities and equipment system. Specific considerations include controls over the final filtration of the processed gas and filter integrity testing. Gas used as a component in a finished product may include the nitrogen gas overlay for oxygen-sensitive products.

- The GMP auditor should verify that gases that come in direct contact with product/primary container surfaces are of appropriate chemical, particulate, and microbial quality.
- The GMP auditor should verify all relevant parameters, including oil and water content, are specified, taking into account the use and type of the gas, the design of the gas generation system, and, where applicable, comply with the current monograph of the relevant pharmacopoeia or the product quality requirement.
- The GMP auditor should verify that gases used in aseptic processes are filtered through a sterilizing grade filter (with a nominal pore size of a maximum of 0.22 μm) at the point of use.
- What happens where the filter is used on a batch basis (e.g., for filtration of gas used for overlay of aseptically filled products) or as a product vessel vent filter?
- Is the filter integrity tested and are the results reviewed?
- The GMP auditor should verify that any transfer pipework or tubing that is located after the final sterilizing grade filter are sterilized.
- How should microbial monitoring of the gas be performed periodically at the point of use?

Where backflow from vacuum or pressure systems poses a potential risk to the product, there should be a mechanism(s) to prevent backflow when the vacuum or pressure system is shut off.

Heating and Cooling and Hydraulic Systems

- The GMP auditor should verify that major items of equipment associated with hydraulic, heating and cooling systems where possible, are located outside the filling room. The GMP auditor should verify that there are appropriate controls to contain any spillage and/or cross-contamination associated with the system fluids.

Any leaks from these systems that would present a risk to the product should be detectable (e.g., an indication system for leakage).

Prewashed/Ready-to-Sterilize Closures

The cGMP regulation (21 CFR 211.94(c)) states that, where indicated, containers and closures must be processed to remove pyrogenic substances. Many manufacturers of small-volume parenterals purchase stoppers that are ready-to-sterilize, (i.e., they are pyrogen-free). No washing or depyrogenation is done by the dosage form manufacturers but these firms are still responsible for ensuring the stoppers are of acceptable quality for use in manufacturing. The pyrogen requirements should be included in the specifications for the stoppers and, if the manufacturer does not conduct testing of each incoming lot for pyrogen/endotoxin, they should establish the reliability of the supplier's test results by qualification of the vendor, followed by periodic testing.

3.26.18 Microbiological and Endotoxin Testing of Component, Container, and Closure

21 CFR 211.84(d) specifies that each lot of component, container, or closure that is liable to microbial contamination, which is objectionable in view of its intended use, shall be subjected to microbiological tests before use. GMP auditors should evaluate the firm's system to determine if microbiological or endotoxin testing is required and the rationale for setting acceptance criteria.

- The GMP auditor should review test data to verify that the materials meet test criteria and, if not, verify that investigations were conducted to determine the cause and corrective actions were implemented.

3.26.19 Verification of Container and Closures

The physical and chemical characteristics of containers and closures can be critical to the sterility and stability of the finished product. Many containers and closures look alike (color and dimensions), but are made of different materials or have a different surface treatment such as silicone on stoppers and ammonium sulfate on type I glass.

- The GMP auditor should evaluate the firm's procedures for ensuring containers and closure consistently meet appropriate specifications.
- The GMP auditor should determine what tests and examinations are done to verify the containers and closures are made of the correct materials with the correct dimensions (critical to ensuring continuing container/closure integrity) and are free of critical defects.

3.26.20 Container/Closure Integrity

It is vital that the container/closure system is kept intact to guarantee that every drug product unit is kept sterile during transportation, storage, and usage. Product contamination is caused by leaky containers or closures.

The GMP auditor should evaluate the tests and studies performed to demonstrate the integrity of container/closure systems for all sterile drugs.

- The GMP auditor should verify that all incoming container/closure components meet specifications, including all appropriate dimensions.
- The GMP auditor should determine whether studies adequately simulate the stress conditions of the sterilization process, handling, and storage.
- The GMP auditor should verify that the units tested in validation are appropriate (e.g., for terminally sterilized drug products, the units selected should be exposed to the maximum sterilization cycles using the production process).
- The GMP auditor should verify sensitivity of the test is specified.
- The GMP auditor should verify container/closure integrity is demonstrated during validation and as part of the stability program (in lieu of sterility testing), over the shelf life of the product.

3.26.21 Production

Production practices and conditions can have a direct and significant adverse impact on drug sterility assurance. Coverage of critical elements, which are typically defined by the firm of the production system, should be part of all audits of aseptically processed sterile drug manufacturers.

The risk of contamination posed by an operation depends greatly on the design of the overall manufacturing operation. Observation of manufacturing is a critical part of evaluating the adequacy of an aseptic processing operation. The following should be carefully observed by the GMP auditor:

- Adequacy of aseptic technique
- Personnel behavior and practices in the cleanroom
- Movement of people and materials before and during the aseptic operation
- Robustness of production process design (e.g., process performance, validation, the impact of equipment configuration on ergonomics of aseptic manipulations)
- Disinfection practices
- More specifically, the audit must include real-time observation of the higher-risk operations including but not limited to (these are examples and not an all-inclusive list):
- Set-up of filling lines, especially difficult-to-assemble lines (e.g., powder filling lines), and lines that require multiple aseptic assemblies or do not employ SIP of the product pathway.
- Cleaning and disinfection of the line and room to ensure all difficult-to-access surfaces are consistently and properly cleaned and disinfected.
- Protection of critical contact surfaces to ensure their sterility throughout operations and post sterilizations.
- Aseptic technique and cleanroom behavior during operations, including handling equipment jams and stoppages.
- Personnel flow in relation to microbial control of the environment.
- Material flow (e.g., whether materials are moved from a lesser controlled area to a cleaner area without disinfection), including a number of staff members and their activities in the aseptic filling room.
- Filling operations, especially personnel gowning technique, gown integrity, strict adherence to SOPs), the nature and frequency of interventions (interventions are also performed during the media fill simulations), and overall condition of the critical filling area.
- Atypical interventions associated with unplanned events (e.g., operator attempts to change the filling pump during operations).
- Extra manipulations during filling operations for assembly of sterile filtration apparatus that is not SIP sterilized.
- Handling (transfer, storage, loading) of partially stoppered vials in lyophilization processes. The GMP auditor should note that for lyophilized products, vials of sterile products are stoppered but not fully sealed until the lyophilization process is completed. The sterile

product is exposed to the environment during filling, half-stoppering, transport, loading of the lyophilizer, and the lyophilization cycle. Complete seating of stoppers typically occurs in the chamber after the cycle is completed. All of these manipulations must be performed under Class 100 conditions.
- Preparation of equipment for sterilization (cleaning, the type of wrapping to ensure protection while still allowing for penetration as part of the validated sterilization cycle with defined loading patterns).
- Environmental monitoring (while the monitoring program is considered a laboratory system, the audit should include observation of the actual monitoring operations and rationale for sample site locations).
- Proper placement and sealing of stoppers on vials as applicable, and the capping (aluminum crimp) is performed in a protected area under unidirectional flow.
- Production of sterile suspensions and sterile bulk powders (e.g., antibiotics) where sterile filtration of the final bulk is not feasible. These are typically formulated and manufactured under aseptic conditions. This requires the sterilization of large pieces of production equipment (e.g., tanks, reactors, dryers, and associated lines) and assurance that these pieces of equipment retain their integrity and remain sterile.

3.26.22 Media Fills or Process Simulations

Media fills are used to validate aseptic processing operations, including those employing newer technologies, such as isolators, BFS, or RABS. Media fills representing manually intensive aseptic operations should equal or approach the size and duration of a commercial production lot. In contrast, a process conducted in an isolator is designed to have a lower risk of microbial contamination because of the lack of direct human intervention and can be simulated with a lower number of units as a proportion of the overall operation. All media fills should closely simulate manufacturing operations, incorporating, as appropriate, worst-case activities and conditions as well as operator interventions.

- The GMP auditor should verify media fills represent actual manufacturing operations by comparing observed operations to those documented in media fill batch records.
- The GMP auditor should determine if media fills are conducted semi-annually for each processing line. The activities and interventions representative of each shift should be included in the semi-annual media fill program. This may require more than one media fill per line every 6 months if aseptic processing is performed during more than one shift. With the exception of isolator operations, at least one semi-annual media fill is performed per line per shift.
- The GMP auditor should determine if the aseptic filling of all types of containers is supported by the media fills performed. If a matrix approach is used, evaluate the firm's justification for selecting the worst-case container/closure configurations for each line.
- The GMP auditor should determine the accountability of all filled units (units filled vs. units incubated).
- The GMP auditor should verify that all units that were discarded during and after filling have a reasonable and assignable cause for rejection (e.g., rubber stopper missing, aluminum cap missing).
- The GMP auditor should determine that cracked and leaking units found after incubation are investigated, counted, and all rejected units properly justified (e.g., is there an assignable cause that is reasonable for the rejection?).
- The GMP auditor should determine how and who examines units after incubation. If the examination is not performed by a microbiologist, determine if it is overseen by the Quality Unit and if the operators doing the exam are properly trained by a microbiologist.

3.26.23 Disinfection

The disinfection of cleanrooms is particularly important. GMP auditors should verify that they are cleaned and disinfected thoroughly in accordance with a written program.

- The GMP auditor should verify that for disinfection to be effective, prior cleaning to remove surface contamination is performed.
- The GMP auditor should verify that cleaning programs effectively remove disinfectant residues.
- The GMP auditor should verify that more than one type of disinfecting agent is employed to ensure that where they have different modes of action, their combined usage is effective against bacteria and fungi.
- The GMP auditor should verify that disinfection includes the periodic use of a sporicidal agent.
- To evaluate the efficacy of the disinfection program and identify alterations in the types of microbial flora (e.g., organisms resistant to the disinfection regime currently in use), the GMP auditor should confirm that regular monitoring is carried out.
- The GMP auditor should verify that the disinfection process is validated.
- The GMP auditor should confirm that validation studies support the in-use expiry periods of prepared solutions and show the suitability and effectiveness of disinfectants on the type of surface material and in the specific manner in which they are used, or representative material if warranted.
- The GMP auditor should verify that disinfectants and detergents used in grade A and grade B areas are sterile prior to use.
- The GMP auditor should verify that disinfectants used in grades C and D are sterile where determined in the CCS.
- How are disinfectants and detergents diluted/prepared?
- Is this done in a manner to prevent contamination?
- Are they monitored for microbial contamination?
- Where are dilutions kept?

In the event that the detergents and disinfectants are provided "ready-made," the certificates of conformance or analysis findings may be accepted, provided that the relevant vendor qualification is completed successfully.

- Are the effectiveness of any fumigation agent and dispersion system understood and validated?

3.26.24 Sterilization

- The GMP auditor should verify that finished products are terminally sterilized, using a validated and controlled sterilization process, as this provides a greater assurance of sterility than a validated and controlled sterile filtration process and/or aseptic processing.

Where it is not possible for a product to undergo terminal sterilization, consideration should be given to using post-aseptic processing terminal heat treatment, combined with the aseptic process to give improved sterility assurance.

- The GMP auditor should verify that the selection, design and location of the equipment, and cycle/program used for sterilization are based on scientific principles and data which demonstrate the repeatability and reliability of the sterilization process.
- How are parameters defined, controlled, monitored, and recorded?
- The GMP auditor should verify that all sterilization processes are validated.

The GMP auditor should confirm that the product composition, storage circumstances, and maximum amount of time between the beginning of a product's production and sterilization are all taken into account in validation tests.

Prior to implementing any sterilization procedure, it is important to confirm that it is suitably validated for the product and equipment, as well as that it consistently achieves the desired sterilizing conditions in every section of each type of load that needs to be processed. This can be done, for example, using physical measurements and, when necessary, biological indicators.

- The GMP auditor should verify that particular attention is given when the adopted product sterilization method is not described in the current edition of the pharmacopoeia, or when it is used for a product that is not a simple aqueous solution.

Where possible, heat sterilization is the method of choice.

- The GMP auditor should verify that validated loading patterns are established for all sterilization processes and that load patterns are subject to periodic revalidation.
- The GMP auditor should verify that maximum and minimum loads are also considered as part of the overall load validation strategy.
- The GMP auditor should verify that the validity of the sterilizing process is reviewed and verified at scheduled intervals based on risk.
- The GMP auditor should verify that heat sterilization cycles are revalidated with a minimum frequency of at least annually for load patterns that are considered the worst case.
- The GMP auditor should verify that other load patterns are validated at a frequency justified in the CCS.
- The GMP auditor should verify that routine operating parameters are established and adhered to for all sterilization processes, e.g., physical parameters and loading patterns.
- The GMP auditor should verify that there are mechanisms in place to detect a sterilization cycle that does not conform to the validated parameters.
- The GMP auditor should verify that any failed sterilization or sterilization that deviated from the validated process (e.g. having longer or shorter phases such as heating cycles) is investigated.
- The GMP auditor should verify that suitable BIs placed at appropriate locations are considered as an additional method to support the validation of the sterilization process.
- The GMP auditor should verify that BIs are stored and used according to the manufacturer's instructions.
- What type of controls are tested where BIs are used to support validation and/or to monitor a sterilization process (e.g., with ethylene oxide)?
- If Bis are used, what type of precautions are taken to avoid transferring microbiological contamination to the manufacturing or other testing processes?

BI results in isolation should not be used to override other critical parameters and process design elements.

The reliability of BIs is important.

- The GMP auditor should verify that suppliers are qualified and that transportation and storage conditions are controlled in order that BI quality is not compromised.
- The GMP auditor should verify that prior to the use of a new batch/lot of BIs, the population, purity, and identity of the indicator organism of the batch/lot are verified.

For other critical parameters, e.g., D-value and Z-value, the batch certificate provided by the qualified supplier can normally be used.

- The GMP auditor should verify that there are clear means of differentiating products, equipment and components, which have not been subjected to the sterilization process from those which have.
- The GMP auditor should verify that equipment such as baskets or trays used to carry products, other items of equipment, and/or components are clearly labeled (or electronically tracked) with the product name and batch number and an indication of whether or not it has been sterilized.
- What types of indicators are used to indicate whether or not a batch has passed through a sterilization process?
- The GMP auditor should verify that sterilization records are available for each sterilization run.
- The GMP auditor should verify that each cycle has a unique identifier.
- The GMP auditor should verify that their conformity is reviewed and approved as part of the batch certification/release procedure.
- The GMP auditor should verify that materials, equipment, and components are sterilized by validated methods appropriate to the specific material.
- What types of protection are applied after sterilization to prevent recontamination?
- What happens if sterilized items are not used immediately after sterilization? How about maximum holding time?
- GMP auditor should verify that the packaging process is undertaken prior to sterilization where protection is achieved by containment in sealed packaging.

In cases where materials, equipment, components, or ancillary items need to be sterilized in closed and/or sealed packages and then transferred to a grade A area, it is essential that this process and disinfection be carried out with appropriate validated methods.The use of fast transfer ports has also gained importance, and it must be proven that both grade A and grade B areas are sufficient to control the potential risk of contamination. In addition, it must be proven that the disinfection method brings the contamination in the packaging to acceptable levels if the product is to be transferred from grade B to grade As. For example, if materials, equipment, components, or ancillary items will be sterilized in a closed and/or sealed package or container, the selected package or container should be in a structure that can keep particle, microbial, endotoxin/pyrogen, or chemical contamination to a minimum and is compatible with the sterilization method.

- The GMP auditor should verify that the packaging sealing process is validated. The GMP auditor should verify that the validation considers the integrity of the sterile protective barrier system, the maximum hold time before sterilization, and the maximum shelf life assigned to the sterilized items.
- The GMP auditor should verify that the integrity of the sterile protective barrier system for each of the sterilized items is checked prior to use.

For materials, equipment, components, and ancillary items that are not direct or indirect product contact parts and are necessary for aseptic processing but cannot be sterilized, an effective and validated disinfection and transfer process should be in place. These items, once disinfected, should be protected to prevent recontamination. These items, and others representing potential routes of contamination, should be included in the environmental monitoring program.

Sterilization by Heat

- The GMP auditor should verify that each heat sterilization cycle is recorded either electronically or by hardcopy, using equipment with suitable accuracy and precision.
- The GMP auditor should verify that the system has safeguards and/or redundancy in its control and monitoring instrumentation to detect a cycle not conforming to the validated

cycle parameter requirements and abort or fail this cycle (e.g., by the use of duplex/double probes connected to independent control and monitoring systems).

- The GMP auditor should verify that the position of the temperature probes used for controlling and/or recording is determined during the validation and selected based on system design and in order to correctly record and represent routine cycle conditions.
- The GMP auditor should verify that validation studies are designed to demonstrate the suitability of system control and recording probe locations and include the verification of the function and location of these probes by the use of an independent monitoring probe located at the same position during validation.
- The GMP auditor should verify that the whole of the load reaches the required temperature before the measurement of the sterilizing time period starts.
- What happens when sterilization cycles are controlled by using a reference probe within the load?
- What happens after the completion of the high-temperature phase of a heat sterilization cycle?

Any cooling liquid or gas that comes into contact with the product or sterilized material should be sterilized.

Moist Heat Sterilization

- How is moist heat sterilization achieved?
- The GMP auditor should verify that the items to be sterilized, other than products in sealed containers, are dry and packaged in a protective barrier system that allows the removal of air and penetration of steam and prevents recontamination after sterilization.
- The GMP auditor should verify that time, temperature, and pressure are used to monitor the process and be recorded for porous cycles (hard goods).
- The GMP auditor should verify that each sterilized item is inspected for damage, packaging material integrity, and moisture on removal from the autoclave.

Any item found not to be fit for purpose should be removed from the manufacturing area and an investigation performed.

- The GMP auditor should verify that the temperature should be recorded at the chamber drain throughout the sterilization period for autoclaves capable of performing pre-vacuum sterilization cycles.

Load probes may also be used where appropriate, but the controlling system should remain related to the load validation. The GMP auditor should verify that the temperature is recorded at appropriate condensate drain locations throughout the sterilization period for steam-in-place systems.

- The GMP auditor should verify that validation of porous cycles includes a calculation of equilibration time, exposure time, correlation of pressure and temperature, and the minimum/maximum temperature range during exposure.
- The GMP auditor should verify that validation of fluid cycles includes temperature, time, and/or F0.
- The GMP auditor should verify that critical processing parameters are subject to defined limits (including appropriate tolerances) and be confirmed as part of the sterilization validation and routine cycle acceptance criteria.
- The GMP auditor should verify that leak tests on the sterilizer are carried out periodically (normally weekly) when a vacuum phase is part of the cycle or the system is returned, post-sterilization, to a pressure lower than the environment surrounding the sterilizer.

- The GMP auditor should verify that there is an adequate assurance of air removal prior to and during sterilization when the sterilization process includes air purging (e.g., porous autoclave loads, lyophilizer chambers).

For autoclaves, this should include an air removal test cycle (normally performed on a daily basis) or the use of an air detector system.

- The GMP auditor should verify that loads to be sterilized are designed to support effective air removal and be free draining to prevent the build-up of condensate.
- The GMP auditor should verify that distortion and damage of non-rigid containers that are terminally sterilized, such as containers produced by blow–fill–seal or for–fill–seal technologies, are prevented by appropriate cycle design and control (for instance, setting correct pressure, heating and cooling rates, and loading patterns).
- The GMP auditor should verify that the system is appropriately designed and validated to ensure all parts of the system are subject to the required treatment where steam-in-place systems are used for sterilization (e.g., for fixed pipework, vessels, and lyophilizer chambers).
- The GMP auditor should verify that the system is monitored for temperature, pressure, and time at appropriate locations during routine use to ensure all areas are effectively and reproducibly sterilized.
- The GMP auditor should verify that these locations are demonstrated as being representative of, and correlated with, the slowest to heat locations during initial and routine validation.

Once a system has been sterilized by steam-in-place, it should remain integral and where operations require, maintained under positive pressure or otherwise equipped with a sterilizing vent filter prior to use.

- The GMP auditor should verify that the heated water consistently reaches all of the required contact points in fluid load cycles where superheated water is used as the heat transfer medium.
- The GMP auditor should verify that initial qualification studies include temperature mapping of the entire load.
- The GMP auditor should verify that there are checks on the equipment to ensure that nozzles (where the water is introduced) are not blocked and drains remain free from debris.
- The GMP auditor should verify that validation of the sterilization of fluids loads in a superheated water autoclave includes temperature mapping of the entire load and heat penetration and reproducibility studies.
- The GMP auditor should verify that all parts of the load heat up uniformly and achieve the desired temperature for the specified time.
- The GMP auditor should verify that routine temperature monitoring probes are correlated to the worst-case positions identified during the qualification process.

Dry Heat Sterilization

Dry heat sterilization utilizes high temperatures of air or gas to sterilize a product or article. Dry heat sterilization is of particular use in the thermal removal of difficult-to-eliminate thermally robust contaminants such as endotoxin/pyrogen and is often used in the preparation of components for aseptic filling.

- The GMP auditor should verify that the combination of time and temperature to which product, components, or equipment are exposed produces an adequate and reproducible level of lethality and/or endotoxin/pyrogen inactivation/removal when operated routinely within the established limits.

The process may be operated in an oven or in a continuous tunnel process, e.g., for sterilization and depyrogenation of glass containers.

- The GMP auditor should verify that dry heat sterilization/depyrogenation tunnels are configured to ensure that airflow protects the integrity and performance of the grade A sterilizing zone by maintaining appropriate pressure differentials and airflow through the tunnel.
- The GMP auditor should verify that air pressure difference profiles are assessed.
- The GMP auditor should verify that the impact of any airflow change is assessed to ensure the heating profile is maintained.
- The GMP auditor should verify that all air supplied to the tunnel passes through at least a HEPA filter and periodic tests (at least biannually) should be performed to demonstrate air-filter integrity.
- The GMP auditor should verify that any tunnel parts that come into contact with sterilized components are appropriately sterilized or disinfected.

Some of the Critical Process Parameters

- The GMP auditor should verify that validation studies are performed to demonstrate that the process provides a suitable F_h value and results in a minimum 3 $\log_{10}$ reduction in endotoxin concentration when a thermal process is used as part of the depyrogenation process for any component or product contact equipment/material.

Table 3.58 presents critical process parameters during validation and/or routine processing.

When this is attained, there is no additional requirement to demonstrate sterilization in these cases.

- The GMP auditor should verify that containers spiked with endotoxin are used during validation and should be carefully managed with a full reconciliation performed.
- The GMP auditor should verify that containers are representative of the materials normally processed (with respect to the composition of the packaging materials, porosity, dimensions, and nominal volume).
- The GMP auditor should verify that endotoxin quantification and recovery efficiency are demonstrated.

Dry heat ovens are typically employed to sterilize or depyrogenate primary packaging components, starting materials, or active substances, but may be used for other processes.

TABLE 3.58
Critical Process Parameters to Consider during Validation and/or Routine Processing

- Belt speed or dwell time within the sterilizing zone
- Temperature – minimum and maximum temperatures
- Heat penetration of the material/article
- Heat distribution/uniformity
- Airflows determined by air pressure difference profiles correlated with the heat distribution and penetration studies

TABLE 3.59
Critical Process Parameters to Consider during Qualification and/or Routine Processing

- Temperature
- Exposure period/time
- Chamber pressure (for maintenance of overpressure)
- Airspeed
- Air quality within the oven
- Heat penetration of material/article (slow to heat spots)
- Heat distribution/uniformity
- Load pattern and configuration of articles to be sterilized/depyrogenated including minimum and maximum loads

- The GMP auditor should verify that they are maintained at a positive pressure relative to lower-grade clean areas throughout the sterilization and post-sterilization hold process unless the integrity of the packaging is maintained.
- The GMP auditor should ensure that all air entering the oven passes through a HEPA filter.

Some of the Critical Process Parameters
Table 3.59 presents critical process parameters during qualification and /or routine processes

Sterilization by Radiation
The purpose of using radiation sterilization is that the materials or products are heat sensitive. UV irradiation is not accepted. The source for this process is Annex 12.

- The GMP auditor should verify that validation procedures ensure that the effects of variation in the density of the product and packages are considered.

Sterilization with Ethylene Oxide
This method should only be used when no other method is practicable.

- The GMP auditor should verify that during process validation, it is shown that there is no damaging effect on the product and that the conditions and time allowed for degassing result in the reduction of any residual ethylene oxide (EO) gas and reaction products to defined acceptable limits for the given product or material.
- The important thing to consider is to ensure that there is direct contact between the gas and microbial cells.The GMP auditor must verify that precautions are in place to prevent microorganisms that may be present in substances such as crystals or dried proteins. The structure, porosity, and amount of packaging material used affect this process.
- The GMP auditor must verify that the environment reaches the required temperature and humidity level for the process before the materials are treated with gas.
- How the quality of steam is evaluated and tested should be considered.
- The GMP auditor should verify that each sterilization cycle is monitored with suitable BIs, using the appropriate number of test units distributed throughout the load at defined locations that have been shown to be worst-case locations during validation.

Some of the Critical Process Parameters
Table 3.60 presents critical control parameters of sterilization process validation and routine monitoring.

TABLE 3.60
Critical Process Parameters to consider As Part of Sterilization Process Validation and Routine Monitoring

- EO gas concentration
- Pressure
- Amount of EO gas used
- Relative humidity
- Temperature
- Exposure time

- The GMP auditor should verify that the load should be aerated to allow EO gas and/or its reaction products to desorb from the packaged product to predetermined levels after sterilization.

Aeration can occur within a sterilizer chamber and/or in a separate aeration chamber or aeration room.

- The GMP auditor should verify that the aeration phase is validated as part of the overall EO sterilization process validation.

Filter Sterilization of Products That Cannot Be Sterilized in Their Final Container
If the product cannot be sterilized in its final container, solutions or liquids should be sterilized by filtration through a sterile sterilizing grade filter (with a nominal pore size of a maximum of 0.22 μm that has been appropriately validated to obtain a sterile filtrate) and subsequently aseptically filled into a previously sterilized container.

- The GMP auditor should verify that the selection of the filter used ensures that it is compatible with the product and as described in the marketing authorization.

Suitable bioburden reduction prefilters and/or sterilizing grade filters may be used at multiple points during the manufacturing process to ensure a low and controlled bioburden of the liquid prior to the final sterilizing filter.

- The GMP auditor should verify that an additional filtration through a sterile sterilizing grade filter, as close to the point of fill as possible, is considered as part of an overall CCS due to the potential additional risks of a sterile filtration process, as compared with other sterilization processes.
- How are the components for the filtration system and their interconnection and arrangement within the filtration system, including prefilters, selected?

Table 3.61 presents the reason behind the design of a filtration system.

- The GMP auditor should verify that sterile filtration of liquids is validated in accordance with relevant pharmacopoeia requirements.

Validation can be grouped by different strengths or variations of a product.

- How is the validation performed?
- The GMP auditor should verify that the rationale for grouping is justified and documented.
- The GMP auditor should verify that the product to be filtered is used for bacterial retention testing of the sterilizing grade filter during filter validation, wherever possible.

TABLE 3.61
Why Is a Filtration System Designed?

- Allow operation within validated process parameters.
- Maintain the sterility of the filtrate.
- Minimize the number of aseptic connections required between the final sterilizing grade filter and the final filling of the product.
- Allow cleaning procedures to be conducted as necessary.
- Allow sterilization procedures, including sterilization-in-place, to be conducted as necessary.
- Permit in-place integrity testing, of the 0.22 μm final sterilizing grade filter, preferably as a closed system, both prior to and following filtration as necessary. In-place integrity testing methods should be selected to avoid any adverse impact on the quality of the product.

Where the product to be filtered is not suitable for use in bacterial retention testing, a suitable surrogate product should be justified for use in the test.

- The GMP auditor should verify that the challenge organism used in the bacterial retention test is justified.
- The GMP auditor should verify that the filtration system minimizes the generation of fibers and particles, does not cause or contribute to unacceptable levels of impurities, or possess characteristics that otherwise alter the quality and efficacy of the product.
- How does the management ensure that the filter characteristics are compatible with the fluid and not adversely affected by the product to be filtered?
- The GMP auditor should verify that the adsorption of product components and extraction/leaching of filter components are evaluated.

Table 3.62 presents filtration parameters for validation.

- The GMP auditor should verify that routine process controls are implemented to ensure adherence to validated filtration parameters.

TABLE 3.62
Filtration Parameters to Consider and Establish during Validation

- The wetting fluid used for filter integrity testing:
  - It should be based on the filter manufacturer's recommendation or the fluid to be filtered. The appropriate integrity test value specification should be established.
  - If the system is flushed or integrity tested in situ with a fluid other than the product, appropriate actions are taken to avoid any deleterious effect on product quality.
- Filtration process conditions include:
  - Fluid pre-filtration holding time and effect on bioburden.
  - Filter conditioning, with fluid if necessary.
  - Maximum filtration time/total time filter is in contact with the fluid.
  - Maximum operating pressure.
  - Flow rate.
  - Maximum filtration volume.
  - The time is taken to filter a known volume of bulk solution and the pressure difference to be used across the filter.

- The GMP auditor should verify that results of critical process parameters are included in the batch record, including but not limited to the minimum time taken to filter a known volume of bulk solution and the pressure difference across the filter.
- The GMP auditor should verify that any significant differences from critical parameters during manufacturing are documented and investigated.
- The GMP auditor should verify that the integrity of the sterilized filter assembly is verified by integrity testing before use (pre-use post sterilization integrity test, or PUPSIT) to check for damage and loss of integrity caused by the filter preparation prior to use.
- The GMP auditor must verify that the filter used to filter liquids has completed the non-destructive integrity test before being removed from the filter housing after use.
- The GMP auditor should verify that the integrity test process is validated and test results correlate to the microbial retention capability of the filter established during validation.

Bubble point, diffusive flow, water intrusion, or pressure hold testing are examples of tests that may be used.

PUPSIT may not always be used after sterilization. The reason for this is that the sample involved in the process is limited by the small volume of the filtered solution. In such cases, the root cause should be determined, and it should be determined whether there are alternative approaches through a detailed risk assessment.

Table 3.63 presents elements to be considered in a risk assessment.

- The GMP auditor should verify that the integrity of critical sterile gas and air vent filters (that are directly linked to the sterility of the product) are verified by testing after use, with the filter remaining in the filter assembly or housing.
- The GMP auditor should verify that the integrity of noncritical air or gas vent filters is confirmed and recorded at appropriate intervals.

Where gas filters are in place for extended periods, the GMP auditor should verify that integrity testing should be carried out at installation and prior to replacement.

- The GMP auditor should verify that the maximum duration of use is specified and monitored based on risk (e.g., considering the maximum number of uses and heat treatment/sterilization cycles permitted as applicable).

TABLE 3.63
Points to Consider in a Risk Assessment

- In-depth knowledge and control of the filter sterilization process to ensure that the potential for damage to the filter is minimized.
- In-depth knowledge and control of the supply chain to include:
  - Contract sterilization facilities.
  - Defined transport mechanisms.
  - Packaging of the sterilized filter, to prevent damage to the filter during transportation and storage.
- In-depth process knowledge such as:
- The specific product type, including particle burden and whether there exists any risk of impact on filter integrity values, such as the potential to alter integrity testing values and therefore prevent the detection of a nonintegral filter during a post-use filter integrity test.
- Pre-filtration and processing steps, prior to the final sterilizing grade filter, which would remove particle burden and clarify the product prior to the sterile filtration

If the sterilizing filtration process has been validated as a system consisting of multiple filters to achieve sterility for a given fluid, the filtration system is considered to be a single sterilizing unit.

- The GMP auditor should verify that all filters within the system satisfactorily pass integrity testing after use.

In a redundant filtration system (where a second redundant sterilizing grade filter is present as a backup but the sterilizing process is validated as only requiring one filter), the GMP auditor should verify that the post-use integrity test of the primary sterilizing grade filter is performed and if demonstrated to be integral, then a post-use integrity test of the redundant (backup) filter is not necessary.

However, in the event of a failure of the post-use integrity test on the primary filter, the GMP auditor should verify that the post-use integrity test on the secondary (redundant) filter is performed, in conjunction with an investigation and risk assessment to determine the reason for the primary filter test failure.

- The GMP auditor should verify that bioburden samples are taken from the bulk product and immediately prior to the final sterile filtration.
- The GMP auditor should verify that systems for taking samples are designed so as not to introduce contamination.
- The GMP auditor should verify that liquid sterilizing grade filters are discarded after the processing of a single batch and that the same filter is used continuously for more than one working day unless such use has been validated.

Table 3.64 presents things a filter user must do in this process.

3.26.25 Form–Fill–Seal (FFS)

- The GMP auditor should verify that the conditions for FFS machines used for terminally sterilized products comply with the environmental requirements of paragraphs 8.3 and 8.4 of Annex 1.
- The GMP auditor should verify that the conditions for FFS machines used in aseptic manufacture comply with the environmental requirements of paragraph 8.10 of Annex 1.
- The GMP auditor should verify that contamination of the packaging films used in the FFS process is minimized by appropriate controls during component fabrication, supply, and handling.

TABLE 3.64
What the Filter User Does When the Campaign Manufacture of a Product Has Been Appropriately Justified in the CCS and Validated

- Assess and document the risks associated with the duration of filter used for the sterile filtration process for a given fluid.
- Conduct and document effective validation and qualification studies to demonstrate that the duration of filter used for a given sterile filtration process and for a given fluid does not compromise the performance of the final sterilizing grade filter or filtrate quality.
- Document the maximum validated duration of use for the filter and implement controls to ensure that filters are not used beyond the validated maximum duration. Records of these controls should be maintained.
- Implement controls to ensure that filters contaminated with fluid or cleaning agent residues, or considered defective in any other way, are removed from use.

TABLE 3.65
Controls of FFS Should Be in Alignment with CCS

- Determination of the boundaries of the critical zone
- Environmental control and monitoring, both of the machine and the background in which it is placed
- Personnel gowning requirements
- Integrity testing of the product filling lines and filtration systems (as relevant)
- Duration of the batch or filling campaign
- Control of packaging films, including any requirements for film decontamination or sterilization
- Cleaning-in-place and sterilization-in-place of equipment as necessary
- Machine operation, settings, and alarm management (as relevant)

Due to the criticality of packaging films, the GMP auditor should verify that procedures are implemented to ensure that the films supplied meet defined specifications and are of the appropriate quality, including material thickness and strength, microbial and particulate contamination, integrity, and artwork, as relevant.

- The GMP auditor should verify that the sampling frequency, the bioburden and, where applicable, endotoxin/pyrogen levels of packaging films and associated components are defined and controlled within the PQS and considered in the CCS.
- The GMP auditor should verify that particular attention is given to understanding and assessing the operation of the equipment, including set-up, filling, sealing, and cutting processes, so that critical process parameters are understood, validated, controlled, and monitored appropriately.
- The GMP auditor should verify that any product contact gases, e.g., those used to inflate the container or used as a product overlay, are appropriately filtered, as close to the point of use as possible.
- The GMP auditor should verify that the quality of gases used and the effectiveness of gas filtration systems are verified periodically in accordance with paragraphs 6.18 and 6.19.

Table 3.65 depicts controls of FFS.

- The GMP auditor should verify that appropriate procedures for the verification, monitoring, and recording of FFS critical process parameters and equipment operation are applied during production.

TABLE 3.66
Critical Process Parameters of FFS during Equipment Qualification?

- Settings for uniform package dimensions and cutting in accordance with validated parameters
- Setting, maintenance and monitoring of validated forming temperatures (including pre-heating and cooling), forming times, and pressures as relevant
- Setting, maintenance and monitoring of validated sealing temperatures, sealing temperature uniformity across the seal, sealing times, and pressures as relevant
- Environmental and product temperature
- Batch-specific testing of package seal strength and uniformity
- Settings for correct filling volumes, speeds, and uniformity
- Settings for any additional printing (batch coding), embossing, or debossing to ensure that unit integrity is not compromised
- Methods and parameters for integrity testing of filled containers

Table 3.66 presents critical process parameters of FFS during equipment qualification.

- The GMP auditor should verify that operational procedures describe how forming and sealing issues are detected and rectified. The GMP auditor should verify that rejected units or sealing issues are recorded and investigated.
- The GMP auditor should verify that appropriate maintenance procedures are established based on risk, and include maintenance and inspection plans for tooling critical to the effectiveness of unit sealing.
- The GMP auditor should verify that any issues identified that indicate a potential product quality concern are documented and investigated.

3.26.26 Blow–Fill–Seal

- The GMP auditor should verify that blow–fill–seal equipment used for the manufacture of products that are terminally sterilized is installed in at least a grade D environment.
- The GMP auditor should verify that the conditions at the point of fill comply with the environmental requirements of paragraphs 8.3 and 8.4 of Annex 1.

For shuttle-type equipment used for aseptic filling, the GMP auditor should verify that the parison is open to the environment and therefore the areas where parison extrusion, blow molding, and sealing take place meet grade A conditions at the critical zones.

- The GMP auditor should verify that the filling environment is designed and maintained to meet grade A conditions for viable and total particle limits both at rest and when in operation.

For rotary-type equipment used for aseptic filling, The GMP auditor should verify that the parison is generally closed to the environment once formed, and the filling environment within the parison is designed and maintained to meet grade A conditions for viable and total particle limits both at rest and when in operation.

- The GMP auditor should verify that the equipment is installed in at least a grade C environment, provided that grade A/B clothing is used.
- The GMP auditor should verify that the microbiological monitoring of operators wearing grade A/B clothing in a grade C area is performed in accordance with risk management principles, and the limits and monitoring frequencies applied with consideration of the activities performed by these operators.

Due to the generation of particles from polymer extrusion and cutting during operation, and the restrictive size of critical filling zones of BFS equipment, in-operation monitoring of total particles for BFS equipment is not expected.

- The GMP auditor should verify that data are available to demonstrate that the design of the equipment ensures that critical zones of the filling process environment would meet grade A conditions in operation.
- The GMP auditor should verify that viable environmental monitoring of BFS processes is risk based, and designed in accordance with section 9 of this Annex 1.
- The GMP auditor should verify that in operation viable monitoring is undertaken for the full duration of critical processing, including equipment assembly.

For rotary-type BFS equipment, it is acknowledged that monitoring the critical filling zone may not be possible.

- The GMP auditor should verify that the environmental control and monitoring program takes into consideration the moving parts and complex airflow paths generated by the BFS process and the effect of the high heat outputs of the process (e.g., through the use of airflow visualization studies and/or other equivalent studies).
- The GMP auditor should verify that environmental monitoring programs are considered factors such as air-filter configuration, air-filter integrity, cooling systems integrity equipment design, and qualification.
- The GMP auditor should verify that air or other gases that make contact with critical surfaces of the container during extrusion, formation or sealing of the molded container undergo appropriate filtration.
- The GMP auditor should verify that the quality of gas used and the effectiveness of gas filtration systems are verified periodically in accordance with paragraphs 6.18 and 6.19 of Annex 1.
- The GMP auditor should verify that particulate and microbial contamination of the polymer granulate is prevented by appropriate design, control, and maintenance of the polymer granulate storage, sampling, and distribution systems.
- The GMP auditor should verify that the capability of the extrusion system to provide appropriate sterility assurance for the molded container is understood and validated.
- The GMP auditor should verify that the sampling frequency, the bioburden and, where applicable, endotoxin/pyrogen levels of the raw polymer are defined and controlled within the PQS and considered in the CCS.
- The GMP auditor should verify that interventions requiring cessation of filling and/or extrusion, molding and sealing, and, where required, resterilization of the filling machine are clearly defined and described in the filling procedure, and included in the APS as relevant.

Table 3.67 displays controls of BFS during the qualification.

Table 3.68 depicts critical process parameters of BFS during equipment qualification.

- The GMP auditor should verify that appropriate maintenance procedures are established based on risk, and include maintenance and inspection plans for items critical to unit sealing, integrity, and sterility.
- The GMP auditor should verify that the molds used to form containers are considered critical equipment, and that any changes or modifications to molds result in an assessment

TABLE 3.67
Controls of BFS Should Be in Alignment with CCS

• Determination of the boundaries of the critical zone
• Environmental control and monitoring, both of the machine and the background in which it is placed
• Personnel gowning requirements
• Integrity testing of the product filling lines and filtration systems (as relevant)
• Duration of the batch or filling campaign
• Control of polymer granulate, including distribution systems and critical extrusion temperatures
• Cleaning-in-place and sterilization-in-place of equipment as necessary
• Machine operation, settings, and alarm management (as relevant)

TABLE 3.68
Critical Process Parameters of BFS during Equipment Qualification

- Clean-in-place and sterilization-in-place of product pipelines and filling needles (mandrels)
- Setting, maintenance and monitoring of extrusion parameters, including temperature, speed, and extruder throat settings for parison thickness
- Setting, maintenance, and monitoring of mold temperatures, including the rate of cooling where necessary for product stability
- Preparation and sterilization of ancillary components added to the molded unit, e.g., bottle caps
- Environmental control, cleaning, sterilization and monitoring of the critical extrusion, transfer and filling areas as relevant
- Batch-specific testing of package wall thickness at critical points of the container
- Settings for correct filling volumes, speeds, and uniformity
- Settings for any additional printing (batch coding), embossing, or debossing to ensure that unit integrity and quality are not compromised
- Methods and parameters for integrity testing of 100% of all filled containers
- Settings for cutters or punches used to remove waste plastic surrounding filled units (flash removal)

of finished product container integrity, and where the assessment indicates, should be supported by validation.

- The GMP auditor should verify that any issues identified that indicate a potential product quality concern are documented and investigated.

3.26.27 Terminal Sterilization

- The GMP auditor should determine what type of sterilization cycles are used (bioburden-based or overkill).
- The GMP auditor should review validation/revalidation or periodic evaluation of terminal sterilization cycles for representative types of products.
- For selected products, the GMP auditor should verify that the parameters and loading patterns used in production are the same as those used in validation studies.
- The GMP auditor should determine the minimum acceptable cycle allowed in the SOP (as opposed to the nominal or routine cycle) and compare that to the validated cycle (using BI) to verify it has been properly qualified.
- The GMP auditor should determine how sterilization cycles are documented, monitored and reviewed.
- The GMP auditor should review deviations or atypical data from sterilization operations that indicate inconsistencies in process performance.
- The GMP auditor should verify that the preparation of components and materials is performed in at least a grade D cleanroom in order to limit the risk of microbial, endotoxin/pyrogen, and particle contamination so that the product is suitable for sterilization.
- What happens when the product is at a high or unusual risk of microbial contamination?
- The GMP auditor should verify that the preparation of ointments, creams, suspensions, and emulsions is carried out in at least a grade C environment before terminal sterilization.

Specific guidance regarding terminally sterilized veterinary medicinal products can be found within Annex 4 of the GMP guidelines.

- The GMP auditor should verify that primary packaging containers and components are cleaned using validated processes to ensure that particle, endotoxin/pyrogen, and bioburden contamination is appropriately controlled.

- The GMP auditor should verify that the filling of products for terminal sterilization is carried out in at least a grade C environment.
- What happens when the CCS identifies that the product is at an unusual risk of contamination from the environment because, for example, the filling operation is slow, the containers are wide necked, or are necessarily exposed for more than a few seconds before closing?
- The GMP auditor should verify that the processing of the bulk solution includes a filtration step with a microorganism retaining filter, where possible, to reduce bioburden levels and particles prior to filling into the final product containers.

Table 3.69 presents the functions of different grades of cleanroom.

Parametric release of the terminally sterilized drug product is defined as a sterility assurance release based on demonstrated control of the sterilization process. It refers to the use of predefined critical process control data without sterility testing to meet the requirements of FDA 21 CFR Part 211.167 (a).It is allowed only for products that are terminally sterilized by heat and it must be identified in the appropriate regulatory filing as the release method. Parametrically released products must have an approved application.

- If encountered during an audit, the GMP auditor should verify the parametric release method has been submitted and approved in the appropriate drug application. If the drug is not the subject of an approved application, the GMP auditor should collect pertinent information and validation data for evaluation.

A 100% inspection of injectable products including cracks, visible particles, and other significant defects should occur.

- The GMP auditor should verify the firm has written procedures that define the defects to be removed from the lot and actions to take if the number of critical defects exceeds a predetermined level.
- Significant defect categories should be identified. The results of the inspection of each batch should be compared to established action levels.
- The GMP auditor should evaluate the appropriateness of and the rationale or justification for predetermined action levels.
- The GMP auditor should evaluate the firm's investigation into the cause of rejects, including units rejected for cracks and visible particulates (e.g., foreign matter).
- The GMP auditor should observe the inspection process.
- The GMP auditor should challenge visual/manual inspection rates through observation.
- The GMP auditor should evaluate the adequacy of written procedures for visual inspection.
- The GMP auditor should evaluate personnel qualification and requalification and equipment qualifications according to established procedures. The GMP auditor should evaluate personnel qualification including the use of reference samples for qualification.

If a manual system is used, the GMP auditor should determine if employees are trained and qualified to verify they can recognize and remove defects under actual or simulated production conditions.

TABLE 3.69
Functions of Cleanroom Grades

Grade A	• Filling of products, when unusually at risk
Grade B	• Preparation of solutions, when unusually at risk • Filling of products
Grade C	• Preparation of solutions and components for subsequent filling

If an automated or semiautomated system is used, the GMP auditor should determine the equipment is qualified and the that software program or equipment settings have been validated for all types of products being inspected (e.g., clear vials, amber vials, colored solution, suspensions). If the equipment is an automatically controlled computer-based system, an assessment of the system and validation is warranted.

- The GMP auditor should evaluate the firm's program for sampling and examination of inspected vials and evaluate the effectiveness of inspection and action taken if the reject level is reached.
- The GMP auditor should evaluate the firm's assessment of units rejected during filling operations (any separate inspection prior to the 100% inspection stage), establish alert/action limits, and perform investigations where appropriate.

3.26.28 Personnel (Gowning, Training, Aseptic Techniques)

- The GMP auditor must verify that the sterile product manufacturer has personnel with the necessary education, training, skills, and experience in the production and testing stages, as well as in all operations used for production.
- The GMP auditor should verify that only the minimum number of personnel required is present in cleanrooms.
- The GMP auditor should verify that the maximum number of operators in cleanrooms is determined, documented, and considered during activities such as initial qualification and APS, so as not to compromise sterility assurance.
- The GMP auditor should verify that all personnel including those performing cleaning, maintenance, and monitoring, and those that access cleanrooms receive regular training, gowning qualification, and assessment in disciplines relevant to the correct manufacture of sterile products.
- The GMP auditor must verify that the training provided focuses on rules to be followed in clean rooms, contamination prevention and control, aseptic production techniques, and protection of sterile products from contamination, and includes basic microbiology and hygiene. Examples include entering grade B clean rooms or training operators who intervene in grade A areas.
- The GMP auditor should verify that the level of training is based on the criticality of the function and area in which the personnel is working.
- The GMP auditor should verify that the personnel accessing grade A and B areas are trained for aseptic gowning and aseptic behaviors.
- The GMP auditor should verify that compliance with aseptic gowning procedures is confirmed by assessment and periodic reassessment at least annually, and should involve both visual and microbial assessment using monitoring locations such as gloved fingers, forearms, chest, and hood (facemask/forehead).
- The GMP auditor should verify that the unsupervised access to the grade A and grade B areas where aseptic operations are or will be conducted is restricted to appropriately qualified personnel, who have passed the gowning assessment and have participated in a successful APS.
- The GMP auditor should verify that unqualified personnel do not enter grade B cleanrooms or grade A in operation.
- What happens if unqualified personnel is brought into the grade B and A areas?
- The GMP auditor should verify that an authorized person from the manufacturer supervises the unqualified personnel during their activities and assess the impact of these activities on the cleanliness of the area.

- The GMP auditor should verify that there are systems in place for the disqualification of personnel from working.
- The GMP auditor should verify that once disqualified, retraining and requalification are completed before permitting the operator to have any further involvement in aseptic practices.

For operators entering grade B cleanrooms or performing intervention into grade A, this requalification should include consideration of participation in a successful APS.

- The GMP auditor should verify that high standards of personal hygiene and cleanliness are essential to prevent excessive shedding or increased risk of introduction of microbial contamination.
- The GMP auditor should verify that personnel involved in the manufacture of sterile products are instructed to report any specific health conditions or ailments that may cause the shedding of abnormal numbers or types of contaminants and therefore preclude cleanroom access.

Staff need to confirm that in situations where there is a risk of excessive microbial contamination, such as illness, there is a designated person to manage the situation and procedures to be implemented.

- The GMP auditor should verify that wristwatches, make-up, jewelry, other personal items such as mobile phones, and any other nonessential items are not allowed in clean areas.
- What happens if electronic devices, mobile phones, and tablets, that are supplied by the manufacturer, are used in cleanrooms?
- The GMP auditor should verify that cleanroom gowning and hand washing follow a written procedure designed to minimize contamination of cleanroom clothing and/or the transfer of contaminants to the clean areas.
- The GMP auditor should verify that the clothing and its quality are appropriate for the process and the grade of the working area.
- Some activities that the GMP auditor must verify include not allowing wristwatches, jewelry, mobile phones, tablets, or any personal items into or out of clean areas.The GMP auditor must verify that the garments worn by the staff are produced specifically for clean areas and that the hand washing procedure is carried out in a way that prevents contamination.The GMP auditor must verify that the garments worn are produced in accordance with the class of work area.The GMP auditor must verify that the garment is worn to prevent product contamination.The GMP auditor must verify that the selected garment is designed to provide two-way protection.The GMP auditor must verify that garments have been checked for integrity before and after wear.
- The GMP auditor should verify that gown integrity is also checked upon exit.
- What kind of particular attention should be taken for sterilized garments and coverings? Are they sterilized? What is their hold time? Is this time validated? What happens if the damage is identified in reusable garments?
- How are the garments qualified? What is the URS? How are DQ, IQ, OQ, and PQ managed?
- The GMP auditor should verify that clothing is chosen to limit shedding due to operators' movement.

What are the gowning requirements for each cleanliness grade?
Details are given in Annex 1 parts 7.13–7.18.

The type of gowns and personal protective equipment (PPE) worn by employees should be appropriate for the areas in which they work. There should be detailed written procedures that describe the gowning requirements for each processing area. The GMP auditor should evaluate the following:

- For aseptic processing, the GMP auditor should determine the gowns (which typically include face masks, hoods, protective goggles, gloves, and boots) are sterilized and made of nonparticle shedding material. Ensure that the gowns cover all skin, hair, and facial hair.
- The GMP auditor should review how the incoming sterile gowns/garb are accepted or rejected for use.
- The GMP auditor should evaluate the firm's program for training, testing, and qualifying and requalifying employees who work in the controlled areas, especially those who set up and operate aseptic processing lines.
- The GMP auditor should evaluate the aseptic techniques of employees by observing aseptic processing operations.
- For selected employees, the GMP auditor should verify the training, testing, qualifying, and requalifying were done as specified in procedures.
- The GMP auditors should verify the training is done on a continuing basis.

Closed Systems

The use of closed systems can reduce the risk of microbial, particle, and chemical contamination from the adjacent environment.

- GMP auditors should verify that closed systems are always designed to reduce the need for manual manipulations and the associated risks.

- The GMP auditor should verify that the design and selection of any closed system used for aseptic processing ensuring maintenance of sterility.
- The GMP auditor should verify that the connection of sterile equipment (e.g., tubing/pipework) to the sterilized product pathway after the final sterilizing grade filter is designed to be connected aseptically (e.g., by intrinsic sterile connection devices).
- The GMP auditor should verify that appropriate measures are in place to ensure the integrity of components used in aseptic connections.
- Which means should be identified and recorded in CCS?
- The GMP auditor should verify that appropriate system integrity tests are considered when there is a risk of compromising product sterility.
- The GMP auditor should verify that supplier assessment includes the collation of data in relation to potential failure modes that may lead to a loss of system sterility.

The background environment in which closed systems are located should be based on their design and the processes undertaken.

For aseptic processing and where there are any risks that the system integrity may be compromised, the GMP auditor should verify that the system is located in grade A.

If the system can be shown to remain integral at every usage (e.g., via pressure testing and/or monitoring), then a lower classified area may be used.

- The GMP auditor should verify that any transfer between classified areas is thoroughly assessed.

If the closed system is opened (e.g., for maintenance of a bulk manufacturing line), then the GMP auditor should verify that this is performed in a classified area appropriate to the materials (e.g., grade C for terminal sterilization processes, or grade A for aseptic processing) or is subject to further cleaning and disinfection (and sterilization in case of aseptic processes).

3.26.29 Environmental Monitoring

Environmental and process monitoring activities at pharmaceutical and biotechnology manufacturing sites should act as part of the established CCS and should have the sole purpose of minimizing the risk of microbial or particle contamination. Evaluating the monitoring system individually, such as live, non-viable, and APS, instead of a holistic approach, may pose reliability risks, and evaluating it separately for asepsis would not be the right approach.

Table 3.70 presents elements of the environmental monitoring program.

- The GMP auditor should verify that the information from these systems is used for routine batch certification/release and for periodic assessment during process review or investigation.

This applies to both terminal sterilization and aseptic processes, however, the criticality of the impact may differ depending on the product and process type.

Environmental and Process Monitoring

GMP auditors should verify that an environmental monitoring (EM) program is established and documented.

Table 3.71 presents the EM program's purpose.

- The GMP auditor should verify that risk assessments are performed in order to establish a comprehensive environmental monitoring program, i.e., sampling locations, frequency of monitoring, monitoring methods, and incubation conditions (e.g., time, temperature(s), aerobic, and/or anaerobic conditions).

TABLE 3.70
Elements of Environmental Monitoring Program

- Environmental monitoring – total particle
- Environmental and personnel monitoring – viable particle
- Temperature, relative humidity, and other specific characteristics
- APS (aseptically manufactured product only)

TABLE 3.71
Purpose of Environmental Monitoring Program

- Provide assurance that cleanrooms and clean air equipment continue to provide an environment of appropriate air cleanliness, in accordance with design and regulatory requirements.
- Effectively detect excursions from environmental limits triggering investigation and assessment of risk to product quality.

- The GMP auditor should verify that these risk assessments are conducted based on detailed knowledge of the process inputs and final product, the facility, equipment, the criticality of specific processes and steps, the operations involved, routine monitoring data, monitoring data obtained during qualification, and knowledge of typical microbial flora isolated from the environment.
- The GMP auditor should verify that the risk assessment includes the determination of critical monitoring locations, those locations where the presence of microorganisms during processing may have an impact upon product quality (e.g., grade A, aseptic processing areas, and the grade B areas that directly interface with the grade A area).
- The GMP auditor should verify that consideration of other information, such as air visualization studies, is included.
- The GMP auditor should verify that these risk assessments are reviewed regularly in order to confirm the effectiveness of the site's environmental monitoring program.
- The GMP auditor should verify that the monitoring program is considered in the overall context of the trend analysis and the CCS for the site.
- The GMP auditor should verify that routine monitoring of cleanrooms, clean air equipment, and personnel are performed in operation throughout all critical stages of processing, including equipment set up.
- The GMP auditor should verify that other characteristics, such as temperature and relative humidity, are controlled within ranges that align with product/processing/personnel requirements and support the maintenance of defined cleanliness standards (e.g., grade A or B).
- The GMP auditor should verify that the monitoring of grade A demonstrates the maintenance of aseptic processing conditions during critical operations.
- The GMP auditor should verify that monitoring is performed at locations posing the highest risk of contamination to the sterile equipment surfaces, containers, closures, and product.
- The GMP auditor should verify that the selection of monitoring locations and the orientation and positioning of sampling devices are justified and appropriate to obtain reliable data from the critical zones.
- The GMP auditor should verify that sampling methods do not pose a risk of contamination to the manufacturing operations.
- The GMP auditor should verify that appropriate alert levels and action limits are set for the results of viable and total particle monitoring.

The maximum total particle action limits are described in Table 3.73 and the maximum viable particle action limits are described in Table 3.74.

- The GMP auditor should verify that more stringent action limits are applied based on data trending, the nature of the process, or as determined within the CCS.

TABLE 3.72
Trend Expectations

- Increasing numbers of excursions from action limits or alert levels.
- Consecutive excursions from alert levels.
- Regular but isolated excursions from action limits that may have a common cause (e.g., single excursions that always follow planned preventive maintenance).
- Changes in microbial flora type and numbers and predominance of specific organisms. Particular attention should be given to organisms recovered that may indicate a loss of control, deterioration in cleanliness, or organisms that may be difficult to control such as spore-forming microorganisms and molds.

- The GMP auditor should verify that both viable and total particle alert levels are established based on results of cleanroom qualification tests and periodically reviewed based on ongoing trend data.
- The GMP auditor should verify that alert levels for grade A (total particle only) grade B, grade C, and grade D are set such that adverse trends (e.g., numbers of events or individual events that indicate a deterioration of environmental control) are detected and addressed.

Table 3.72 presents trend expectations.

- The GMP auditor should verify that the monitoring of grade C and D cleanrooms in operation is performed based on data collected during qualification and routine data to allow effective trend analysis.

The requirements of alert levels and action limits will depend on the nature of the operations carried out. Action limits may be more stringent than those listed in the tables.

If action limits are exceeded, the GMP auditor should verify that the operating procedures prescribe a root-cause investigation, an assessment of the potential impact on the product (including batches produced between the monitoring and reporting), and requirements for corrective and preventive actions.

If alert levels are exceeded, the GMP auditor should verify that operating procedures prescribe assessment and follow-up, which should include consideration of an investigation and/or corrective actions to avoid any further deterioration of the environment.

Environmental Monitoring – Total Particle

- The GMP auditor should verify that a total particle monitoring program is established to obtain data for assessing potential contamination risks and to ensure the maintenance of the environment for sterile operations in a qualified state.

The limits for environmental monitoring of airborne particle concentration for each graded area are given in Table 3.73.

However, if there are consecutive or regularly low levels counted, it should be examined in detail and investigated whether this situation is a result of contamination. Situations like this can often be caused by malfunctions of room air supply filters or equipment in the area, or may even be an indication that the machines are installed incorrectly.

For grade A, the GMP auditor should verify that particle monitoring is undertaken for the full duration of critical processing, including equipment assembly.

- The GMP auditor should verify that the grade A area is monitored continuously (for particles ≥0.5 and ≥5 μm) and with a suitable sample flow rate (at least 28 liters [1 ft^3] per minute) so that all interventions, transient events, and any system deterioration is captured.
- The GMP auditor should verify that the system frequently correlates each individual sample result with alert levels and action limits at such a frequency that any potential excursion can be identified and responded to in a timely manner.
- The GMP auditor should verify that alarms are triggered if alert levels are exceeded.
- The GMP auditor should verify that procedures define the actions to be taken in response to alarms including the consideration of additional microbial monitoring.

It is recommended that a similar system be used for the grade B area, although the sample frequency may be decreased.

- The GMP auditor should verify that the grade B area is monitored at such a frequency and with a suitable sample size that the program captures any increase in levels of contamination and system deterioration.

TABLE 3.73
Maximum Permitted Total Particle Concentration for Monitoring

Grade	Maximum Limits for Total Particle ≥0.5 µm/m³		Maximum Limits for Total Particle ≥5 µm/m³	
	At Rest	In Operation	At Rest	In Operation
A	3,520	3,520	29	29
B	3,520	352,000	29	2,930
C	352,000	3,520,000	2,930	29,300
D	3,520,000	Not predetermined[a]	29,300	Not predetermined[a]

[a] For grade D, "in operation" limits are not predetermined. GMP auditor should verify that the manufacturer establishes "in operation" limits based on a risk assessment and on routine data, where applicable.

Note 1: The particle limits given in the table for the "at rest" state should be achieved after a short "clean-up" period defined during qualification (guidance value of less than 20 minutes) in an unmanned state, after the completion of operations.

Note 2: The occasional indication of macro particle counts, especially ≥5 µm, within grade A may be considered to be false counts due to electronic noise, stray light, coincidence loss, etc.

If alert levels are exceeded, the GMP auditor should verify that alarms are triggered.

- The GMP auditor should verify that the selection of the monitoring system takes into account any risk presented by the materials used in the manufacturing operation (e.g., those involving live organisms, powdery products, or radiopharmaceuticals) that may give rise to biological, chemical, or radiation hazards.

In the case where contaminants are present due to the processes involved and would potentially damage the particle counter or present a hazard (e.g., live organisms, powdery products and radiation hazards), the frequency and strategy employed should be such to ensure the environmental classification both prior to and after exposure to the risk.

- The GMP auditor should verify that an increase in viable particle monitoring is considered to ensure comprehensive monitoring of the process.
- The GMP auditor should verify that monitoring is performed during simulated operations.
- The GMP auditor should verify that such operations are performed at appropriate intervals.
- The GMP auditor should verify that the approach is defined in the CCS.

The size of monitoring samples taken using automated systems will usually be a function of the sampling rate of the system used. It is not necessary for the sample volume to be the same as that used for the formal classification of cleanrooms and clean air equipment.

- The GMP auditor should verify that monitoring sample volumes is justified.

Environmental and Personnel Monitoring – Viable Particle

When aseptic operations are performed, the GMP auditor should verify that microbial monitoring is frequent using a combination of methods such as settle plates, volumetric air sampling, glove, gown, and surface sampling (e.g., swabs and contact plates).

- The GMP auditor should verify that the method of sampling used is justified within the CCS and is demonstrated not to have a detrimental impact on grade A and B airflow patterns.

- The GMP auditor should verify that cleanroom and equipment surfaces are monitored at the end of an operation.
- The GMP auditor must verify that particle monitoring activities are carried out in cleanrooms during periods when there are no routine production operations, for example after disinfection or after shutdown, and must also verify that this process is carried out in unused rooms.

In case of an incident:

- The GMP auditor should verify that additional sample locations are used as a verification of the effectiveness of corrective action (e.g., cleaning and disinfection).
- The GMP auditor should verify that continuous viable air monitoring in grade A (e.g., air sampling or settle plates) are undertaken for the full duration of critical processing, including equipment (aseptic set-up) assembly and critical processing.
- The GMP auditor should verify that a similar approach is considered for grade B cleanrooms based on the risk of impact on the aseptic processing.
- The GMP auditor should verify that the monitoring is performed in such a way that all interventions, transient events, and any system deterioration would be captured, and any risk caused by interventions of the monitoring operations is avoided.
- The GMP auditor should verify that a risk assessment evaluates the locations, type, and frequency of personnel monitoring based on the activities performed and the proximity to critical zones.
- The GMP auditor should verify that monitoring includes a sampling of personnel at periodic intervals during the process.
- The GMP auditor should verify that the sampling of personnel is performed in such a way that it will not compromise the process.
- The GMP auditor must verify that special attention is given to personnel performing critical interventions and monitoring personnel exiting each grade B area.

Where monitoring of gloves is performed after critical interventions, the GMP auditor should verify that the outer gloves are replaced prior to the continuation of activity.

If monitoring of gowns is required after critical interventions, the GMP auditor should verify that the gown is replaced before further activity in the cleanroom.

- The GMP auditor should verify that microbial monitoring of personnel in the grade A and grade B areas is performed.

Where operations are manual in nature (e.g., aseptic compounding or filling), the GMP auditor should verify that the increased risk leads to the enhanced emphasis placed on microbial monitoring of gowns and justified within the CCS.

- The GMP auditor should verify that the adoption of suitable alternative monitoring systems such as rapid methods is considered by manufacturers in order to expedite the detection of microbiological contamination issues and to reduce the risk to the product.
- The GMP auditor should verify that the sampling methods and equipment used are fully understood and procedures are in place for the correct operation and interpretation of results obtained.

Table 3.74 presents maximum action limits for viable particle contamination.

The GMP auditor should verify that supporting data for the recovery efficiency of the sampling methods chosen are available.

- The GMP auditor should verify that microorganisms detected in the grade A and grade B areas are identified to species level and the potential impact of such microorganisms on product quality (for each batch implicated) and the overall state of control should be evaluated.

TABLE 3.74
Maximum Action Limits for Viable Particle Contamination

Grade	Air Sample (CFU/m³)	Settle Plates (diameter 90 mm) CFU/4 hours[a]	Contact Plates (diameter 55 mm) CFU/plate[b]	Glove Print Including 5 Fingers on Both Hands CFU/glove
A	No growth[c]			
B	10	5	5	5
C	100	50	25	—
D	200	100	50	—

[a] The GMP auditor should verify that settle plates are exposed in grade A and B areas for the duration of operations (including equipment setup) and changed as required after a maximum of 4 hours (exposure time should be based on validation including recovery studies and the GMP auditor should verify that it does not have any negative effect on the suitability of the media used). The GMP auditor should verify that for grade C and D areas, exposure time (with a maximum of 4 hours) and frequency are based on QRM. Individual settle plates may be exposed for less than 4 hours.

[b] Contact plate limits apply to equipment, room and gown surfaces within the grade A and grade B areas. Routine gown monitoring is not normally required for grade C and D areas, depending on their function.

[c] It should be noted that for grade A, any growth should result in an investigation.

Note 1: It should be noted that the types of monitoring methods listed in the table are examples, and other methods can be used provided they meet the intent of providing information across the whole of the critical process where the product may be contaminated (e.g., aseptic line setup, aseptic processing, filling, and lyophilizer loading).

Note 2: Limits are applied using CFU throughout the document. If different or new technologies are used that present results in a manner different from CFU, the manufacturer should scientifically justify the limits applied and where possible correlate them to CFU.

- The GMP auditor should verify that consideration is given to the identification of microorganisms detected in grade C and D areas (for example where action limits or alert levels are exceeded) or following the isolation of organisms that may indicate a loss of control, deterioration in cleanliness, or that may be difficult to control such as spore-forming microorganisms and molds and at a sufficient frequency to maintain a current understanding of the typical flora of these areas.

Batch Records

- The GMP auditor should review environmental and personnel monitoring data, as well as other data relating to the acceptability of support systems (e.g., HEPA/HVAC, WFI, steam generator) and manufacturing equipment. This review is considered essential to batch release decisions. The batch record should include documentation that ensures this type of holistic review is done before the release of a lot for distribution.
- For aseptic processing, the GMP auditor should verify interventions in critical areas (Class 100/ISO 5) are documented so they can be reviewed and evaluated by the Quality Unit.
- The GMP auditor should review batch records to verify they include complete information for all sterilization processes.

3.26.30 Packaging and Labeling

Areas of special concern for sterile products include:

- The GMP auditor should determine that packaging and labeling operations do not introduce risk to product integrity (for example, damage to the container or closure that could affect the integrity of the unit).

- The GMP auditor should determine that the container, closure, and packaging systems provide adequate protection against foreseeable external factors in storage, shipment, and use that can cause contamination or deterioration (e.g., cracked vials during shipment if not properly protected; pinhole leaks in bags, frozen drug products, tears or holes in overwraps of sterile bulk antibiotics and large-volume parenteral; and unseating of stoppers in aluminum cans containing sterile bulk APIs due to pressure changes during shipment by air).
- It is not unusual for filled containers of sterile products to be stored unlabeled for a period of time. The firm must have adequate controls to ensure proper identification of the unlabeled product at all times.
- Tracking of refrigerated or temperature-controlled units for room temperature exposure times (e.g., warm-up of refrigerated units prior to label application).
- Tracking and investigation (as specified and appropriate) of rejected units culled during packaging and labeling operations.

3.26.31 Laboratories

Auditing of sterile drug manufacturers should also cover microbiology laboratories. Quality control tests (sterility and Limulus amebocyte lysate, or LAL test) and the collection of environmental and personnel monitoring samples should be observed to verify that acceptable techniques are used and written procedures are followed. The audit of the microbiology laboratory should evaluate the following:

- Sterility testing, including the collection of samples that are representative of the entire lot and processing conditions; adequate control and monitoring of the testing environment; validation of the method for specific products; growth promotion testing of the media; and incubation times and temperature. It is important to note that increasing the number of samples or the number of tests does not greatly increase the probability of detecting contamination if it is present at a very low level in a lot.
- LAL testing, including product-specific validation; collection of representative samples of raw materials, components/containers, in-process, and finished product, where appropriate; and adequate laboratory facilities for conducting the tests. The GMP auditor should verify the rationale of the sample size of the endotoxin test relative to the production batch.
- Environmental monitoring, which includes a well-defined written program that covers all production shifts and includes air, floors, walls, equipment surfaces, and, in aseptic process operations, critical surfaces that come in contact with sterile products, containers, and closures; establishment of appropriate alert and action levels, use of sampling (contact plates, swabs, active air samplers) and testing methods (media, plate exposure times, incubation times and temperatures) that are designed to detect environmental isolates. Evaluation of the validity of the sampling locations and sampling methods is needed.
- Personnel monitoring which includes a routine program for daily/shift monitoring of operator's gloves and an appropriate schedule for monitoring gowns; establishment of limits that are based on the contamination risk to the product; and investigations of results that exceed the established levels or demonstrate an adverse trend. Personnel monitoring is important in all sterile product operations, but it is especially critical in aseptic processing, and inspectional emphasis should be risk based, focusing on those operations that require employees to enter the critical areas of the processing line.
- Efficacy of disinfectants, including assessment of the suitability, efficacy and limitations of the disinfecting agents used in the controlled area, production equipment, and laboratories. The firm's assessment typically includes laboratory studies that test the effectiveness of agents on different surface materials. Material coupons are usually used with surface types as found in production. The studies should be done with the same disinfecting agents and

contact times (which should be clearly defined in written procedures). It is also important to understand that disinfectants have limitations and most are not effective against every type of microorganism.

- Identification of microorganisms, including procedures that require identification of organisms found in positive sterility tests, media fills, and environmental monitoring (environmental and personnel) samples as specified by the firm. The program should ensure routine identification of microorganisms found in samples taken in critical areas, surrounding areas, and from personnel in the production area. The GMP auditor should review the procedures, equipment, and controls used in the identification activities of the contaminants.
- Microbiological media, including the preparation, sterilization, and growth promotion testing of the media used in performing tests (sterility tests, raw material testing, prefiltration bioburden, environmental monitoring, media fills, etc.). Where appropriate, inactivating agents for disinfectants or product residuals should be added to allow the detection of contaminants.
- BIs and biological cultures used in sterilization validation studies should be used and stored under appropriate conditions. Typically, the conditions are described in the literature received with the BI, if supplied by a vendor. The microbial population should be confirmed by testing each lot. Spore counts should be verified prior to use in validation studies. The D-value should be determined for each lot of a BI if it is used in a way not described by the vendor. If used specifically as directed, the D-value supplied by the vendor can be accepted if the reliability of the Certificate of Analysis has been established, but the D-value of incoming batches should be periodically verified.
- Microorganisms (e.g., ATCC) are used for growth promotion tests of media. Organisms isolated from environmental monitoring samples can also be used to perform growth promotion tests.
- Monitoring, calibration, and maintenance programs for microbiology laboratory equipment, such as incubators.
- Training of microbiologists and evaluation of microbiologists or technicians that perform sterility, LAL, and environmental monitoring tests.
- Documented investigations into out of specification results. The GMP auditor should evaluate positive sterility tests and media fill and LAL failures investigations. Also, the GMP auditor should review environmental/personnel monitoring results at alert and action levels to identify and determine the firm's response to significant incidents or trends. Because of the limited sensitivity of the sterility tests to detect batch contamination, any positive is a serious issue and should be thoroughly investigated by the firm with Quality Unit oversight and approval. The investigation followed should be reviewed during audits to assess the decision-making process. An initial positive can be found invalid only if there is clear documented evidence that the microbial growth was unequivocally a laboratory error.

21 CFR 211.180(e) requires that records be maintained in such a manner that the data can be used to evaluate adherence to quality standards. The evaluation of data generated by the microbiology lab plays an integral role in establishing the sterility assurance of the finished product. The audit should determine if the firm generates and reviews testing data and product quality-related data (e.g., trend reports) to make timely, informed, and science-based decisions to ensure an ongoing state of control.

- The GMP auditor should verify that there is personnel available with appropriate training and experience in microbiology, sterility assurance, and knowledge of the processes to support the design of the manufacturing activities, environmental monitoring regime, and any investigation assessing the impact of microbiologically linked events to the safety of the sterile product.

- The GMP auditor should verify that specifications for raw materials, components, and products include requirements for microbial, particulate, and endotoxin/pyrogen limits when the need for this has been indicated by monitoring and/or by the CCS.
- The GMP auditor should verify that the bioburden assay is performed on each batch for both aseptically filled products and terminally sterilized products, and that the results are considered part of the final batch review.
- The GMP auditor should verify that there are defined limits for bioburden immediately before the final sterilizing grade filter or the terminal sterilization process, which are related to the efficiency of the method to be used.
- The GMP auditor should verify that samples are taken to be representative of the worst-case scenario (e.g., at the end of hold time).

If overkill sterilization parameters are set for terminally sterilized products, the GMP auditor should verify that bioburden is monitored at suitable scheduled intervals.

For products authorized for parametric release, the GMP auditor should verify that a supporting presterilization bioburden monitoring program for the filled product prior to initiating the sterilization cycle is developed and the bioburden assay should be performed for each batch.

- The GMP auditor should verify that the sampling locations of filled units before sterilization are based on worst-case scenarios and representative of the batch.
- The GMP auditor should verify that any organisms found during bioburden testing are identified and their impact on the effectiveness of the sterilizing process determined.

Where appropriate, the GMP auditor should verify that the level of endotoxin/pyrogen is monitored.

- The GMP auditor should verify that the sterility test applied to the finished product is only regarded as the last in a series of critical control measures by which sterility is ensured.

It cannot be used to ensure the sterility of a product that does not meet its design, procedural, or validation parameters.

- The GMP auditor should verify that the test is validated for the product concerned.
- The GMP auditor should verify that the sterility test is performed under aseptic conditions.
- The GMP auditor should verify that samples taken for sterility testing are representative of the whole of the batch but include samples taken from parts of the batch considered to be most at risk of contamination.

Table 3.75 presents several risks.

TABLE 3.75
Risks

- For products that have been filled aseptically, samples should include containers filled at the beginning and end of the batch. Additional samples, e.g., taken after critical interventions should be considered based on risk.
- For products that have been heat sterilized in their final containers, samples taken should be representative of the worst-case locations (e.g., the potentially coolest or slowest to heat part of each load).
- For products that have been lyophilized, samples were taken from different lyophilization loads.

Note: Where the manufacturing process results in sub-batches (e.g., for terminally sterilized products), then the GMP auditor should verify that sterility samples from each sub-batch are taken and a sterility test for each sub-batch performed. The GMP auditor should verify that consideration is given to performing separate testing for other finished product tests.

For some products, it may not be possible to obtain a sterility test result prior to release because the shelf life of the product is too short to allow completion of a sterility test.

In these cases:

- The GMP auditor should verify that the additional considerations of the design of the process and additional monitoring and/or alternative test methods required to mitigate the identified risks are assessed and documented.
- The GMP auditor should verify that any process (e.g., vaporized hydrogen peroxide, ultraviolet) used to decontaminate the external surfaces of sterility samples prior to testing does not negatively impact the sensitivity of the test method or the reliability of the sample.
- The GMP auditor should verify that the media used for product testing is quality control tested according to the related pharmacopoeia before use.
- The GMP auditor should verify that media used for environmental monitoring and APS is tested for growth promotion before use, using a scientifically justified and designated group of reference microorganisms and including suitably representative local isolates.
- The GMP auditor should verify that media quality control testing is normally performed by the end user.
- The GMP auditor should verify that any reliance on outsourced testing or supplier testing of media is justified, and transportation and shipping conditions should be thoroughly considered in this case.
- The GMP auditor should verify that environmental monitoring data and trend data generated for classified areas are reviewed as part of product batch certification/release.
- The GMP auditor should verify that a written procedure is available that describes the actions to be taken when data from environmental monitoring are found out of trend or exceeding the established limits.

For products with short shelf life, the environmental data for the time of manufacture may not be available; in these cases, the GMP auditor should verify that the compliance includes a review of the most recent available data.

- The GMP auditor should verify that manufacturers of these products consider the use of rapid/alternative methods.

If rapid and automated microbial methods are used for general manufacturing purposes, the GMP auditor should verify that these methods are validated for the product(s) or processes concerned.

3.26.32 Sampling

Samples of sterile drug products should be collected to document suspected contamination, adulteration, or misbranding encountered during an inspection. The samples can be physical or documentary. In the event of a for-cause assignment where in-process samples are being requested, collect the samples aseptically at points where such contamination might occur. These samples should be collected by the firm under the observation of the auditor. Great care shall be taken to prevent the possibility of contamination and/or compromising the integrity of the sample for all physical samples of raw materials, in-process samples, or finished drug products. A production lot with an initial sterility failure result that was invalidated by the firm may be considered a good candidate for sampling.

An analysis of the physical sample may not be necessary to document deficiencies that do not meet the requirements specified in cGMP. In such cases, document samples can be presented as evidence.

In general, samples will be submitted to your district's designated servicing laboratory, except as requested in a special assignment or by your supervisor.

3.26.33 Analysis

For sterility and bacterial endotoxin testing:

Sterility testing methods should be based on current editions of USP <71> Sterility Tests.
Bacterial endotoxin testing methods should be based on current editions of USP <85> Bacterial Endotoxin Test.
Particulate matter testing methods should be based on current editions of USP <78>.
Other microbiological examinations should be based on appropriate sections of USP.

Table 3.76 presents a group of questions about moist heat sterilization.

Dry Heat Sterilization/Depyrogenation

Dry heat sterilizers are used, primarily, to sterilize and depyrogenate glass containers for injectable drug products. Both dry heat ovens and dry heat tunnels are used. An oven is a batch process and, at the end of the cycle, the sterile vials are manually removed from the oven, transported, and then manually loaded onto the filling line. A dry heat tunnel uses a continuous, integrated process. In a tunnel, the vials move on a belt from the washing processing through a heating zone where the vials are sterilized and depyrogenated to a cooling zone and then directly into the Class 100 area of a filling line. Typically, firms validate the depyrogenation of the glassware rather than the sterilization cycle. This is based on the fact that it is generally more difficult to remove or denature pyrogenic substances than microorganisms. Components that have undergone heat depyrogenation are rendered sterile. Table 3.77 presents a group of questions about dry heat sterilization.

Table 3.78 presents various questions about sterile drug products.

3.27 AUDITING CLEANROOMS

Although I have mentioned cleanrooms, RABS, and isolators in sterile medical product manufacturing facilities, I would like you to know that I give special importance to cleanroom audits. As you know, technology has started to produce state-of-the art equipment and even if there is one person in the cleanrooms, doubts begin to arise about whether he or she is clean. While working at a German pharmaceutical company, the factory director used exactly this expression about cleanrooms that I really liked: "Even if there is only one person in the cleanrooms, that room is no longer clean in my opinion."

The auditor should request a map(s) of the facility that details cleanroom grades, process and equipment flows, and pressure differentials. The auditor can gain an impression of the contamination controls in place. The cleanroom design should take into consideration the flow of personnel, product, equipment, and components. The most effective facility designs will take into account the flow of all of these from the receipt of raw materials at the warehouse through the facility to the packaging of the product (although the latter takes place outside of a cleanroom, this may fall within the scope of the audit).

- The GMP auditor should verify that cleanrooms and clean air equipment such as unidirectional airflow units (UDAFs), RABS, and isolators, used for the manufacture of sterile products, are qualified according to the required characteristics of the environment.

TABLE 3.76
Moist Heat Sterilization

Moist Heat Sterilization	Question	Comments	Notes
General			
1	Who is the manufacturer of the steam sterilizer (autoclave)?		☐ Fulfilled ☐ Control needed ☐ Action needed [room for notes]
2	What are the model number, age, and internal volume of the autoclave?		☐ Fulfilled ☐ Control needed ☐ Action needed [room for notes]
3	If jacketed, what pressure/temperature is maintained in the jacket as opposed to the chamber?		☐ Fulfilled ☐ Control needed ☐ Action needed [room for notes]
4	What is the sterilizing agent (e.g., steam, air overpressure, superheated water, gamma irradiation)?		☐ Fulfilled ☐ Control needed ☐ Action needed [room for notes]
5	What types of vent filters are used and how often are they integrity tested?		☐ Fulfilled ☐ Control needed ☐Action needed [room for notes]
6	Are vent filters hydrophobic? Are the vent filter housings heated to prevent condensation?		☐ Fulfilled ☐ Control needed ☐ Action needed [room for notes]
7	Are cycles controlled manually or automatically?		☐ Fulfilled ☐ Control needed ☐ Action needed [room for notes]
8	What type of monitoring and controlling sensors are used (e.g., mercury-in-glass thermometer, thermocouple, RTD, pressure gauge)?		☐ Fulfilled ☐ Control needed ☐ Action needed [room for notes]
9	How are these sensors calibrated? Are the standards NIST traceable (or traceable to a National standard for foreign firms), where appropriate?		☐ Fulfilled ☐ Control needed ☐ Action needed [room for notes]
10	Is the autoclave equipped with a steam spreader (more than one steam entry line would be considered in this category)?		☐ Fulfilled ☐ Control needed ☐ Action needed [room for notes]
11	If more than one autoclave is used by the firm, what is the system's capacity for steam production in relation to all autoclaves being in operation at the same time?		☐ Fulfilled ☐ Control needed ☐ Action needed [room for notes]

(Continued)

TABLE 3.76 (CONTINUED)
Moist Heat Sterilization

Moist Heat Sterilization	Question	Comments	Notes
12	What are the sterilization cycle parameters?(Compare Master Process Record/SOP specifications against processing records completed for selected drug products.)		☐ Fulfilled ☐ Control needed ☐ Action needed [room for notes]
13	What are the firm's specifications and observed parameters for: • Time • Temperature • Pressure (psi, inHg)		☐ Fulfilled ☐ Control needed ☐ Action needed [room for notes]
14	Where is the cycle controller sensor located?		☐ Fulfilled ☐ Control needed ☐ Action needed [room for notes]
15	How are each of the sterilization parameters monitored? Are the come-up times for chamber temperature during sterilization reproducible when compared to the come-up time achieved in validation studies?		☐ Fulfilled ☐ Control needed ☐ Action needed [room for notes]
16	Is the slowest to heat spot ("coldspot") in each load monitored during each autoclave cycle?		☐ Fulfilled ☐ Control needed ☐ Action needed [room for notes]
17	Have any changes in the steam sterilization system occurred since the last EI? Have these changes been evaluated for the need for revalidation?		☐ Fulfilled ☐ Control needed ☐ Action needed [room for notes]
18	Is clean steam being used (control of bacterial endotoxin)?		
	Validation		
19	Does the firm have written procedures for validation that include: • Design objectives • Installation qualification (IQ) of equipment • Operational qualification (OQ) of equipment • Performance qualification (PQ) with the product (maximum and minimum load submitted in the application, and any changes after) • Description of circumstances requiring revalidation of the system • Procedures for revalidation		☐ Fulfilled ☐ Control needed ☐ Action needed [room for notes]
20	Does validation documentation include the following? 1. Empty chamber/loaded chamber heat distribution studies: • Number of runs? • Was the cold spot determined? • Allowable variation? • Is actual variation found? • What is the worst-case load? 2. Heat penetration studies • For each type of loading pattern/for each container size utilized? • Number of runs per pattern? • Was the "cold spot" determined for each pattern?		☐ Fulfilled ☐ Control needed ☐ Action needed [room for notes]

(*Continued*)

TABLE 3.76 (CONTINUED)
Moist Heat Sterilization

Moist Heat Sterilization	Question	Comments	Notes
	• Is the established load configuration submitted in the application followed and/or reportable changes submitted as appropriate? 4. Empty chamber/loaded chamber heat distribution studies: • Number of runs? • Was the cold spot determined? • Allowable variation? • Is actual variation found? • What is the worst-case load? 5. Heat penetration studies • For each type of loading pattern/for each container size utilized? • Number of runs per pattern? • Was the "cold spot" determined for each pattern? • Is the established load configuration submitted in the application followed and/or reportable changes submitted as appropriate?		
21	What type of temperature measurement system was used? Does it provide a separate printed reading for each thermocouple?		☐ Fulfilled ☐ Control needed ☐ Action needed [room for notes]
22	What type of temperature sensors were used, and were they calibrated before and after each run?		☐ Fulfilled ☐ Control needed ☐ Action needed [room for notes]
23	If biological indicators were used during validation runs: • Type of indicator (spore strip, inoculated product, ampoule) • Source of indicator • Microorganisms used, including concentration and D value • Were BIs used in an "endpoint" or "count reductions" mode? If any positive BIs were found (when not expected), what was the firm's response?		☐ Fulfilled ☐ Control needed ☐ Action needed [room for notes]
24	In the event a heat distribution or penetration variance was disclosed during the studies, how did the firm correct or allow for it?		☐ Fulfilled ☐ Control needed ☐ Action needed [room for notes]
25	Has the firm determined lag times for all container sizes/mass, product viscosities, etc., and adjusted their cycles accordingly?		☐ Fulfilled ☐ Control needed ☐ Action needed [room for notes]

- The GMP auditor should observe that each manufacturing operation requires an appropriate environmental cleanliness level in the operational state in order to minimize the risk of contamination of the product or materials being handled.
- How are appropriate cleanliness levels in the "at rest" and "operational" states maintained?
- The GMP auditor should verify that cleanrooms and clean air equipment are qualified using the methodology in accordance with the requirements of Annex 15.

TABLE 3.77
Dry Heat Sterilization

Dry Heat Sterilization	Question	Comments	Notes
General			
1	Determine what types of dry heat sterilizers are used (ovens, tunnels)?		☐ Fulfilled ☐ Control needed ☐ Action needed [room for notes]
2	Determine the location of the heat source. The heating element/device that generates the depyrogenation temperature can affect the HEPA filters and their ability to provide the desired filtration for nonviable particles. For example, the high temperatures create expansion and contraction of the HEPA filters and filter housing, which can compromise the integrity and functionality of the HEPA filters.		☐ Fulfilled ☐ Control needed ☐ Action needed [room for notes]
3	How is the heat distribution in the oven or tunnel (fan or convection)?		☐ Fulfilled ☐ Control needed ☐ Action needed [room for notes]
4	Where are the HEPA filters located in the oven or tunnel? What zones in a tunnel are supplied with HEPA-filtered air (supply, heating, cooling zones)? Does the cool-down section of the tunnel provide Class 100 conditions to ensure the sterility of the containers is maintained? What zones are controlled as Class 100 areas?		☐ Fulfilled ☐ Control needed ☐ Action needed [room for notes]
5	Due to thermal dynamics (high temperature of >360°C leads to the expansion and contraction of the filter housing), the HEPA filter normally undergoes integrity testing every 6 months. However, depending on the amount of usage of the depyrogenation tunnel and data obtained from particle monitoring in the vicinity outside the heat tunnel, integrity tests may need to be conducted on a more frequent basis. How does the firmensure the integrity of the HEPA filters? How often are HEPA filters changed?		☐ Fulfilled ☐ Control needed ☐ Action needed [room for notes]
6	Are nonviable particle counts taken in a hot air tunnel? Nonviable particle measurements for the Class 100/ISO-5 zone of the depyrogenation tunnel are not done during routine operation because of high temperature. However, periodic measurements of nonviable in ISO 5 area should be verified and they are normally measured prior to and subsequent to the manufacture of a batch/campaign and at ambient temperature.		☐ Fulfilled ☐ Control needed ☐ Action needed [room for notes]
7	Are the sterilization/depyrogenation cycles controlled manually or automatically?		☐ Fulfilled ☐ Control needed ☐ Action needed [room for notes]
8	What type of monitoring and controlling sensors are used (e.g., thermocouple, RTD, pressure gauge, belt speed indicators)? How often are they calibrated?		☐ Fulfilled ☐ Control needed ☐ Action needed [room for notes]
9	What are the sterilization/depyrogenation cycle parameters or equipment settings? Compare master production record/SOP specifications and processing records for specific representative products.		☐ Fulfilled ☐ Control needed ☐ Action needed [room for notes]

(Continued)

TABLE 3.77 (CONTINUED)
Dry Heat Sterilization

Dry Heat Sterilization	Question	Comments	Notes
10	What are the firm's specifications for time, temperature, belt speed, and pressures if applicable?		☐ Fulfilled ☐ Control needed ☐ Action needed [room for notes]
11	What are the critical parameters? How were they established? How does the firm ensure the critical parameters are met for each lot or cycle?		☐ Fulfilled ☐ Control needed ☐ Action needed [room for notes]
12	Where are the sensors for the cycle controllers located?		☐ Fulfilled ☐ Control needed ☐ Action needed [room for notes]
13	How are each of the above parameters monitored? Are any of the parameters not monitored? Are the critical parameters alarmed in continuous systems, such as those used in tunnels? What is done when there is an alarm?		☐ Fulfilled ☐ Control needed ☐ Action needed [room for notes]
14	How does the firm ensure that all critical parameters are met for each cycle in an oven or for the continuous run of the tunnel?		☐ Fulfilled ☐ Control needed ☐ Action needed [room for notes]
15	Do dry heat tunnels have alarms to alert operators if critical parameters (heat, belt speed, pressure) are not met? Are the alarms documented? Did the conditions that caused the alarm require the implementation of change controls and did they impact the validated process?		☐ Fulfilled ☐ Control needed ☐ Action needed [room for notes]
16	Have any changes in the dry heat sterilization/depyrogenation system occurred since the last inspection? Have these changes been evaluated for the need for revalidation?		☐ Fulfilled ☐ Control needed ☐ Action needed [room for notes]
	Validation		
17	Does the firm have written procedures for qualifying the dry heat sterilizers that include: • Design objectives • Installation qualification of equipment • Operational qualification of equipment • Performance qualification with product • Description of circumstances requiring revalidation and procedures		☐ Fulfilled ☐ Control needed ☐ Action needed [room for notes]
18	Does validation documentation include heat distribution/penetration studies? • What is the firm's allowable temperature variation and the actual variation found in production? • What type of temperature measurement system was used? • Was calibration performed before and after the validation? • If the dry heat sterilizer is an oven, was the slowest to heat area determined?		☐ Fulfilled ☐ Control needed ☐ Action needed [room for notes]

(Continued)

TABLE 3.77 (CONTINUED)
Dry Heat Sterilization

Dry Heat Sterilization	Question	Comments	Notes
19	How was the depyrogenation cycle validated? • Was a known amount of endotoxin added to vials? • Was the endotoxin allowed to dry on the vials? • Was validation conducted with recovery tests to ensure the spiked endotoxin can be recovered? Were the vials challenged with enough endotoxin to allow the calculation of a 3-log reduction? • What cycle or equipment settings were used in the validation runs? • Was a 3-log reduction of endotoxin achieved?		☐ Fulfilled ☐ Control needed ☐ Action needed [room for notes]
20	Are the depyrogenation cycles for all vials validated or does the firm use a matrix approach? If a matrix approach (vial sizes/ mass) is used, what criteria are used to select the worst-case challenges to the system? Are all vial types and sizes bracketed?		☐ Fulfilled ☐ Control needed ☐ Action needed [room for notes]
21	Have the dry heat sterilizer cycles or equipment settings changed since the validation studies were completed? Compare validation studies, current SOPs, and processing records for recent batches.		☐ Fulfilled ☐ Control needed ☐ Action needed [room for notes]
22	In the event a heat distribution or penetration variance was disclosed during the studies, how did the firm correct or allow for it?		☐ Fulfilled ☐ Control needed ☐ Action needed [room for notes]
23	Has the firm determined lag times for all container sizes/mass, product viscosities, etc. and adjusted their cycles accordingly?		☐ Fulfilled ☐ Control needed ☐ Action needed [room for notes]

- The GMP auditor should verify that cleanroom qualification (including classification) is clearly differentiated from operational environmental monitoring.
- The GMP auditor must be trained to know that cleanroom and clean air equipment qualification is the overall process of assessing the level of compliance of a classified cleanroom or clean air equipment with its intended use.
- The GMP auditor must be trained to know that cleanroom classification is part of the cleanroom qualification and is a method of assessing the level of air cleanliness against a specification for a cleanroom or clean air equipment by measuring the total particle concentration.
- The GMP auditor should verify that classification activities are scheduled and performed in order to avoid any impact on process or product quality.

Table 3.79 presents elements to be considered as part of EU GMP Annex 15.

How should initial classification be performed during simulated operations?

- GMP auditor should verify that for cleanroom classification, the total of particles equal to or greater than 0.5 and 5 μm are measured. The auditor also may want to see that this measurement is performed both at rest and in simulated operations in accordance with the limits specified in Table 3.80.

TABLE 3.78
Various Questions About Sterile Drug Products

1	When observing personnel during production, do operators practice proper cleanroom behaviors as specified in established cleanroom SOPs?	☐ Fulfilled ☐ Control needed ☐ Action needed [room for notes]
2	Can you observe the aseptic filling processes without going into the clean room (i.e., through a window or TV monitor)? A. If yes, watch/observe the aseptic filling processes from the preparation of the bulk liquid product to filling and sealing of final dosage form, including the environmental monitoring performed in critical areas during actual production. B. If not, consider observing the aseptic filling inside the clean room, for example, via a Class 10,000 (ISO-7) area. Prior to entry, consult the district. • Do personnel enter and perform interventions at the critical Class 100 (ISO 5) areas of the filling lines? If so, how are such entries recorded? • How is this done (full body entry, hands only)? • Why? • Is this done routinely or infrequently? • Are proper aseptic techniques used? [Reference: FDA 2004 Aseptic Processing Guidance, Section V.A.] • Do hands or arms go over open vials for sterile components? If yes, are the vials discarded?	☐ Fulfilled ☐ Control needed ☐ Action needed [room for notes]
3	Does the firm have written procedures describing the aseptic filling of drug products? Does it include discussions of proper aseptic techniques and acceptable techniques for performing interventions in the Class 100 areas?	☐ Fulfilled ☐ Control needed ☐ Action needed [room for notes]
4	Review trend reports of the nonviable, viable, and personnel monitoring data. • Have any trends been identified by the firm? If yes, what is done? • Are investigations done of out-of-limit results?	☐ Fulfilled ☐ Control needed ☐ Action needed [room for notes]
	Aseptic Filing Validation	
5	How does the process used for media fill compare to the aseptic filling of commercial drug products? Does the firm accurately evaluate the production operation on a routine basis (changes over time) against the media fill design? Does the firm have detailed procedures that describe the media fill process, including frequency, challenge conditions, personnel participation, container/closures, interventions, duration of fill, reconciliation of vials, acceptance criteria, incubation, examination after incubation, actions to take if positive growth is found, etc.?	☐ Fulfilled ☐ Control needed ☐ Action needed [room for notes]
6	Request a summary of all media fills conducted since the last inspection, including lot identification, fill dates, production lines, number of units filled, number of units incubated, number of vials with no growth, number of vials with positive growth, microbial identification for any containers with growth, disposition of media fill units, etc. • Have media fills been done at the frequency described in the procedure? • Have any media fills shown positive growth? • Have any media fills failed to meet acceptance criteria?	☐ Fulfilled ☐ Control needed ☐ Action needed [room for notes]
7	Are procedures followed when positive growth is found in media-filled vials? Are investigations conducted of all positive growth found in media-filled vials? What is done when a media fill fails to meet acceptance criteria? Review investigations of any positive growth.	☐ Fulfilled ☐ Control needed ☐ Action needed [room for notes]

(Continued)

TABLE 3.78 (CONTINUED)
Various Questions About Sterile Drug Products

8	Are media fills performed on all shifts (e.g., representative of shift operations in a campaign)? Do media fills include shift changes and breaks that occur during routine production?	☐ Fulfilled ☐ Control needed ☐ Action needed [room for notes]
9	When can media fills be aborted or units not incubated? Would production lots be rejected if the same conditions existed during filling?	☐ Fulfilled ☐ Control needed ☐ Action needed [room for notes]
10	Are all personnel included in the media fill program? Does this include set-up personnel and mechanics who work on the aseptic filling lines? What system does the firm have for assuring all personnel are included?	☐ Fulfilled ☐ Control needed ☐ Action needed [room for notes]
11	If end-line filters are used in actual manufacture, are they also used during media fills?	☐ Fulfilled ☐ Control needed ☐ Action needed [room for notes]
12	What size vials or ampoules are used for media fills? Evaluate justification for selecting worst-case challenge if the firm does not perform media fills using all the container/closure configurations filled on the line.	☐ Fulfilled ☐ Control needed ☐ Action needed [room for notes]
13	Are vials inverted before incubation to assure media touches all inner surfaces?	☐ Fulfilled ☐ Control needed ☐ Action needed [room for notes]
14	How does the firm conduct media fills for products that are filled into amber or opaque containers?	☐ Fulfilled ☐ Control needed ☐ Action needed [room for notes]
15	How does the firm assure that all integral vials are incubated? How does the firm handle filled vials removed during the run because of interventions?	☐ Fulfilled ☐ Control needed ☐ Action needed [room for notes]
16	Do media fills include interventions that occur during routine production? Are there written procedures specifying the removal of media fill units (type of intervention and number of units removed) corresponding to the practice of a routine production run? Did you observe any interventions during the filling of routine production lots that are not included in media fills?	☐ Fulfilled ☐ Control needed ☐ Action needed [room for notes]
17	Does the duration of a media fill approximate the duration of a routine production lot? If not, evaluate the firm's justification for running shorter media fills?	☐ Fulfilled ☐ Control needed ☐ Action needed [room for notes]
18	What is the microbial growth medium?	☐ Fulfilled ☐ Control needed ☐ Action needed [room for notes]
19	Are growth promotion studies performed on each type of medium used? Are growth promotion studies performed on each type of medium used? Are growth promotion studies conducted every time a media fill is done? When are the growth promotion studies performed (before/after filling, after incubation, etc.)? What organisms are used to perform the growth promotion tests? Are any environmental organisms used?	☐ Fulfilled ☐ Control needed ☐ Action needed [room for notes]

(Continued)

TABLE 3.78 (CONTINUED)
Various Questions About Sterile Drug Products

20	What temperatures and incubation times are used to incubate media-filled vials?	☐ Fulfilled ☐ Control needed ☐ Action needed [room for notes]
21	Are microorganisms from positive vials identified to genus and species? Are such microorganisms correlated to those found during environmental monitoring?	☐ Fulfilled ☐ Control needed ☐ Action needed [room for notes]
22	Is an investigation done when cracked vials are found after incubation begins?	☐ Fulfilled ☐ Control needed ☐ Action needed [room for notes]
23	When and where are the media-filled vials examined?	☐ Fulfilled ☐ Control needed ☐ Action needed [room for notes]
24	Who performs the examination of media-filled vials? If production employees perform the examination, have they been trained to recognize all types of microbial growth? Is a microbiologist present when the examination is done?	☐ Fulfilled ☐ Control needed ☐ Action needed [room for notes]
25	What incubators are used to incubate media-filled vials? How is the temperature controlled and monitored? Have tests been done to determine if the temperature is uniform throughout the incubator?	☐ Fulfilled ☐ Control needed ☐ Action needed [room for notes]
	Lyophilization General	
1	Who is the manufacturer of the lyophilizer?	☐ Fulfilled ☐ Control needed ☐ Action needed [room for notes]
2	Describe the heating and cooling systems used in the lyophilizer, the vacuum system, the gas that is used to break the vacuum and whether it is sterile, and the temperature-controlling system.	☐ Fulfilled ☐ Control needed ☐ Action needed [room for notes]
3	How are the vials transported from the filling line to the lyophilizer? How are Class 100 (ISO 5) conditions maintained during the transport, loading, and unloading of the lyophilizer?	☐ Fulfilled ☐ Control needed ☐ Action needed [room for notes]
4	Are the loading and unloading of the vials automated or manual?	☐ Fulfilled ☐ Control needed ☐ Action needed [room for notes]
5	How is the sealing of vials (final seating of stoppers) performed?	☐ Fulfilled ☐ Control needed ☐ Action needed [room for notes]
6	If the stoppering is performed automatically in the chamber at the end of the cycle, is it under vacuum? If not under vacuum, what gas is used and how is it sterilized?	☐ Fulfilled ☐ Control needed ☐ Action needed [room for notes]

(Continued)

TABLE 3.78 (CONTINUED)
Various Questions About Sterile Drug Products

7	If the vials are stoppered outside of the chamber, how is the lyophilized product protected from contamination during transport to the stoppering station and during the stoppering operation?	☐ Fulfilled ☐ Control needed ☐ Action needed [room for notes]
8	Describe the chamber clean-up procedures between batches of the same product and between different products (including sterilant/ cleaning agents used, monitoring for sterilant residue where appropriate, and exposure cycle).	☐ Fulfilled ☐ Control needed ☐ Action needed [room for notes]
9	Past history has shown that the lyophilizer condenser can be a source of contamination and an assessment of the lyophilization process should include an inspection of the condenser. How is the lyophilization cycle monitored/ documented and reviewed?	☐ Fulfilled ☐ Control needed ☐ Action needed [room for notes]
	Lyophilization Validation – Sterilization of Chamber	
1	How is the lyophilization chamber sterilized? How are all surfaces, such as all surfaces of moving shelves, exposed to steam?	☐ Fulfilled ☐ Control needed ☐ Action needed [room for notes]
2	When is the lyophilization chamber sterilized? What is the explanation/justification if it is not done between each batch?	☐ Fulfilled ☐ Control needed ☐ Action needed [room for notes]
3	How are sterilization cycles controlled (manual, programmed)? How are cycles monitored during production?	☐ Fulfilled ☐ Control needed ☐ Action needed [room for notes]
4	Has the sterilization of the lyophilization chamber been validated? Have the slowest-to-heat surfaces in the chamber been determined and challenged in the validation runs?	☐ Fulfilled ☐ Control needed ☐ Action needed [room for notes]
5	Review current procedures and sterilization records for production lots. Are the cycle parameters the same as those used during validation? Do production cycles meet the validated cycle parameters?	☐ Fulfilled ☐ Control needed ☐ Action needed [room for notes]
	Lyophilization Validation – Aseptic Handling	
1	Is the aseptic handling of lyophilized products validated by media fills?	☐ Fulfilled ☐ Control needed ☐ Action needed [room for notes]
2	In the aseptic process, is lyophilization simulation performed during media fill?	☐ Fulfilled ☐ Control needed ☐ Action needed [room for notes]
3	Is the maximum amount of time the vials are held prior to lyophilization simulated during media fills? If vials are not sealed in the lyophilization chamber, is the maximum hold time prior to stoppering simulated in media fills?	☐ Fulfilled ☐ Control needed ☐ Action needed [room for notes]
4	During validation, what level of vacuum is pulled on the lyophilization chamber?	☐ Fulfilled ☐ Control needed ☐ Action needed [room for notes]

(*Continued*)

TABLE 3.78 (CONTINUED)
Various Questions About Sterile Drug Products

5	How long do media fill vials remain in the lyophilization chamber under vacuum? How does this compare to commercial lots?	☐ Fulfilled ☐ Control needed ☐ Action needed [room for notes]
6	Does the process simulation result in the freezing of the media? Note that this process simulation should not include freezing of the media.	☐ Fulfilled ☐ Control needed ☐ Action needed [room for notes]
7	Is environmental monitoring performed during the loading of the lyophilizer both during production and as well as during validation?	☐ Fulfilled ☐ Control needed ☐ Action needed [room for notes]
8	Does the firm have data on the growth promotion of the media? Are growth promotion tests done on vials after incubation is completed?	☐ Fulfilled ☐ Control needed ☐ Action needed [room for notes]
9	Is environmental monitoring performed during unloading of the chamber during production as well as during media fill validation?	☐ Fulfilled ☐ Control needed ☐ Action needed [room for notes]
10	What is used to break the vacuum during media fills (nitrogen, air, other gas)?	☐ Fulfilled ☐ Control needed ☐ Action needed [room for notes]
	Lyophilization Validation – Process	
1	Has the firm validated the lyophilization cycle (e.g., time, rate of heat input, temperatures, and eutectic melting point) for each product? Review validation records for selected drug products with different physical and chemical characteristics.	☐ Fulfilled ☐ Control needed ☐ Action needed [room for notes]
2	Review lyophilization production records for the same products. Are the cycle parameters and observed results within the validated cycles?	☐ Fulfilled ☐ Control needed ☐ Action needed [room for notes]
3	What are the firm's criteria for acceptable vs. unacceptable runs, including general appearance, cake appearance, melt back, reconstitution time, moisture, etc.?	☐ Fulfilled ☐ Control needed ☐ Action needed [room for notes]
4	Has the firm performed equipment qualification, preventive maintenance, critical instrument calibration, and cleaning validation?	☐ Fulfilled ☐ Control needed ☐ Action needed [room for notes]
	Lyophilization of Sterile API	
1	The aseptic process may include either a manual or automated transferring process or a combination of both, to transfer the lyophilized API out of the lyophilizer and into a SIP holding vessel or transfer tank. It is equally important to observe and assess the manual operations of the aseptic lyophilization process and assure that similar operations are performed during the media fill processing steps for the API.	☐ Fulfilled ☐ Control needed ☐ Action needed [room for notes]

(Continued)

TABLE 3.78 (CONTINUED)
Various Questions About Sterile Drug Products

	Isolators Methods (e.g., hydrogen peroxide, peracetic acid) used to decontaminate isolator barriers are capable of rendering surfaces free of viable organisms but lack the capabilities of steam sterilization. While these agents do not effectively penetrate obstructed or protected surfaces, validated systems are highly effective at ensuring contamination-free internal isolator surfaces.	
1	Determine: • Model of isolator or barrier, and materials of construction • Type of isolator (open or closed) • Airflow (turbulent, unidirectional) • Classification of the surrounding room environment • Number and location of gloves or half-suits • Attire worn by operators (e.g., are sterile gloves worn under isolator gloves) • Operating parameters (pressures, air velocities, temperature, humidity)	☐ Fulfilled ☐ Control needed ☐ Action needed [room for notes]
2	Is a written maintenance program in place that requires routine documented checks or tests of gloves, half-suits, door seals, etc. for integrity? What type of tests/checks are done and how frequently?	☐ Fulfilled ☐ Control needed ☐ Action needed [room for notes]
3	Do written procedures specify glove replacement frequency? If so, what is the frequency, and is the SOP followed?	☐ Fulfilled ☐ Control needed ☐ Action needed [room for notes]
4	Does the isolator maintain continuous positive pressure at sufficient levels?	☐ Fulfilled ☐ Control needed ☐ Action needed [room for notes]
5	How are materials transferred into and out of the isolator? How robust are the transfer mechanisms?	☐ Fulfilled ☐ Control needed ☐ Action needed [room for notes]
6	Are the equipment and surfaces that have direct contact with sterile products and components sterilized by heat? Does it achieve a minimum of a six-log reduction of the BI spores?	☐ Fulfilled ☐ Control needed ☐ Action needed [room for notes]
7	What method is used to decontaminate the inner surface of the isolator barrier (e.g., vapor hydrogen peroxide, steam hydrogen peroxide, chlorine dioxide, etc.)? Determine decontamination parameters.	☐ Fulfilled ☐ Control needed ☐ Action needed [room for notes]
8	Did the surface decontamination validation study sufficiently address the ability of the sterilant to disperse throughout the chamber and reach all surfaces? Did the decontamination process include the use of chemical indicators (CI) to determine the presence or absence of the VHP on the work surfaces and/or the worst-case locations to decontaminate? The CI can assist in the evaluation process by providing useful qualitative data. Were replicate BIs placed throughout the isolator, including the most difficult-to-reach locations (e.g., underneath any items remaining in isolator during sterilization)? Are the most difficult-to-sterilize materials evaluated?	☐ Fulfilled ☐ Control needed ☐ Action needed [room for notes]

(Continued)

TABLE 3.78 (CONTINUED)
Various Questions About Sterile Drug Products

9	What is the isolator decontamination frequency and is it justified by validation data?	☐ Fulfilled ☐ Control needed ☐ Action needed [room for notes]
10	Is a decontamination cycle performed after a power failure or pressure reversal or other unanticipated breach of system integrity?	☐ Fulfilled ☐ Control needed ☐ Action needed [room for notes]
11	How often is the isolator decontamination cycle revalidated?	☐ Fulfilled ☐ Control needed ☐ Action needed [room for notes]
12	Does the written environmental monitoring program include routine tests for nonviable particles, as well as an appropriate number of microbial tests (e.g., active air and surface samples; gloves samples) during each campaign? Evaluate the tests performed and the testing frequencies.	☐ Fulfilled ☐ Control needed ☐ Action needed [room for notes]
13	Have any false positives been identified in the laboratory isolator barrier (this should be an exceedingly rare occurrence)? Were the false positives investigated to determine and correct the cause?	☐ Fulfilled ☐ Control needed ☐ Action needed [room for notes]
	Environmental Monitoring – Nonviable	
1	Is the air supplied to critical areas (exposed product/filling areas) filtered through HEPA filters under positive pressure?	☐ Fulfilled ☐ Control needed ☐ Action needed [room for notes]
2	Is the airflow in critical areas unidirectional when delivered to the point of use? At what velocity? Is velocity determined at the critical work height and at the filter face? Airflow pattern evaluations (smoke studies) under dynamic conditions are performed to visualize and demonstrate unidirectional airflow within a designated area/room (e.g., Class 100 & 10,000 [ISO5 and ISO7]). The smoke studies may also reveal air turbulences and air eddies that could be a vector assisting the dissemination of microbial and/or nonviable contamination within critical manufacturing areas.	☐ Fulfilled ☐ Control needed ☐ Action needed [room for notes]
3	How is the air filtered that is supplied to critical areas (where unsterilized product, in-process materials, and container/closures are prepared)?	☐ Fulfilled ☐ Control needed ☐ Action needed [room for notes]
4	What are the firm's air quality classifications for the following areas: • Compounding • Equipment preparation • Any area where product or sterilized components are exposed • Areas where aseptic connections are made • Filling lines • Room surrounding filling lines • Capping (crimping) area	☐ Fulfilled ☐ Control needed ☐ Action needed [room for notes]
5	How often is HEPA filter integrity tested? What test method is used? What is done if leaks are found? If done by an outside firm, are results reviewed by on-site personnel including the Quality Unit?	☐ Fulfilled ☐ Control needed ☐ Action needed [room for notes]

(*Continued*)

TABLE 3.78 (CONTINUED)
Various Questions About Sterile Drug Products

6	How often are airflow velocities checked for each HEPA filter? What are the air velocity specifications? What is done if velocity readings are out of specification?	☐ Fulfilled ☐ Control needed ☐ Action needed [room for notes]
7	Does the firm have a written monitoring program for classified areas that includes scientifically sound sampling schedules; descriptions of sampling locations and frequency of sampling? How were the locations selected?	☐ Fulfilled ☐ Control needed ☐ Action needed [room for notes]
8	What type of instrument is used to check nonviable particle counts in the classified areas? Is the air sampled continuously? If continuous, is there an alarm when counts exceed pre-set limits or detect cleanroom doors open for an extended time? What is done in response to the alarms? Is there an alarm log? Are permanently installed sensors used or portable units that are taken into and out of the critical areas? What is done if counts meet or exceed alert and action limits?	☐ Fulfilled ☐ Control needed ☐ Action needed [room for notes]
9	What are the pressure differential requirements in the sterile core area? Does this ensure an air cascade from the most clean area to the least clean air (by air classification)?	☐ Fulfilled ☐ Control needed ☐ Action needed [room for notes]
10	How are pressure differentials monitored? Is a continuous monitoring system used? If yes, does it include alarms so operators are aware of excursions? Are alarm conditions documented (alarm log)? What is done in response to an alarm? How long must a condition exist before an alarm is sounded?	☐ Fulfilled ☐ Control needed ☐ Action needed [room for notes]
11	How are temperature and humidity monitored? What is the acceptable range? What is done if the reading is outside of the range?	☐ Fulfilled ☐ Control needed ☐ Action needed [room for notes]
12	How are environmental excursions (readings that meet or exceed alert and action levels) handled? Are the excursions investigated to determine the impact on the product, root cause, and needed corrective actions?	☐ Fulfilled ☐ Control needed ☐ Action needed [room for notes]
	Environmental and Personnel Monitoring Air	
1	Does the firm have an effective Environmental Monitoring (EM) Program in place? What is the purpose and scope of the EM Program? Are EM sampling locations strategic, based on the product contamination risk of the process and operation? Are the microbial alert and action levels based on the historical EM data derived from the manufacturing operations, support utilities and personnel practices performed at the manufacturing site? What is the frequency of microbiological sampling of air using "active" samplers (systems that sample a known volume of air) in various locations, such as: • Areas where product or sterile components are exposed to the environment • Filling areas • Loading areas for lyophilizers • Surrounding areas	☐ Fulfilled ☐ Control needed ☐ Action needed [room for notes]
2	What are the established microbial alert and action levels or limits for quantitative air samples? What is the length of the sampling period? Is sampling done during production or at rest?	☐ Fulfilled ☐ Control needed ☐ Action needed [room for notes]

(*Continued*)

TABLE 3.78 (CONTINUED)
Various Questions About Sterile Drug Products

3	What type of active air sampling equipment is used (centrifugal, impaction, membrane)? Is the sampling equipment calibrated? What is the efficiency of the active air sampler?	☐ Fulfilled ☐ Control needed ☐ Action needed [room for notes]
4	Does the firm have data on the ability of these samplers to recover organisms without deleterious effects on survivability such as through impact or desiccation of the media?	☐ Fulfilled ☐ Control needed ☐ Action needed [room for notes]
5	What is the actual volume of air sampled per location?	☐ Fulfilled ☐ Control needed ☐ Action needed [room for notes]
6	Are settling plates used? What is the length of the exposure period? Sampling frequency? Locations (including proximity to critical operations)? Microbial limits?	☐ Fulfilled ☐ Control needed ☐ Action needed [room for notes]
	Surfaces	
1	Are there written procedures that describe the monitoring of surfaces in the clean rooms? Does it describe locations, frequency and sampling techniques?	☐ Fulfilled ☐ Control needed ☐ Action needed [room for notes]
2	What surfaces are sampled in the critical (Class 100/ISO 5) areas? Are critical surfaces included (those that touch the sterile product or sterile components)?	☐ Fulfilled ☐ Control needed ☐ Action needed [room for notes]
3	When are surface samples taken? Are critical surfaces only sampled at the end of the operation?	☐ Fulfilled ☐ Control needed ☐ Action needed [room for notes]
4	What is the frequency of sampling surfaces?	☐ Fulfilled ☐ Control needed ☐ Action needed [room for notes]
5	What types of samples are collected at each location (RODAC plate, swab samples)?	☐ Fulfilled ☐ Control needed ☐ Action needed [room for notes]
6	What are the alert and action levels/limits for the microbial sampling of critical surfaces, surfaces in Class 100 areas and surfaces in other classified areas? How were the levels/limits established? What is done when the samples meet or exceed alert and action levels?	☐ Fulfilled ☐ Control needed ☐ Action needed [room for notes]
	Personnel	
1	Are there specific programs for training personnel who work in classified and aseptic processing areas? Does the program include microbiological testing or qualifying employees before they are allowed to work in these areas? Does the program include prequalification requirements?	☐ Fulfilled ☐ Control needed ☐ Action needed [room for notes]
2	How often is monitoring performed on filling room personnel? Is it done at least upon exiting the cleanroom (e.g., per shift)? How often are gloves (hands) sampled? How often are gowns sampled? Is any sampling done after interventions are performed?	☐ Fulfilled ☐ Control needed ☐ Action needed [room for notes]

(*Continued*)

TABLE 3.78 (CONTINUED)
Various Questions About Sterile Drug Products

3	Who performs the sampling of personnel? Is it self-sampling? Is the sampling performed by another production employee, a microbiologist, a technician, or another quality control employee?	☐ Fulfilled ☐ Control needed ☐ Action needed [room for notes]
4	What are the firm's alert and action levels/limits for personnel monitoring? How were the levels/limits established? What is done if the sample results reach or exceed the alert and action limits?	☐ Fulfilled ☐ Control needed ☐ Action needed [room for notes]
5	Do employees spray their hands with sanitizer or disinfectant before the sample is taken?	☐ Fulfilled ☐ Control needed ☐ Action needed [room for notes]
	General	
1	What are the microbiological growth media used for environmental and personnel monitoring samples?	☐ Fulfilled ☐ Control needed ☐ Action needed [room for notes]
2	Are the media used in the viable monitoring program shown to be capable of detecting molds and yeasts as well as bacteria by means of growth promotion tests?	☐ Fulfilled ☐ Control needed ☐ Action needed [room for notes]
3	Is anaerobic monitoring ever performed? When?	☐ Fulfilled ☐ Control needed ☐ Action needed [room for notes]
4	Are inactivators used for antibiotics or other bactericidal/bacteriostatic substances? Has the firm shown that these are effective? Are records available?	☐ Fulfilled ☐ Control needed ☐ Action needed [room for notes]
5	When are recovered microorganisms identified? To what level (genus, species)?	☐ Fulfilled ☐ Control needed ☐ Action needed [room for notes]
6	What are the incubation times and temperatures?	☐ Fulfilled ☐ Control needed ☐ Action needed [room for notes]
7	How are environmental and personnel monitoring data trended? How often are trend reports prepared? How often are they reviewed? Who reviews the trend reports? What type of action is taken based on the review of trend reports?	☐ Fulfilled ☐ Control needed ☐ Action needed [room for notes]
8	Are environmental and personnel monitoring samples that exceed action levels investigated to determine product impact, cause, and needed corrective actions? What is done when an alert level is exceeded?	☐ Fulfilled ☐ Control needed ☐ Action needed [room for notes]
9	Review and assess the EM trending data, which will provide a good indication of the viable and nonviable particles are maintained within the established levels or drifting out of control. What are the causes of the aberrant events? Were corrective actions and preventive measures taken to preclude the reoccurrence of the viable and nonviable particle anomalies?	☐ Fulfilled ☐ Control needed ☐ Action needed [room for notes]

(Continued)

TABLE 3.78 (CONTINUED)
Various Questions About Sterile Drug Products

	Biological Indicators	
	In sterile drug manufacturing, BIs are typically used to validate the cycles used for terminal sterilization, and sterilization of equipment and components. BIs are also used to validate SIP systems used to sterilize lyophilizers, processing tanks, sterilizing filters, and product lines and systems used to decontaminate surfaces in isolators.	
1	What type of indicator is used (e.g., inoculated carrier, inoculated product, inoculated simulated product, etc.)? Are BIs inoculated into the components (i.e., stoppers) used whenever possible? What are their corresponding D values? Is the sterilization cycle adequate corresponding to the D value of the BI?	☐ Fulfilled ☐ Control needed ☐ Action needed [room for notes]
2	If the source of the indicator is commercial, what is the brand name and manufacturer? What labeling is received with the BI? If the BI is prepared in-house, determine the supplier of the organism, how the BI is propagated and stored, and the method of preparation.	☐ Fulfilled ☐ Control needed ☐ Action needed [room for notes]
3	What organism is used (genus, species)? Is it the appropriate microorganism used for sterilization?	☐ Fulfilled ☐ Control needed ☐ Action needed [room for notes]
4	What is the challenge level of the biological indicator prior to exposure to sterilant?	☐ Fulfilled ☐ Control needed ☐ Action needed [room for notes]
5	Does the firm verify viable spore count on each lot of BIs before use in validation?	☐ Fulfilled ☐ Control needed ☐ Action needed [room for notes]
6	Does the firm or the BI labeling claim to meet USP performance criteria for steam or ETO biological indicators?	☐ Fulfilled ☐ Control needed ☐ Action needed [room for notes]
7	Does the firm perform USP testing on each lot of BIs received?	☐ Fulfilled ☐ Control needed ☐ Action needed [room for notes]
8	What is the approximate D value of the biological indicator? Is this verified prior to validation?	☐ Fulfilled ☐ Control needed ☐ Action needed [room for notes]
9	How many BIs are used per sterilizer load?	☐ Fulfilled ☐ Control needed ☐ Action needed [room for notes]
10	What procedure is used to assay the indicators after exposure? What growth media is used? What are the optimal and actual incubation time and temperature for the BI (compare to COA received with the BI)?	☐ Fulfilled ☐ Control needed ☐ Action needed [room for notes]
11	How are the BIs prepared for sterilization?	☐ Fulfilled ☐ Control needed ☐ Action needed [room for notes]

(Continued)

TABLE 3.78 (CONTINUED)
Various Questions About Sterile Drug Products

12	Are biological indicators located in the most difficult-to-sterilize product sites (explain)? How are these locations determined?	☐ Fulfilled ☐ Control needed ☐ Action needed [room for notes]
13	Is there a diagram of the distribution of biological indicators in the loading pattern(s) used?	☐ Fulfilled ☐ Control needed ☐ Action needed [room for notes]
14	What is the elapsed time (hrs.) between removing indicators from the sterilizer and testing? Are there time limits established for this period? What happens if they are exceeded?	☐ Fulfilled ☐ Control needed ☐ Action needed [room for notes]
15	What is done if positive BIs are found after sterilization?	☐ Fulfilled ☐ Control needed ☐ Action needed [room for notes]
16	Describe biological indicator storage conditions: A. Type of room, cabinet, etc. (if stored in freezer or refrigerator, state if frost-free) B. Temperature C. Relative humidity (if known) D. Compare to literature received with BI or written procedures	☐ Fulfilled ☐ Control needed ☐ Action needed [room for notes]
17	Does the firm use chemical process monitor(s) to indicate cycle exposure or to measure one or more cycle parameters?	☐ Fulfilled ☐ Control needed ☐ Action needed [room for notes]

TABLE 3.79
Elements to be Considered As Part of EU GMP – Annex 15

- Installed filter system leakage and integrity testing
- Airflow tests – volume and velocity
- Air pressure difference test
- Airflow direction test and visualization
- Microbial airborne and surface contamination
- Temperature measurement test
- Relative humidity test
- Recovery test
- Containment leak test

For the classification of the cleanroom, the minimum number of sampling locations and their positioning can be found in ISO 14644 Part 1.

TABLE 3.80
Maximum Permitted Total Particle Concentration for Classification

	Maximum Limits for Total Particle ≥0.5 µm/m³		Maximum Limits for Total Particle ≥5 µm/m³	
Grade	At Rest	In Operation	At Rest	In Operation
A	3,520	3,520	Not specified[a]	Not specified[a]
B	3,520	352,000	Not specified[a]	2,930
C	352,000	3,520,000	2,930	29,300
D	3,520,000	Not predetermined[b]	29,300	Not predetermined[b]

- The GMP auditor should verify that the aseptic processing area and the background environment (the grade A and grade B areas, respectively), additional sample locations are considered, and all critical processing areas such as the point of fill and container closure feeder bowls are evaluated.
- The GMP auditor should verify that critical processing locations are determined by documented risk assessment and knowledge of the process and operations to be performed in the area.

Table 3.80 presents the maximum permitted total particle concentration for classification.

The definition of an "at rest" state is the condition whereby the installation of all the utilities is complete including any functioning HVAC, with the main manufacturing equipment installed as specified but not operating and without personnel present in the room.

The definition of "in operation" state is the condition where the installation of the cleanroom is complete, the HVAC system fully operational, and equipment installed and functioning in the manufacturer's defined operating mode with the maximum number of personnel present performing or simulating routine operational work.

- The GMP auditor should verify that the total particle limits given in Table 3.80 for the "at rest" state are achieved after a "clean up" period on completion of operations and line clearance/cleaning activities.
- The GMP auditor should verify that the total particle limits given in Table 3.80 for the "in operation" state are achieved after a "clean up" period on completion of operations and line clearance/cleaning activities.
- The GMP auditor should verify that the "clean up" period (guidance value of fewer than 20 minutes) is determined during the qualification of the rooms, documented, and adhered to in procedures to reinstate a qualified state of cleanliness if disrupted during operation.
- The GMP auditor should verify that the speed of air supplied by unidirectional airflow systems is clearly justified in the qualification protocol including the location for air-speed measurement.
- The GMP auditor should verify that air speed is designed, measured and maintained to ensure that appropriate unidirectional air movement provides protection of the product and open components at the working position (e.g., where high-risk operations occur and where product and/or components are exposed).
- The GMP auditor should verify that unidirectional airflow systems provide a homogeneous air speed in a range of 0.36–0.54 m/s (guidance value) at the working position unless otherwise scientifically justified in the CCS. The GMP auditor should verify that airflow visualization studies correlate with the air-speed measurement.

TABLE 3.81
Maximum Permitted Microbial Contamination Level during Qualification

Grade	Air Sample (CFU/m³)	Settle Plates (diameter 90 mm) CFU/4 hours[a]	Contact Plates (diameter 55 mm) CFU/plate	Glove Print Including 5 Fingers on Both Hands CFU/glove
A	No growth			
B	10	5	5	5
C	100	50	25	-
D	200	100	50	-

- The GMP auditor should verify that the microbial contamination level of the cleanrooms is determined as part of the cleanroom qualification. The GMP auditor should verify that the number of sampling locations is based on a documented risk assessment and the results obtained from room classification, air visualization studies, and knowledge of the process and operations to be performed in the area.

The maximum limits for microbial contamination during qualification for each grade are given in Table 3.81. Qualification should include both "at rest" and "in operation" states.

- The GMP auditor should verify that the maximum time interval for requalification of grade A and B areas is 6 months, and the maximum time interval for requalification of grade C and D areas is 12 months.
- The GMP auditor should verify that a requalification, including at least the tests highlighted above, has been carried out following completion of remedial action to address non-conformance of equipment or improvement of conditions at the facility, or after procedural changes have been made to the facility or processes.

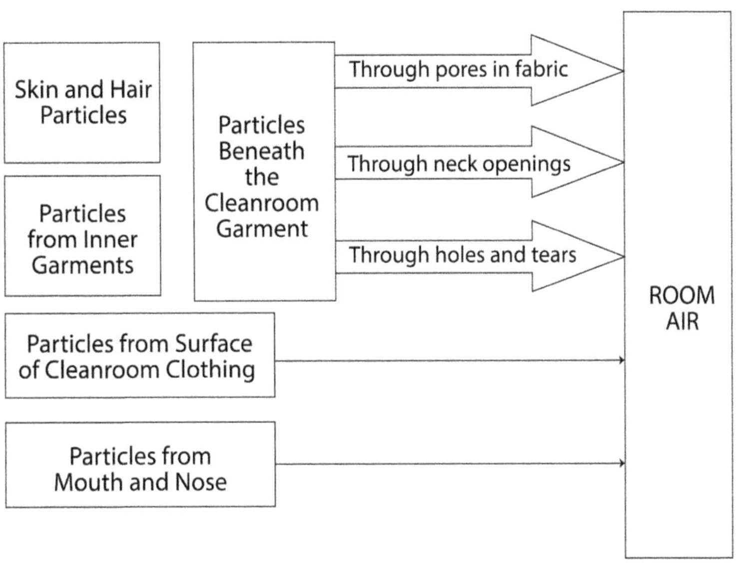

FIGURE 3.37 What type of contaminants are shedded from people?

When questioning change management, a GMP auditor must verify whether the change was made in accordance with the relevant procedures. Some of the changes that the GMP auditor should pay attention to are:

i. Cutting off air movement that would disrupt the operation of the installation
ii. A design change anywhere in the cleanroom or a parameter change in the HVAC system
iii. Special maintenance requirements that will disrupt the operation of the installation (e.g., replacement of filters)

In Figure 3.37, a GMP auditor can review typical contaminants shedded from people in cleanrooms.

New ISO 14644-1:2015 Instrument Calibration Requirements

B.2.2 Instrument Calibration

The instrument shall have a valid calibration certificate, and the frequency and method of calibration should be based on current accepted practice. ISO 14644-1:1999

A.22 Instrument Calibration

Particle counters used in cleanrooms must be designed to meet the requirements of ISO 21501-4. Calibration frequencies based on this standard are very important and should definitely be questioned in GMP audits.

Note: Some particle counters may be calibrated outside the requirements specified in the ISO standard. This situation should not be ignored.

What does this mean to you?

You are expected to have instruments calibrated in accordance with ISO requirements.

Use of non-ISO 21501-4 compliant instruments will require additional and undesirable explanations to authorities.

ISO 21501-4 is the worldwide-recognized standard for optical particle counter calibration.

This standard was introduced by the ISO organization in 2007, and it is based on the following purpose.

An important result of the ISO 14644-1:2015 revision once again shows the need for the ISO 21501:4 standard that regulates particle counters. Pharmaceutical and biotechnology companies with cleanrooms will occasionally need ISO 21501 for Current Good Manufacturing (CGMP), EU, GMP, ISO 14644-1, and other requirements.

This change to ISO 14644-1 represents an important step in improving the accuracy of clean room contamination evaluation and aseptic process control.

ISO 14644-2

- Monitoring plan
- Risk assessment
- Periodic classification
- Alarms

3.28 AUDITING DISINFECTION

- Environmental control and cleaning of the manufacturing or packaging area and equipment used to manufacture the antimicrobial chemical agent.
- Control and disinfection or sterilization of the antimicrobial chemical agent packaging containers.

- Documentation and review of antimicrobial chemical agent production processing activities.
- For aseptically filled agents, the EM program data, including alert and action levels, trending, corrective actions taken, and the use of neutralizing agents for the EM media used.
- For agents labeled as sterile, sterility testing data, and qualification of the sterilization process.
- Water systems and the quality of water used in the manufacturing process.
- Package or container integrity studies.
- For double- and triple-bagged containers, disinfection of filled containers, and overwrapping integrity.
- For double- and triple-bagged containers where a claim of sterility is made for inner bags, qualification of the sterilization process is used.
- Handling and storage of finished product containers or work in progress.
- Study results to support the label claim of an agent.
- Documentation related to regulatory approval of agent.
- Change control: customer notification of ingredient changes or process changes that would affect the finished product, for example, wrapping, irradiation, and sterilization.

3.29 ASEPTIC OPERATORS AUDIT

- What is the "curriculum" for an aseptic operator?
- Is this curriculum a "controlled document," signed by the departmental manager?
- Is training planned? What is the plan?
- How often is retraining performed? For what reasons?
- If personnel work in more than one area, are there any measures to prevent cross-contamination when moving from one area to another?
- How is the effectiveness of training measured?
- Is there any feedback to the employees on their aseptic performance?
- How long have employees been working in the pharmaceutical industry? And in cleanrooms?
- Are duties and responsibilities grouped into more experienced and less experienced personnel?
- Look at connections and transfers and evaluate "routine."
- Do new personnel receive sufficient practical training in outdoor areas before performing production in clean areas?
- What are the rules for good personal hygiene?
- Are the operators trained in proper gowning techniques? How often is retraining?
- Is there a cheat sheet in the gowning area? If so, is it in line with the SOP?
- How is the qualification of gowning measured?
- How is it ensured that only qualified people enter the areas?
- Are all make-up and all jewelry really removed?
- Is there a mirror in the locker room where staff can fully check their clothes before entering the clean areas?
- How often are personnel cleaning, disinfecting, and changing their gloves?
- How is monitoring personnel trained? What is their "curriculum"?
- Is understanding the sources of contamination involved with sampling part of that training for monitoring personnel?
- When are the operator samples taken?
- What are typical conclusions on investigations related to operators exceeding limits/showing adverse trends?

3.30 AUDITING MEDIA FILL

- What are the results of media fills split up by operator?
- Is there any adverse trend on an operator level? And (how) was this investigated?
- What happens if a media fill fails with a specific shift or team?
- What are operators doing before they pass a media fill?
- Evaluate the aseptic process simulation (media fill or broth fill)
- Know the key points in designing a media fill protocol to meet regulatory expectations
- Learn how failures have to be evaluated and which consequences they have
- To discuss in a team

3.30.1 Principles of Media Fill

Why is the validation of the aseptic process required by pharmaceutical regulations?

A "sterile product" is defined as "free of viable organisms." As it is not practical to examine every unit for confirmation of sterility, all efforts are made to minimize the risk of contamination (finishing, HVAC, pressure differentials cleaning procedure, monitoring program). Despite such measures, contamination is an ever-present danger because aseptic processing is a process being operated in an uncontrolled – but not sterile – environment and sample numbers are too small, so that only gross contamination is likely to be detected.

- Pharmaceutical regulations
  - FDA "Guidance for industry, sterile drug products produced by aseptic processing – cGMP"
  - EU GMP Part I Annex 1
  - PIC/S PI 007-2 "Recommendations on the validation of aseptic process"
- Although media fills must duplicate aseptic manufacturing conditions, it is not possible for them to be conducted in exactly the same way as the manufacture of a production batch of a pharmaceutical product.

What medium? How many units? How long?

- In aseptic processing, the greatest risk comes from the personnel working in the clean room; the operators have to participate in media fills.

Which qualifications do the operators need and when can operators be considered qualified?

- Environmental monitoring activities are required during aseptic filling operations.

Are there additional monitoring activities necessary or not?

- It is usual to include the worst-case conditions that can occur in production runs

What kind of interventions have to be considered?

3.30.2 Key Elements to be Taken into Account

- Number and frequency of runs
- Medium culture (to replace the product)
- Number of units filled
- Container (vial) size
- Fill volume
- Line speed (or filling speed)
- Duration of fill

- Operators shifts
- Monitoring activities
- Interventions (both routine and non-routine)
- Incubation method
- Acceptance criteria

3.30.3 Behavior

- Are people behaving relatively slowly and deliberately?
- Are they touching their clothing/face?
- Are they touching sterile surfaces, equipment, products, etc. with gloves or gowns?
- How are they manipulating, if needed, related to the airflow?
- What happens to people who have an infectious disease?
- Is this described in an SOP?
- Are, e.g., athlete's foot and heavy dandruff also included?
- Is there any restriction for smokers working in the aseptic core?

Periodic verification of the effectiveness of the controls in place for aseptic processing should include an APS using a sterile nutrient media and/or surrogate in place of the product.

- The GMP auditor should verify that the APS is not considered as the primary means to validate the aseptic process or aspects of the aseptic process. Table 3.82 presents critical manufacturing steps in APS.
- The GMP auditor should verify that the effectiveness of the aseptic process is determined through process design, adherence to the pharmaceutical quality system and process controls, training, and evaluation of monitoring data.

TABLE 3.82
What Critical Manufacturing Steps Should Be Included in APS?

- The APS should assess all aseptic operations performed subsequent to the sterilization and decontamination cycles of materials utilized in the process to the point where the container is sealed.
- For non-filterable formulations, any additional aseptic steps should be assessed.
- Where aseptic manufacturing is performed under an inert atmosphere, the inert gas should be substituted with air in the process simulation unless anaerobic simulation is intended.
- Processes requiring the addition of sterile powders should use an acceptable surrogate material in the same containers as those used in the process under evaluation.
- Separate simulations of individual unit operations (e.g., processes involving drying, blending, milling, and subdivision of a sterile powder) should be avoided. Any use of individual simulations should be supported by a documented justification and ensure that the sum total of the individual simulations continues to fully cover the whole process.
- The process simulation procedure for lyophilized products should represent the entire aseptic processing chain including filling, transport, loading, a representative duration of the chamber dwell, unloading, and sealing under specified, documented, and justified conditions representing worst-case operating parameters.
- The lyophilization process simulation should mimic all aspects of the process, except those that may affect the viability or recovery of contaminants. For instance, boiling-over or actual freezing of the solution should be avoided. Factors to consider in determining APS design include, where applicable:
  - The use of air to break vacuum instead of nitrogen or other process gases.
  - Replicating the maximum interval between sterilization of the lyophilizer and its use.
  - Replicating the maximum period of time between filtration and lyophilization.
  - Quantitative aspects of worst-case situations, e.g., loading the largest number of trays, replicating the longest duration of loading where the chamber is open to the environment.

- The GMP auditor should verify that the selection of an appropriate nutrient media and/or surrogate is made based on the ability of the media and/or surrogate to imitate physical product characteristics assessed to pose a risk to product sterility during the aseptic process.

If processing stages may indirectly impact the viability of any introduced microbial contamination (e.g., aseptically produced semisolids, powders, solid materials, microspheres, liposomes, and other formulations where the product is cooled or heated or lyophilized), the GMP auditor should verify that alternative procedures that represent the operations as closely as possible are developed.

If surrogate materials, such as buffers, are used in parts of the APS, the surrogate material should not inhibit the growth of any potential contamination

- The GMP auditor should verify that the APS takes into account various aseptic manipulations and interventions known to occur during normal production as well as worst-case situations.
  - The GMP auditor should verify that inherent and corrective interventions representative of the routine process are performed in a manner and frequency similar to that during the routine aseptic process.
  - The GMP auditor should verify that the inclusion and frequency of interventions in the APS are based on assessed risks posed to product sterility.
- The GMP auditor should verify that APS is not used to justify practices that pose unnecessary contamination risks.

Table 3.83 presents elements to be considered in APS.

- The GMP auditor should verify that for sterile active substances, the batch size is large enough to represent a routine operation, simulate intervention operation at the worst case, and cover all surfaces that may come into contact with the sterile product.
- The GMP auditor should verify that, in addition, all the simulated materials (surrogates or growth medium) are subjected to microbial evaluation.
- The GMP auditor should verify that the simulation materials are sufficient to satisfy the evaluation of the process being simulated and do not compromise the recovery of microorganisms.
- The GMP auditor should verify that APS is performed as part of the initial validation, with at least three consecutive satisfactory simulation tests that cover all working shifts that the aseptic process may occur in, and after any significant modification to operational practices, facilities, services, or equipment that are assessed to have an impact on the sterility assurance of the product (e.g., modification to the HVAC system, equipment, changes to process, number of shifts and number of personnel, major facility shut down).
- The GMP auditor should verify that APS (periodic revalidation) is repeated twice a year (approximately every six months) for each aseptic process, each filling line, and each shift.
- The GMP auditor should verify that each operator participates in at least one successful APS annually.
- The GMP auditor should verify that consideration is given to performing an APS after the last batch prior to shutdown, before long periods of inactivity, or before decommissioning or relocation of a line.
- If manual operation (e.g., aseptic compounding or filling) occurs, the GMP auditor should verify that each type of container, container closure, and equipment train are initially validated with each operator participating in at least three consecutive successful APS and revalidated with one APS approximately every six months for each operator.

TABLE 3.83
Elements to Consider in APS Plan

- Identification of worst-case conditions covering the relevant variables, such as container size and line speed, and their impact on the process. The outcome of the assessment should justify the variables selected.
- Determining the representative sizes of container/closure combinations to be used for validation. Bracketing or matrix approach may be considered for validation of the same container/closure configuration for different products where process equivalence is scientifically justified.
- Maximum permitted holding times for sterile products and equipment exposed during the aseptic process.
- The volume filled per container, which should be sufficient to ensure that the media contacts all equipment and component surfaces that may directly contaminate the sterile product. The volume used should provide sufficient headspace to support potential microbial growth and ensure that turbidity can be detected during inspection.
- The requirement for substitution of any inert gas used in the routine aseptic manufacturing process by air unless anaerobic simulation is intended. In these situations, the inclusion of occasional anaerobic simulations as part of the overall validation strategy should be considered.
- The selected nutrient media should be capable of growing a designated group of reference microorganisms as described by the relevant pharmacopoeia and suitably representative local isolates.
- The method of detection of microbial contamination should be scientifically justified to ensure that contamination is reliably detected.
- The process simulation should be of sufficient duration to challenge the process, the operators that perform interventions, shift changes, and the capability of the processing environment to provide appropriate conditions for the manufacture of a sterile product.
- Where the manufacturer operates different or extended shifts, the APS should be designed to capture factors specific to those shifts that are assessed to pose a risk to product sterility, for example, the maximum duration for which an operator may be present in the cleanroom.
- Simulating normal aseptic manufacturing interruptions where the process is idle (e.g., shift changeovers, recharging dispensing vessels, introduction of additional equipment).
- Ensuring that environmental monitoring is conducted as required for routine production and throughout the entire duration of the process simulation.
- Where campaign manufacturing occurs, such as in the use of barrier technologies or the manufacture of sterile active substances, consideration should be given to designing and performing the process simulation so that it simulates the risks associated with both the beginning and the end of the campaign and demonstrating that the campaign duration does not pose any risk.
- The performance of "end of production or campaign APS" may be used for additional assurance or for investigative purposes; however, their use should be justified in the CCS and should not replace routine APS. If used, it should be demonstrated that any residual product does not negatively impact the recovery of any potential microbial contamination

- The GMP auditor should verify that the APS batch size mimics that used in the routine aseptic manufacturing process.
- The GMP auditor should verify that the number of units processed (filled) for APS is sufficient to effectively simulate all activities that are representative of the aseptic manufacturing process.
- The GMP auditor should verify that justification for the number of units to be filled is clearly captured in the CCS.

Typically, a minimum of 5,000 to 10,000 units are filled. For small batches (e.g., those under 5,000 units), the number of containers for APS should at least equal the size of the production batch.

- The GMP auditor should verify that filled APS units are agitated, swirled, or inverted before incubation to ensure contact of the media with all interior surfaces in the container.

- The GMP auditor should verify that all integral units from the APS are incubated and evaluated, including units with cosmetic defects or those that have gone through nondestructive in-process control checks.

If units are discarded during the process simulation and not incubated, these should be comparable with units discarded during a routine fill, and only if production SOPs clearly specify that units must be removed under the same circumstances (i.e., type of intervention; line location; specific number of units removed).

In no case should more units be removed during a media fill intervention than would be cleared during a production run. Examples may include those that must be discarded during routine production after the set-up process or following a specific type of intervention. To fully understand the process and assess contamination risks during aseptic set-up or mandatory line clearances, these units would typically be incubated separately and would not necessarily be included in the acceptance criteria for the APS.

If processes include materials that contact the product contact surfaces but are then discarded (e.g., product flushes), the discarded material should be simulated with nutrient media and be incubated as part of the APS unless it can be clearly demonstrated that this waste process would not impact the sterility of the product.

- The GMP auditor should verify that filled APS units are incubated in a clear container to ensure visual detection of microbial growth.

If the product container is not clear (e.g., amber glass, opaque plastic), clear containers of identical configuration may be substituted to aid in the detection of contamination.

When a clear container of identical configuration cannot be substituted, the GMP auditor should verify that a suitable method for the detection of microbial growth is developed and validated.

- The GMP auditor should verify that microorganisms isolated from contaminated units are identified to the species level when practical, to assist in the determination of the likely source of the contaminant.
- The GMP auditor should verify that filled APS units are incubated without unnecessary delay to achieve the best possible recovery of potential contamination.
- The GMP auditor should verify that the selection of the incubation conditions and duration are scientifically justified and validated to provide an appropriate level of sensitivity for the detection of microbial contamination.

3.30.4 What Happens after Incubation?

- The GMP auditor should verify that filled APS units are inspected by personnel who have been appropriately trained and qualified for the detection of microbiological contamination. The GMP auditor should verify that the inspection is conducted under conditions that facilitate the identification of any microbial contamination.
- The GMP auditor should verify that samples of the filled units undergo positive control by inoculation with a suitable range of reference organisms and suitably representative local isolates.

Table 3.84 depicts actions to be taken in case of any failures in APS.

3.30.5 What Happens If Target Zero Growth Is Exceeded?

- The GMP auditor should verify that all APS runs are fully documented and include a reconciliation of units processed (e.g., units filled, incubated, and not incubated).

TABLE 3.84
Actions to Take after APS Failure

- An investigation to determine the most probable root cause(s).
- Determination and implementation of appropriate corrective measures.
- A sufficient number of successful, consecutive repeat APS (normally a minimum of 3) should be conducted in order to demonstrate that the process has been returned to a state of control.
- A prompt review of all appropriate records relating to aseptic production since the last successful APS.
  a) The outcome of the review should include a risk assessment of potential sterile breaches in batches manufactured since the last successful APS.
  b) All other batches not released to the market should be included in the scope of the investigation. Any decision regarding their release status should consider the investigation outcome.
- All products that have been manufactured on a line subsequent to a process simulation failure should be quarantined until a successful resolution of the process simulation failure has occurred.
- Where the root cause investigation indicates that the failure was related to operator activity, actions to limit the operator's activities, until retrained and requalified, should be taken.
- Production should resume only after the completion of successful revalidation.

- The GMP auditor should verify that justification for filled and nonincubated units is included in the documentation.
- The GMP auditor should verify that all interventions performed during the APS are recorded, including the start and end time of each intervention and the involved person.
- The GMP auditor should verify that all microbial monitoring data as well as other testing data are recorded in the APS batch record.
- The GMP auditor should verify that an APS run is aborted only under circumstances in which written procedures require commercial lots to be equally handled. The GMP auditor should verify an investigation is documented in such cases.

The GMP auditor should verify that an aseptic process is subject to a repeat of the initial validation when:

i. The specific aseptic process has not been in operation for an extended period of time.
ii. There is a change to the process, equipment, procedures, or environment that has the potential to affect the aseptic process or addition of new product containers or container/closure combinations.

In Figure 3.38, A GMP auditor can review a standard autoclave function.

Jacket – maintains heat
Door – seals the chamber
Vacuum pump – removes air
Baffle plates – distribute steam
Probes – monitor temperature and pressure

3.31 AUTOCLAVE AUDIT CONSIDERATIONS

- When auditing, look at:
  - Was validation conducted?

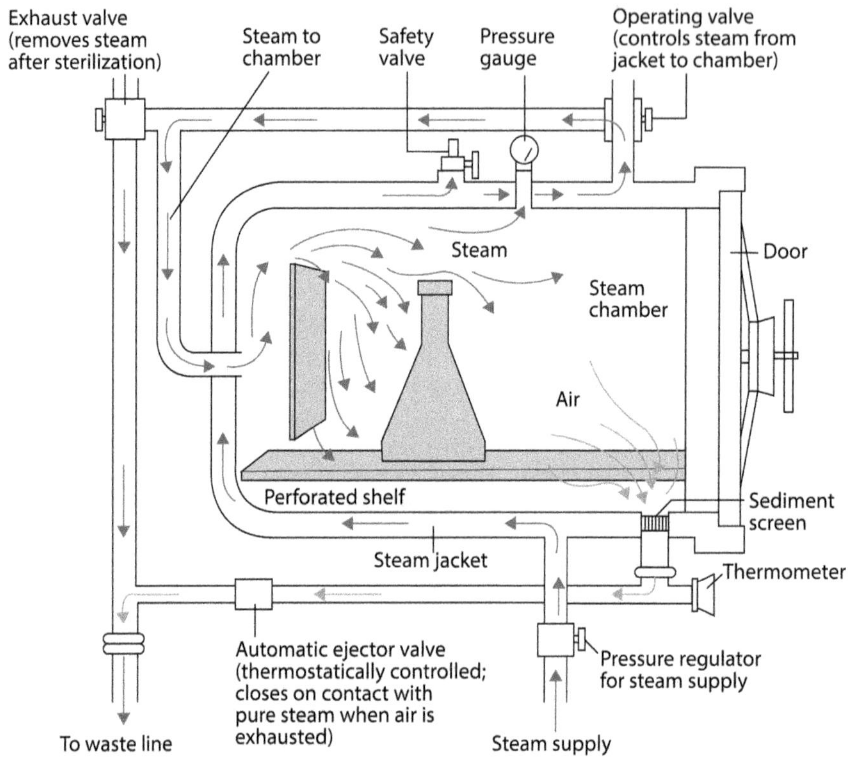

FIGURE 3.38 How an autoclave works.

- Mainly at PQ only if the unit is >15 years of age. If newer, the whole validation package. Also you want to see a preventive maintenance program, SOPS, and leak rate test data.
- Are all expiry dates current?
- Cycle time – Is it sufficient for tested D values?
- Was validation equipment within calibration (pre- and post-use for thermocouples)?
- Compare temperature traces for validation and the most recent cycle. Discrepancies?
- Are vacuum cycles used appropriately?
- Are the coolest and warmest positions clearly stated in the validation report?
- Examine the largest (bulkiest) loading pattern. Was the validation acceptable? Is anything not listed on the loading pattern present in the autoclave? Is there enough room for steam to circulate through the chamber?
- Were there any deviations from the protocols? Are conclusions valid and justified?
- Can the site show that the product that has been terminally sterilized has retained efficacy?
- Could also check that the steam system is validated and that dye bathing (or some other integrity checking) occurs post sterilization.

In Figure 3.41, a GMP auditor can review a standard display of an autoclave.

3.32 QUESTIONS TO AUDITORS

- You are auditing the API facility for compliance with GMP.
    - Identify at least eight major deficiencies experienced in GMP of active pharmaceutical ingredients manufacture.

- A vaccine manufacturer needs to audit the supplier of its fetal bovine serum. The supplier is located 400 km away. A team of three people set out in the morning and return that evening. At some point in the evening, the final report is written and signed off. There were no significant findings. There is no list of the documents reviewed or the operational areas seen. A good portion of the checklist is marked N/A. In this case, you are a regulatory auditor reviewing this company's QMS system.
  - Based on the conduct of this audit, what is your impression and what will you write in your report to the company?
- You are asked to visit a factory producing the following product lines: injections in ampoules and vials, including insulin, vaccines, and heat-stable pharmaceuticals sterile eye ointment
  - Describe the type of facility you would expect to find.
  - List the typical rooms, their purpose, and air classification.
- You received a complaint that an injectable product (WFI, 10 ml ampoule) is possibly contaminated with microorganisms. You have to organize an audit of the company in question.
  - What type of audit would be performed?
  - Will the audit be announced or unannounced?
  - Who will be part of the audit team? Who would you take with you?
  - What will you consider in preparation for the audit?
- You have been assigned to perform data integrity audits in your company. At the laboratory you stop by first, the analysis certificate of the finished tablet product abc123, standing on a bench, catches your eye. You start to examine whether the tablet hardness value written in the analysis certificate is really the measured value, with your data integrity experiences and GMP auditor awareness.
  - Which departments should you get help from when evaluating how this value is obtained in terms of data integrity?
  - Which documents should you review?
  - What verifications do you think you need to do?
  - Who do you think you should talk to?

BIBLIOGRAPHY

Active Pharmaceutical Ingredients Committee (APIC) Auditing Guide

Active Pharmaceutical Ingredients Committee (APIC) Supplier Qualification & Management Guideline, December 2009

Active Pharmaceutical Ingredients Committee (APIC) Practical Risk-Based Guide for Managing Data Integrity, Version 1, March 2019

Achana Pandit, *B.Sc. Microbiology Microbiology Laboratory Audit, a Checklist for the Pharmaceutical Industry*, GMP Scientific Inc

Adriana G. Lopes, Andrew Brown, *Practical Guide to Single-Use Technology: Design and Implementation*, Smithers RAPRA, 2016

Andrew Teasdale, David Elder, Raymond W. Nims, *ICH Guidelines an Implementation Guide*, Wiley, 2018

A. Hamid Mollah, Mike Long, Harold S. Baseman, *Risk Management Applications in Pharmaceutical and Biopharmaceutical Manufacturing*, Wiley, 2013

Considerations for a Corporate Data Integrity Program a Concept Paper by ISPE GAMP Community of Practices, March 2016

Current Good Manufacturing Practice for Medical Gases Guidance Industry Draft Guidance 2017

C.K. Moorthy, *Good Cleanroom Practices a Manual for Cleanroom Personnel Center for GMP*, 2007

David M. Bliesner, *Laboratory Control System Operations in a GMP Environment*, Wiley, 2020

David M. Bliesner, *Validating Chromatographic Methods a Practical Guide*, Wiley Interscience, 2006

David Nettleton, Janet Gough, *Electronic Record Keeping, Achieving and Maintaining Compliance with 21 CFR Part 11 and 45 CFR Parts 160, 162 and 164*, Interpharm/CRC, 2004

David Stokes, *Testing Computer Systems for FDA/MHRA Compliance*, Interpharm/CRC, 2004
Donald C. Singer, *A Laboratory Quality Handbook of Best Practices*, ASQ Quality Press, 2001
Douglas J. Pisano, David S. Mantus, *FDA Regulatory Affairs: A Guide for Prescription Drugs, Medical Devices, and Biologics*, Informa Healthcare, 2008
Explanatory Document for ISPM 15 (Regulations of Wood Packaging Material in International Trade), February 2017
Field Compliance Programs: Inspections of Licensed Biological Therapeutic Drug Products, 7356.002M
Food and Drug Administration Compliance Program Chapter 56 – Drug Quality Assurance 7356.002C
Food and Drug Administration Compliance Program Chapter 56 – Drug Quality Assurance 7356.002A
Gary Walsh, Brendan Murphy, *Biopharmaceuticals, An Industrial Perspective*, Kluwer Academic Publishers, 1999
Gary Walsh, *Biopharmaceuticals Biochemistry and Biotechnology*, Second Edition, 2003
George Verghese, Nancy Kaiser, *Cleaning Agents and Cleaning Chemistry Steris*, 2009
Graham P. Bunn, *Good Manufacturing Practices for Pharmaceuticals*, Seventh Edition, CRC Press, 2019
Guide to Inspections of Microbiological Pharmaceutical Quality Control Laboratories
Guide to Inspections of Oral Solid Dosage Forms Pre/Post Approval Issues for Development and Validation
International Standard ISO 19011 Guidelines for Auditing Management Systems, Third Edition, 2018-07
IPEC Europe How-To Document European Commission Guidelines on the Formalized Risk Assessment for Ascertaining the Appropriate Good Manufacturing Practice for Excipients of Medicinal Products for Human Use (OJ 2015/C 95/02)
Ira R. Berry and Daniel Harpaz, *Validation of Active Pharmaceutical Ingredients*, Second Edition, Informa Healthcare, 2001
IVT Network, Essential Microbioloy for GXP Compliance Special Edition
Jeanne Moldenhauer, *System Based Inspections for Pharmaceutical Manufacturers*, PDA, 2006
Jeanne Moldenhauer, *Contamination Control in Healthcare Product Manufacturing*, PDA, 2013
Jeanne Moldenhauer, *Assuring Data Integrity for Life Sciences*, PDA, 2016
Jeanne Moldenhauer, *Disinfection and Decontamination a Practical Handbook*, CRC Press, 2019
James Agalloco, Phil DeSantis, Anthony Grilli, and Anthony Pavell, *Handbook of Validation in Pharmaceutical Processes*, Fourth Edition, 2022
James L. Vesper, *Quality and GMP Auditing Clear and Simple*, Informa Healthcare, 2010
John Andrews, *Validating Pharmaceutical Systems, Good Computer Practice in Life Science Manufacturing*, Taylor & Francis, 2006
John M. Centanni, Michael J. Roy, *Biotechnology Operations Principles and Practices*, Second Edition, CRC Press, 2017
John W. Helgeson, *The Software Audit Guide*, ASQ Quality Press, 2010
Kim Huynh-Ba, *Handbook of Stability Testing in Pharmaceutical Development*, Springer, 2009
Kursad Turksen, *Stem Cells and Good Manufacturing Practices Methods, Protocols, and Regulations*, Humana Press, 2015
Ludwig Huber, *Validation of Computerized Analytical and Networked Systems*, Informa Healthcare, 2007
Ludwig Huber, *Validation and Qualification in Analytical Laboratories*, Second Edition, Informa Healthcare, 2007
Luis Jimenez, *Microbial Contamination Control in the Pharmaceutical Industry*, Marcel Dekker Inc., 2004
Maik W. Jornitz, *Filtration and Purification in the Biopharmaceutical Industry*, Third Edition, CRC Press, 2020
Mario L. Rocci Jr., Stephen Lowes, *Regulated Bioanalysis: Fundamentals and Practice*, Springer, 2017
Michael J. Akers, *Sterile Drug Products Formulation, Packaging, Manufacturing, and Quality*, Informa Healthcare, 2010
Michael E. Swartz, İra S. Krull, *Handbook of Analytical Validation*, CRC Press, 2012
Milton A. Anderson, *GLP Essentials a Concise Guide to Good Laboratory Practices*, Second Edition, Informa Healthcare, 2002
Myo Min Aung, Yoon Seok Chang, *Cold Chain Management*, Springer, 2023
Nigel A. Halls, *Achieving Sterility in Medical and Pharmaceutical Products*, Marcel Dekker Inc., 1994
Oliver Schmidt, *Pharmaceutical Quality Systems*, Informa Healthcare, 2010
Orlando Lopez, *21 CFR Part 11: Complete Guide to International Computer Validation Compliance for the Pharmaceutical Industry*, Interpharm/CRC, 2005
Orlando Lopez, *Data Integrity in Pharmaceutical and Medical Devices Regulation Operations Best Practices Guide to Electronic Records Compliance*, CRC Press, 2017
Orlando Lopez, *E-Records Integrity Requirements*, Nova, 2022

Orlando Lopez, *EU Annex 11 Guide to Computer Validation Compliance for the Worldwide Health Agency GMP*, CRC Press, 2015
Parag Kolhe, Mrinal Shah, Nitin Rathore, *Sterile Product Development Formulation, Process, Quality and Regulatory Considerations*, Springer 2013
Paul L. Pluta, First Impressions, the plant tour, *Journal of GXP Compliance*, Winter 2011
Pharmaceutical Auditing Pharmaceutical Quality Group Monograph No. 5 (Revised)
Pharmaceutical Inspection Convention PI-025–2 Aide Memoire Inspection of Medicinal Gases
Pharmaceutical Inspection Convention PI-009–4 Aide Memoire Inspection of Utilities
Pharmaceutical Inspection Convention PI-024–3 Aide Memoire Inspection of Biotechnology Manufacturers
Pharmaceutical Inspection Convention PI-028–2 Aide Memoire GMP Inspection Related to Packaging
Pharmaceutical Inspection Convention PI-041–1 PIC/S Guidance Good Practices for Data Management and Integrity in Regulated GMP/GDP Environments
Pharmaceutical Inspection Convention PI-008–3 PIC/S Guidance Document for Inspectors PIC/S Guide to Inspections of Source Plasma Establishments and Plasma Warehouses
PIC/S Guidance Document for Inspectors PI-008-4 PIC/S Aide Memoire to Inspections of Blood Establishments and Plasma Warehouses
Ronald Blank, *The Basics of Quality Auditing*, CRC Press, 1999
R.D. McDowall, *Data Integrity and Data Governance*, Royal Society of Chemistry, 2019
Regine Eibl, Dieter Eibl, *Single-Use Technology in Biopharmaceutical Manufacture*, Second Edition, Wiley 2019
Robert A. Nash, Alfred H. Watcher, *Pharmaceutical Process Validation An International*, Third Edition, revised and Expanded 2003
RX-360 Supply Chain Security Template-Requirements for Third Party Logistics Providers 6 June 2012
Rx-360 Audit Guide for Pharmaceutical Packaging Materials for Medicinal Products 10 May 2012
Rx-360 Remote Audit Best Practice Guide Version 1.0 Published August 2020
Rx-360 Selection and GMP Auditing of Software and Hardware Vendors Version 1.0 Issue Date October 2019
Rx-360 GMP Audit Manual Data Governance and Data Integrity Published March 2018
Rx-360 White Paper Managing Critical Vendors December 2015
Sarfaraz K. Niazi, *Biosimilars and Interchangeable Biologics Tactical Elements*, CRC Press, 2016
Sidney H. Willig, *Good Manufacturing Practices for Pharmaceutical a Plan for Total Quality Control from Manufacturer to Consumer*, Fifth Edition, Revised and Expanded Marcel Dekker Inc., 2001
Single-Use System Audit Guide BioPhorum September 2021
Siri H. Segalstad, *International IT Regulations and Compliance Quality Standards in the Pharmaceutical and Regulated Industries*, Wiley, 2008
Terry Jacobs, Andrew A. Signore, *PE Good Design Practices for GMP Pharmaceutical Facilities*, Second Edition, Taylor & Francis, 2017
Tim Sandle, Madhu Raju Saghee, *Cleanroom Management in Pharmaceuticals and Healthcare Euromed Communications*, Euromed Communication, 2017
Tim Sandle, *Pharmaceutical Microbiology Essentials for Quality Assurance and Quality Control*, Elsevier, WP, 2016
Thomas Catalano, *Essential Elements for a GMP Analytical Chemistry Department*, Springer 2013
Vinaj Bhatt, *GMP Compliance, Productivity, and Quality*, Taylor & Francis, 2010

4 Auditing Pharmaceutical Quality Management System

DOI: 10.1201/9781003284970-4

4.1 AUDITING PHARMACEUTICAL QUALITY MANAGEMENT SYSTEM

4.1.1 Complaints

Every effort is made to ensure that drug products that reach the market are safe, pure, effective, and of the highest quality. However, a company may receive complaints from a patient, pharmacist, or physician about a drug product not meeting its specifications. The company or site must know how to manage these complaints.

The goal of this section is to describe how to audit a complaint system using the appropriate good manufacturing practice (GMP) standards. Some of the principles outlined in this module can apply to a quality management system based on ISO 9001 or other similar systems as well as GMPs. Although regulatory agencies do not specify a standard complaint-handling system, the GMP requirements specify certain actions that shall be included in any system.

Manufacturers shall:

- Document, review, evaluate and file all complaints
- Formally designate a unit or individual to perform these activities
- Determine if an investigation is necessary
- Record the reason if no investigation is made
- Assign responsibility for deciding when not to investigate

The site should have an approved standard operating procedure (SOP) for managing complaints. It should contain the roles and responsibilities of the Quality Unit regarding complaints, the method for recording and retaining complaint information, and the criteria for determining if the complaint should be investigated. The Quality Unit should review the complaint and the subsequent investigation or document the decision not to perform an investigation.

The complaint system is important because it acts as an indicator for the regulators to determine if the manufacturer's quality system and corrective actions are adequate. It also provides valuable information to the manufacturer regarding the quality and performance of distributed products. The site should also have a system in place to monitor for trends periodically and to notify upper management of significant incidents and trends.

4.1.1.1 Notification of Complaint

A finished drug product manufacturer may be notified of a complaint via telephone, facsimile, or written correspondence by sales representatives, pharmacists, patients, and physicians. Active pharmaceutical ingredient (API) manufacturers may also be notified using the same methods.

When the complaint is received, the following information must be recorded, if available:

- Name of complainant and demographics
- Batch number
- Product name/strength (local product code is useful in the US for finished dosage forms)
- Description of the problem

Each complaint record should have a unique number or code assigned to it. The site should have a method to track the status of the complaint and to trend complaints.

Authorities need to be notified of major quality issues, which are required by EU GMP, for medical devices and for comparator manufacturers for research and development (R&D). All complaint records must be readily available for review by regulatory agencies. If the complaint is recorded in an electronic system, the system must be validated.

4.1.1.2 Types of Complaints

The company may receive many types of complaints. These complaints can be classified into four categories:

1) Supply and logistics complaints
2) Product quality complaints
3) Adverse drug events
4) Medication errors

The key parameters of auditing a complaint system prior to the audit are to:

- Determine the site's main functions, (manufacturing, packaging, etc.)
- Request an organizational chart
- Review the chart for appropriate separation of Quality and Production Units
- If the contractor is a manufacturing site, review the site's record for meeting complaint investigation deadlines

For investigations, obtain both the sponsor's reference number and the contractor's number.

4.1.1.3 During the Audit

- Ensure that Quality has the responsibility for all product quality complaint investigations.
- Ensure that the site has an approved SOP used to manage drug quality complaints.
- Ensure that the site SOP includes the following topics:
    - Where the complaint files are maintained.
    - The system used to track complaints.
- What information should the author look for to find evidence?
- Ensure that there is a system in place to report, evaluate, and investigate all adverse events.

- Ensure that the adverse event reporting, evaluation, and investigation system mandates notification of Quality Assurance for all adverse events that may involve a product quality issue.
- Ensure that all adverse event complaints with potential quality issues are investigated by Quality Assurance in accordance with the approved SOP that is used to manage drug quality complaints.
- Ensure that all product quality complaints determined to be not justified include documented evidence to support this decision.
- Ensure that received product quality complaints are reviewed on a regular basis to identify any trends according to complaint type, manufactured lot, or other defined categories. If the site categorizes complaints by type, verify that the SOP defines what the types are.
- Ensure that employees who receive complaints by phone are trained to gather the necessary information from the caller.
- Ensure that each complaint file includes, at minimum, the:
  - Name of complainant and demographics
  - Batch number
  - Description of the reported product defect
  - Product name/strength (local product code)

All correspondence between the complaint source and the company, including letters and summaries of verbal communications

The contractor responsible for sending a letter to the complainant, if appropriate

The documented decision to investigate or not investigate

The investigation report, if appropriate.

- Review representative complaints from different types.
- Verify that the complaints were fully investigated.
- Determine if the "root cause" may impact other product lots.
- Verify that the conclusions in the investigation were appropriate and valid.
- Verify that appropriate corrective actions were taken.
- Ensure that systems are in place to define the retention of samples and documentation to include specific time periods.
- If an electronic system is used to track and trend complaints, ensure that the system has been validated.
- Ensure for formulated contractors, where the distribution is the responsibility of the contractor, or for internal audits that there is a local procedure for recall.
- Verify that distribution records are in sufficient detail to provide traceability.
- Verify that distribution records are in sufficient detail to provide traceability for exported products.
- Ensure the effectiveness of the recall procedure has been evaluated.
- Verify that the procedure can be instigated promptly and at any time.

4.1.2 Change Control

Planned changes may be recommended for starting materials, raw materials, excipients, process equipment, process media, production or testing methods, packaging material used in drug production, or any change that affects product quality or consistency. There should be written procedures covering all activities if there is a possibility of change. Consider if the procedures describing change control activity involve a revision of a process. In that case, they must ensure that sufficient supporting data is generated to demonstrate that this revision will result in a product of the desired quality. In other words, all changes that will affect the repeatability of the process, especially product quality, must be officially requested, documented, and accepted. The possible impact of facilities, systems, and equipment on the product is evaluated through risk analysis. The qualification and requalification requirements and scope should be determined, and the rationale should be documented.

4.1.2.1 Basic Components of a Change Management System

If a piece of equipment or a process needs to be modified, it should follow an established and approved change control process as outlined in a detailed approved procedure.

This procedure should include direction and guidance and answer the following questions:

- What is the proposed change?
- How will it be implemented?
- Who are the necessary parties to approve it?
- How will it be tested or validated once the change has been made?
- What is the rationale for the change?
- What is the timing of the change?
- What is the impact on quality once the change is implemented?

A change management request should be issued and circulated to designated parties to review and approve the change. Each site should have a formal and documented change management system that evaluates the effectiveness of each change proposed, defines clear roles and responsibilities, and assesses the impact of change on the validation/stability and/or registration status of the change.

4.1.2.2 Audit Strategy

To determine the effectiveness of the change management program, one strategy is to identify a completed change that has taken place at the site and follow it through the change management process. Depending on the actual change, several GMP systems may be impacted.

The site's change management program should be established as part of the quality system. Under the quality system, the site's requirements for change control should be described.

Any change made as part of the change management program should be:

- Documented
- Evaluated
- Approved
- Tracked to completion
- Assessed for impact
- Closed

If the change is a production change, within the production GMP system, it should be part of the change management system and assessed to determine if it requires validation, a stability study, and communicated to or approved by the regulatory authorities. For each site or contractor, it is important to determine if the proposed change requires global approval.

4.1.2.3 Prior to the Audit

Review the inventory list of the site's critical equipment, instruments, and systems and determine if there have been changes to them.

Determine if there are any outstanding corrective actions from either internal audits or regulatory inspections involving change management.

4.1.2.4 During the Audit

- Ensure that any change made to facilities, equipment, materials, processes, documents, or informational technology is part of a site's change management program.
- Verify that there is a written policy, guidance, or procedure in place that identifies all aspects of the change management program.
- If more than one change management system exists, determine what system governs what type of change and how they are connected.
- Determine if the site has different levels of changes.
- Ensure that there is a procedure that identifies the activities for each level of change.
- Ensure that there is appropriate documentation for each level of change.
- Determine if there is a standard method of approval for documents and that the approval process is documented.
- Ensure that if changes are made to the proposed original change after the review process, the revised changes are documented.
- Verify that GMP changes are approved by the Quality group.
- Verify that no change, with potential regulatory impact, is implemented until it has been approved by Regulatory Affairs.
- Verify that changes requiring regulatory approval do not progress prior to receiving regulatory approval.
- Ensure that there is a list of approvers or approver titles for various levels of change.
- Ensure the supplier or contractor, as part of the change management process, includes a step for approval or notification to the sponsor company of changes to manufacturing, specification, or testing requirements.

If approvers have delegated their authority to another individual, ensure that there is a memo reflecting this.

- Ensure that if changes are made to the change approval process they are documented.
- Ensure that there is a method to coordinate a change in all affected documents. If the same product is produced at multiple sites, determine how each site is notified and the change is implemented. If a change has been made to an electronic system, ensure that training on the system has taken place.
- Determine what the site's procedure is for bundling changes.
- Verify that the status of changes is tracked and accessible.
- Determine if there is more than one level of approving bodies for a change, e.g., global changes.
- Determine who the approvers are at each level.
- Determine the criteria for approvals at more than one level.
- Ensure that before implementation of a change requiring Regulatory Affairs approval takes place, it has been approved by Regulatory Affairs.
- Verify that the user area is formally notified of the approval.
- Ensure that all personnel are trained on the change being effected, if appropriate.
- Verify that all training is documented.
- Verify temporary employees and consultants are trained if working on or with equipment or processes that have changed.
- Review the quality system.
- Verify that there is an established and approved quality system/quality policy that is endorsed by management.

4.1.3 Training Records

Within the pharmaceutical industry, it is agreed that people are the most important element in any pharmaceutical operation. The number of people working at a site needs to be sufficient to ensure that the drug product manufactured, processed, packaged, or held is compliant with all necessary GMP and regulatory requirements. Personnel must be qualified and have the right attitude, training, and supervision to produce good quality products.

4.1.4 Organization

There must be an organizational chart and a clearly defined reporting structure in the company or facility. There should be no gaps or unexplained overlap in the duties and responsibilities of employees implementing GMPs. There should be a quality unit in the organization that is independent from the production department.

4.1.5 Personnel Responsibilities

Personnel must assume responsibility for producing a quality product and should understand the impact of their activities on the product and, ultimately, the patient. All personnel are responsible for following company and site policies, procedures, practices, company standards, and quality requirements and regulations.

These responsibilities include but are not limited to:

- Following established dress codes within the work area
- Maintaining good hygiene and health that will not compromise a pharmaceutical product
- Completing documentation correctly
- Advising management of problem situations
- Understanding and following SOPs for specific job functions
- Undergoing training when necessary
- Refraining from eating, smoking, drinking, or chewing within a production area
- Assuring that there is adequate supervision for all activities related to drug production
- Incorporating regulatory requirements into their particular job function
- Preventing contamination of drug products
- Avoiding direct contact with the exposed products as well as with any part of the equipment that comes in contact with the product
- Protecting product quality

Personnel working at a pharmaceutical company are expected to be qualified in the operations that are assigned to them. This qualification can be achieved through training that focuses on job tasks (job skills training) and regulatory requirements, experience, and education, or a combination of the three. Employees are considered qualified when they can demonstrate satisfactory proficiency in job performance.

4.1.5.1 Training

To create qualified employees, companies need to have an effective GMP training system in place.
Training should be:

- Competency-based
- Conducted by a qualified trainer
- Evaluated for effectiveness

- Focused on particular job skills and regulatory requirements
- Documented
- Joint responsibility among the trainer, trainee, and management

4.1.5.2 Role of Management in Training

Management should be involved with training. Some of the responsibilities of management include:

- Identifying personnel training needs
- Developing and establishing a robust training system
- Providing adequate resources to successfully accomplish the training
- Providing support for employee training
- Evaluating the effectiveness of the training

One of the most important duties expected from managers is to define appropriate tasks for the working personnel and determine a work schedule that is fully compatible with those tasks.

4.1.5.3 Prior to the Audit

- Determine the site's main functions (manufacturing, packaging, etc.).
- Request an organizational chart.
- Review the chart for appropriate separation of Quality and Production Units.

4.1.5.4 During the Audit

- Ensure there is a procedure defining the responsibilities of Quality and Production.
  - While touring the plant, observe the employees.
  - Verify that employees are complying with the area dress code.
  - Verify that employees are following approved and accepted procedures.
  - Verify that employees are completing documents concurrently with the production activities.
- Ensure there is a documented training system in operation at the site by reviewing actual training documentation for selected employees.
- Ensure training records are compliant with GMPs and use good documentation practices.
  - Verify that the records are complete and accurate.
  - Verify that cGMP training is as indicated in the site's training plan or SOP.
  - Verify that training in the training record is appropriate for the job description.
  - Verify that no chronological gaps appear in the training records unless documented.
- Ensure that training materials meet their stated objectives and are developed to support specific job functions as defined in the job description.
  - Verify that the materials have been approved by the functional area and/or quality.
  - Verify that the materials, if used on a repetitive basis (i.e., task training on equipment), are reviewed and updated on a regular basis and follow some form of change management.
- Ensure that a training plan/curriculum has been developed for each job classification and that it is recorded.
  - Verify that the training plan matches the job description.
  - Verify that it contains both ongoing GMP and job skills training.
  - Verify that it contains training on procedures.
- Ensure that there is a method to track training.
  - Verify that there is a provision within the training system that ensures that personnel who have missed scheduled training receive the missed training.
  - Verify that training information is readily accessible.

- Ensure that trainers are qualified to conduct specific training according to site procedures.
  - Verify that there is a system to determine the trainer's competency.
  - Verify that there is an SOP in place defining what is included in the training system.
  - Verify that there is an assessment method in place to determine employee competency.
- Ensure that contractors/consultants have the necessary education, training, and experience to perform their assigned job function.
  - Verify that the training program incorporates requirements for temporary employees and consultants.
- Ensure that training is conducted with sufficient frequency to ensure that employees remain familiar with applicable regulations.
- Ensure that laboratory personnel are receiving both GMP and job skills training that is documented.
- Ensure that all personnel involved in the pharmaceutical products' manufacturing, packaging, testing, and quality control activities have training records.

These should meet the GMP requirement of assuring that staff is appropriately trained to perform their job function.

4.2 AUDITING CONSULTANTS

When the US Food and Drug Administration (FDA) sends a warning letter to a facility after its inspections, it usually underlines that one of the essential parts of the company's actions is to work with a consultant or a group of consultants. The use of consultants does not mean that all duties and responsibilities are delegated to these individuals and should not exempt the company from all responsibilities. Although the FDA supports and encourages working with the consultant, it knows very well that the firm should have the last word.

Although it is very important to work with consultants, experts should also have a number of features.

First, before deciding to work with a consultant, the firm should ask itself a number of questions to determine the characteristics that a consultant should have, decide on a risk-based basis why a consultant is needed for which project, and question the validity of some expertise documents that the consultant will present.

Although working with a consultant is not very similar to the process of procuring an API, I think that more or less similar actions and documentation can be operated.

First:

- How do you look for a consultant?
- How will you understand that the consultant can respond to your expectations and wishes?
- How would you like to receive the consultant's service?
- Will you have a daily or weekly service request at your facility remotely and monthly?
- Did you do a gap analysis before working with the consultant?
- What aspects do you think you are missing?
- Will you sign a confidentiality agreement before hiring the consultant on a project basis?
- Will you have a quality agreement with the consultant where roles and responsibilities are clearly defined?
- Before deciding to work with a GMP consultant, how will you test that person's past experience, success, and failure in GMP?
- How do you plan to use chapter 7 of the EU GMP, which details outsourcing, effectively?
- Do you plan to get to know the person or people closely by sending a list of questions before you start working with them?
- What information do you want to be included in the questionnaire?

- Do you intend to audit the consultant's office based on the response to the questionnaire?
- With what expectation do you expect consultants to fill this gap since they do not have a GMP document yet?

A consultant working in accordance with the GMP must at least be a sole proprietor or a member of the consulting firm. If we think bigger, a consultancy firm is expected to meet the aforementioned criteria before being decided by you as a solution partner.

In some cases, you may obtain key information from a friend or colleague in the industry who has previous knowledge of this company or person. Internet searches can often be misleading, and it shouldn't be surprising to learn that a consulting firm claiming to work with a team of dozens doesn't even have an office.

Assuming that we are progressing on the basis of a consulting firm, I openly support that this firm should at least have a quality management system standard such as ISO 9001. In this way, you can have information about the quality system of the company and have the first view of the way it manages its processes.

For example, let's say there is a contract manufacturer with more than 100 dosage forms. This company wants to add 20 more new tablet forms to its product portfolio but does not know how and from where to start cleaning validation. There is no competent and experienced employee in the company in this regard. In this case, the company urgently needs to get a consultancy service and complete the cleaning validations on time.

- So what will the manufacturer do?
- What kind of service should it get and from whom?
- Which company is experienced in this field?

EU GMP 2.23 is allocated to consultants and consultants are expected to have sufficient education, training, and experience in this field for whichever service they are hired for, but are pharmaceutical companies really researching and looking for people who meet these criteria or are they only half-heartedly reviewing information given by some people or organizations? Or is it the first contact with these people based on information found on the internet? FDA 21 CFR Part 211.34 summarizes the necessary information about consultants, but unlike the FDA, EU GMP, specific naming of consultants' service areas, such as production, quality control, packaging, and labeling, is an approach that I quite like.

If we continue with the example I gave:

- What kind of experience could the person who describes themself as a consultant have in cleaning validation?
- What kind of training have they received on the subject? What kind of project have they managed or contributed to on behalf of which pharmaceutical company?
- In fact, which health authority might have concluded that the counseling of this person was sufficient?
- From which school could the candidate have graduated and with what kind of diploma?

Another example is a person who works as a consultant in the field of microbiology. We do not know that there are departments in universities where careers can be gained in many fields such as biology, microbiology, and medical biology. But when it comes to pharmaceutical microbiology, how do you find someone with double expertise in pharmaceutical microbiology and data integrity? An expert who can detect and prevent data integrity breaches in endotoxin testing. Where are they?

The FDA also makes another striking request about the GMP consultants it recommends at the end of its warning letters. The FDA expects the consultant to inspect the facility against GMP guidelines before the relevant inspection or warning letter finding is improved.

As an expert and auditor who has been actively working in the sector for 25 years, I can say that these people do not even have the proper GMP knowledge, let alone succeed in the GMP audit.

I've passed everything, and GMP wants us to keep the records of these consultants. I'm talking about the records that contain key information such as the full address of the consultant or the consultancy firm, the services it provides, qualification documents, and features.

In that case, how will you find consultants suitable for all your demands and expectations and suitable for evaluating the services of the health authority from a GMP perspective?

You must first prepare the consultant user requirement specification (URS). A detailed URS containing some information such as what qualifications a GMP consultant should have and how they should respond to your expectations can be useful at the beginning of everything. For consultants who claim to meet your criteria in the URS, you must then initiate a design qualification (DQ), installation qualification (IQ), operational qualification (OQ), and performance qualification (PQ). After declaring the qualities you expect from a GMP consultant in the design qualification plan or protocol, you can complete your DQ report by finding a consultant who meets these expectations.

Once a GMP consultant arrives after the DQ report as you ordered in the IQ protocol, you need to clarify these points on how to adapt and set up your facility. After the IQ report, you should set up tests in your OQ plan where you can see if your GMP consultant meets the test criteria that you have previously determined and approved. The OQ report will be the stage where the GMP consultant who successfully passes this test is approved. In your PQ protocol, it is a phase where you are looking for evidence that the GMP consultant can provide the services they offer at all times and without interruption, and you should predetermine this period for the PQ and test the adequacy of the consultant, for example, for a period of six months.

You have to carry out the DQ, IQ, OQ, and PQ stages with a risk-based approach. At this point, you can take advantage of ICH Q9, the new version of which has been released.

After all these stages and after auditing the GMP consultant's office, you will have reached a very good point in the qualification of your GMP consultant. The qualification process of consultants will not only complete the inspection or qualification process, but as I mentioned, it will be quite safe for the services they provide to be accepted by at least three health authorities. In the qualification process, the speed of the consultant's response to the questions you ask and the accuracy of their answers stand out as important factors.

Well, who will test whether the service provided by the GMP consultant in pharmaceutical or biotechnology companies is in compliance with GMP?

First, it is to your advantage that this person is better, more experienced, and more educated than the hired GMP consultant. If you have such a well-educated and skillful staff, then why do you hire a person whose skills, knowledge, and experience are questionable as a consultant? Time … Your most important constraint is time. I suggest you also consider the costs of testing the GMP consultant.

4.3 DOCUMENTATION SYSTEM

Good documentation is essential for providing assurance that drug products are safe, pure, effective, and of the highest quality. It is expected from any pharmaceutical process. Through the use of established documentation systems, the company and its suppliers are able to provide a complete written record or proof of all activities used in producing a quality drug product. This module focuses on written documentation, but electronic data processing systems, photographs, or other reliable means may record the data. If electronic systems are used, the requirements for the documentation in principle are the same.

Good documentation practices should be universally used in all documents considered part of GMP documentation.

4.3.1 General Documentation Practices

All required data should be recorded directly into production batch records, applicable logbooks, forms, or worksheets. Before signing a document or confirming that an activity was performed, the person signing should both read and understand the steps in the process or the statement. Initials and signatures of individuals working in that department should be maintained. Entries may not be made in advance or for another person. Entries may not be backdated. Unfamiliar terms should be defined within the document.

4.3.2 Types of Documentation Systems That Are Audited

Documentation systems that should be routinely audited are:

- Equipment assembly, cleaning, and use of log
- Component, drug product container, closure, and labeling records
- Master production and control records
- Batch production and control records
- Production record review
- Laboratory record systems including electronic data systems and analytical methods
- Distribution records
- Annual product reviews
- Complaint files
- Validation and qualification systems of processes and equipment
- Equipment calibration and maintenance records
- Environmental monitoring records
- Training records

4.3.3 Components of Documentation System

All documentation systems should have approved and established SOPs that govern how the particular documentation system is managed.

The topics that should be included, at a minimum, are:

- Document control and distribution
- Approving bodies and date of approval
- Timeline for review and revision
- A tracking and follow-up method
- Assurance of training regarding documentation
- A description of the purpose of the system
- Storage and retrieval of information
- Directions for handling atypical occurrences within the system

4.3.4 Prior to the Audit

- Review the types of documentation systems in use at the site.
- Determine if there are any outstanding corrective actions from either internal audits or regulatory inspections involving documentation.

4.3.5 During the Audit

- Ensure that the documentation system is operating in full compliance.
- Verify that all documentation within a system applies good documentation practices.

For production, laboratory, or training documentation:

- Verify that personnel activity indicated in current SOPs is actual practice.
- Verify that data added to process documents is correct and current with the process.
- Verify that the system is working correctly by asking to retrieve a previously generated document.
- Verify that there is a procedure in place for double-checking data if required for that documentation system.
- Verify that access to documentation is restricted to authorized personnel.
- Verify that all documentation contains approvals by the appropriate personnel.
- Verify that completed documentation is reviewed by appropriate personnel including a quality representative.
- Verify that document distribution and circulation are controlled.
- Verify that documents are identified with appropriate control and access statements.
- Verify that the site has a method for assuring that only current versions of documents are used.
- Verify that documents are reviewed or revised either on a scheduled frequency or as needed, whichever is the shorter time interval.
- Verify that there are SOPs for managing each documentation system. This should include a review and approval process for documents including those that require change control management.
- Verify that personnel understands the differences when signing in different capacities.
- Determine if documentation discrepancy/deficiency classifications are used by an audited site.
- Determine the criteria for each classification.
- Verify that the same criteria apply if electronic systems are used.
- Review the quality system.
- Verify that there is an established and approved quality system/quality policy that is endorsed by management.
- Verify that the Quality Unit is independent of manufacturing.
- Verify that one of the Quality Unit's responsibilities is to approve all appropriate quality-related documents.
    - Ensure that personnel has received adequate training on new or revised versions of documents.
    - Ensure that training is documented.
- Verify that the training program incorporates requirements for temporary employees and consultants.
    - Ensure that documents are revised according to a written procedure.
- Verify that superseded documents are removed when a new procedure is issued.
- Verify that revisions to documents are only made by authorized personnel.
- Verify that documents are reviewed and/or revised according to an established frequency.
- Ensure that documents are retained according to a site retention schedule.
    - Verify that the site has a retention schedule.
- Verify that documents can be easily retrieved.
- Verify that manufacturing records are held for a minimum of one year past the expiration date.

4.4 QUALITY RISK MANAGEMENT AUDIT

Quality risk management (QRM): How to apply – How to audit.

- Revised version is relatively new for both sides – companies and regulators.
- It is a challenge for all participants.

- Key criteria – for the inspection and application of QRM can directly be developed from the current guidelines.
- Personal view on how to go about it.

4.4.1 QRM – Requirements I

- Quality risk management is part of the quality system.
- To achieve the quality target, the pharmaceutical quality system implemented must be designed comprehensively.
- Quality risk management is a systematic approach to assessing, controlling, effectively communicating, and reviewing risks to the quality of the drug product.

There must be a systematic approach!

4.4.1.1 QRM – What Auditors Are Looking For

- Is quality risk management part of the quality system?
    - Is there a high-level controlled document addressing QRM (e.g., policy, Site Master File, quality manual)?
    - Is there a document explaining the methods used in risk management, which are applied in the company?
    - Are QRM principles incorporated into GMP training?
- Is QRM systematically used (e.g., as checklists within the change control system)?
    - Is it part of the qualification/validation system (e.g., in the validation master plan [VMP])?
    - Is there an internal requirement to use it in investigations?

4.4.2 QRM – Requirements II

QRM is part of the quality system that ensures the process performance and quality monitoring and review systems are based on risk.

- An appropriate level of root-cause analysis should be applied to investigate deviations, suspected product defects, and other problems. This strategy can be determined using QRM.
- The quality system should ensure processes are in place to control outsourced activities and the quality of purchased materials. These processes should incorporate QRM principles.
- Quality management system (QMS), GMP, and QRM are interrelated. QRM must be visible in different areas!

4.4.2.1 QRM – What Auditors Are Looking For

- Where and when is QRM applied and used?
- How is it embedded in GMP?
- Evaluation of changes within the change control system.
- Definition of actions before implementing changes.
- Definition of qualification and validation work.
- Definition of maintenance activities.
- Definition of process monitoring.
- Performance of investigations, root cause, and impact analysis (to other products/patients) within the deviation and complaint-handling process.
- Definition of the depth of supplier qualification.
- Determination of the internal/external audit program.

4.4.3 QRM – Requirements III

- In order to evaluate the quality risk, it is essential to have scientific knowledge and experience related to the process and, as a result, to protect the patient.
- QRM activities are often managed by interdisciplinary teams. When creating teams, representatives not only from the quality unit but also from units such as engineering, production, laboratory, operations, legal, and supply chain should be included. In addition, each person included in the team is expected to have both experience and continuing education records in QRM.

4.4.3.1 QRM – What Auditors Are Looking For

- Who is involved in quality risk assessments and analysis?
  - Are experts in the relevant process involved?
  - An experienced employee or a new one?
  - Is there at least one individual who is specifically trained in QRM?
  - If external resources are used, what is their level of knowledge and do they know the process?
- Is it linked to the protection of the patient?
  - Is there someone on the team who addresses the clinical aspects?
  - Is the possible impact on the patient an aspect that is addressed in the QRM?

4.4.4 QRM – Requirements IV

- The purpose of applying QRM should be compatible with the status, documentation, and level of the risk.
- It may not always be appropriate or necessary to use a formal risk management process.
- Appropriate and correct use of QRM may facilitate, but not hinder, the pharmaceutical industry's health authority requirements and expectations.

The management and depth of QRM must be appropriate and sufficient.

4.4.4.1 QRM – What Auditors Are Looking For

- Is the method chosen suitable/appropriate?
  - For example, for the identification of risks related to a new process, a fishbone analysis might be better than a failure mode and effects analysis (FMEA).
  - Is the source of data and information used correct and reliable?
- Is the level of effort adequate?
  - Is one person or a team more adequate (depending on complexity)?
- Is the level of formality adequate?
  - Simple checklist of certain questions or a complex FMEA?
- Is QRM misused, e.g., to justify that no qualification of a piece of equipment is necessary?
- Are the measures to mitigate the risks appropriate?
- Is the level/category of risk plausible?
  - Is it really not critical?
  - Is the potential impact on the patient the focus?
- Are the conclusions adequate?
- Is an aspect missing?
- Are other business risks addressed?
  - Bears the danger that they compromise quality!
- Plausibility check!

4.4.5 QRM – What Auditors Might Say (Example)

- New equipment/machine.
  - It is the fifth that has been ordered in the last three years,
- According to the risk assessment done, no URS/DQ was performed, with the justification that it is the same piece of equipment as the four machines installed before, documented in the VMP.
  - Why not when it can be shown that they are identical (same supplier, same machine generation, same IT version)?
  - Reference to former URS/DQ.
- Certain OQ tests were skipped with the same justification.
  - Not acceptable!
- Determination of how many products/runs must be included in the PQ.
  - Yes, a formal risk assessment might be applied here.
  - Historical data on the former PQs and problems from the routine should be taken into account.

4.4.6 QRM – Tips and Tricks Auditing QRM

- Inspecting QRM should be taken into account while planning the inspection.
- Enough time should be reserved within the inspection in order to evaluate QRM aspects.
- When inspecting QRM for the first time in detail, start with the system and a manageable, clearly limited example.
- Be careful if exclusively external resources are used for risk assessments. Involvement of the internal personnel is strictly recommended!

4.4.7 QRM – Audit Criteria – Conclusion

- Is there evidence that QRM is part of the quality system?
- Are the right people involved (with knowledge and experience)?
- Is QRM appropriately used?
  - A plausibility check will be applied.
  - Common sense.
- Be careful if exclusively external resources are used for risk assessments. Involvement of the internal personnel is strictly recommended!
- It is not possible to inspect a complex risk assessment in detail.

4.5 DEVIATIONS

During manufacturing, unplanned events or incidents may occur. These events are considered deviations from established processes, procedures, and policies possibly placing the product out of compliance with regulatory requirements and jeopardizing the safety, purity, and effectiveness of the drug product. Deviations may be referred to by many different names such as atypical events, discrepancies, problems, abnormal occurrences, events, or incidents. There must be a deviation management system in place to determine the extent of the deviation and the impact of the deviation, and to investigate why the deviation occurred and what can prevent it from reoccurring. The goal of this section is to describe how to audit a deviation management system using the appropriate GMP standards. The GMP requirements detailed in this module should be applied to the supplier on a sliding scale, the scale being dependent on the product or service being audited. In addition, the auditor should consider the risks associated with an unplanned incident that would potentially result in the product not meeting customer specifications or regulatory requirements.

A "nonconformance" or "nonconforming" product can be:

- A product not meeting specifications.
- Customer rejects or complaints.
- Results of inspections or in-process testing failures.
- An unexplained event or observation.

Refer to ISO 9001 standards for additional nomenclature and requirements of ISO quality systems for managing nonconformances.

4.5.1 Your Role as an Auditor

As a GMP auditor, your job is to determine if there is a robust system for handling deviations if deviations have been sufficiently investigated according to the site's procedure and if a cause has been determined to prevent the deviation from recurring. Typically this determination will include reviewing examples of deviations. Consider that all deviations may not need to be investigated to the same level. The level of investigation may depend on the significance of and risk associated with the incident, problem, or deviation.

4.5.1.1 Prior to the Audit

- Determine what manufacturing processes are used at the site.
- If available, review the list of deviations generated at the site or company.
- Review complaints to ensure that corresponding investigations are in place, as applicable.

4.5.1.2 During the Audit

- Ensure that the site has a robust system described in an approved SOP for management and investigation of manufacturing deviations and OOS test results.
- Ensure that the SOP includes:
  - How investigation tracking numbers are generated and assigned.
  - What the time limit is for the investigation (usually working days).
  - How and when Quality is notified, usually within 24 hours, once a deviation is confirmed.
  - What steps to take if an investigation is not completed within the defined time frame.
  - The criteria and requirements for a preapproved plan for retesting.
  - Guidance in determining what information must be examined to determine the root cause.
  - All GMP areas including in-process material and finished product.
  - A review of deviation data on a periodic basis, within a defined time frame, to identify trends.
- Request a list of deviations/OS, including trend reports.
- Review process behavior charts and deviation key performance indicators (KPIs), if available, to give insight into key deviation issues.
- Ensure the SOP is followed by verifying through a review of a few deviation investigations that:
  - A unique tracking number is assigned to each investigation.
  - Deviations found in all GMP areas including in-process material and finished products are investigated according to SOP.
  - Quality is notified in a timely manner (within a specific time frame, e.g., 24 hours) as to lots affected.
  - A procedure is clearly described to manage the deviation investigation.
  - The investigation has a 30-working-day time limit for completion.

  - Investigations not completed within the directed timeframe followed the SOP.
  - Other manufactured product lots or products that might be affected by the deviation were investigated.
  - Identify any trends and determine how the site addresses them. Verify that the date the deviation occurred and the date the deviation was detected was minimal time.
- Ensure that the investigation report includes:
- A written, factual description of the event, including the product name and lot number involved in the event, the date and time of the event, and the extent which the product lot was affected.
  - An assessment of other affected products and any lots that could potentially be impacted.
  - The true root cause, while eliminating, with proof, other causes. If the true root cause is unable to be determined, then the most probable cause is investigated.
  - Clear corrective and preventive actions based on the conclusion of the root cause.
  - Assignment of a responsible party and completion date for both the corrective action and preventive action.
  - Disposition information about the lot is supported through data.
  - A due date for completion of the investigation activities.
  - Supporting documentation.
  - Approval by a quality representative as well as other appropriate parties.
- Ensure that there is a system in place that verifies that corrective actions have been implemented by their due date.
- Ensure that before the batch is released and the investigation has been approved by Quality.

4.5.1.3 Reviewing the Out of Specification (OOS) Investigation

- Verify that averaging practices for laboratory testing data do not hide data variability and conform to regulatory expectations.
- Verify that lab data is invalidated based on sound scientific rationale.
- Verify that retesting was appropriate and that an appropriate test plan was generated and followed. It is important that this test plan be clearly defined in the SOP and followed without variation.
- Verify that laboratory personnel are aware of potential problems and do their best to prevent them.

4.6 AUDITING VALIDATION AND QUALIFICATION

Validation is proving, per the principles of GMP, that any procedure, process, equipment, material activity, or system consistently leads to the expected results. Documented evidence guarantees that a specific system, equipment, or process will always produce a product meeting its predetermined specifications and quality attributes. Validation is nothing more than proving that a process works.

4.6.1 What Should Be Validated?

We use the terms validation and qualification to cover the documented verification of a spectrum of GMP activities including:

- Facilities
- Equipment used in manufacturing
- Equipment used to control the environment(s) where the product is manufactured or stored
- Utilities with product contact (e.g., water systems, compressed gases, air)
- HVAC systems

- Alarm systems that monitor utilities and air handling for process and storage areas
- Analytical methods
- Analytical instruments
- Computerized systems (e.g., computerized training system, documentation control system)
- Cleaning processes
- Manufacturing processes

Storage facilities, air, water, and systems that come in contact with the product, can have a lasting impact on the quality and efficacy of the product.

Regulations require that all aspects of facilities and utilities that may directly or indirectly alter or affect products including laboratories, warehouses, cleanrooms, water, ventilation, temperature, and humidity systems, be assessed and appropriately validated.

4.6.2 Validation and Qualification

Qualification differs from validation in that it involves components or elements of a system whereas validation involves the entire process for a product or system. The amount of documentation and level of detail may vary depending on the type of material to be manufactured/tested, the criticality of GMP, and the system's complexity. In the excipient industry, full validation programs are not always set up as typical in the pharmaceutical industry. However, most excipient manufacturers have evaluated their processes using process capability studies that usually contain much of the appropriate information.

4.6.3 Risk Assessment

In validation, risk assessment is used to know the scope and extent of validation, e.g., an initial impact assessment should be performed to determine the extent of qualification required. Risk assessment is a systematic approach to identifying potential process failures and quantifying the risk they present to the company and its customers.

HACCP (hazard analysis critical control points) may be used. The HACCP concept is an eight-step process designed to identify, characterize, control, and confirm the control of hazards to a process. The hazard analysis involves the preparation of a flow chart detailing the individual steps in a process. At each step, the hazards to the process are identified and the associated risk and severity are assessed respectively. This may be done using FMEA where risk priority numbers are calculated taking into account the severity, probability, and likelihood of detection of the failure. Appropriate systems are implemented to ensure that the theoretical hazards do not present a hazard to the process. The appropriate limits are set for these systems. If the risk assessment has been properly carried out and documented, parts of a process that are low risk may be eliminated or paid minimal attention during validation.

4.6.4 Validation Master Plan

The complexity of validation requires that each element be described in the validation master plan (VMP). VMP provides an overview of the validation, organizational structure, and content. The VMP contains the items' names, descriptions, and planning schedules to be validated. All validation activities in critical technical operations related to products and processes in a pharmaceutical company should be included in the VMP.

4.6.4.1 Prior to the Audit

- Review the critical manufacturing equipment and utilities in use at the site.
- Review the types of processing performed at the site.

4.6.4.2 During the Audit

- Walk through the area with copies of the P&ID (piping and instrumentation diagram) and HVAC blueprint to determine:
  - The field conditions are the same as indicated in the drawings.
  - The location of the piping and HVAC is correctly indicated.
- Ensure that the site has an approved established procedure for performing validation and qualification including a VMP.
- Ensure the extent of qualification/validation is justified.
- Ensure that the supplier's documentation has been reviewed for:
  - Completion of activities.
  - Certification that their work has met the design specification.
  - Verify that a validation protocol exists for each validation project, which identifies qualification activities for DQ/IQ/OQ/PQ and their locations.
  - If DQ is performed as a separate qualification, verify that there is a system to manage the setup of a preventive maintenance and change control program for new equipment. This management system may be part of the IQ/OQ process or part of the site's change control system.
- Ensure that the site has performed the following checks during the IQ phase.
  - For electrical and instrumentation completion:
    - Verify that there is documentation of loop checks.
    - Verify that the software version installed is documented.
- For installation and instrumentation:
  - Verify that the delivered equipment and instrumentation meet the defined specifications of the equipment order.
  - For utility completion:
    - Verify that there is documentation (e.g., a utilities checklist) of the connection of utilities to equipment.
- For piping systems completion:
  - Verify that the system "walk down" was completed.
  - Verify that any red line changes to the P&IDs have been submitted.
  - Verify that the following documentation is correct:
    - As-built engineering and installation drawings
    - Calibration documentation
    - Weld documentation
    - Material of construction certifications
  - Ensure that the operational qualification is complete.
    - Verify that the valve sequencing (e.g., API tanks) is correct by comparing it against the piping and installation drawing.
    - Verify that ranges for process and safety alarms have been set according to process parameters.
    - Verify that critical alarms are present by comparing the alarms against a list of critical parameters.
    - Verify that the range tested in the protocol encompasses the current alarm condition.
    - Verify that worst-case conditions have been tested for the equipment and ancillary systems.
  - Ensure that the documentation system is operating in full compliance.
    - Verify that when any protocol has been approved, any changes to the protocol follow the site's change control process.
  - Ensure that the performance qualification is complete.
    - Verify that the testing performed is within established acceptance criteria.

- Ensure that appropriate SOPs and batch documentation are developed for the equipment and systems being qualified.
- Ensure that any restrictions to use are implemented.
- Ensure process validation has been performed and that manufacturing is performed within validated criteria.
- Ensure that an appropriate level of cleaning validation has been performed and that the validated cleaning process is implemented and conformed to.
- Ensure that analytical methods used are appropriately validated for the intended purpose.
- Ensure that individuals involved in the qualification activities are adequately trained and that their training is documented.
- Ensure validation reports are written in a timely manner, that acceptance criteria are met, and that results are concluded.
- Ensure validated systems are subject to change control.
- Ensure there is a process for periodic review of the validated system and that requalification/revalidation is performed where required.

A number of issues that are identified by inspectors as concerns with validations are listed next:

- Trying to "validate" a process that clearly is poorly developed.
- Poorly organized documentation.
- Validation requirements remain unfulfilled.
- Illogical validation priorities – important work left out.
- Validation approach inconsistent.
- Not adhering to validated processes.
- Poor or even nonexisting change control.

4.6.4.3 What Are the Challenges?

Inspection readiness for validation is an area that is often overlooked and not well managed in the industry. Most companies typically put more focus on inspection readiness for manufacturing, quality management systems, and quality control. Inspection readiness for validation tends to be inadequately structured. When the validation team is not well prepared, the risk of compliance violations is high. Validation is a critical area in the organization that must be adequately prepared for regulatory audits.

This discussion will address the challenges with the traditional approach to validation inspection readiness. It will provide strategies to correct deficiencies and effectively manage the preparatory process.

The traditional approach to inspection readiness for validation is not typically based on a strategy that enables adequate and logical knowledge of the critical areas that will be covered during a regulatory inspection. The approach is also not integrated with related functional areas at the site.

The traditional approach is usually based on disorganized and poorly managed internal audits. Inadequate preparation fails to meet the intent of the audit process. These issues cause inadequate knowledge transfer, audit inefficiencies, business constraints, negative regulatory inspections, compliance observations, and business risks. Developing an effective validation inspection readiness strategy can help to mitigate these challenges.

4.6.5 How to Implement an Inspection Strategy

The validation readiness strategy at the manufacturing site should be a carefully developed and documented plan. An SOP addressing the validation inspection that incorporates an understanding of those areas, topics, and employees involved is highly recommended. Example topics, processes, and questions that will most likely be addressed during a regulatory inspection should be highlighted. Every employee who may be directly or indirectly involved in an audit must be familiar with the SOP content.

Validation audit readiness ensures that critical issues are successfully conveyed to the auditor without implying a lack of knowledge and understanding of the facility validation program.

- Validation of hot topics
- Validation history and knowledge management
- Validation program and procedures
- Sterilization validation program and procedures
- Process validation program and procedures
- Cleaning validation program and procedures
- Critical deviations with potential impact on the validation program

4.6.6 Hot Topics

Validation inspection hot topics may vary between different companies and manufacturing facilities within the same general business. There are some commonalities, however, depending on the type of manufacturing operation performed at the facility. For example, hot topics may be different if the company is a sterile manufacturing facility, a traditional small molecule nonsterile pharmaceutical company; or a biotechnology company. Validation inspection readiness hot topics include those areas that most likely will be reviewed by regulatory agencies. Some example areas may include the following:

- Deviations with sterilization validation impact
- Deviations with impact on equipment qualification
- Deviations with process validation impact
- Deviations with cleaning validation impact
- Legacy systems validation
- Critical change controls with validation impact

The validation inspection readiness process must include strategies for adequately preparing validation personnel to present and defend these validation hot topics. For example, deviations and nonconformities that impact critical processes require adequate inspection preparation. Planning for discussion on these areas requires expert awareness in:

- Chronology of events that triggered the deviation
- Validation impact
- Validation deliverables executed because of the deviation
- Root cause
- Validation program gaps contributing to the deviation

Process deviations require a significant amount of preparation to enable knowledge and a logical approach to adequately defend subsequent actions. The chronology of the events that triggered the deviation must be clearly understood. This requires an understanding of the activities and unit operations performed in the equipment prior to the event that triggered the deviation. Regulatory auditors will inquire about the events preceding the deviation. An adequate understanding of the event that triggered the deviation and a logical strategy for discussing and defending the potential impact on the validation program is required.

4.6.7 Validation and Knowledge Management

One of the biggest challenges during regulatory inspections is knowledge management related to the validation history. This is a challenge that mostly impacts legacy systems, their validation, and issues encountered during the system life cycle. This challenge is typically created due to attrition and changes in the validation organization. These challenges create a knowledge vacuum that creates compliance and business risk.

Companies need to implement a strategy that includes adequate tools and processes related to the validation history of legacy systems to mitigate these risks. These strategies should include databases or tools that capture critical validation history information related to legacy systems above and beyond what is included in a validation plan. The following are critical validation topics:

- Deviations
- Corrective and preventive actions (CAPAs)
- Change control
- Validation document high-level summaries
- Critical protocol deviations

Inspection readiness activities must focus significantly on the validation history of legacy systems to enable adequate knowledge transfer to newer validation participants who were not involved in the initial validation. This is critical because legacy systems often yield numerous regulatory observations driven by inadequate knowledge transfer. Companies need to implement inspection readiness processes and procedures that enable adequate knowledge transfer and management of critical validation history information.

4.6.8 What Is a Validation Plan?

Validation inspection readiness activities need to include a strategy for presenting and defending the validation program and related procedures. The strategy needs to focus on critical procedures such as sterilization, process, and cleaning validation. In addition, critical program documents such as VMPs, global standards, and SOPs must also be reviewed prior to the audit. These program-level documents must be aligned and periodically assessed against current industry standards and regulatory requirements. Activities related to validation programs and procedures require qualified subject matter experts (SMEs) who have a high level of ownership and understanding of these documents. The SMEs must participate in inspection readiness activities that include mock audits conducted internally.

Inspection readiness activities related to the validation program and procedures are critical to ensure a successful outcome during regulatory agency inspections.

4.6.9 What Are the Roles and Responsibilities?

- Validation inspection readiness requires an integrated cross-functional effort that includes SMEs from the following departments:
- *Quality Compliance.* This function must support mock audits and provide strategic support during inspection readiness activities.
- *Quality Assurance Validation* – This function requires a process SME who can address procedure and progression as well as oversight of all validation activities and link to the validation history.
- *Engineering*– This technical SME brings a significant amount of technical and validation knowledge of validated systems.
- *Validation* – The owner of the program, procedures, processes, and practices.

These experts must all be involved in the validation inspection readiness activities, including knowledge management and transfer, participating in mock audits, and any remediation activities necessary because of internal observations and gaps.

4.6.10 What Should a GMP Auditor Verify?

- Is there a validation policy covering the procedure for preparing, reviewing, approving, and executing the validation plan?
- Is a validation master plan in place that details the scope, responsibility, and approach of validation?
- Is the validation approach in line with current industry practices and with the latest guidelines?
- Are all facilities and utilities, equipment, computer systems, and processes, including cleaning processes and lab instruments, validated?
- Are validation records properly documented, assembled, reviewed, and easily tracked?
- Are validation studies conducted per predefined protocols?
- Are validation protocols clear and easy to follow?
- Is validation properly documented, with VMP, validation protocols, and summary reports included?

4.6.11 Process Validation

- Are all critical process parameters identified, monitored, and validated?
- Are all critical quality attributes identified, monitored, and validated?
- Has the regulatory-favored validation type been used?
- Is validation rationale/strategy scientifically supported?
- Were process validation prerequisites met prior to the execution?
- Were CAPAs properly handled during execution?
- Is the current master batch procedure based on the validation?
- Has a life-cycle approach been applied as recommended by FDA Guidance for Industry Process Validation: General Principles and Practices, Health Canada, and European Medicines Agency (EMA) Guidelines on Process Validation?
- Has a risk-based approach been applied as recommended by ICH Q9 on quality risk management?

4.6.12 Cleaning Validation

- Does the cleaning procedure have sufficient detail to follow?
- Is the cleaning procedure validated with the worst-case products?
- Are the validation limits established with sound scientific rationale?
- Is the residue analytical method validated?
- Has the visual inspection been validated if it is used as the sole acceptance criterion?
- Have critical process parameters been identified, monitored, and validated?
- Have other variables been identified and validated?
- Has campaign length been established based on a cleaning validation study?
- Has the clean hold time been challenged and established?
- Has the dirty hold time been challenged and established?
- Have newly introduced products and new equipment/parts been properly assessed?
- Has the revalidation criteria been properly defined?
- Has the cleaning procedure been monitored at a reasonable interval?

4.7 AUDITING CALIBRATION

Calibration of critical items, preventive maintenance, and housekeeping, while not appearing to be the most important parts of GMP, are vital to keeping a site in compliance and producing quality products. If an audit is being performed on calibration, preventive maintenance, or housekeeping, the GMP systems should be reviewed during the audit.

4.7.1 Quality System

The site audited should have SOPs in place describing the overall calibration and maintenance program including:

- How critical items requiring calibration and maintenance are identified (a risk/GMP assessment should be in place for defining what is critical and not)
- Establishing calibration and maintenance requirements and review during item's life cycle
- Description of the workflow for calibration and maintenance
- Management of "out of tolerance" conditions and remedial actions
- Requirements for trend reviews (e.g. calibration failures, maintenance frequencies)
- Prevention from use of "uncalibrated" and "out of service" items
- Acceptance of items as suitable for use
- Use of contractor calibration and maintenance providers

4.7.2 Documentation Requirements

There are cleaning requirements when reintroducing/removing equipment to/from sterile areas or areas where hazardous materials are handled.

4.7.3 Calibration and Preventive Maintenance Audit

Equipment and instrumentation routinely used in the manufacturing and QA laboratory areas need to be within calibration at all times during normal operation. To ensure this, each site should have a written approved calibration program that includes specific directions, schedules, calibration requirements, and provisions for remedial action if accuracy and/or precision limits are not met. Calibration requirements for each piece of equipment, instrument, or system should be documented. The limits and parameters for the calibration should be established through a combination of:

- Process requirements (including accuracy/regulatory)
- Duty and location of instruments
- Qualification work
- Recommendations from the manufacturer
- Experience, history

The frequency of calibration should be recorded and determined using the same criteria as limits and parameters. Once a frequency is determined it should be part of an approved maintenance and calibration schedule. If there is a question as to the interval of frequency, the smallest interval should be chosen that precludes any risk to the product.

Allowable time frames for the calibration should be defined (e.g., monthly + three days, quarterly ± seven days) and compliance with the schedule verified. This schedule should be part of a procedure to manage maintenance and calibration.

The calibration status of equipment and instruments should be easily identified. Status may be indicated through a label visible on the instrument or equipment, through a database, or through other means. If an external contractor is performing calibration work for the site, the contractor must work with the Site Engineering and Maintenance group. Site personnel should also review and approve the calibration documentation and/or records generated by the contractor in order to ensure that jobs have been completed per established procedures and that equipment is within established tolerances.

If possible, calibration should be performed in a calibration laboratory, free of conditions that may impact calibration such as:

- Vibration
- Variable line voltage
- Dust
- Fumes
- Extreme temperature ranges or other environmental extremes that could impact equipment functionality

All calibration work must be recorded. Calibration records should be kept in a secure document filing and archive system. Calibration data should be reviewed for trends. All work performed on equipment or instruments should be documented and use traceable standards. Computer systems for scheduling, tracking, and reporting of calibration and preventive maintenance jobs should be developed and implemented to defined quality standards with appropriate qualifications performed.

4.7.4 Preventive Maintenance

Preventive maintenance is considered planned maintenance. This maintenance is performed on equipment, instruments, or systems to:

- Ensure equipment is able to operate normally within its specifications before an unplanned repair is needed
- Extend the operating life of the equipment or instrument
- Replace worn parts before they affect equipment or instrument performance, which will affect drug quality
- Provide routine or recommended maintenance such as lubrication

All preventive maintenance should be scheduled at sufficient frequency to ensure that equipment, instrumentation, and systems are within their operating limits under normal operating conditions. The frequency may be established through consideration and review of the following criteria:

- Work performed during qualification
- Frequency of use
- Experience and application
- Process requirements
- Criticality of the device with respect to GMP impact
- Duty and environment of instruments
- Historical trend data
- Required accuracy of the process
- Regulatory aspects
- Maintenance recommendations from the manufacturer

The rationale for the frequency of preventive maintenance must be documented. The frequency should be on the conservative side by ensuring that no product is at risk because of infrequent maintenance. The site should have an approved preventive maintenance schedule. A written approved procedure to manage calibration and maintenance according to the schedule should be in place. In the case of unplanned maintenance, the site should have a written procedure defining how to appropriately perform and document such work. The impact of unplanned maintenance on routine manufacturing needs to be assessed and documented, and remedial actions taken.

4.7.5 Use of Contractors for Calibration and Maintenance

Site procedures should define the approach to the management of calibration and maintenance performed by external contractors and calibration service suppliers, including requirements for contracts/service-level agreements/shared SOPs. The roles and responsibilities of the contract acceptor and contract giver need to be defined in an appropriate written agreement. Delineation of responsibilities should be agreed upon. SOPs should describe the provision of calibration and maintenance records or certificates by external contractors. This should include a review of work performed to verify that it has been completed to an acceptable standard, and approval by the site indicating that the appropriate acceptance criteria have been met.

Prior to the audit, review the inventory list of the site's critical equipment, instruments, and systems. Determine if there are any outstanding corrective actions from previous audits or regulatory inspections involving calibration and/or preventive maintenance of critical equipment, instruments, or systems, or housekeeping citations.

During the audit, check the facility/area for appearance and cleanliness. When arriving at the site look carefully around to get a first impression of maintenance and housekeeping and check for proper and controlled segregation of materials, i.e., waste not next to components or raw materials. Determine if the facility has any obvious problems (e.g., missing ceiling tiles, holes in walls, peeling paint, "stained" floors, trash overflowing). Ensure that critical equipment and instrumentation within the facility have stickers or labels to indicate calibration status or have another system in place to indicate calibration status. Determine if the facility appears well-maintained and visibly clean. Determine if there is any evidence of vermin infestation.

Review the quality system.

- Verify that there is an established and approved quality system/quality policy that is endorsed by management.
- Verify that Quality Unit/Quality Assurance is independent of manufacturing.
- Verify that one of the responsibilities of the Quality Unit/Quality Assurance is to review out of tolerance results and determine the impact on product quality.
- Ensure that personnel have received adequate training.
- Verify that there is a documented training and qualification program for any person responsible for conducting calibration, preventive maintenance, or housekeeping.
- Verify that detailed job-specific training is conducted and documented for each employee.

If external personnel are used to perform the calibration, preventive maintenance, or housekeeping function, ensure that they have been qualified and are competent.

- Verify that the training program incorporates requirements for temporary employees and consultants.
- Ensure that there is a calibration program in place at the site.

For calibration of equipment, instrumentation, and systems, ensure that there is a written and approved program for managing calibration.

This program should contain the following information:

- An inventory of critical equipment, instruments, and systems (including computer systems) requiring calibration should be available and include appropriate acceptance criteria.
- Specific calibration requirements including limits and parameters for each piece of equipment, instrument, or system. The accuracy, precision, and linearity of the calibration need to be relevant to the critical instrument and scale of measurements.
- Frequency of calibration for each piece of critical equipment, instrumentation, or system.
- Review the system for issuing the calibration schedule.
- Determine the time frame within which calibration must be performed.
- Determine what will happen if calibration does not take place as scheduled.
- Standards and testing equipment to be used in each calibration, which can be traced, back to a primary standard.
- Methods to be used in each calibration.
- Documentation for recording calibration results.
- Calibration schedules.
- Procedures detailing what to do in the case of an out of tolerance result.
- Procedures for performing trends on calibration data.
- A system for retention and retrieval of calibration history and data.
- Requirements for using contractors.
- Documentation that computer systems for calibration program management are validated.

Ensure that there is a preventive maintenance program in place at the site.

For maintenance of equipment, instrumentation, and systems, ensure that there is a written and approved program for managing maintenance.

Verify that the program contains the following information:

- An inventory of critical equipment, instruments, and systems (including computer systems) requiring maintenance.
- Specific maintenance requirements including limits and parameters for each piece of equipment, instrument, or system.
- Frequency of maintenance for each piece of critical equipment, instrumentation, or system.
- Methods to be used for maintenance.
- Documentation for recording maintenance results.
- Maintenance schedules.
- Procedures for cleaning equipment after maintenance (and calibration) if necessary,
- Procedures for performing trends on maintenance data.
- A system for controlling the spare parts inventory.
- The system used to communicate the maintenance schedule.
- The timeframe within which maintenance must be performed.
- Verify that computer systems for preventive maintenance program are validated.
- Verify that in the course of preventive maintenance if lubrication is required that food grade lubrication is used.

Ensure that there is a housekeeping program in place at the site.

- Verify that there is a system for managing facility cleaning.
- Verify that there are written procedures that include cleaning schedules, methods, equipment-approved sanitization, and cleaning agents.
- Verify that cleaning is documented.
- Verify that the facility has a documented program for the prevention of infestation from rodents, birds, insects, or other vermin.
- Verify that the site has a list of approved infestation prevention agents and uses only those.

- Verify that the site locations for the prevention infestation (pest control) stations are placed logically, documented, and reviewed on a regular basis for activity and that any relevant recommended actions are implemented.
- Verify if a contractor is used for the prevention of infestation (pest control), that appropriate agreements are placed, and any visit reports are reviewed and approved.
- Verify that the site has adequate personnel, procedures, and methods to ensure timely removal of trash and refuse from the site.
- Verify that there is a calibration program, described in a sufficiently detailed written procedure(s).
- Verify that there is a mechanism, such as an equipment master list, for differentiating critical from noncritical equipment/instrumentation.
- Verify that there is a requirement that equipment/instrumentation be calibrated prior to initial use.
- Verify that there is a system for identifying equipment (e.g., ID number).
- Verify that there is a mechanism for establishing calibration tolerances, which takes into consideration the allowable or existing process tolerance and the capability of the equipment being calibrated.
- Verify that there is a system for assigning calibration frequencies. This should include the time window within which the calibration must be performed.
- Verify that frequencies are documented on the equipment master list or in individual calibration procedures.
- Verify that there is a requirement for the use of suitably recognized standards and the requirement for maintaining certificates of traceability for all reference standards used in calibrations.
- Verify that there is a requirement that there be a calibration procedure for each piece or type of equipment/instrument.
- Verify that each procedure includes:
  - Standard(s) to be used
  - Step-by-step calibration instructions
  - Calibration tolerance
  - Provisions for adjustments, if necessary
  - Requirement for recording actual measurements found both before and after adjustment
- Verify that there is a format for the calibration record, including appropriate provisions for approval(s) of data.
- Verify that there is provision for timely investigation of any out of tolerance (OOT) conditions, including:
  - An assessment of all production/work performed using the OOT equipment instrumentation since the previous acceptable calibration or verification/accuracy check.
  - A review of that work to evaluate the possible impact on product quality.
  - A documented rationale, approved by the quality function, as to why the product/work was or was not impacted.
- Verify that there is a requirement that any OOT equipment/instrumentation be labeled "Out of Service" and not be used until it is adjusted and found to be within the required tolerances.

The GMP auditor should review that following appropriate supervisory review and approval, calibration records are retained on a schedule commensurate with site record retention policies.

The GMP auditor should review that records are retained at least as long as other records associated with products manufactured using the equipment/instrumentation.

The GMP auditor must ensure that an appropriate change control system covers:

- Change in a calibration tolerance
- Change in calibration frequency
- Change in a calibration procedure
- Addition/deletion of equipment/instrumentation to/from the calibration program
- Changes in the location of equipment/instrumentation
    - GMP auditors should verify that the relevant procedures are being followed correctly, by appropriately trained personnel. The audit's scope must be defined precisely.
    - GMP auditors should verify that procedures and systems are being operated and completed successfully.

The audit report should:

1. Serve as the formal record of the audit and its findings
2. Provide a major input when determining corrective action

The audit report should present an accurate objective record of what was audited and the findings.

References made in the audit report to documentation examined during the audit should be unambiguous (e.g., by title, reference, date, version, copy number, and author). An audit report normally will contain:

1. An introduction
2. Scope information
3. Organization of audit, including agenda, criteria, representatives
4. Detailed findings
5. Record of the closing meeting
6. Conclusions

Audit reports should be responded to with agreed actions and retained as part of the overall validation for a process.

4.7.6 Outsourced Calibration Service Audit

When considering a supplier to provide calibration contract services, a clear understanding of their capability should be achieved before any work is undertaken.

This can be achieved by a:

- Formal supplier audit
- Postal audit
- Combination of formal supplier and postal audits

4.8 UTILITY AUDIT

- HVAC
- Water
- Steam
- Gas

To produce either a finished pharmaceutical, an API, an excipient, a medical device, or packaging materials, the environment within the plant has to be considered. One of the least noticed but important parts of producing a product free of contamination is the plant utility system. The system includes the heating, ventilation, and air-conditioning (HVAC) system, the water system, the compressed gas system, and the steam system if used in the plant. Each of these systems must be qualified/validated.

4.8.1 HVAC System

Air systems should be designed to prevent contamination and cross-contamination. To achieve this protection, the air handling system must be carefully designed with many factors taken into account.

4.8.1.1 Phases of an Air Handling System

A conventional air handling system has four different phases.

The first phase manages the incoming/intake air or "fresh" air. This phase may include a coarse filter that eliminates large contaminants such as soot, pollution, or large visible particles from the external environment. Some systems may recirculate air to "clean" it. When air is recirculated, it continuously passes through filters. During this process some air is "lost," so some fresh air or "make-up" air needs to be added. The second phase is the conditioning phase. Air enters the central air handling unit (AHU) and is conditioned there (heated, cooled, humidified or dehumidified and filtered). The cooling system in this phase can be a potential source of mold contamination. In the AHU, fresh air and recirculated air, if any, will be mixed. The third phase supplies the air to the actual room or area. The air is blown with a fan/blower through an air diffuser into the specific area. Before the air enters the room it may be pushed through another filtration system (e.g., HEPA or ULPA) and then dispersed through a diffuser. The fourth phase exhausts air from the area. This system contains the air returns found in a room. An air return allows the air to be drawn back into the duct system. Air returns may be located on the ceiling or on the wall. The design should take the movement of air as well as how to make the best environmental conditions in the room into account. Returns on the wall are thus more common in cleanrooms. The return should exhibit negative pressure.

A well-designed air handling system should start with a good design, be installed properly, be qualified, and be maintained. At a minimum, it should have sealed ducts to prevent unfiltered air from entering the system, and tight filters for the same reason. Different areas with different air cleanliness requirements should have separate air handling systems. The level of filtration should be appropriate for the level of air cleanliness that is to be achieved. The percentage of recirculation air to make-up air will vary depending on the level of cleanliness desired. Risks for cross-contamination should also be considered. No recirculation of air minimizes the risk of cross-contamination and should be used when handling some types of substances or products. Air filtration, dust collection, and exhaust systems should be used in production areas where the product may be exposed and according to regulatory requirements. The filter should be located in the system where it will do the most good. Intake and air returns should not be located next to each other. Dust extractors and bag filters should be located upstream of the final filters. Temperature and humidity requirements should be established to keep the product free from contamination. Depending on the product and the environment to be achieved, air changes should be consistent with industry standards with more air changes for cleaner environments. (See ISO 14644, Cleanrooms and associated controlled environments – Part 1: Classification of air cleanliness for cleanroom standards.)

Pressure differences between cleanrooms and areas of different cleanliness are means of maintaining the environment and the principle is to have higher pressure in cleaner areas. To minimize the risk for cross-contamination, which is important in dust-generating activities, such as in tablet manufacturing, and when containment is desirable, the principle will be to have the highest pressure in the corridor or in the airlocks and a lower pressure in the cleanroom to ensure the air is not spread to the environment. Pressure differentials should be maintained according to design criteria and tested and/or monitored on a frequent basis. An approved procedure and backup power supply should be in place for responding to HVAC failures or loss of power. Ductwork should run in long straight sections, which allow the air to flow freely. The ducts should be designed so that they can be easily and effectively cleaned. An alarm system should be in place to notify the site if there are unexpected drops in pressure. Depending on the environment, there should be an established airflow direction. There should be adequate controls and monitoring devices for air pressure, microorganisms, dust, humidity, and temperature in finishing areas. Air handling units for dry processing areas (i.e., drying, milling, and packaging areas) should be independent.

4.8.1.2 Air Pressure Differentials

In areas where air changes and differential pressure are critical, the system must be monitored and documented. Air pressure must be measured and correct pressure differentials maintained. If the pressure differentials are not maintained, there should be a procedure in place defining what action to take. Blowers and fans should be on a routine preventive maintenance schedule. Ductwork should be checked at an established interval for leaks. Preventive maintenance must be established based on written specifications. Filters are one of the main components of an HVAC system since they determine the size of airborne particles that pass through them, and thus the hygiene class. Depending on the type of room air requirements, different filters will be used. In an area where there are powders, often the exhaust will have a bag filter attached to it to catch the escaping powder.

4.8.1.3 Cleanroom Air Requirements

Cleanrooms, because of the desired high level of cleanliness, require HEPA or ULPA filters to prevent microbial and particulate contamination. For manufacturing injectables and sterile products, it is recommended that the filters be placed in a terminal position. The reasons are that they provide better product protection. Any problem arising from the ducts is eliminated and it is the preferred method for cleanroom classes with high requirements. HEPA or ULPA filters may be mounted either vertically or horizontally in the cleanroom, depending on what the room will be used for. For the air exhaust, in the case of a vertical unit, a low return is better, as the air is better distributed in the room. It is important to monitor airflow velocities for each HEPA filter according to a program of established intervals because significant reductions in velocity can increase the possibility of contamination, and changes in velocity can affect the laminarity of the airflow. The differential pressure on filters is an indication of the clogging of the filters: With the charging of dust on the filters, the differential pressure will increase.

4.8.2 Water Systems

Water is the most widely used starting ingredient or raw material in the manufacture of active pharmaceutical ingredients and medicinal products. Its uses in the production environment range from starting material to cleaning agent to steam source for sterilization processes. Because of this, controlling the chemical and microbiological quality of water and steam is critical. Special care should be taken in the processing, storage, and distribution of water. The higher the risk to the pharmaceutical product, the purer the water should be.

4.8.2.1 General Considerations for Water Systems

The rationale for appropriate water quality for each process should be documented. The water quality depends on the stage of its use in the manufacturing process and the intended route of administration for the finished product. Water used in production must be routinely sampled and tested according to a predefined and approved plan.

This includes the potable water supply. For any water treated to achieve an established quality level, a validated, qualified, and maintained treatment process and distribution system must be used.

Water should circulate in a closed loop system, not open to the external environment. Fittings and valves should be of a sanitary type. There should be no threaded pieces where bacteria can enter and multiply. Fittings and valves can be taken apart and sterilized if needed. Water that is held in tanks should either be used within 24 hours of manufacture or circulated.

4.8.2.2 Validation of Water Systems

Water systems should take a full year to validate because seasonal environmental conditions will affect the quality of the feed water coming into the plant. Validation should follow the same criteria and produce equivalent documents as other types of validation.

4.8.2.3 Contamination in Water

Water can support microbial growth. Water is susceptible to bacterial contamination for many reasons. Causes of contamination in water can be:

- Stagnant or noncirculating water
- Improperly processed water
- Water circulating at too low a temperature
- Water circulating too slowly
- Dead spots or dead legs in the system

However, a significant part of good water system design is minimizing microbial contamination. Microbial contamination can occur due to the colonization of surfaces and stagnant areas by aquatic bacteria with biofilm formation. The battle is almost won if the bacteria can be prevented from sticking to the surfaces. Smooth surfaces, high temperatures, moving water, and no dead spots are all good design elements.

4.8.2.4 Types of Water

The type of water used will depend on the requirements of the product during various stages of production, e.g., purified water for tablets and water for injection (WFI) for sterile products. The way the medicine will be used in the body actually determines which water will be used. Sterile products like injectables need to be produced with the highest quality water (WFI) since these products go directly into the body bypassing the body's defense system. Water has unique chemical properties, and is able to dissolve, adsorb, absorb, or suspend many compounds. These contaminants need to be removed before water can be used as an ingredient.

Figure 4.1 presents different types of water, Table 4.1 presents water systems for sterile products, and Table 4.2 presents water systems for nonsterile products.

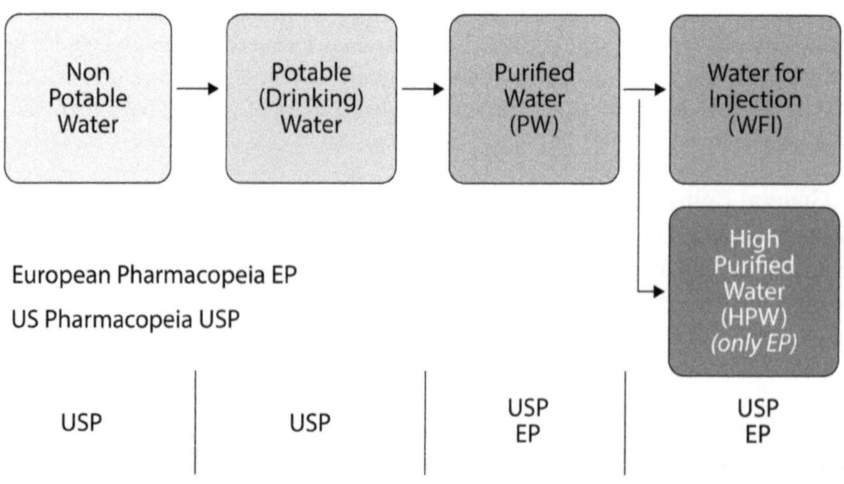

FIGURE 4.1 Different water qualities according to pharmacopoeias.

TABLE 4.1
Water Systems: Sterile Medicinal Products

Sterile Medicinal Products	Minimum Acceptable Quality of Water
Biologics (including vaccines and ATMP)	WFI
Parenteral	WFI
Ophthalmic	Purified water
Hemofiltration solutions, hemodiafiltration solutions	WFI
Peritoneal dialysis solutions	WFI
Irrigation solutions	WFI
Nasal/ear preparations	Purified water
Cutaneous preparations	Purified water

Four different categories of water are common. They are:

- Potable
- Purified
- Highly purified water
- Water for Injection

Water Sources

- City water
- Company-generated potable water
- Well water

In Figure 4.1 are the different types of water according to valid pharmacopoeias that should be known by a GMP auditor.

TABLE 4.2
Water Systems: Nonsterile Medicinal Products

Nonsterile Medicinal Products	Minimum Acceptable Quality of Water
Vaccines for nonparenteral use	Purified water*
Oral preparations	Purified water
Nebulizer solutions	Purified water**
Cutaneous preparations	Purified water***
Nasal/ear preparations	Purified water
Rectal/vaginal preparations	Purified water

*According to the outcomes of the risk assessment, WFI may be required in some cases for the manufacture of nonsterile vaccines where in order to ensure the vaccines' safety and quality (avoiding the introduction of microorganisms undesirable in the specific finished product formulation) greater microbiological purity of water is needed.

**In certain disease states (e.g., cystic fibrosis), medicinal products administered by nebulization are required to be sterile and nonpyrogenic. In such cases, WFI should be used.

***For some products, such as veterinary teat dips, it may be acceptable to use potable water where justified and authorized taking account of the variability in chemical composition and microbiological quality.

TABLE 4.3
Water Used for Cleaning/Rinsing

Cleaning/Rinsing of Equipment, Containers, Closures	Product Type	Minimum Acceptable Quality of Water
Initial rinse	Intermediates and AS	Potable water
Final rinse	AS	Use the same quality of water as used in the AS manufacture
Initial rinse including CIP* of equipment, containers and closures, if applicable	Medicinal products – nonsterile	Potable water
Final rinse including CIP* of equipment, containers and closures, if applicable	Medicinal products – nonsterile	Purified water or use the same quality of water as used in the manufacture of medicinal products, if higher quality than purified water
Initial** rinse including CIP* of equipment, containers and closures, if applicable	Sterile products	Purified water
Final rinse including CIP* of equipment, containers, and closures, if applicable	Sterile nonparenteral products	Purified water or use the same quality of water as used in the manufacture of medicinal products, if higher quality than purified water
Final rinse including CIP* of equipment, containers and closures, if applicable	Sterile parenteral products	WFI***

*CIP, clean in place.

**Some containers, e.g., plastic containers for eyedrops, may not need an initial rinse. Indeed this may be counterproductive since particulate counts could be increased as a result. In some cases, e.g., blow–fill–seal processes, rinsing cannot be applied.

***For injection for veterinary use, purified water may be used for cleaning/rinsing of equipment, containers, and closures, if the preparation falls into the exempted category from the test for bacterial endotoxins or test for pyrogens in the European Pharmacopoeia monograph *Parenteral Preparations*.

In this case, a risk-based approach to justify the use of purified water instead of WFI must be implemented as part of an overall control strategy, particularly to ensure sterility and to avoid the introduction of contaminants and the growth of microorganisms in the final product.

Water Qualities for Pharmaceutical Purposes

- Drinking water
- Bulk purified water
- Bulk highly purified water
- Bulk water for injections
- Other grades of water

Table 4.3 presents water used for cleaning/rinsing.

Water Testing Methods

- Conductivity
- TOC
- pH
- Ozone
- Aerobic bacteria
- Endotoxin
- Heavy metals/nitrates

Distribution Systems

- Dead-end system
- Loop system
  - Hot
  - Ambient
  - Cold

4.8.2.5 Qualification of Water Systems

Figure 4.2 presents the water system qualification phases and can help the GMP auditor ask questions about the permission and prohibition of water use in manufacturing.

- Installation qualification
  - System installed as planned, redline drawings, loop check, material, gaskets
- Operational qualification
  - Integrity of the system, general functionality, cleaning, sanitization, sterilization, water complies to specification, SOPs
- Performance qualification
  - Constant water of the required quality
    - Test period I: Two to four weeks daily
    - Test period II: Monitoring results from one year

4.8.2.5.1 Test Points during Qualification

- OQ and PQ period I
  - Before and after critical steps of generation
  - Begin and end of the loop
  - Every point of use
- PQ period II
  - Begin and end of the loop
  - Critical use points (same as monitoring)

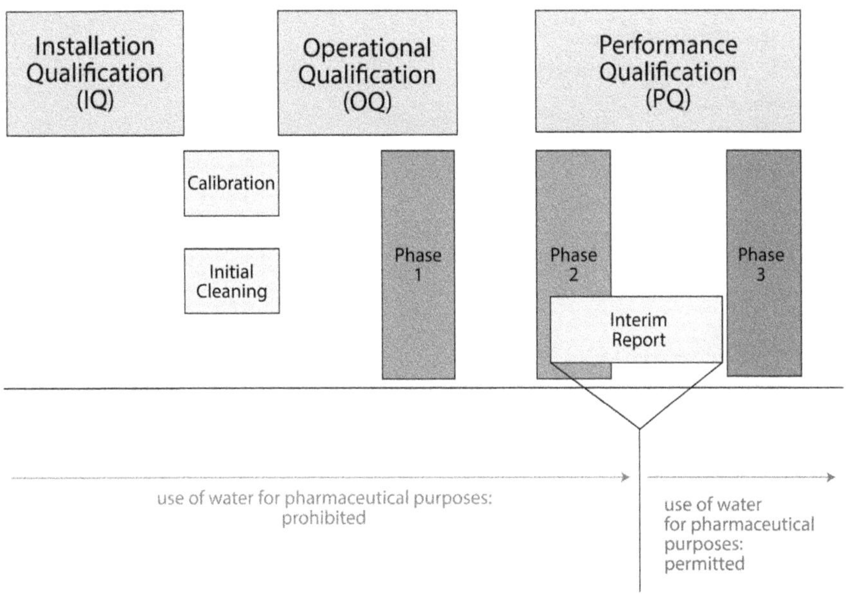

FIGURE 4.2 Water system qualification phases and reports.

4.8.2.5.2 Sanitization

- Hot sanitization
  - Hot water (constant/periodically)
  - Steam/sterilization (periodically)
- Chemical sanitization
  - Ozone (periodically)
  - Chemical substances (periodically; concentration of the agent, removal of the substance)
- Physical sanitization
  - UV irradiation (constant)
  - Filtration (constant)

4.8.2.5.3 Monitoring

- Tests
  - Microorganism/endotoxin
  - TOC/conductivity/pH
- Frequency
  - Daily/weekly/monthly/quarterly
- Sampling points
  - Critical use points
  - Begin and end of the loop
  - Critical generation steps

4.8.2.5.4 Documentation

- SOPs for
  - Operation
  - Sanitization
  - Monitoring
  - Maintenance
- Trend analysis/quality review
- Relevant technical documentation
  - First installation
  - Change control
  - Calibration

4.8.2.5.5 Special Points

- Maintenance program
  - Filter change
  - Sanitization
  - Gaskets
  - Calibration
  - Integrity
- Labeling
  - Content/flow direction
  - Use points/sampling points
- Heat exchangers
  - Integrity (double tubesheet heat exchanger, double seals, differential pressure)

Users should consider conducting routine audits and self-inspection of established water systems.

Those recommendations can be used as the basis of an audit and inspection. A tour of the water system, treatment system, storage and distribution system, as well as visible pipework and user points, should be performed to ensure that the system is appropriately designed, installed, qualified, validated, maintained, and monitored.

The audit of the water system should include:

- An examination of valid and approved water systems showing all components comprising the system from the entrance to the points of use, sampling points, and their markings
- A physical check to ensure that the system matches the P&ID
- Approved piping drawings (e.g. orthographic and/or isometric)
- Qualification and validation protocols, reports, and results
- A sampling and monitoring plan with a drawing of all sample points with evidence of sample management, sample preparation, testing, and results
- A training program for sample collection and testing
- The setting and monitoring of alert and action levels
- Monitoring of results and evaluation of trends
- Absence of leaks
- A review of any changes made to the system since the last audit or inspection
- A review of deviations recorded and their investigation
- A general inspection of the system for status and condition
- A review of maintenance, failure, and repair logs
- A check of calibration and standardization of critical instruments and devices
- A review of the performance capability of the system
- Procedures and records for sanitization

4.8.3 Steam

Pure, pyrogen-free steam (called clean steam) must be used where steam can come into contact with the product, or "product contact surfaces," e.g., sterilization-in-place (SIP) equipment. It must:

- Have the same chemical quality as water for injections
- Have no pyrogens or endotoxins
- Have no volatile additives such as amines or hydrazines
- Be produced by stills without condensation
- Have a limit on noncondensable gases

Clean steam should be tested periodically.

Water systems should be scheduled for routine maintenance. Parts of the system that should be included are:

- Replacement of resins for deionizer/softener
- Inspection and replacement of heat exchanger parts, stills, vent filters, fittings, membranes, valves

4.8.4 Compressed Gases

Compressed gases such as nitrogen (N), argon (Ar), and compressed air (CA) may be used in the production of drug products and contact with the products. Depending on the use of the gas, qualification may be required. Gases may be

- Manufactured and distributed by the site
- Purchased from an outside vendor and distributed through a site distribution system
- Purchased from an outside vendor and used at the point of source (gas cylinders)

In all cases, the gas must meet specifications. These specifications are determined based on the use of the gas and the environment in which it is used. If the gas is supplied by an outside vendor, a certificate of analysis should be required to prove that the gas meets identity and other specifications.

If the gas is to be circulated throughout a plant distribution system, the system should be qualified both at the source and points of use. Qualification should include testing of points of use farthest from the source, identification of the gas, testing for hydrocarbons, and particulate and microbial monitoring. A qualification program should demonstrate that the distribution system for the gas does not adulterate the quality of the gas from the controlled source to the point of contact in manufacture. If the gas will be used within a controlled environment, it should meet both the particulate and microbial requirements of that environment. Qualification samples should be taken when the system is in use by multiple users, at various points of use, and "worse case" points of use where the system does not have long stretches of straight piping. The system should be checked for leaks. If the site is manufacturing its own compressed air and/or nitrogen for use in a controlled environment, it should use an oil-free compressor. A monitoring program should be based on the criticality of the gas to the product. The monitoring program should provide information on the quality of the gas being monitored, and serve to demonstrate that the distribution system has not deteriorated or degraded the quality of gas being distributed. Monitoring frequency should be based on the variability of the gas supply to demonstrate that the supply at the point of use is consistent across a period of time. Test samples taken for the monitoring program should be taken under identical conditions as used in the process. If production flushes the point of use for three minutes, the sample should be taken after a flush of three minutes. If a filter is installed at the point of use, the sample should be taken from a point-of-use filter. If maintenance work is performed on the system, test samples should be taken to verify that the system has not been compromised through the work. All routine monitoring should be documented as appropriate.

4.8.5 Factors Contributing to Quality Products

The manufacturing environment is critical for product quality. Factors to be considered include:

- Light
- Temperature
- Relative humidity
- Air movement
- Microbial contamination
- Particulate contamination

An uncontrolled environment can lead to product degradation, product contamination (including cross-contamination), and loss of product and profit.

4.8.5.1 What Is Contamination?

Contamination is the undesired introduction of impurities (chemical/microbial/foreign matter) into or on to starting material or intermediate – during sampling, production, packaging or repackaging. Impurities could include products or substances other than the product manufactured, foreign products, particulate matter, microorganisms, endotoxins (degraded microorganisms), etc.

Figure 4.3 presents differences between contamination and cross-contamination and Figure 4.4 presents airflow patterns.

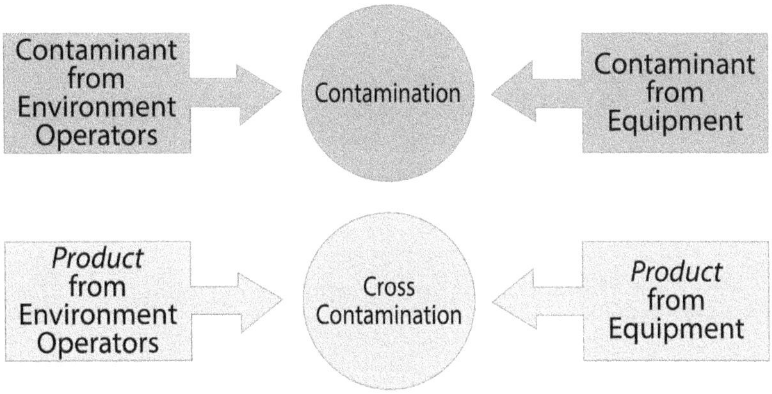

FIGURE 4.3 Cross-contamination.

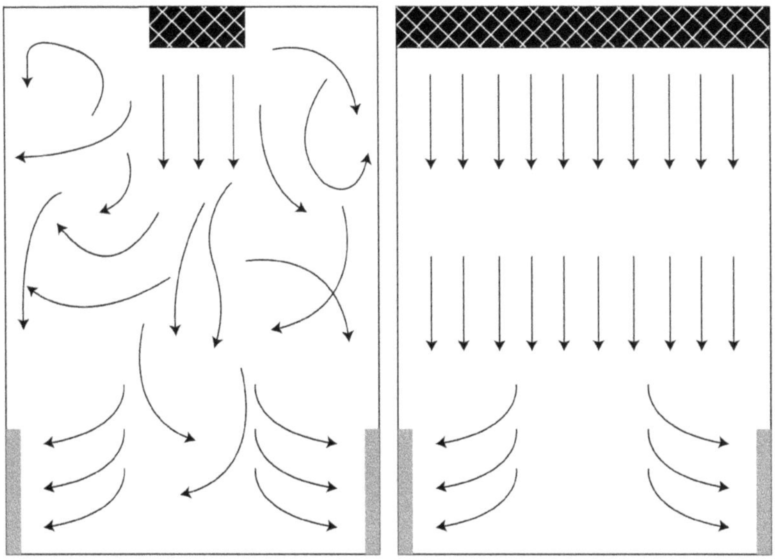

FIGURE 4.4 Airflow patterns. Turbulent dilution of dirty air (left) versus unidirectional/laminar displacement of dirty air (right).

4.8.5.2 What Is Cross-Contamination?

Cross-contamination is contamination of a starting material, intermediate product, or finished product with another starting material or product during production.

Cross-contamination can result from, e.g.:

- Poorly designed, operated, or maintained air handling systems and dust extraction systems
- Inadequate procedures for, and movement of, personnel, materials, and equipment
- Insufficiently cleaned equipment

4.8.6 Microorganisms

- Microbiological
    - Open product
    - Doors/openings to lower cleanliness classes
    - Contamination sources (personnel)

- Particle
  - Open product
  - Contamination sources (personnel/machines)
  - 30 cm to 30 cm grid (below HEPA filter and working high)
- Airflow
  - 30 cm to 30 cm grid (below HEPA filter and working high)
- Airflow visualization
  - Class A

4.8.7 HEPA Filter Test

- Challenge test
  - Test aerosol upstream
  - Aerosol photometer/particle counter
  - Full filter sampling
- Leak test
  - Normal or challenging condition
  - Filter frame/gasket
  - Sampling grid

4.8.8 Monitoring and Requalification

- Microbiological tests
- Particle tests
- Room pressure
- Filter Integrity
- Airflow/speed
- Filter differential pressure
- Temperature/humidity

4.8.9 Documentation

- General SOPs for operation
- SOPs for personnel and gowning
- SOPs for cleaning and sanitization
- Monitoring
- Maintenance/filter change
- Calibration

4.8.10 Gases (Steam, Compressed Gas, etc.)

4.8.10.1 Specification

- Clean steam: (condensate) according to water for injection
- Compressed air: according to pharmacopoeia if part of the drug
- In all other cases: must be defined by the user
- Compressed air: Content of water and oil

4.8.10.2 Qualification

- Installation Qualification (IQ)
  - Installed according to the approved drawings, material, (gaskets), filtration, labeling

- Operational Qualification (OQ)
    - Cleaning of the system, identification, microorganisms, quality (moisture, oil, particle, assay)
- Performance Qualification (PQ)
    - n.a.

4.8.10.3 Generation

- Delivered from external companies
- Certificate of analysis
- Self-generated
- Clean steam: generated from purified water, overheated steam, saturated steam
- Compressed gases: oil-free compression, moisture reduction
- Filtration

4.8.10.4 Monitoring

- Microbial count (excluding clean steam)
- Particle
- Oil/water content (compressed gases)
- Endotoxin (clean steam)
- Noncondensable gases (clean steam)

4.8.10.5 Documentation

- Monitoring
- Maintenance
- Calibration

4.8.10.6 Special Points

- Rust from the piping system
- Particles

4.8.11 Engineering Tasks

- Design qualification/specification of new systems
- Qualification IQ/OQ
- Maintenance
- Calibration
- Repair
- Operation of supporting systems

4.8.11.1 Internal/External Engineers

- Company engineers
    - Working according to SOPs
    - Trained in general GMP aspects
    - Part of the whole GMP system
- External engineers
    - Well-trained for the special job
    - Not trained in GMP
    - Documentation not in accordance with the GMP guidelines

4.8.11.2 General Requirements

- All engineers/technical staff have to be qualified/trained for the work, the environment, and GMP purposes.
  - Following instructions
  - Documentation requirements
  - Change management
  - Deviation management
  - Hygiene
  - Medical examination

4.8.11.3 Qualification

- The engineering department is involved/responsible for DQ, IQ, and OQ.
  - Preparing the technical specification
  - Involved in the DQ
  - Writing/discussing of protocols for IQ and OQ
  - Testing of rooms, equipment, and systems
    - According to the protocol
    - Documentation of results
  - Final reports

4.8.12 Maintenance

- Writing maintenance protocols
- Documentation of the work
- Scheduling/tracking the work
- Writing deviation reports in case of deviations with possible influence on the product quality

4.8.13 Calibration

- Calibration is testing and documentation.
- Calibration is not repairing or adjusting.
- Before repairing/adjusting, the instrument should be calibrated.

4.8.13.1 Calibration Program

- List of instruments to calibrate
  - Tolerances
  - Calibration points
  - Frequency
  - Test equipment
  - Procedure
- Documentation
- Labeling
- Tracking (next recalibration)
- Deviation management (out of tolerance/time)

4.8.14 Gases Defined

Process gases are defined as gases that can affect product quality. This chapter considers the following gases:

- Nitrogen
- Oxygen
- Argon
- Carbon dioxide
- Compressed air

4.8.14.1 Process Gases: Typical Applications

The biopharmaceutical and life science industries use compressed gases in a wide range of applications including:

- Pneumatic systems
- Chemical synthesis
- Controlled atmospheres and blanketing
- Cell cultivation
- Chilling/freezing
- Material Extraction

Nitrogen is generally utilized as an inert gas for conditioning pharmaceutical products in storage and transportation. In specific applications, argon or carbon dioxide may be used instead of nitrogen. These gases (nitrogen, argon, and carbon dioxide) are also used during the manufacture of APIs and in the production of finished pharmaceutical products. Gases such as oxygen, compressed air, and carbon dioxide are often used for the cultivation of mammalian cells and for fermentation.

4.8.14.2 Prior to the Audit

- Determine the relevant utilities to be covered.
- Determine if there are any risks of cross-contamination needing specific attention.
- Request drawings for the site's water systems, steam systems (if applicable), HVAC systems, and compressed gas distribution systems (if applicable).

4.8.14.3 During the Audit

For any utility system, ensure that:

- All piping or ductwork is correctly labeled with the utility name (clean steam, distilled water, compressed air, etc.).
- Directional arrows are in place to indicate flow.
- Valves, pressure gauges, sampling points, etc. are identified with unique IDs.
- Minimum and maximum operating rates are shown.

4.8.15 HVAC

For a walkthrough specifically focused on the HVAC system, ensure that:

- The room air returns are not blocked.
- The room air supplies are not blocked.
- The pressure differential gauges and/or manometers arc operating properly.
- Cross-contamination is minimized by not recirculating air if required, correcting pressure differences as designed.
- Ensure that each heating, air-conditioning, and ventilation system has been qualified and operates according to the necessary requirements and specifications.
- Verify that preventive maintenance is scheduled on a regular basis and that:
- Pressure differential manometers on AHUs are monitored.

- Fans and motors are checked on a scheduled basis to determine if they are operating properly.
- Ducts are checked for leaks from the external environment.
- ULPA filters are monitored through pressure differential measurements and the system is alarmed if an over-limit is obtained.
- Ensure that critical filters within the HVAC are monitored for drops in pressure.
- Ensure that the following documentation is in place:
- A complete description of installation (drawings, installed elements, description of the elements, etc.).
- An SOP that establishes action and alert levels and what actions to take when these are reached.
- An SOP defining how to operate the system and what performance factors need to be monitored.
- A maintenance SOP for the system.
- Maintenance records and the preventive maintenance schedule showing that the system has been properly maintained.
- Ensure that adequate and appropriate training for all personnel working with the HVAC is in place.
- Verify that personnel have been qualified in their jobs.
- Verify that appropriate training has been received.

4.8.16 Water

- Determine what material the water system of the plant is composed of.
- Verify that an acceptable material, 316L stainless steel or polypropylene/PTFE, is used.
- Verify that internal welds in the stainless steel have been electropolished either by inspecting the welds or inspection of records.
- Verify that if polypropylene or polytetrafluoroethylene (PTFE; e.g., Teflon) is used for purified water and systems, it has been specifically designed for this use.
- Ensure that any fittings for the purified or WFI systems are sanitary fittings with no threaded portions.
- Ensure that there is no direct connection to drains or sewers.
- Ensure that the system has non-return valves that prevent water from backing up into the system.
- Ensure that the piping system and holding tanks appear intact.
- Verify that there are no stains and leaks that could indicate problems.
- Verify that the tank is kept at an acceptable temperature and that the water is circulating.
- Ensure that the valve positions indicated on the drawing of the water system correlate with the actual positions, i.e., if the valve is marked "normally open" it should be physically open.
- Ensure that valves are stainless steel ball valves with no gaskets. Check where detachable hoses are used these are removed immediately after use to drain freely.
- Ensure that the appropriate monitoring is performed on each plant water system.
- Verify that potable water meets at least the World Health Organization (WHO) drinking water standards.
- Verify that the following tests are performed on purified water:
  - Total microbial count
  - Chemical tests
- Verify that the following tests are performed on WFI:
  - Total microbial count and typing if required
  - Chemical tests
  - Pyrogen and endotoxin testing

- Ensure that water sampling is performed by trained employees who practice good aseptic techniques when sampling.
- Verify that sample containers are clean and, for WFI, sterile.
- Verify that there is a water testing SOP that includes sampling points, frequency of sampling, testing requirements, specifications, and tests to be performed for each type of water, sampling technique to be used, and additional tests based on special products being manufactured.
- Verify that the testing laboratory is equipped to conduct microbial limits testing, total microbial counts, objectionable organism testing, and US Pharmacopoeia (USP) chemical water tests.
- Verify that water samples are held under refrigeration if not tested immediately.
- Ensure that all personnel working with a water system are trained in both their GMPs and job duties.
- Verify that they are trained to take "clean" or aseptic samples.
- Verify that they are trained in responding to system alarms.
- Ensure that there is documentation for qualification, maintenance, and monitoring.
- Verify that there is an SOP for maintenance of the water system and that it includes the frequency of:
  - Maintenance
  - Filter replacement
  - Sanitization
  - Maintenance of stills and heat exchangers
- Verify that the maintenance documentation includes records for:
  - Passivation of the system
  - Regeneration of the deionizer
  - Sanitization
  - Replacement of UV light if UV sterilization method is used
- Verify that validation records are complete and that qualification takes place over a year to determine what seasonal and environmental conditions affect the water.
- Ensure that the steam system operates correctly.
- Ensure critical operating parameters (temperature, pressure, etc.) are monitored.
-

4.8.16.1 Compressed Gases

- Ensure that the diagrams of the compressed gas system, if used at the site, are the same as observed.
- Verify that a certificate of analysis accompanies all compressed gases. If purchased from a vendor ensure that a certificate of analysis accompanies the cylinder.
- Ensure that the site's compressed gas distribution system has been qualified.
- Verify that during qualification the following were completed:
- Points of use farthest from the source were tested.
- Use of the system by multiple users at various points of use was tested.
- "Worse case" points of use where the system does not have long stretches of straight piping were tested.
- Particulate and microbial tests were conducted to ensure that the gas meets the same environmental conditions as the room or location.
- Monitoring is in place to determine if the gas is not changing the particulate level in the environment.
- Ensure that the gas is tested for particulates, identity, total hydrocarbon content, moisture content, and microbial content. Ensure that all test results are documented.

4.9 QUESTIONS TO AUDITORS

- You are auditing a pharmaceutical company for compliance with GMP.
  - What do you think will be the key qualification and validation issues to arise during an audit?
  - Suggest at least five observations.

- While in the Quality Control department, you were reviewing some of the cleanroom's statistical process control (SPC) control charts. The charts tracked critical environmental parameters in several rooms. The charts identify the upper and lower control limit points. On a number of charts, these points were exceeded. These charts displayed several extended periods of time during which the operation continued even after they exceeded the limit points. When asked about this situation the technician said it was OK to run outside the limits for some periods of time as long as things didn't get "too bad" and that "these problems usually get corrected when we do our monthly machine maintenance." Besides, the technician says, these control points are only "alarm" points, and the situation isn't considered nonconforming until the levels of particulates are much higher.

- Is this a nonconformance?

Y_______ N_________ What element?_____________________

- In the Regulatory Affairs department you were examining records for the internal audits that were performed over the last six months. The audit reports were all signed off by the Regulatory Affairs manager and there was a listing of other auditors. The audit reports identified nonconforming observations. Looking at the follow-up reports, 2 of the 22 areas audited had not provided information about their planned corrective action. The Regulatory Affairs manager said that he had talked to the area managers but had not been successful in obtaining adequate corrective action plans. The manager plans to bring this situation to the president's attention in the next management review meeting.

- Is this a nonconformance?

Y_______ N_________ What element?_____________________

- You are auditing the ampoule filling area in a sterile manufacturing facility. At a particular station, there is a poster showing various abnormalities that can occur during production. It is stamped "For Reference Only." The supervisor indicates that because it is marked as a reference only, it does not have to be in the document control system. While interviewing employees, several indicated that they regularly use the poster to identify the abnormalities they discover. They further use the poster to help determine why the abnormality occurred, and how to make the necessary adjustments to prevent recurrence.

- Is this a nonconformance?

Y_______ N_________ What element?_____________________

- Suppose you are auditing a facility that produces sterile medical products. While examining the pure water tanks during the facility tour, you notice that there is a puddle on the ground. When you ask the technical department manager about the source of the water, he says that one of the pumps may have burst. You also note that one of the three pumps leaks water, but the technical manager adds that it is beyond his knowledge and that the general manager's approval should be obtained in order to take immediate action, even if the leak originates from the pump.

- Is this a nonconformance?

Y_______ N_________ What element?______________________

REFERENCES

EMA. Guideline on the quality of water for pharmaceutical use. EMA/CHMP/CVMP/QWP/496873/2018.
EMA. ICH guideline Q9 on quality risk management. EMA/CHMP/ICH/24235/2006.
Eudralex – Volume 4 – Good Manufacturing Practice (GMP) guidelines.
Chapter 1 Pharmaceutical Quality System
Chapter 2: Personnel
Chapter 3: Premises and Equipment
Chapter 4: Documentation
Chapter 5: Production
Chapter 6: Quality Control
Chapter 7 Outsourced Activities
Chapter 8: Complaints, Quality Defects and Product Recalls
Chapter 9: Self-Inspection
Annex 1: Manufacture of Sterile Medicinal Products
Annex 2: Manufacture of Biological active substances and Medicinal Products for Human Use
Annex 11: Computerized Systems
Annex 15: Qualification and Validation
ISO 14644-1, ISO 14644-2, ISO 14644-3, ISO 14644-4, ISO 14644-5.

5 Alternative Auditing Methods

5.1 ALTERNATIVE AUDITING METHODS

5.1.1 eAudit

An eAudit is an audit that uses electronic means to remotely obtain audit evidence to determine the extent of conformity to the audit criteria.

5.1.2 Remote Audit

Remote audits leverage technology such as video teleconferencing and shared file folders to facilitate interviews and share documents and records when an on-site audit is not viable or desired.

 DOI: 10.1201/9781003284970-5

5.1.3 Virtual Audit

As defined by ISO 19011, a virtual audit is a set of audit activities in a virtual environment. A virtual environment may be composed of digital and/or nondigital activities using technological assets (software, hardware, sensors, programmable logic controllers, automated devices) taking some or all decision(s) in the process(es).

5.1.4 On-Site Audit

An on-site audit is a supplier audit that is conducted on-site and in-person at the auditee site. The audit is focused on a specific area identified by the pharmaceutical manufacturer or audit sponsor.

5.2 COVID-19 AND AUDIT CONSTRAINTS

5.2.1 Why Perform Audits?

When it is necessary to ensure that suppliers performing activities such as GxP, (i.e., cGMP, GLP, GCLP, GCP), comply with legal and international legislation, the necessary compliance audits of these companies are inevitable.

> 21 CFR Part 200.10: Contract facilities (including consulting laboratories) utilized as extramural facilities by pharmaceutical manufacturers. The Food and Drug Administration (FDA) is aware that many manufacturers of pharmaceutical products utilize extramural independent contract facilities, such as testing laboratories, contract packers or labelers, and custom grinders, and regards extramural facilities as an extension of the manufacturer's own facility.

5.2.2 Why Remote Audits?

Few might have anticipated how our lives would change in 2019 with the COVID-19 pandemic. But, the work of qualifying and monitoring GxP vendors must continue to ensure contracted work is performed in a compliant manner, as these essential companies contribute to drug development and/or provide much-needed clinical and commercial drugs to patients.

Social distancing guidelines, travel restrictions, and efforts to eliminate nonessential personnel on sites forced us to be creative and perform effective remote audits now and for the foreseeable future.

Companies must perform risk assessments of their suppliers to determine which remote audits should be performed as a priority. If possible, audits of critical suppliers should not be postponed.

Qualification and monitoring audits must stay on track!

5.3 REMOTE AUDIT CONDUCT

Good planning is the key to an effective audit. Is the supplier willing, able, and ready to host a remote audit? Do they have the IT infrastructure/capability to share a large number of documents electronically in a secure manner?

A detailed audit letter with an agenda must include the scope of the audit, what will be covered specifically and at what times. A questionnaire and detailed list of documents that you want to review (including SOPs) should be shared at least two to three weeks prior to the audit date(s).

Do your homework! You want to find out as much as possible about the company you will be auditing before the audit date. If it will be a repeat/surveillance audit, review past reports so you can zero in on previous observations and confirm they have been addressed and fully resolved.

TABLE 5.1
Audit Agenda – Overseas Remote Audit

Time	Schedule
Day 1	
06:00 A.M. EDT, 1:00 P.M. (GMT +3)	Opening meeting • Introductions; review agenda and audit objectives • Organizational overview • Brief company/facility presentation • Virtual tour of the warehouse, GMP suites, and QC and Microbiology • Media fill SOP and record review and discussions
06:30–09:30 A.M. EDT, 1:30–4:30 P.M. (GMT+3)	• Document review, SMEs discussions on training, quality systems, and equipment • Site master plan • Validation master plan • Quality systems SOPs and quality metrics • Management review • Deviations, OOSs, CAPAs • Change control • Equipment records and SOPs • Training and qualification of staff including operators and visual inspection • Environmental monitoring and trend reports • Computerized systems
09:30 A.M. EDT, 4:30 P.M. (GMT+3)	• Brief check-in discussion and planning of Day 2
09:45 A.M.–3:30 P.M. EDT	• Document review
Day 2	
06:00 A.M. EDT, 1:00 P.M. (GMT+3)	• Brief call to discuss comments and questions from Day 1, audit plan for Day 2
06:15–07:00 A.M. EDT, 1:15–2:00 P.M. (GMT+3)	• Quality agreement review and discussion
07:00–09:30 A.M. EDT, 2:00–4:30 P.M. (GMT+3)	• Document review, SME discussions • Review of laboratory controls • Material and product controls • Internal and external audits and supplier controls
09:30–09:45 A.M. EDT, 4:30–4:45 P.M. (GMT+3)	• Auditor review of notes and findings
09:45–10:00 A.M. EDT, 4:45–5:00 P.M. (GMT+3)	• Audit close-out meeting

What should your letter, agenda, and document request list look like? Table 5.1 depicts an example remote audit agenda.

5.3.1 Audit Letter Excerpt

The purpose of this audit is to assess the cGMP compliance (21 CFR 210 and 211) of your quality systems, production controls, facilities, equipment, organization, personnel, and technical capabilities related to the aseptic filling/manufacturing, visual inspection, packaging, release testing, and stability storage and testing of phase 1 clinical trial material batches of ABC drug product. The scope of this audit will include an assessment of the overall compliance with the current good manufacturing practices and related documentation.

In order to prepare for an efficient audit, please forward any pre-audit preparation package that may exist for the facility. As permitted by your procedures, please provide a copy of the following by XX/XXXX:

1. Organizational chart
2. Floor plan
3. List of licenses and certifications
4. Controlled document index
5. History of regulatory inspections (5 years) including a summary of findings
6. Quality manual

5.3.2 A Detailed List of Documents to Review

Tip 1: Request a standard operation procedures (SOPs) index prior to the audit and provide the list of SOPs you want to review during the audit so they are waiting for you when you start!

Tip 2: Be as specific and comprehensive as possible when generating the list of documents you want to review. Although you will have some time to request other documents during the audit, it will make for a more efficient and smooth audit if you can anticipate and request all documents for review *before* the audit.

5.3.2.1 Detailed List of Documents to Review – Part 1

1. Organization chart
2. Layout
3. SOP list
4. Quality manual
5. Validation master plan
6. Site master file
7. Equipment list
8. Risk assessment/report for the introduction of new products into your facility
9. List of computerized systems and each system's function and validation (21 CFR part 11) status
10. Equipment validation/qualification SOPs
11. Regulatory inspection history summary from 2020 to the present including inspection reports, observations, and responses
12. SOPs for sterility testing, endotoxin, and bioburden
13. Deviation and CAPA
14. OOS/OOT
15. Change control
16. Supplier qualification
17. Approved supplier list
18. Data integrity
19. Disaster recovery and business continuity SOPs/plans for supply chain, personnel, etc.
20. Aseptic process validation (media fill) SOP
21. Gowning qualification SOP and qualification documentation for a current technician/operator
22. Media fill schedule that contains vials filled, vial, stopper, and seal type and results
23. Internal audit SOP and proof of completion for audits performed in 2021–2022
24. Training SOP, training requirements, and training files for at least three employees including one in QA, QC, and Manufacturing

25. Visual inspection
26. Qualification of visual inspector's SOP and representative qualification documentation for a current inspector
27. Environmental monitoring and WFI trend reports for 2021 and any 2022 reports completed to date
28. Complaints
29. Recalls
30. Batch record review

5.3.3 Virtual Tour

- Nice to have but certainly not all companies are willing and able to provide a virtual tour. Some companies have uploaded facility walkthroughs to their website and these are readily accessible.

5.3.4 Before the Audit Commences

- Double-check to ensure you have all the nondisclosure/confidentiality documentation executed.
- Make sure you can access the document "share site" prior to the audit if possible to ensure it is accessible and functioning properly. Check with the auditees to determine how long you will have access to the documents.
- Make sure you have downloaded whatever cloud platform will be used during the audit: Zoom, WebEx, or Microsoft Teams. Have a list of questions that you can discuss if necessary after the opening presentation.
- Create a detailed game plan for document review beforehand.
- At the beginning of the audit, provide some information about yourself during the introductions.
- State the purpose and scope of the audit.
- Create a communication plan for the audit: Who can you email or call if you have questions?
- Set up call times to check in (at least twice during one audit day).
- Set up the audit wrap-up meeting date and time so that the auditees can have all the necessary personnel present for the close-out.
- Request a copy of their opening presentation for your records.

5.3.5 During the Audit

Stick to the agenda but remain flexible to accommodate discussions when SMEs (subject matter experts) are available.

I recommend you start with regulatory inspections (483s, EIRs, and company responses to inspections) to gauge the compliance track record for a company. Go back at least three to five years to ensure that there are no repeat 483 observations. Keep detailed notes and start a table of documents reviewed during the audit that will become an attachment to the audit report (write as much during the audit to save time with the report).

Be transparent. Don't wait until the close-out meeting to share observations. You want to be sure you understand any deficiencies that will require discussions with the auditee, allowing them to correct misimpressions if needed.

5.3.6 Audit Close-Out Meeting

At an audit close-out meeting is a must to present:

- What systems and documents were covered (high-level)
- What observations you've identified (don't classify them until the report; more on this later)
- Thank the auditees for their time and for hosting the audit. Communicate when the audit report or list of observations will be presented to them (determine who will be the respondent) and request that they respond in writing within 30 calendar days.

5.3.7 Audit Reporting

The deliverable for the audit – the audit report – should look the same as if you conducted an on-site audit.

The audit report or list of observations should be presented as a signed PDF report within 30 calendar days.

Any *critical* observations should be immediately communicated.

Send a thank-you email the next day to thank the audit team for hosting the audit.

5.3.8 Challenges to Consider

- *Time zone differences* – You need to be in their time zone as much as possible.
- *No in-person communication* – Use video sharing as much as possible for check-ins in addition to opening and closing meetings.
- *Time* – There never seems to be enough of it during any audit. Your audit review game plan must be generated before the audit. Efficiency and prep work are key.
- *Auditor burnout* – There are no lunches or tours to break up the audit day(s). If possible, take frequent short breaks, perform as much of the audit standing, and take a walk when you normally would be having lunch at the audit site.
- *Communication* – Use screen sharing to review a document with the auditees if needed as opposed to playing email ping-pong.

5.3.9 Final Thoughts

Although remote audits cannot replace on-site, in-person audits, they are an effective means of performing a compliance assessment of a company. The IT infrastructure of the company being audited is of critical importance in document review. Patience and professionalism are important. Take your time while speaking on Zoom calls and allow each participant to speak and be heard. I think that the ability to perform remote audits as "the new norm" won't go away even post-pandemic, as they are quite cost-effective with no travel time or expenses.

5.3.10 Pros and Cons of Remote Audits

Remote auditing – desperate times call for desperate measures.

The conversation about new audit approaches, the use of technology, and remote auditing techniques has gone on for several years. However, in various polls of companies conducted over the past two years, few firms report implementing remote auditing beyond the test scale.

The emergence of COVID-19 and its related worldwide travel restrictions, combined with the need to conduct regulatory-, legal-, or immediate need-driven audits, are renewing the conversation and focusing internal audit efforts on finding alternatives to the traditional face-to-face audit that can be implemented now. Remote auditing may be a good go-to alternative, especially because most companies have restricted travel to business-critical functions, and many jurisdictions in the world temporarily closed borders during the pandemic.

The example presented below identifies all kinds of obstacles and difficulties encountered in remote auditing and provides tips on how to overcome difficulties at each stage, including conducting audits from the planning stage, opening and closing speeches, including document checks, facility tours, and interviews with employees. In fact, application examples are included in remote audits in the COVID-19 environment.

5.3.11 Considerations for Remote Auditing

As readers will note, remote audits are similar in structure to face-to-face or on-site audits. The differences are the audit stages and the workload within each stage. This section highlights things to consider, based on the lessons learned by three contributing companies that utilized remote audits as part of their response to COVID-19.

5.3.11.1 Planning

Audit planning and scoping are critical in every audit. Because it is more difficult to pivot at the moment from a remote location, the stakeholder planning meeting is very important. This meeting includes a discussion of the scope and schedule, and auditors must allow adequate time to explain the remote audit approach to the participants. This should include an explanation of anticipated similarities and differences between face-to-face audits that stakeholders are accustomed to and remote audits.

Examples of the types of information that should be covered include how and when the information will be shared, what technology will be used (everything from cameras to drones to telepresence support), what authorizations need to be obtained in advance to collect video and still photography, and what confidential or restricted areas need to be considered or avoided.

It is also important to discuss from the outset the limitations of remote auditing and explain that future on-site work may be required based on the findings of the remote audit or once the barriers to a physical site visit are removed (e.g., travel bans are lifted).

Based on the additional content and explanation, experience indicates that internal auditors should allocate about twice as much time for a remote audit planning meeting compared to what is needed for a traditional audit. The use of video teleconferencing, as well as PowerPoint or other visual guides, is also helpful for this meeting.

5.3.11.2 Document Review

Remote document reviews are in many ways analogous to reviewing documents at the facility, with a couple of major caveats.

It may take significantly more time for the facility to prepare and upload documents to a file-sharing platform (SharePoint, shared drives, etc.) than it would to provide access to a file drawer or binder on-site. Depending on the documentation method (paper records, database storage system, etc.), the facility personnel will need to take the time to convert these records into a reviewable format (such as PDF) and upload the files.

Auditors should be open to receiving and reviewing information in whatever format is most easily obtained so that the burden on the facility is minimized. If possible, consideration should be given to the accessibility of the digital file system used by the facility where records are stored. Often, direct access can be temporarily granted, solely for the duration of the audit.

Careful consideration should be given to building effective strategies for reviewing data remotely. For example, depending on the number of records to be reviewed, sampling may be the best option. Whether reviewing all or part of the available data, the strategy should be discussed with auditees to ensure they provide the correct information to support the sampling strategy.

As it is known, in face-to-face audits, the desired question is asked by the GMP auditor during the examination of the records and answers can be received simultaneously. However, in remote audits, questions regarding records may not always be answered simultaneously. In such cases, the

GMP auditor should quickly take notes and be able to derive questions regarding these notes. If the issue of simultaneous questioning and asking questions is the main agenda of the GMP auditor, a solution such as video conferencing should be discussed. Thus, the questions will be asked and the other party will be able to respond by presenting evidence simultaneously.

5.3.11.3 Site Reconnaissance

The remote site reconnaissance is perhaps the most challenging aspect of the remote audit. The companies that contributed to this knowledge brief have tried various technology solutions to accomplish an adequate site reconnaissance.

One approach was to use live, two-way communication technology, including live streaming and two-way smart glasses. Although the idea of a live walk-through was compelling, the following limitations were encountered:

- Most facilities did not have facility-wide Wi-Fi. Even those facilities that claimed to have facility-wide Wi-Fi often did not have good coverage of remote areas of the site (such as tank farms, docks, large warehouses, and storage areas).
- Covered facilities also happened to be in remote locations or were within old buildings constructed like bunkers where cellular service was limited or the signal strength was poor, degrading the quality of the live video feed.
- There was little value in remote observations traveling from point to point, as the video feed provided tunnel vision and peripheral observations were not possible.
- Ambient noise and (conversely) the noise-canceling properties of the technology prevented remote personnel from hearing interview responses.

An alternative to live, two-way communication is video and still digital photographs taken using company cellular phones. The advantage of this approach is that these devices are readily available and compatible with most company computer storage solutions (e.g., SharePoint). Additionally, signal strength at the point of collection does not negatively impact video and photograph quality on such devices. As mentioned earlier, the use of digital devices often requires preplanning and special permissions (such as in sensitive areas or in environments where an explosion hazard exists).

During the document review, auditors should compile a list of areas of concern that they want to capture with video and still photographs. Examples of areas of focus for chemical security audits include:

- Active facility access points
- Secure and restricted areas
- Videos of vehicle and personnel screening
- Videos of CCTV monitoring stations, including the use of cameras

The photos and videos are reviewed by the audit manager and compiled into an album, which is then reviewed by each auditor. Auditors take notes and prepare questions to be asked during the remote interviews.

5.3.11.4 Remote Interviews

Remote interviews are conducted in much the same way as in-person interviews and can be conducted by scheduling video calls with key individuals using any number of readily available technologies (e.g., Microsoft Teams, Skype, and Zoom). Plan on interviews of 30 to 90 minutes with the facility program owners. Shorter, 15-minute interviews can be conducted with designated personnel who have implementation responsibilities, and short (10-minute) interviews can be conducted with selected general facility employees who have responsibilities ancillary to the focus of the audit. These help the internal auditor gain knowledge of the general culture. Video calls are preferred over voice-only calls because nonverbal cues are an important part of communication and are often lost without video.

Preparing for remote interviews takes additional time for the auditor. Each auditor should come ready with a list of questions and points regarding what additional information is needed, based on information from the document review.

When more than one auditor is participating in an interview, care must be taken to avoid talking over either the interviewee or other auditors. Keep in mind that many people may not be comfortable chatting by video, especially auditees who do not regularly do so (for example, a skilled-trade supervisor). While this is unavoidable, try to set a comfortable tone and be aware that the video alone may change body language or perception.

5.3.12 Closing Meeting

The closing meeting for a remote audit is much the same as the closing meeting for an in-person audit. It is suggested to schedule the closing meeting one to two days following the remote interviews. This allows the audit team members to review their own notes and findings, as well as have a remote audit team meeting to compile preliminary draft audit results.

The audit closing meeting is an opportunity to present these preliminary draft audit results to the stakeholder audience, resolve any questions or concerns, and discuss the path forward to audit result finalization and continuous improvement.

5.3.13 Advantages and Disadvantages of Remote Auditing

The following is a summary of the advantages and limitations discovered as the three companies implemented remote audits.

5.3.13.1 Advantages of Remote Audits

A short list of positives discovered while implementing remote audits includes:

- Restoration of a much-needed sense of normalcy.
- Reduced travel costs. For an audit program with multiple annual audits, remote audits can provide significant savings.
- An expanded pool of available auditors. The retiree in Tennessee, the part-timer in Colorado, or even the auditor in Van Nuys (Los Angeles, California) can expand your audit staff.
- Expanded coverage. Remote audits allow for more coverage when competing priorities of volume and time limitations occur.
- Expanded use of specialists. Specialists can connect remotely for selected interviews or parts of audit planning, and they need not be present for a full audit.
- Improved document reviews. Remote reviews of plans and documentation, at the auditor's own pace, contribute to a higher quality review and a deeper dive into the documentation.
- Improved use of available technology strengthens documentation and reporting. Facility personnel's use of technology to capture video and photographic information contributes to improving their understanding and use of available technology. This contributes to better documentation of facility conditions, an improved ability to report incidents and conditions to remote corporate personnel, and increased opportunities for future remote training tools.
- The daily workload on the audited facility can be reduced. The time and effort spent during the preparation and digitization of both documents and visuals such as videos, pictures, figures, diagrams, etc. can be reduced by spreading them over time, unlike the workload encountered in on-site audits.

- Improved organization and confirmation of required documentation. Because facility personnel have to review and assemble the required documents, remote auditing provides an opportunity to organize and confirm that all documentation required for a regulatory inspection is readily available.

5.3.13.2 Limitations of Remote Audits

Some limitations of the remote audit approach as executed by the companies include:

- First-hand observations cannot be replaced. There is nothing like seeing processes first-hand, observing body language, or even noticing a smell that should not be there. Online auditing also has limits under certain conditions, such as where operations are secure, highly restricted, or in sensitive environments. You cannot walk through a cleanroom with a video device, nor can you take it into many chemical plants or refineries.
- Remote auditing makes it hard to build rapport with auditees. Opportunities to provide hints, tips, and observations for improvement are lost. It is hard to identify best practices or describe things that others may benefit from, outside of the documentation process. Good auditors do this, and these are often the most useful things that auditees get from the exercise.
- The lack of in-person interaction opens opportunities for fraud**.** The opportunity to present doctored documents and to omit relevant information is increased. This may call for additional planning, some additional/different audit procedures, or a follow-up once the barriers to a traditional audit are lifted.

Table 5.2 presents the remote audit risk assessment.

5.3.14 How Should I Measure Audit Effectiveness?

- Are auditors able to effectively utilize technology to audit?
- Do your audit results align with stakeholder expectations for the use of technology?
- Do you get the same responsiveness from leadership even though you (the auditor) may not be physically on-site?
- Do savings in travel balance against any extra time required in planning/conducting virtual audits? How much travel will occur post-COVID-19?
- Are auditees and auditors satisfied with the use of technology?

5.3.15 Pre-Audit Questionnaire

When conducting a remote audit, preparation and the ability to review documents before visual and/or audio interviews are essential to the success and timeliness of the audit. Some examples of primary components of a pre-audit survey or questionnaire include:

General company information:

- Address, key point contacts, and geographic awareness.
- List of materials or products produced at the audit site.
- Are any materials or processes outsourced to another organization?
- Are there any documents that cannot or will not be shared during the audit?
- Will all documents reviewed be in English or the language of origin?
- Facility map or layout, including room classification.

Personnel and training information:

- Does the audit site have training manuals and training programs in place for all employees?
- Are training records readily available?

TABLE 5.2
Remote Audit Risk Assessment

Potential Failure Mode	Potential Effect(s) of Failure	Potential Cause(s) of Mechanisms of Failure	Recommended Action
Broadband performance	Slow unable to support VoIP, video; visual review of the manufacturing floor will be impacted	High-speed internet not used or intermittent service	Use phone lines Only have necessary cameras on when necessary
Collaborative program	Room is hacked; uninvited persons can view information during the audit The room does not contain the necessary tools to observe evidence	• Collaborative conference room is not encrypted • Security protocols are not supported – shareware • Password protections not used • No internet security, firewall software installed or updates not up to date	Define the tool set required for auditing Use an encrypted collaborative program Use password protection Keep the internet security firewall updated
Technical difficulties	Poor visibility of objective evidence and poor sound quality	Bandwidth issues, the configuration of (aged) computers and mobile devices	Technology checks should be used to determine Wi-Fi weakness, software and equipment performance prior to the audit
Digital integration	Not all hardware interfaces have the collaborative program or software	• Dated equipment • Equipment does not interface with Apple • Mobile device does not connect to the collaborative program	Technology checks should be used to ensure hardware works with current software programs

- Does the supplier have established communicable disease processes for employee health screening and workplace exclusion where necessary?
- Gain an understanding of the breakdown of employees (regular vs. temps/contractors) and whether the same curriculum would apply to an employee regardless of their employment status (e.g., regular vs. temp/contractor).

Product-specific production:

- How do you define an individual batch?
- Is this product manufactured by individual batches or is manufacturing continuous?
- Is the product manufactured in dedicated equipment?
- Is the original manufacturer's name printed on each packaging unit?
- A schedule of product launches from the site.
- A general process flow of the production process in order to understand what activities occur at the site.

Material source – product-specific:

- Are only nonbiological origin materials used as raw materials during the manufacturing process?
- Is the product or any of its raw materials derived from GMOs (genetically modified organisms)?

Residual solvents and elemental impurities:

- Are solvents excluded from or consistently removed during the manufacturing process?

Cross-contamination control:

- Which cross-contamination controls are in place?

- Are production materials handled and stored in a manner to prevent degradation, contamination, and cross-contamination?
- What controls have been established to prevent cross-contamination?

Regulatory:

- List of recent regulatory inspections at the site. If there were observations associated with the inspection, what is the current status of the corrective actions, and are they completed as per the commitment?
- Please list regulatory correspondence impacting the site within the last five years (i.e., warning letters, CEP suspension, import alerts, etc.).
- Is the site registered with any government regulatory agency (FDA registration, GMP certification, etc.)?

Computerized systems:

- Does the facility utilize computerized systems for managing GxP activities or data?
- Do computer systems that could impact quality have controls for operation, maintenance, and prevention of unauthorized access?
- Do you document changes and deviations of validated systems?

Quality:

- What quality management system is utilized on-site?
- Which regulatory initiatives does the site follow/comply with?
- Provide an organizational chart. Is the quality department independent?
- A copy of the quality manual and organizational charts are required in order to understand the level of quality maturity and relative size of organization/reporting structure conflicts.

Quality and laboratory control:

- Which written information about laboratory instruments is/are available for review?
- What testing is performed on-site? Were any external labs used for testing?

Sampling:

- Are samples stored, shipped, and handled the same as the provided product?
- Are samples pulled in a statistically significant method to provide an adequate representation of a product batch?
- Are there any environmentally controlled requirements for sample management?

Materials management:

- Are returnable shipping containers used for the finished product?
- Do you have written procedures that describe finished goods storage?

5.4 QUESTIONS TO AUDITORS

- Do you think that remote audits will be replaced with on-site audits sooner or later?
- If you cannot see the pre-uploaded documents necessary on a shared platform between the auditor and auditee site to perform a robust remote audit, what would your first reaction be?
- Let's assume that the auditee site does not accept a real-time facility tour via a smartphone or a tablet and recommends pre-recorded videos and pictures of the site. What would your reaction be?
- Imagine that you will be auditing a distribution center in a developing country where the internet infrastructure and network do not work properly to connect to the site. What would your decision be?
- Take a closer look at the following list that shows potential expenses and lost income. Can you think of other losses for your situation?

5.4.1 Lost Income

- Travel to airport – 1 hour
- Wait at airport – 2 hours
- Flying time – 5 hours
- Travel to rental car – 1 hour
- Travel to client site – 1 hour

5.4.2 Expenses

- $600 airfare (5-hour flight)
- $200 hotel
- $150 rental car
- $100 food

BIBLIOGRAPHY

J.P. Russell, Shauna Wilson, *eAuditing Fundamentals Virtual Communication and Remote Auditing* ASQ Quality Press 2013

Julie Barnhill, *Ph.D. Remote Compliance Auditing During the Covid-19 Pandemic*, Meridian BioGroup LLC

Rx-360 Remote Audit Best Practice Guide Version 1.0, Published August 2020

6 Audit Question Examples

6.1 CHECKLISTS FOR VARIOUS AUDITS

Table 6.1 presents a simple data integrity audit checklist.

Table 6.2 presents a water systems audit checklist.

Table 6.3 presents a brief list of questions for the qualification of water systems.

Table 6.4 presents an environmental monitoring checklist.

The purpose of Table 6.5 is to help determine if a computer system complies with the FDA Rule 21 CFR 21 Part 11 for electronic records and electronic signatures. This audit questionnaire applies to systems that meet the definition of a "closed" system as defined in Section 11.3 (b)(4) of the rule and that do not utilize biometrics identification methods. Table 6.6 presents an EU, PIC/S GMP Annex 11 audit checklist.

6.1.1 Biopharmaceutical Manufacturing Checklist

6.1.1.1 MCB and MWCB Storage Conditions

For the Master Cell Bank (MCB) and Manufacturer's Working Cell Bank (MWCB):

- Are the storage conditions (for the MCB and MWCB) clearly defined?
- If storage conditions are clearly defined, can you ensure the storage conditions have been maintained?
- If the storage conditions specify temperature limitations (e.g., DO NOT FREEZE), can you verify there is documentation to ensure the recording of temperature excursions?
- Is there a working alarm system to alert one if the temperature excursion is outside the specified limits?

6.1.1.2 MCB and MWCB Identification Procedures

For the Master Cell Bank (MCB) and Manufacturer's Working Cell Bank (MWCB):

- Has the MWCB been characterized and met the written specifications before use?
- If not, was an OOS investigation satisfactorily completed – if MWCB was released for use – that shows the MWCB would have met specifications?

DOI: 10.1201/9781003284970-6

TABLE 6.1
Data Integrity Checklist

	Question	Comments	Notes
Attributable Paper	• Does your company maintain a signature log for employees who work in GxP areas? • Is staff trained in good documentation practices outlining that GxP records must be initialed and dated? • Is the use of scribes prevalent in your company? • Are digital images of a person's handwritten signature permitted at your company?		☐ Fulfilled ☐ Control needed ☐ Action needed [room for notes]
Attributable Electronic	• Does the system use unique user logins with electronic signatures? • Are there audit trials in place recording the identity of operators entering, changing, confirming, or deleting data? • Does the system identify and record the person releasing or certifying the batches? Is an electronic signature used? • Is staff trained on the fundamentals of data integrity, which emphasizes never disclosing their username or passwords to other staff?		☐ Fulfilled ☐ Control needed ☐ Action needed [room for notes]
Legible Paper	• Are controls in place to ensure data is recorded using permanent, indelible ink? • Is the use of correction fluid, pencils, and erasures prohibited? • Is there controlled issuance of bound, paginated notebooks for GMP activities? • Are archiving of paper records performed by an independent, designated archivist? • Are operators trained to use single-line cross-outs accompanied by an initial and date when recording changes to a record?		☐ Fulfilled ☐ Control needed ☐ Action needed [room for notes]
Legible Electronic	• Is your stored data checked periodically for readability? • Are audit trails convertible to a generally intelligible form? • Can general users switch off the audit trail? • Is archived data checked periodically for readability? • Is data backed up in a manner permitting the reconstruction of an activity?		☐ Fulfilled ☐ Control needed ☐ Action needed [room for notes]
Contemporaneous Paper	• Is staff trained in good documentation practices emphasizing the importance of recording data entries at the time of activity? • Is staff trained in good documentation practices emphasizing that it is improper to backdate or forward date a record?		☐ Fulfilled ☐ Control needed ☐ Action needed [room for notes]
Contemporaneous Electronic	• Does your system automatically generate a timestamp when data is entered? • Do electronic signatures contain an automatically generated timestamp? • Are users able to change the timestamps applied to records? • Are general users able to gain access and change the system clock or time zone settings? • Is data saved to unauthorized storage locations such as USB sticks? • Is there sufficient availability of user terminals at the location where a GxP activity takes place?		☐ Fulfilled ☐ Control needed ☐ Action needed [room for notes]

(Continued)

TABLE 6.1 (CONTINUED)
Data Integrity Checklist

	Question	Comments	Notes
Original Paper	• Are sticky notes or other unofficial notepads permitted in GMP areas of the facility? • Are qualification/validation activities performed on original preapproved protocols? • Is there a controlled and secure area for archiving records? • Are original records readily available for inspection?		☐ Fulfilled ☐ Control needed ☐ Action needed [room for notes]
Original Electronic	• Is it possible to print out batch release records, showing any data that has been changed since the original entry? • Are your electronic signatures permanently linked to their respective record? • Does the person processing the data have the ability to influence what data is reported or how it is presented? • Does the system prevent the deletion of original data? • Is it possible to take screenshots and use snipping tools to manipulate data? • Is metadata periodically reviewed?		☐ Fulfilled ☐ Control needed ☐ Action needed [room for notes]
Accurate Paper	• Are forms, logbooks, and notebooks formatted to easily allow for the entry of correct data? • Are procedures in place to independently review original paper records? • Are deviations and out-of-specification results investigated? • Are laboratory instruments calibrated and maintained? • Are secondary checks performed to check the accuracy of critical data? • Is staff pressured into meeting production targets, leading to compromised accuracy of records?		☐ Fulfilled ☐ Control needed ☐ Action needed [room for notes]
Accurate Electronic	• Do interfaces contain built-in checks for the correct and secure entry and processing of data? • Does your system perform a check on the accuracy of critical data and configurations? • Are systems periodically reviewed? • Are interfaces validated to demonstrate security and no corruption of data? • Is archived data protected against the unauthorized amendment?		☐ Fulfilled ☐ Control needed ☐ Action needed [room for notes]
+ Complete	• Is any data left out or deleted? All data should be present including repeat/reanalysis.		☐ Fulfilled ☐ Control needed ☐ Action needed [room for notes]
+ Consistent	• Is the chronology of events reproducible? Data, metadata, and sequential timestamps must be consistent.		☐ Fulfilled ☐ Control needed ☐ Action needed [room for notes]
+ Enduring	• Are data recorded in a durable medium and readable for as long as it is retained? Unacceptable are pencil/Post-it notes, singatures.		☐ Fulfilled ☐ Control needed ☐ Action needed [room for notes]
+ Available	• Are data available/retrievable for the whole data life cycle (including storage backup/restore, archiving, business continuity, audits, inspections, reviews, etc.)?		☐ Fulfilled ☐ Control needed ☐ Action needed [room for notes]

TABLE 6.2
Water Systems Audit Checklist

Section	Question	Answer Yes ☐ No ☐ N/A ☐	Comments
Feed Water	What is the source of the plant water (city or private supply; well groundwater or surface water)?		
	Is source water entering the facility, whether from a municipal supply or a private well, tested for microbiological contamination, and what are the specifications?		
	Is coliform bacteria testing performed according to 40 CFR 141.14 and 141.21?		
	Is source water entering the facility tested for chemical contamination, or are municipality reports provided?		
	Does the water meet the EPA specifications for potable water?		
	What is the frequency of testing?		
	If the results are provided by the water supplier, are they reviewed and approved by a suitably qualified user?		
	What actions are taken if specifications are not met?		
	If the water is being tested by the user, is a written SOP describing sampling and testing available?		
General Water System Requirements	What water purification system is used?		
	Is the system a one-way flow (i.e., not recirculating)? These are more problematic and difficult to maintain as they are basically a "dead leg."		
	What is the velocity of the circulating system (6 fps recommended)?		
	What is the temperature range of the water in a recirculating system (normally 65°C to 80°C), and how is this monitored?		
	What is the temperature of the water in the tank?		
	Is the tank storage time less than or equal to 24 hours for room-temperature batch processing?		
	Is there a continuous temperature recorder and controller for each storage tank?		
	How often is the tank vent filter integrity tested and checked for condensate blockage?		
	Is the vent filter a hydrophobic 0.2 µm?		
	What type of tubing is used? (316L stainless steel piping is common in WFI and purified water. Some plastics, e.g., PVDF, can also be used but must be checked for compatibility.)		
	Is there acceptable documentation/video of the stainless steel welding and unique identification of each welded joint?		
	Is there a copy of the welder's certification and the welding procedure on file?		
	Are there any dead legs or potential areas where air can become trapped or where water can stagnate?		
	Does the system contain screw-threaded fittings instead of the required sanitary fittings?		
	Does the system contain ball valves or other fittings that can possibly retain water from the main system and cause potential microbial problems?		

(*Continued*)

TABLE 6.2 (CONTINUED)
Water Systems Audit Checklist

Section	Question	Answer Yes ☐ No ☐ N/A ☐	Comments
	Does the heat exchanger, with the exception of a double concentric tube or double tubesheet, have greater pressure on the water system side than the coolant?		
	Do any of the use points in the system have 0.2 μm filters (FDA prefers no in-line filters)? Actual water microbial counts may be "masked" by filters.		
	Review water sampling results for at least six months preceding and two months following the manufacture of lots selected during the audit. If specifications were not met, review investigation and corrective actions. Are the results, investigations, and corrective actions acceptable?		
	Are the sampling locations and frequency of testing suitable for the system?		
	Who performs sampling for the chemical and microbial samples?		
	Check that the sampling personnel is adequately trained and the training is appropriately documented.		
	Are chemical and microbial results being trended? Are the results acceptable?		
	Have changes been made to the water system since the last audit? Since the system was initially qualified? If major system changes have occurred, have the changes been evaluated for the need for requalification?		
	Are any pumps only used periodically? These can be a source of bacterial contamination from stagnant water.		
	Are thorough and complete records of the system cleaning, passivation, and maintenance maintained? The records should include who performed and supervised the cleaning, the date, cleaning agents used, pH, conductivity, and microbial results.		
	What chemical sanitization is used and is it adequate?		
	How often are the reverse osmosis membrane seals checked for integrity to prevent bypass and contamination?		
	Is there a maintenance program for reverse osmosis to prevent membrane fouling and integrity failure?		
	If resins are regenerated on-site, what SOP is followed?		
	Are the ion exchange beds tested microbiologically and chemically after regeneration?		
	What is the quality of the regenerant chemicals? Are they released by QC before use?		
	Is regeneration of the cation/anion/mixed-bed resins documented?		
	If the resins are regenerated off-site, does the manufacturer have written certification that the units have been used for the treatment of water in only systems such as this?		
	Are cation/anion/mixed-bed resins regenerated, or are virgin resins always used?		
	Has the correlation between in-line and laboratory testing for conductivity and TOC been established?		

(Continued)

TABLE 6.2 (CONTINUED)
Water Systems Audit Checklist

Section	Question	Answer Yes ☐ No ☐ N/A ☐	Comments
	What is the minimum output of the UV light in the system before replacement is required? (Normally 40%–50% of original.)		
	What wavelength of light is used (254 nm; germicidal, 185 nm; TOC reduction)?		
	Is there a maintenance program for the UV lamp, especially cleaning the lens to maintain effectiveness?		
	Are the established specifications and corresponding action and alert levels suitable and based on historical/statistical data?		
	Have any vendors supplying products for the system been qualified?		
	Are any of the components controlled/monitored by a computer or programmed logic controller? If so, has the system been validated?		
Water for Injection (WFI)	Is the water prepared by distillation or reverse osmosis?		
	Is clean steam (free of additives) used to generate the WFI?		
	How is the feed water treated prior to the distillation? Is the source water chlorinated, carbon treated, or deionized?		
	Is endotoxin testing performed on the tank, pre-/post-final treatment step, and points of use?		
	What is the level of endotoxins from the feed water, and is it appropriate to feed the still or reverse osmosis? (Stills normally only affect a 2.5 to 3 log reduction in endotoxin content.)		
	Is the system in a state of control and producing water of a quality suitable for its intended use?		

- If not, what product lots were affected by this MWCB? What is the fate of those product lots?
- Can you confirm the records to show the origin of history, number of passages, and other pertinent features of the MCB and MWCB characterization/identification?

6.1.1.3 MCB and MWCB Handling

- Can you confirm the records for inventory and handling of the MWCB to ensure they are adequate to protect the integrity of the cells?
- Can the firm produce records to show which MWCB was used to initiate a particular production batch?
- If the firm has generated a new MCB from an MWCB, is this procedure allowed under the approved BLA or supplements?
- If so, does the firm have records showing adequate characterization of the second-generation MCB?

6.1.1.4 Fermentation/Bioreactors

- Fermentation may include early grow-up phases in open containers under laminar flow hoods, while later stages of production are performed in closed vessels. Ensure that procedures are in place for the proper characterization testing of the MWCB prior to use in production.

TABLE 6.3
Brief List of Questions of Qualification of Water Systems

	Question	Comments	Notes
Installation Qualification	Approved protocol? Meet acceptance criteria?		☐ Fulfilled ☐ Control needed ☐ Action needed [room for notes]
Operational Qualification	Approved protocol? Meet acceptance criteria including daily sampling after each step of the purification process at each point of use for two to four weeks? Are the operational SOPs approved? Repeat testing as above.		☐ Fulfilled ☐ Control needed ☐ Action needed [room for notes]
Performance Qualification	Approved protocol? Meet acceptance criteria including WFI sampling one point of use each day with all points tested weekly? Complete a year of testing and meet specifications?		☐ Fulfilled ☐ Control needed ☐ Action needed [room for notes]

- Are written procedures in place for the handling and release of grow-up material for further production use?
- Do the procedures include microbial contamination ID testing?
- Are there written procedures in place for the transfer of grow-up phase material into the production vessel?
- Written procedures and ways of how the firm identifies this.
- For lots used outside of the prescribed procedures, how are these aberrations handled?
- What key parameters are monitored during fermentation? Are these all identified in the batch records?
- How is growth monitored?
- For batch-fed reactions that may involve induction, what concentration of material is used and how is this level determined? Is the level validated for the ranges noted in the batch records?
- How is the fermentation reaction determined to be concluded? Are these data chronicled in the batch records?
- Are the fermentation parameters currently used the same as what is described in the application?
- If the changes are not detailed in any update to the application, note when the changes were put into effect and what lots are impacted.

6.1.1.5 Disruption and Harvest

- Disruption may be by mechanical, chemical, or enzymatic methods. Determine the method and the parameters. Were the parameters optimized and validated for the operations?
- If a key calculation is required to determine precise amounts of the enzyme needed for disruption, is this critical step double-check/double signature in the batch records?
- Are the disruption and harvest steps and parameters currently used the same as what is described in the application?
- If the changes are not detailed in any update to the application, note when the changes were put into effect and what lots are impacted.

TABLE 6.4
Environmental Monitoring

	Question	Comments	Notes
Facility Design and Operation	• Is there a facility qualification protocol and test report for the current facility? • Is there a current floor plan? • Is there a current mechanical drawing for the HVAC? • Has the HVAC been validated? • Is there a written preventive maintenance agreement for the HVAC system? • Has preventive maintenance (PM) been performed and documented? If so, when? ______________________ • Does the work completed conform to the signed agreement? • Are the PM records audited by Quality Assurance (QA)? If so, when were the records last audited? • Were all adverse findings satisfactorily resolved? • Is there a current "zone map" indicating what air handlers are designated for what rooms/areas? • Is the HVAC design adequate to provide separation of clean from less clean activities? • Are there startup and shutdown procedures in shared production areas? • Is there a written changeover procedure to ensure adequate cleaning and inspection of shared areas prior to initiating a new job? • Are the walls, floors, and ceilings in good condition? (For example, is there any peeling paint, exposed wood, cracks, open joints, damaged tiles, walls, or ceiling tiles?) • Do production areas have pipework, light fixtures, ventilation ducts, or other fixtures that are difficult to clean? • Are drains constructed in a manner that allows ease of cleaning? Is there routine cleaning of drains that is documented? • Is the facility properly constructed to support the environmental control program (ECP)? • Is the facility designed to allow proper cleaning, maintenance, and other necessary operations? • Are there procedures in place for routine inspection and documentation of the facility's general condition? • Does the facility appear clean and well-maintained? • Are there areas of clutter where cleaning would be difficult? • Are items stored off of the floor to allow ease of cleaning? • Do corners appear clean? • Is there visible dirt or dust on ledges, window seals, door jams, or similar surfaces? • Is there a documented pest control program? • Is the facility designed and maintained in a manner to prevent the entry of insects and animals? • Are pesticides described in the standard operating procedure (SOP)? Are the pesticides approved for use in food or other FDA-regulated environments? • Does the designated individual who reviews the pest control service provided ensure that only approved agents are used? Is this review documented? • Is there an established procedure for monitoring and recording the temperature in the areas used for storage, handling, and processing product? • Has acceptance criteria for temperature been established? • Are there data to demonstrate that the products will not be affected by the temperature that falls below or rises above the established range? • Are data available that determine what other environmental conditions must be controlled, such as light, humidity, ventilation, etc.? • Do areas used for production and quality control (QC) have restricted access to prevent individuals not working in those areas from entering? • Are restrooms and breakrooms separate from production and QC areas? • Are there procedures in place to prohibit preventive maintenance activities, repair, or construction while production is in process, or at times when there may be a potential risk of product contamination? • Are there documented inspections of environmental controls that indicate systems are functioning properly?		☐ Fulfilled ☐ Control needed ☐ Action needed [room for notes]

(*Continued*)

TABLE 6.4 (CONTINUED)
Environmental Monitoring

	Question	Comments	Notes
Utilities (e.g., water, air)	• Are there written procedures for routine monitoring of water at all points of use? • Are there additional sampling locations, such as before and after key system components? • Is water sampled in a manner consistent with the use of the water from that port? • Is microbial testing performed at least weekly? • Are the microbes detected submitted for identification? • Are results from the water testing routinely evaluated and trended? • Are there written procedures that describe the evaluation and trending that is performed? • Are there written procedures for the reporting of deviations and nonconformances of water test results? • Are there established acceptance levels for microbial contamination? • Are there written procedures for action to be taken when the levels are exceeded? • Are nonconformances reported to the corrective and preventive action (CAPA) program with documented investigations, corrective action, and follow up? • Are there written preventive maintenance procedures? • Does the PM include resin and filter changes? How often are the filters changed? • Are there contractor SOPs on-site that describe the treatment of resin? • Are the results from the PM documented? • Is there a written contract with the service provider? • Has the contractor been qualified and routinely monitored under written purchasing control policies and procedures? • Is the water system routinely sanitized? • How often is the sanitization performed? • Is this performed by the service contractor or by in-house personnel? • Are there procedures for sanitization? • Does the procedure ensure that all surfaces in the water system are sanitized? • Was the sanitization procedure validated for effectiveness and removal of the sanitizing agent? • Does the procedure include a check for the removal of the sanitizing agent? • Is the result from the check documented? • Is there a procedure that describes the water system design, including all components, preventive maintenance procedures, sampling points, sanitization procedure, and their frequency? • Are there "dead legs" that have the potential to contaminate the water system? • Are hoses attached to the points of use? • Are there procedures in place for the removal, replacement, and sanitization of the hoses? • Has the current water system been validated? • Did the validation account for seasonal variations? When was the validation completed? • Have there been changes to the water system since the original validation was completed, and if so, has the water system been revalidated?		☐ Fulfilled ☐ Control needed ☐ Action needed [room for notes]
Manufacturing Processes (e.g., product exposure)	• Has a risk assessment been performed to determine the potential impact of the environment on the product? • Are bulk containers left uncovered during processing or storage? • Are personnel working in areas where the product is uncovered or exposed to the environment properly gowned? • For example, is all hair completely covered? • Are arms exposed? • Are beard covers used? • Are lab coats fully buttoned? • Are shoe covers worn? • Is the product ever exposed to the outside environment, such as open doors, windows, etc.? • Are food, plants, or animals kept in areas where the product is handled, stored, or tested?		

(Continued)

TABLE 6.4 (CONTINUED)
Environmental Monitoring

	Question	Comments	Notes
	• Are product, personnel, and waste flows designed and practiced in a manner to minimize the potential for cross-contamination? • Is there a process for reporting deviations that may potentially have an adverse impact on the product from an environmental perspective? • Are these deviations monitored on a regular basis for trends? • Is waste handled, transported, and stored in a manner that will minimize potential product contamination? • Are the outside surfaces of containers cleaned prior to transport into clean areas? For example, are the lids of containers cleaned? • Do production instructions require that a physical check be performed to confirm equipment is visibly clean prior to its use? • Is this check documented in the device history record or batch record? • Is there a maximum time limit specified between equipment use and when it was last cleaned? • Are there written procedures that will ensure that mix-ups, damage, deterioration, contamination, and other adverse effects do not occur during the handling and storage of the product (in-process and final product)?		☐ Fulfilled ☐ Control needed ☐ Action needed [room for notes]
Personnel (e.g., traffic flow, training)	• Have personnel involved with the handling and processing of products been trained in the potential adverse impact of the environment? • Have personnel been trained in proper gowning techniques? Has the training received been documented? • Is there a procedure that prevents personnel who have not received training from working in production? • Are traffic flow patterns established in a manner to minimize potential product mix-up and contamination? • Are there written personnel hygiene procedures in place? • Do these procedures include health, hygiene practice, and gowning requirements? • Are requirements in place that ensure no person with potentially infectious disease or open cuts, wounds, or lesions engages in production activities? • Are food and drink prohibited from production areas? • Are precautions taken to prevent the hands and arms of operators from contaminating the product? • Are personnel instructed to wash their hands prior to entering and after exiting the production areas? • Have the personnel been trained in the proper procedure for washing hands?		☐ Fulfilled ☐ Control needed ☐ Action needed [room for notes]

6.1.1.6 Purification

- Are the purification steps and parameters currently used the same as what is described in the application?
- If the changes are not detailed in any update to the application, note when the changes were put into effect and what lots are impacted.
- Are there critical steps with calculations that require double-check/double signature in the batch records?
- Are there repeated steps for purification based on IPC test data? If so, are these repeated steps noted in the application?
- Are the IPC test methods validated?
- When chromatographic columns are used, are the conditions of use specified in written instructions? For instance, are any columns being used as "once only"?
- If columns are for repeated use, are there written instructions for cleaning, storage, and end-of-life evaluation? Is there validation data for these critical steps?
- If columns are reused, are there written criteria for adequacy of use (e.g., system suitability checks)? If so, are these procedures being followed?

6.1.1.7 Cross-Contamination

Preface: Cross-contamination is a significant concern for multiuse facilities.

- Does the firm use adequate time, equipment, personnel, and space to separate products and activities?
- If equipment is not dedicated, review the cleaning records and cleaning validation.
- If personnel work across products, are the SOPs adequate to prevent contamination?
- Does the firm have SOPs for control of nonconforming products?
- Do these SOPs include an investigation and written evaluation of the investigation?
- Request a list of all failed final and in-process lots and a list of all deviations. Is this available?
- If so, evaluate the list to see if any failed lots were released (which did not meet specifications). Did this occur?
- If so, what lots were affected?
- For biotech products, the firm should perform an investigation when the WCB does not generate the necessary growth when used according to established SOPs or if the user deviates from routine practices.
- Did this occur?
- If so, what was the disposition of the resulting product?

6.1.2 Logistics Service Provider Questions – Procurement (Vehicle, Material, etc.)

- Is there a description of all trucks and cooling compartments used and of materials (drawings, specifications, data sheets) used for the transport?
- How is it made certain that the trucks, cooling compartments, and materials used meet the defined requirements?
- How are packages protected from jarring and jolting?
- Are packages protected from the effects of weather?

6.1.2.1 Receipt of Goods – Transport Begin

- Are there written protocols for the receipt of a refrigerated delivery?
- Are the products to be transported inspected prior to receipt in accordance with defined procedures? (For example, direct transfer from refrigeration?)
- Are there precise rules as to when refrigerated products to be transported are accepted or rejected?
- Are the reasons for the product rejection recorded at the transport inspection?
- Is there precise labeling of the container (with the name of the product, if applicable coding, batch designation, pharmacy) that identifies the products?
- Have the refrigerated trucks and/or the refrigerator units been qualified?
- How is the temperature monitored in the truck?
- How are the measuring sensors calibrated?
- Is there an alarm system in the trucks (failure of refrigeration, failure of alarm system)?
- How are the temperature curves documented?
- Are they always documented?
- What procedures follow on the failure of refrigeration in a truck? Are there rules for this?
- Are there packing/loading rules for the truck (areas in which nothing may be placed)?
- What is the difference between refrigerated products and the cold chain?
- Are there appropriate records on all points involving alarm/temperature detection and qualification?

TABLE 6.5
21 CFR Part 11 Checklist

Section	Question	Answer Yes ☐ No ☐ N/A ☐	Comments
11.10(b)	Are data entries stored in the database tables in an encrypted form?		
11.10(b)	Can data entries be printed directly from the database, then read by a human without a key?		
11.10(b)	Can records be output in common formats such as ASCII, VSAM, XLS, PDF, etc.?		
11.10(e), 11.50	Is every data or signature entry recorded in the database with a date and timestamp?		
11.10(e)	Is an entry that has been made and then later corrected retained in the system?		
11.10(e)	Is a prior entry that has been corrected identified as superseded?		
11.10(e)	If an entry is made by the operator, but is not saved due to a warning about its potential incorrectness retained in the system?		
11.10(e)	Does the system include in its audit trail operator actions that are not based upon a data entry or signature, (e.g., when an operator triggers an action, responds to a message, etc.?		
11.10(d)	Does the system limit who may access the system? Functions?		

6.1.2.2 Storage

- How is the transfer into depots etc. carried out?
- Is there an operating instruction (SOP) for this?
- Are there written protocols for the storage of refrigerated delivery?
- Are the products to be stored inspected prior to receipt in accordance with defined rules (for example, direct transfer from refrigeration)?
- Are there precise rules as to when refrigerated products to be stored are accepted or rejected?
- Have the cold-storage compartments and/or the power units been qualified?
- How is the temperature monitored in the refrigerated compartment?
- How are the measuring sensors calibrated?
- Is there an alarm system in the refrigerated compartments (failure of the cooling, failure of the alarm system)?
- How are the temperature curves documented? Are they always documented?
- What procedure is followed for failure of refrigeration in a storage facility? Are there rules for this?
- Are there storage rules for refrigeration areas (areas in which nothing may be placed)?
- What is the difference between refrigerated products and the cold chain?
- Are there appropriate records on all points involving alarm/temperature detection and qualification?
- Are goods received treated and stored in a manner that prevents damage (such as through moisture, temperature, and ventilation)?
- Are goods received stored in such a manner that there can be no mix-ups?
- Are there rules for warehousing?
- Is access control in place in the storage areas?

TABLE 6.6
EU, PIC/S GMP Annex 11 Checklist

Section	Question	Answer Yes ☐ No ☐ N/A ☐	Comments
2.2.1.5	What is the relationship (or impact) between computerized systems or processes to patient and product safety, or electronic data quality or integrity?		On the basis of this question, critical systems can be identified. These are in the scope of the major objective of the inspection. For example, such systems are controlling production process systems (e.g., reactor supervision, filling lines, blender) or systems used in the nearer production areas (e.g., air handling units/HVAC, CIP/SIP processes, WFI or Aqua Purificata production) or in the area of quality control (HPLC, analytical instruments for IPC, data for batch releases).
2.2.1.6	Which actions have been addressed for risk mitigation/reduction during the risk assessment(s)?		In some cases, all relevant GMP requirements cannot be fulfilled by an existing system for technical reasons (restrictions). Within the scope of a risk control process, there might exist additionally defined actions or the operational range was limited. A replacement of such systems should be initiated (refer to section validation).
2.2.1.7	Which statements (declarations) are defined in higher-level QA documents regarding the identification and evaluation of risks?		The management (handling) of computerized systems must be included in the related quality system.
2.2.1.8	Which exigent and prospective risk reduction actions can be derived from these?		The fundamental management of risk reduction and avoidance should be defined in the QA system.
2.2.1.9	To what extent have GMP-related processes been assessed for the type and range of validation activities by the risk assessment(s)?		During the risk assessment, the direct and indirect impact of the computerized system regarding GMP should be analyzed. Critical processes and functions, that have been identified, might be the subject of a separate/dedicated inspection (part) (refer to section validation).
2.2.1.10	Was a risk assessment executed in the case of a retrospective validation?		At minimum, the following actions are expected for a retrospective validation approach: - Execution of risk analysis in order to evaluate GMP-relevant parts of the system and to define additional actions - Analysis and evaluation of historic data - Test execution of the as-critical ranked GMP-relevant parts (refer to section validation)

(Continued)

TABLE 6.6 (CONTINUED)
EU, PIC/S GMP Annex 11 Checklist

Section	Question	Answer Yes ☐ No ☐ N/A ☐	Comments
2.2.1.11	In which way are risk assessments for computerized systems implemented in the change control system?		Changes should be investigated toward the (potential) risks (by/for the computerized systems).
2.2.1.12	To what extent is risk management implemented in the respective phases of the system life cycle?		Risk management should be applied throughout the entire system life cycle. By the initial assessment, the GMP criticality should be analyzed. Especially the impact on patient safety, data integrity, and product quality should be evaluated. The user requirement specification(s) should be developed on the basis of potential risks. These form the basis of the first initial risk ranking. Complex systems should be based on a detailed risk assessment, resulting in the identification of critical functions. This should assist in addressing all critical functions during validation. Risk management includes the implementation of control strategies and their verification.
2.2.1.13	Does the recognizability of risks have any impact on the overall risk?		Only risks, that have been detected, can result in the reduction of the overall risk.

6.1.2.3 Shipping/Transport

- Are the shipping containers clearly marked so that the contents can be identified?
- Is appropriate transport from the factory premises assured?
- Are there rules on transfer (from depot to truck) and continued transport?

6.1.2.4 Additional Workflows

- Are written processes in place for workflow and organization?
- Are the process steps clearly formulated so that employees understand their duties and rules are unambiguous?
- Are written procedures in place, being followed, and self-checked accordingly (for instance, with name symbols)?
- Is it certain that the customer will be notified early of changes and deviations?
- Are transport procedures documented?
- Are all documents being properly archived for an appropriate time period?
- Are the current rules always available to the person they relate to?
- Are maintenance plans and change control procedures in place?
- Are cleaning rules available?
- Is (proper) cleaning documented?
- Is there a monitoring system for hygiene and air conditioning?
- Have pest control measures been taken?

TABLE 6.7
Contracted Manufacturer Audit Checklist

Section	Question	Answer Yes ☐	No ☐	N/A ☐	Comments
Personnel training and assessment	☐ How are personnel from the production, laboratory, and engineering areas trained?				
	☐ How often does training take place?				
	☐ How is the training documented?				
	☐ Is there a management system for training and participants?				
	☐ Is there an approved training program?				
	☐ How is it ensured that the participants have understood the topic?				
	☐ How are the instructors trained?				
	☐ Is an assessment carried out?				
State of repair of the premises	☐ Is the ceiling sealed?				
	☐ Are the floor tiles damaged?				
	☐ Is there a wall guard and is it sealed?				
	☐ Is a dust-tight ceiling fitted?				
Cleaning and sanitation programs for personnel and facilities	☐ Are there cleaning schedules for production rooms?				
	☐ Are they adhered to and documented?				
	☐ Are there cleaning procedures for all machines?				
	☐ Are there approved cleaning agents and disinfectants?				
	☐ Are the ceilings cleaned regularly?				
	☐ How is cleaning documented?				
	☐ Are regular checks made for the cleaning?				
	☐ How often is protective apparel changed?				
	☐ Are there specific regulations for particular areas?				
	☐ Does regular training of the cleaning staff take place?				
Calibration of balances	☐ How often are the balances calibrated?				
	☐ Is there a calibration instruction?				
	☐ Is the function of the balances tested daily?				
Maintenance and repair measures	☐ Is there an SOP for maintenance?				
	☐ Who monitors the maintenance due dates?				
	☐ How is the maintenance documented?				
	☐ Is there a maintenance schedule?				
	☐ Are the calibration weights certificated regularly?				
Quality management system/safety measures	☐ Is there an SOP system?				
	☐ How often are SOPs updated?				
	☐ Who is responsible for SOP training?				
	☐ How are deviations handled?				
	☐ How are changes controlled?				
	☐ Is there a change control system?				
	☐ Is there a process for out-of-specification results?				
Approval systems	☐ Who is authorized to release a product?				
	☐ Is there a user draft for the release on the computer?				
	☐ Who approves raw materials?				
	☐ Who assigns the expiration date?				
	☐ Is the computer system validated for the release part?				
	☐ Who can release in the event of absence?				

(Continued)

TABLE 6.7 (CONTINUED)
Contracted Manufacturer Audit Checklist

Section	Question	Answer Yes ☐ No ☐ N/A ☐	Comments
Storage	☐ Where are the samples taken from?		
	☐ Who is responsible for sampling?		
	☐ Who trains the staff who carry out the sampling?		
	☐ How are receipts checked?		
	☐ Who creates the labels?		
	☐ Who controls the products supplied?		
	☐ Are all partially filled containers sealed?		
	☐ How is the temperature and humidity monitored in the warehouse?		
	☐ Are samples taken from all containers? If not, what is the basis of the sampling plan?		
	☐ How is container labeling carried out?		
	☐ How is the reprinting of labels carried out?		
Complaints	☐ Are internal and external complaints analyzed?		
	☐ Are there any statistics on complaint trends?		
	☐ Who is responsible for processing complaints?		
	☐ Are the topics of the complaint incorporated into the training program?		
Documentation systems	☐ Who is responsible for compiling the manufacturing instructions?		
	☐ Who compiles the test procedures?		
	☐ Is there a packaging instruction?		
	☐ Is there a variant of the manufacturing instructions?		
	☐ Are there logbooks for all machines and media equipment?		
	☐ Who completes the cleaning documentation?		
	☐ Who can change the bills of materials?		
Job descriptions for key personnel	☐ Are there current job descriptions for staff in charge? In the USA, job descriptions for all personnel performing GMP-related activities should exist.		
	☐ How often are the job descriptions updated?		
	☐ Is responsibility for the release clearly worded?		
Building installation	☐ How are temperature and humidity monitored?		
	☐ What is the air exchange rate?		
	☐ Are the measuring instruments calibrated?		
	☐ Where can the temperatures and pressures be read?		
	☐ Is there an alarm signal in the event of deviations?		
	☐ What happens in the event of deviations?		
	☐ Is the building control system qualified?		
Risk of contamination	☐ How is contamination between machines and areas prevented?		
	☐ How is contamination by staff prevented?		
	☐ Is there specific clothing or air locks?		
	☐ What are the ventilation systems installed?		
	☐ What are the pressure differentials in the rooms?		
Zonal concept	☐ What is the company zonal concept?		
	☐ Is there an SOP for it?		
	☐ Are room plans with personnel and material flow available?		
	☐ How often is monitoring carried out (particle, organism)?		

(Continued)

TABLE 6.7 (CONTINUED)
Contracted Manufacturer Audit Checklist

Section	Question	Answer Yes ☐ No ☐ N/A ☐	Comments
Utilities and supply	☐ Are all measuring instruments calibrated? ☐ Are there maintenance schedules for utilities? ☐ Are there any current drawings? ☐ Is there a qualification for the water treatment system? ☐ Where appropriate, is validation performed?		
Validation of process and analytical procedures	☐ How are validations carried out? ☐ Who is responsible for the validation? ☐ When must revalidation be carried out? ☐ Is a risk analysis carried out? ☐ Are all analytical procedures validated? ☐ Is there a validation master plan? ☐ Are regular checks made that the manufacturer still conforms to the validation data?		
Qualification of facilities and machines	☐ Are all existing facilities qualified? ☐ Are new facilities qualified? ☐ How is a performance qualification carried out? ☐ How is the qualification of old facilities carried out? ☐ Who compiles the user specifications? ☐ Who carries out monitoring of activities?		
Cleaning validation	☐ Is a cleaning validation carried out? ☐ Which product groups are defined? ☐ Is there a cleaning procedure? ☐ Do the cleaning procedures contain the detergent and the concentration? ☐ How is the cleaning result defined? ☐ What is the basis for classification into product groups? ☐ How are new products handled?		
Specifications and test procedures	☐ Are inspections always carried out in accordance with prevailing instructions? ☐ How is raw data documented? ☐ Who transfers the raw data to the certificate of analysis? ☐ Is there a separate material number for each material? ☐ Are complete analyses always carried out? ☐ Are laboratory reagents correctly labeled?		
Labeling of containers and premises	☐ Are there cleaning labels? ☐ How are containers identified? ☐ Does staff sign these labels? ☐ How long is cleaning valid? ☐ Are all rooms labeled and assigned a floor plan?		
Self-inspection	☐ Are self-inspections carried out on a regular basis? ☐ How are the results of the self-inspection implemented? ☐ Who carries out the self-inspection? ☐ Who takes over monitoring of activities?		

TABLE 6.8
Sterilization Supplier Audit Checklist

Question	Comments	Notes
• Is there an approved quality agreement that defines the quality and compliance roles and responsibilities between the contract sterilization service and the client? • Has the contract sterilization service facility been commissioned? Have equipment and processes been qualified by Installation/Operational/Performance Qualifications? • Is there a deviation alert/notification process that includes an investigation process and notification of the customer? • Is there a change control system that identifies and approves changes to validated sterilization processes, equipment, cobalt radiation sources, and package configuration changes (e.g., density, bioburden, packaging, and component), etc.? • Is the client notified of such changes and are client product/material-related changes approved by a representative of the client? • Are qualification/validation studies performed according to preapproved validation protocols with defined acceptance criteria? Are qualification/validation summary reports approved by Quality? • Are there established load patterns (i.e., orientation defined by a controlled and approved diagram) for the product or materials being sterilized? • Is there a review of sterilization documentation prior to product/material release? Are sterilization certificates or other documentation provided to the client upon shipment of sterilized materials? • Are dosimeters used during validation studies and routinely used to monitor absorbed radiation doses? • Has the contract irradiation sterilization service qualified and/or audited the dosimeter manufacturer? • Does reprocessing and additional exposure to sterilization conditions get preapproved by the client? • Are periodic sterilization dose audits, at least annually, carried out to confirm the continued appropriateness of the sterilization dose? • Do ethylene oxide (EtO) processing steps include preconditioning and conditioning to a specified temperature and relative humidity (RH); sterilization cycle; and aeration? • For EtO, has the qualification process included packaging bioburden, load preconditioning, and survival cycle determination? • Are product bioburden samples taken up to the point of sterilization? • Are critical processing parameters controlled, monitored, and recorded in the batch sterilization record; e.g., exposure time, timer setting (batch mode processing), conveyor speed (continuous mode processing), dose measurement, package size, and bulk density, etc.? • Does batch and continuous sterilization process documentation include accountability of product before and after sterilization by batch/lot number/load pattern identification? • Dosimeter placement and retrieval? • Specified minimum and maximum dose? Processing parameters? Sterilization date? Dosimeter readings and analysis? • Deviation and investigations? Identification of operators? • Are process interruptions handled through a deviation initiation and investigation process? For absorbed dose readings outside specified limits? • Are at least two dosimeters placed and exposed in batch mode and continuous mode irradiation processing? • Are there written and approved procedures that cover all phases of the operation (i.e., from receipt of nonsterile materials through release and shipment)?		☐ Fulfilled ☐ Control needed ☐ Action needed [room for notes]

(*Continued*)

TABLE 6.8 (CONTINUED)
Sterilization Supplier Audit Checklist

Question	Comments	Notes
• Are the written procedures sufficiently detailed, in a site-defined format, reviewed and approved for use, and periodically reviewed? • Is employee training covered by a site process/procedure to include new employee requirements, expectations related to procedure updates, and periodic reinforcement? • Are the laboratory methods included in a site documentation program, periodically reviewed and version controlled? • Is there a process to evaluate the impact, if any, if an instrument/device is found to be out of calibration specifications? • What about during routine PM events? • Are dosimeters calibrated on an annual basis? Is the radiation source added, redistributed, or replaced on a weekly basis?		

6.1.2.5 Quality Control

- Are spot checks for temperature carried out if necessary to ensure that the temperature is being maintained?
- Are tests that are performed and the test results properly documented?
- Is every deviation fully recorded and investigated?
- Has the customer/manufacturer been informed?
- Is a system in place to make certain that measuring instruments, control equipment, and testing devices are regularly calibrated?
- Are completed calibrations recorded in writing?
- When the calibration itself is being conducted are these calibration procedures defined in compliance with appropriate standards and are they understandable?
- When calibrations are conducted externally, is the external party tested for competence?
- Are complaints reviewed?
- Are the causes of quality failures investigated?
- Are appropriate actions and measures taken to prevent a repeat occurrence of the failure?

6.1.2.6 Quality Assurance

- Does the company management issue a written description of the quality policy?
- Is there a "quality assurance system" that describes the organizational structure, the accountabilities, and standard operating procedures?
- Is there a quality assurance manual describing the "quality assurance system"?
- Is there an internal quality review program?
- Do self-inspections take place at regular intervals?
- Are they recorded in the log files?
- Is there a complete organogram of the organization's structure with workplace descriptions, staffing plans, and definitions of accountabilities?
- Is there a continuous change service for QA regulations that ensures that outdated versions are no longer used?
- Is there a system in place that ensures validations and qualifications for necessary processes and equipment?
- Is there an appropriate procedure in place to ensure the cleanliness of vehicles, facilities, and containers?
- Is the traceability of products (transport tracking) ensured?

- Is there a system that ensures that only intact packages are shipped?
- How are subcontractors deployed if applicable?
- If subcontractors are used, are they certified and/or qualified and approved?
- Are these subcontractors audited?
- Does the company have a list of approved subcontractors?
- Can it be assumed that employees (potential subcontractors as well!) possess the necessary knowledge and awareness of quality?
- Is an adequate training program in place for all employees in the relevant areas?
- Is there a program for regular refresher training courses/retraining?
- Are the training courses recorded?
- Is attention paid to employee hygiene?
- Are there adequate sanitation facilities?
- Are eating, drinking, and smoking prohibited where necessary?
- When pharmaceutical and biotechnology manufacturers look for a service provider for the storage or distribution of their products, a right business partner is always advantageous. Experienced and well-referenced 3PL companies in healthcare should provide benefits that will make a difference in supply-chain security, which has become increasingly important especially in recent days.
- It is very important to evaluate the capabilities of the relevant candidate company during the selection phase of the 3PL company. Therefore, the company's closeness and compliance with legal and international cGMP and GDP legislation will provide you with the most critical data.

6.2 LESSONS LEARNED FROM 483S, WARNING LETTERS, AND GMDP DATABASE

6.2.1 Case 1

The first case I want to review is a beta-lactam manufacturer. The company repackaged beta-lactam products in a nondedicated facility. In addition, personnel can move freely between beta-lactam and non-beta-lactam areas.

When the matter was discussed with the FDA, the company had to recall its products. Beta-lactam production by the company was stopped. The firm decontaminated the facility at a million-dollar expense and renovated the building to separate the spaces. The FDA has sent a warning letter.

Lesson: It is very difficult and demanding to obtain a safe level for beta-lactam contamination. Exposure to beta-lactam causes severe allergic reactions on the skin. You cannot decontaminate the areas that need to be separated by cleaning only. Due to the complexity and limited capability of analytical methods, you may be inadequate in most cases to trace beta-lactam contamination. The separation seems to be the most logical of the solutions.

6.2.2 Case 2

A company carries out the repackaging of therapeutically different products. Operators change master labels. There is no procedure for manual repackaging. Complaints about the products being below the filling level are increasing. The company is forced to recall products, and the FDA issues a warning letter for inadequate manufacturing conditions and controls, and labeling violations.

Lesson: Failure to repackage a pharmaceutical product can have critical consequences on product effectiveness and quality. Inadequately repackaged products can cause serious side effects. The manufacturer had to urgently meet the requirements of 21 CFR Parts 210, 211.

6.2.3 Case 3

A company has been producing a wide variety of transdermal patches for many years. It has developed a new product using the same adhesion matrix as the others. In its first year in the market, the product received 500 complaints due to its effectiveness and adhesion problems. About 25% of the drug was adhering incorrectly.

In the research conducted by the company, it was revealed that the source of the problem was the drug-adhesive problem. The company continued to market the drug, claiming that there was no problem with the application. The company eventually had to recall its products. The FDA sent a warning letter to the company because of insufficient specifications and the inability to provide proper adhesion. The company had to renew its specifications.

Lesson: You cannot think that a specification or production method that you thought would work for any product would work for every product. You should revise your specifications by reevaluating the data you have obtained in production.

6.2.4 Case 4

The FDA asked a pharmaceutical company about the in-process control results, which are out of limits. As a result of the audit, the company announced that it stopped endotoxin in-process controls.

The company could not avoid receiving a warning letter. The products were recalled after discussion with the FDA. As a corrective action, the company chose to collaborate with the FDA and locate the source of the endotoxin. As a result of detailed root-cause research, it was determined that the endotoxin originated from the raw material.

Lesson: The company concluded that quality cannot be ensured by testing but must be taken into account during the development phase of the product (quality by design).

Raw materials, containers, closures, and manufacturing processes, including excipients critical to product quality, must be robust, and control strategies must be validated.

6.2.5 Case 5

A pharmaceutical company manufactures a drug product containing a material that aids in dissolution. The supplier informs the company that it will no longer produce this aid material. The company starts looking for other sources and wants to buy another product that it claims is the same. The company decides to try for a material change. Testing of products produced with new materials is a time-consuming process. The company decides to market the products manufactured with this material without evaluating all the data coming from the trial batches.

The quality control (QC) test results of lots manufactured and marketed with the new aid material failed. It is decided that all parties made with the new material aid will be destroyed with a loss of millions of dollars. Since the company thinks that it has found a similar previously used aid, it has made enough stock for almost a year and continues to seek a solution to the problem.

Lesson: Never risk your manufacturing. Do not assume that two products that meet the same pharmacopoeia requirements will give the same results. Always have your resources as your backup supplier in order to be protected from risks in the supply chain. Always keep in mind that the supply chain is a multicomponent and complex process.

6.2.6 Case 6

The pharmaceutical company manufactures chewable tablet products. In routine stability tests, it is determined that the active pharmaceutical ingredient (API) result of some tablets is three times higher than normal. It seems that the release tests failed to catch the problem earlier.

The company urgently recalled its products. It is determined that different particle sizes, densities, and flow characteristics cause the mixture to separate during discharge. The company applied these fixes to other products with the same issue.

Lesson: For blending before pressing, the company should evaluate the uniformity of the material entering the press. The root-cause investigation should be extended to other batches, other products, and other parts of the process. It should be kept in mind that release tests may be insufficient to detect process problems.

6.3 HOW TO BECOME AN EXEMPLAR GLOBAL ACCREDITED GMP LEAD AUDITOR

From the first page of this book, I have tried to explain how important GMP audits are for a pharmaceutical or biotechnology company. I also stated that the first requirement of planning for a successful and effective GMP audit, strictly following the plan, asking questions, making observations, and writing reports containing accurate information is to successfully complete a globally recognized certified auditor training.

You may be a senior manager in a company or you may be an inexperienced quality employee who has only been hired for a few years, but be sure that it will take years to become a competent GMP auditor.

But, I also have good news for you. I want you to know that there are some opportunities for you to become an auditor whose GMP auditor certification is valid on a global scale.

When it comes to educating, training, and certifying GMP auditors, I can say that there are a few consultancy and training companies. An internationally recognized organization like Exemplar Global is the most important one that can help you in this regard, in my opinion. Exemplar Global approves companies that offer GMP Lead Auditor training and ensures that they are competent organizations.

The only institution authorized by Exemplar Global is Quality Academia (www.qualityacademia.com), based in Istanbul, Turkey. You can visit its website for details. However, what I want to focus on is the scope and purpose of the GMP Lead Auditor training.

The training was approved by Exemplar Global before the start of COVID-19, but due to the pandemic, the training cannot be offered face-to-face. But it is offered to pharmaceutical and biotechnology companies online. A company can also offer face-to-face training to small groups onsite if there is a demand.

The duration of the training is six days in total. Depending on the number of customers and demand, it can be offered consecutively for three days in the first week and three days in the second week. Training includes all the audit components covered in this book and I can say that it is quite popular.

In addition to quizzes and oral questions during the training, case studies are designed to help participants refresh and improve their GMP knowledge. At the end of each day, there is a written exam covering the day's topics. After the evaluation of the training, successful participants are entitled to become a "Certified GMP Lead Auditor."

By increasing the number of Certified GMP Lead Auditors, the number of companies that want to perform their own audits without outsourcing is increasing day by day.

In addition, one of the topics that the health and agriculture authorities of the countries question in GMP audits is how many GMP auditors a company has and their training and certificates. Even for this reason alone, pharmaceutical and biotechnology companies should increase the number of qualified auditors with this training.

REFERENCES

Quality Academia Training Documents
Quality Academia Training Presentations

Index

B

C

D

O

P

Q

R

S

www.ingramcontent.com/pod-product-compliance
Ingram Content Group UK Ltd.
Pitfield, Milton Keynes, MK11 3LW, UK
UKHW052217190926
475702UK00006B/1057
* 9 7 8 1 0 3 2 2 5 7 3 0 3 *